Everyday Mathematics®

The University of Chicago School Mathematics Project

Student Reference Book

Everyday Mathematics®

The University of Chicago School Mathematics Project

Student Reference Book

Chicago, IL • Columbus, OH • New York, NY

UCSMP Elementary Materials Component
Max Bell, Director, UCSMP Elementary Materials Component; Director, *Everyday Mathematics* First Edition
James McBride, Director, *Everyday Mathematics* Second Edition
Andy Isaacs, Director, *Everyday Mathematics* Third Edition
Amy Dillard, Associate Director, *Everyday Mathematics* Third Edition
Rachel Malpass McCall, Associate Director, *Everyday Mathematics* Common Core State Standards Edition

Authors
Max Bell, Jean Bell, John Bretzlauf, Amy Dillard, James Flanders, Robert Hartfield, Andy Isaacs, Deborah Arron Leslie, James McBride, Kathleen Pitvorec, Peter Saecker

Assistants
Lance Campbell (Research), Adam Fischer (Editorial), John Saller (Research)

Technical Art
Diana Barrie

The *Student Reference Book* is based upon work supported by the National Science Foundation under Grant No. ESI-9252984. Any opinions, findings, conclusions, or recommendations expressed in this material are those of the authors and do not necessarily reflect the views of the National Science Foundation.

everyday**math**.com

STEM McGraw-Hill is committed to providing instructional materials in Science, Technology, Engineering, and Mathematics (STEM) that give all students a solid foundation, one that prepares them for college and careers in the 21st century.

Send all inquiries to:
McGraw-Hill Education
STEM Learning Solutions Center
P.O. Box 812960
Chicago, IL 60681

ISBN: 978-0-07-657650-0
MHID: 0-07-657650-7

Printed in the United States of America.

4 5 6 7 8 9 QVR 17 16 15 14 13 12

The McGraw·Hill Companies

Contents

Geometry and Constructions 87

Measurement 125

Algebra 147

Problem Solving 173

Mathematics... Every Day
Machines that Calculate **185**

Calculators 191

Games 227

CHINA
Bay of Bengal
South China Sea
Philippine Sea
Hong Kong
VIETNAM
PHILIPPINES
INDONESIA
Timor Sea
Indian Ocean
Sea of Japan
TAIWAN
Kuala Lumpur
MALAYSIA
BRUNEI
Manila
Saigon
Hanoi
Bangkok
Rangoon
Phnom Penh
Davao
Wenzhou
Guiyang

About the *Student Reference Book*

A reference book is organized to help people find information quickly and easily. Dictionaries, encyclopedias, atlases, cookbooks, and even telephone books are examples of reference books. Unlike novels and biographies, which are usually read in sequence from beginning to end, reference books are read in small segments to find specific information at the time it is needed.

You can use this *Student Reference Book* to look up and review information on topics in mathematics. It consists of the following sections:

- A **table of contents** that lists the topics covered and shows how the book is organized.
- Essays on **mathematical topics,** such as whole numbers, decimals, percents, fractions, data analysis, geometry, measurement, algebra, and problem solving.
- A collection of **photo essays** called **Mathematics... Every Day,** which show in words and pictures some of the ways that mathematics is used.
- Descriptions of how to use a **calculator** to perform various mathematical operations and functions.
- Directions on how to play **mathematical games** that will help you practice math skills.
- A set of **tables and charts** that summarize information, such as a place-value chart, prefixes for names of large and small numbers, and tables of equivalent measures and of equivalent fractions, decimals and percents.
- A **glossary** of mathematical terms consisting of brief definitions of important words.
- An **answer key** for every Check Your Understanding problem in the book.
- An **index** to help you locate information quickly.

This reference book also contains a **World Tour** section. It is a collection of numerical information about people and places around the world.

How to Use the *Student Reference Book*

Suppose you are asked to solve a problem and you know that you have solved problems like it before. But at the moment, you are having difficulty remembering how to do it. This is a perfect time to use the *Student Reference Book.* You can look in the **table of contents** or the **index** to find the page that gives a brief explanation of the topic. The explanation will often show a step-by-step sample solution.

There is a set of problems at the end of most essays, titled **Check Your Understanding.** It is a good idea to solve these problems and then turn to the answer key at the back of the book. Check your answers to make sure that you understand the information presented in the section you have been reading.

Always read mathematical text with paper and pencil in hand. Take notes, draw pictures and diagrams to help you understand what you are reading. Work the examples. If you get a wrong answer in the Check Your Understanding problems, try to find your mistake by working back from the correct answer given in the answer key.

It is not always easy to read text about mathematics, but the more you use the *Student Reference Book,* the better you will become at understanding this kind of material. You may find that your skills as an independent problem-solver are improving. We are confident that these skills will serve you well as you undertake more advanced mathematics courses.

Whole Numbers

Uses of Numbers

It is hard to live even one day without using or thinking about numbers. Numbers are used on clocks, calendars, car license plates, rulers, scales, and so on. The major ways that numbers are used are listed below.

♦ Numbers are used for **counting.**

Examples The first U.S. Census counted 3,929,326 people.

The population of Copper Canyon is 889.

♦ Numbers are used for **measuring.**

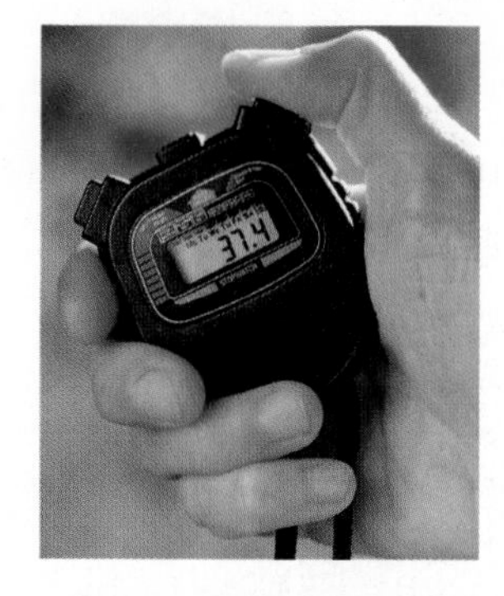

Examples Alice swam the length of the pool in 37.4 seconds.

The package is 25 inches long and weighs $3\frac{1}{4}$ pounds.

♦ Numbers are used to show where something is in a **reference system.**

Examples

Situation	Reference System
Normal room temperature is 21°C.	Celsius temperature scale
Jan was born on May 25, 1998.	Calendar
The time is 6:19 P.M.	Clock time
Detroit is located at 42°N and 83°W.	Earth's latitude and longitude system

♦ Numbers are used to **compare** counts or measures.

Examples There were 3 times as many boys as girls at the game.

The cat weighs $\frac{1}{2}$ as much as the dog.

Did You Know?

The first product with a bar code was scanned at a check-out counter in 1974. It was a 10-pack of Wrigley's Juicy Fruit chewing gum.

♦ Numbers are used for **identification** and as **codes.**

Examples driver's license number: M286-423-2061

ZIP code: 60637 phone number: (709) 555-1212

Kinds of Numbers

The **counting numbers** are the numbers used to count things. The set of counting numbers is 1, 2, 3, 4, and so on.

The **whole numbers** are any of the numbers 0, 1, 2, 3, 4, and so on. The whole numbers include all of the counting numbers and zero (0).

Counting numbers are useful for counting, but they do not always work for measures. Some measures fall between two whole numbers. **Fractions** and **decimals** were invented to keep track of such measures.

Fractions are often used in recipes for cooking and for measurements in carpentry and other building trades. Decimals are used for almost all measures in science and industry. Money amounts are usually written as decimals.

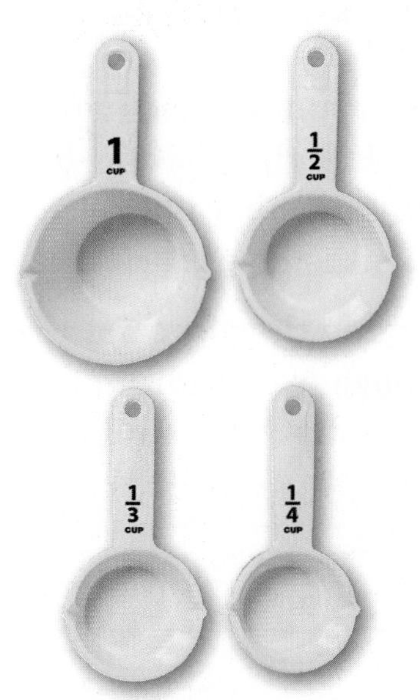

A measuring cup set marked with fractions

Examples The recipe calls for $2\frac{3}{4}$ cups of flour.

The window sill is 2 feet $5\frac{1}{2}$ inches above the floor.

The lunch meat weighs 0.55 pound and costs $2.74.

A food scale showing weight as a decimal

Negative numbers are used to describe some locations when there is a zero point.

Examples A temperature of 10 degrees below zero is written as −10°F.

A depth of 235 feet below sea level is written as −235 feet.

Negative numbers are also used to indicate changes in quantities.

Examples A weight loss of $4\frac{1}{2}$ pounds is recorded as $-4\frac{1}{2}$ pounds.

A decrease in income of $1,000 is recorded as −$1,000.

Place Value for Whole Numbers

Any number, no matter how large or small, can be written using one or more of the **digits** 0, 1, 2, 3, 4, 5, 6, 7, 8, and 9. A **place-value chart** is used to show how much each digit in a number is worth. The **place** for a digit is its position in the number. The **value** of a digit is how much it is worth according to its place in the number.

Study the place-value chart below. Look at the numbers that name the places. As you move from right to left along the chart, each number is **10 times as large as the number to its right.**

10,000s ten thousands	**1,000s** thousands	**100s** hundreds	**10s** tens	**1s** ones
8	1	9	0	3

Example The number 81,903 is shown in the place-value chart above. It is read "eighty-one thousand, nine hundred three."

The value of the 8 is 80,000 (8 * 10,000).
The value of the 1 is 1,000 (1 * 1,000).
The value of the 9 is 900 (9 * 100).
The value of the 0 is 0 (0 * 10).
The value of the 3 is 3 (3 * 1).

In **expanded notation,** 81,903 = 80,000 + 1,000 + 900 + 3.

In larger numbers, look for commas that separate groups of 3 digits. The commas help you identify the thousands, millions, billions, and so on.

Example Last year, the U.S. Mint made 14,413,236,610 pennies.

billions				**millions**				**thousands**				**ones**		
100	10	1	,	100	10	1	,	100	10	1	,	100	10	1
	1	4	,	4	1	3	,	2	3	6	,	6	1	0

Read from left to right. Read "billion" at the first comma, "million" at the second comma, and "thousand" at the last comma.

The number is read as "14 **billion,** 413 **million,** 236 **thousand,** 610."

Check Your Understanding

Read each number to yourself. What is the value of the 5 in each number?

1. 35,104 **2.** 82,500,000 **3.** 71,054 **4.** 3,657,000

Check your answers on page 340.

Powers of 10

Numbers like 10, 100, and 1,000 are called **powers of 10.** They are numbers that can be written as products of 10s. 100 can be written as 10 * 10; 1,000 can be written as 10 * 10 * 10; and so on.

There is a shorthand method for writing products of 10s.
10 * 10 can be written as 10^2.
10 * 10 * 10 can be written as 10^3.
10 * 10 * 10 * 10 can be written as 10^4.

The raised digit is called an **exponent.** The exponent tells how many 10s are multiplied.

This chart shows powers of 10 from ten through one billion.

Note

10^2 is read "10 to the second power" or "10 squared." 10^3 is read "10 to the third power" or "10 cubed." 10^4 is read as "10 to the fourth power."

Powers of 10

Standard Notation	Product of 10s	Exponential Notation
10	10	10^1
100	10*10	10^2
1,000 (1 thousand)	10*10*10	10^3
10,000	10*10*10*10	10^4
100,000	10*10*10*10*10	10^5
1,000,000 (1 million)	10*10*10*10*10*10	10^6
10,000,000	10*10*10*10*10*10*10	10^7
100,000,000	10*10*10*10*10*10*10*10	10^8
1,000,000,000 (1 billion)	10*10*10*10*10*10*10*10*10	10^9

Note

A number written with an exponent, such as 10^3, is in **exponential notation.**

A number written in the usual place-value way, such as 1,000, is in **standard notation.**

Example 1,000 * 1,000 = ?

Use the table above to write 1,000 as 10 * 10 * 10.

1,000 * 1,000 = (10 * 10 * 10) * (10 * 10 * 10)
= 10^6
= 1 million

1,000 * 1,000 = 1 million

Example 1,000 millions = ?

Write 1,000 * 1,000,000 as (10 * 10 * 10) * (10 * 10 * 10 * 10 * 10 * 10). This is a product of nine 10s, or 10^9.

1,000 millions = 1 billion

Did You Know?

A *googol* is the number 1 followed by 100 zeros. The word was invented by a 9-year-old boy who was asked by a mathematician to think up a name for a very big number. A googol is a very large power of 10: 1 googol = 10^{100}

Comparing Numbers and Amounts

When two numbers or amounts are compared, there are two possible results: They are equal, or they are not equal because one is larger than the other.

Different symbols show that numbers and amounts are equal or not equal.

- Use an **equal sign** (=) to show that the numbers or amounts *are equal.*
- Use a **greater-than symbol** (>) or a **less-than symbol** (<) to show that they are *not equal* and to show which is larger.

Examples

Symbol	=	>	<
Meaning	"equals" or "is the same as"	"is greater than"	"is less than"
Examples	$\frac{1}{2} = 0.5$	$7 > 3$	$2 < 4$
	$20 = 4 * 5$	$1.23 > 1.2$	$398 < 1{,}020$
	3 cm = 30 mm	14 ft 7 in. > 13 ft 11 in.	99 minutes < 2 hours
	$5 + 5 = 6 + 6 - 2$	$8 + 7 > 9 + 5$	$2 * (3 + 3) < 4 * 5$

When you compare amounts that include units, be sure to use the *same unit* for both amounts.

Examples Compare 30 yards and 60 feet.

The units are different—yards and feet.

0 1 2 3 4 5 6 7 8 9 10 11 12 13 14 15 16 17 18 19 20 21 22 23 24 25 26 27 28 29 30 31 32 33 34 35 36

0 1 2 3 4 5 6 7 8 9 10 11 12

Change yards to feet, and then compare.
1 yd = 3 ft, so 30 yd = 30 ∗ 3 ft, or 90 ft.
Now compare feet. 90 ft > 60 ft

Therefore, 30 yd > 60 ft.

Or, change feet to yards, and then compare.
1 ft = $\frac{1}{3}$ yd, so 60 ft = 60 ∗ $\frac{1}{3}$ yd, or 20 yd.
Now compare yards. 30 yd > 20 yd

Therefore, 30 yd > 60 ft.

Check Your Understanding

True or false?

1. $9 + 7 < 12$ **2.** 38 in. > 3 ft **3.** $5 * 4 = 80 / 4$ **4.** $13 + 1 > 15 - 1$

Check your answers on page 340.

Arrays and Factors of Counting Numbers

When two numbers are multiplied, the answer is called the **product.** The two numbers that are multiplied are called **factors** of the product.

> **Note**
>
> Whenever you are asked to find the factors of a counting number, those factors *must* be counting numbers.

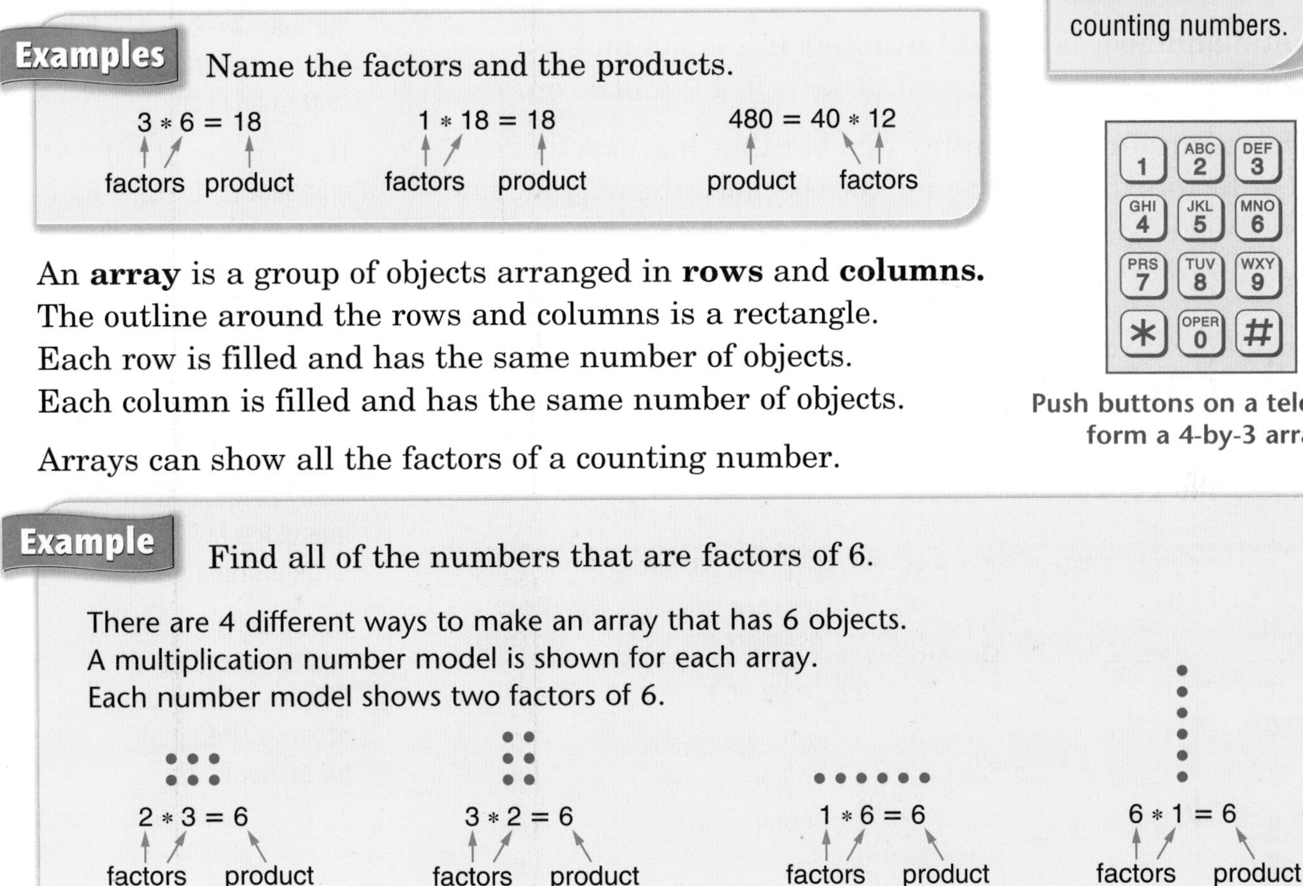

Examples Name the factors and the products.

$3 * 6 = 18$ — factors: 3 and 6; product: 18

$1 * 18 = 18$ — factors: 1 and 18; product: 18

$480 = 40 * 12$ — product: 480; factors: 40 and 12

An **array** is a group of objects arranged in **rows** and **columns.** The outline around the rows and columns is a rectangle. Each row is filled and has the same number of objects. Each column is filled and has the same number of objects.

Arrays can show all the factors of a counting number.

Push buttons on a telephone form a 4-by-3 array.

Example Find all of the numbers that are factors of 6.

There are 4 different ways to make an array that has 6 objects.
A multiplication number model is shown for each array.
Each number model shows two factors of 6.

$2 * 3 = 6$ — factors: 2 and 3; product: 6

$3 * 2 = 6$ — factors: 3 and 2; product: 6

$1 * 6 = 6$ — factors: 1 and 6; product: 6

$6 * 1 = 6$ — factors: 6 and 1; product: 6

The arrays show that 1, 2, 3, and 6 are factors of 6. They are the *only* factors of 6.

Example Find all of the numbers that are factors of 5.

There are only 2 ways to make an array that has 5 objects.
The arrays show that 1 and 5 are the only factors of 5.
There are only 2 ways to multiply two counting numbers and get 5.
$1 * 5 = 5$ and $5 * 1 = 5$. So 1 and 5 are both factors of 5.

$1 * 5 = 5$

$5 * 1 = 5$

1 and 5 are the *only* factors of 5.

Check Your Understanding

Draw arrays and find all of the factors for each number.

1. 12 **2.** 15 **3.** 24 **4.** 11

Check your answers on page 340.

Kinds of Counting Numbers

The **counting numbers** are the numbers 1, 2, 3, 4, 5, and so on.

A counting number is an **even number** if 2 is one of its factors. The even numbers are 2, 4, 6, 8, 10, and so on.

A counting number is an **odd number** if it is not an even number. The odd numbers are 1, 3, 5, 7, 9, 11, and so on.

A **prime number** is a counting number that has exactly 2 different factors. 5 is a prime number because the only factors of 5 are 1 and 5. All of these numbers are prime numbers:

2, 3, 5, 7, 11, 13, 17, 19

A **composite number** is any counting number greater than 1 that is not a prime number. Each composite number has 3 or more different factors. All of these numbers are composite numbers:

4, 6, 8, 9, 10, 12, 14, 15, 16, 18, 20

Facts about the Numbers 1 through 20

Number	Factors	Prime or Composite	Even or Odd
1	1	neither	odd
2	1 and 2	prime	even
3	1 and 3	prime	odd
4	1, 2, and 4	composite	even
5	1 and 5	prime	odd
6	1, 2, 3, and 6	composite	even
7	1 and 7	prime	odd
8	1, 2, 4, and 8	composite	even
9	1, 3, and 9	composite	odd
10	1, 2, 5, and 10	composite	even
11	1 and 11	prime	odd
12	1, 2, 3, 4, 6, and 12	composite	even
13	1 and 13	prime	odd
14	1, 2, 7, and 14	composite	even
15	1, 3, 5, and 15	composite	odd
16	1, 2, 4, 8, and 16	composite	even
17	1 and 17	prime	odd
18	1, 2, 3, 6, 9, and 18	composite	even
19	1 and 19	prime	odd
20	1, 2, 4, 5, 10, and 20	composite	even

Note

The number 0 and the opposites of even numbers (−2, −4, −6, and so on) are usually said to be even.

The opposites of odd numbers (−1, −3, −5 and so on) are usually said to be odd.

Did You Know?

Many persons are interested in identifying large prime numbers.

See http://primes.utm.edu/largest.html for details and for lists of large and small prime numbers.

Note

Prime and composite numbers have at least two different factors. The number 1 has only one factor. So, the number 1 is not prime, and it is not composite.

Multiples

When you skip-count by a number, your counts are the **multiples** of that number. Since you can always count further, lists of multiples can go on forever.

Examples Find multiples of 2, 3, and 5.

Multiples of 2: 2, 4, 6, 8, 10, 12, 14, 16, 18, 20, 22, 24, ...

Multiples of 3: 3, 6, 9, 12, 15, 18, 21, 24, 27, 30, 33, 36, ...

Multiples of 5: 5, 10, 15, 20, 25, 30, 35, 40, 45, 50, 55, ...

Note

The three dots, ..., mean that a list can go on in the same way forever.

Common Multiples

On lists of multiples for numbers, some numbers are on all the lists. These numbers are called **common multiples.**

Example Find common multiples of 2 and 3.

Multiples of 2: 2, 4, **6,** 8, 10, **12,** 14, 16, **18,** 20, 22, **24,** ...

Multiples of 3: 3, **6,** 9, **12,** 15, **18,** 21, **24,** 27, ...

Common multiples of 2 and 3: 6, 12, 18, 24, ...

Least Common Multiples

The **least common multiple** of two numbers is the smallest number that is a multiple of both numbers.

Example Find the least common multiple of 6 and 8.

Multiples of 6: 6, 12, 18, **24,** 30, 36, 42, **48,** 54, ...

Multiples of 8: 8, 16, **24,** 32, 40, **48,** 56, ...

24 and 48 are common multiples. 24 is the smallest common multiple.

24 is the least common multiple for 6 and 8. It is the smallest number that can be divided evenly by both 6 and 8.

Check Your Understanding

Find the least common multiple of each pair of numbers.

1. 6 and 12 **2.** 4 and 10 **3.** 9 and 15

Check your answers on page 340.

Addition Methods

Partial-Sums Method

The **partial-sums method** is used to find sums mentally or with paper and pencil. Here is the partial-sums method for adding 2-digit or 3-digit numbers:

1. Add the 100s. **2.** Add the 10s. **3.** Add the 1s.

4. Then add the sums you just found (the partial sums).

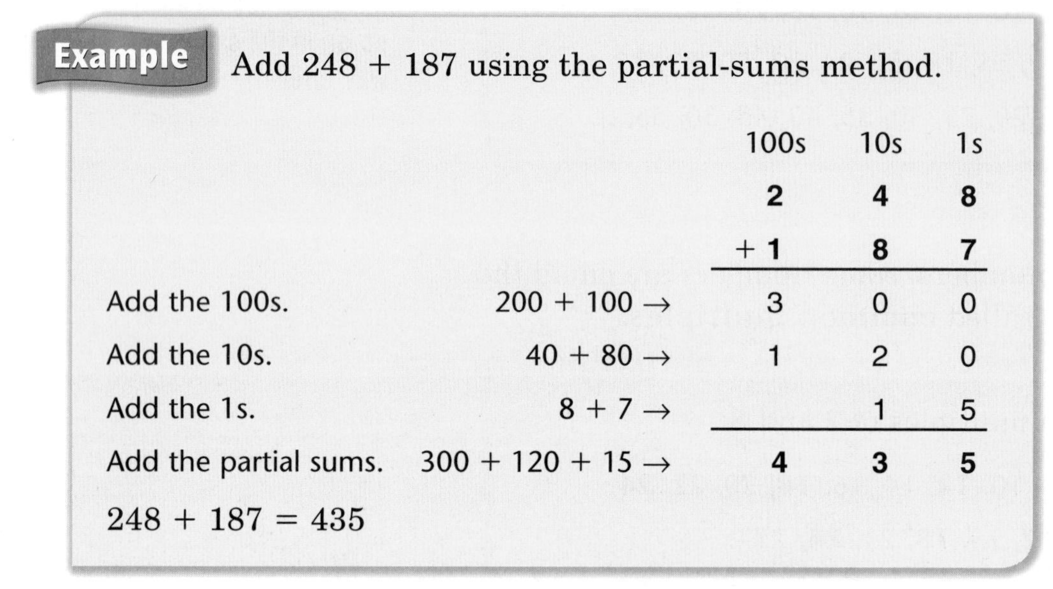

Example Add 248 + 187 using the partial-sums method.

		100s	10s	1s
		2	**4**	**8**
		+ 1	**8**	**7**
Add the 100s.	200 + 100 →	3	0	0
Add the 10s.	40 + 80 →	1	2	0
Add the 1s.	8 + 7 →		1	5
Add the partial sums.	300 + 120 + 15 →	**4**	**3**	**5**

248 + 187 = 435

Note
Larger numbers with 4 or more digits are added the same way.

Use base-10 blocks to show the partial-sums method.

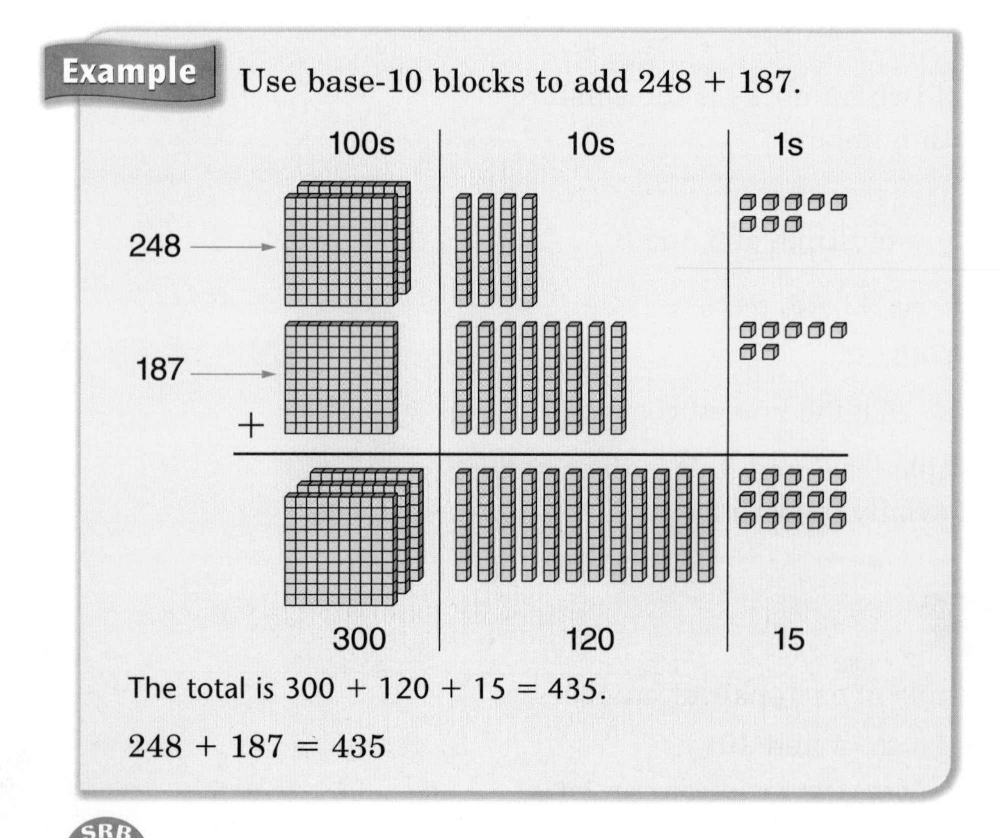

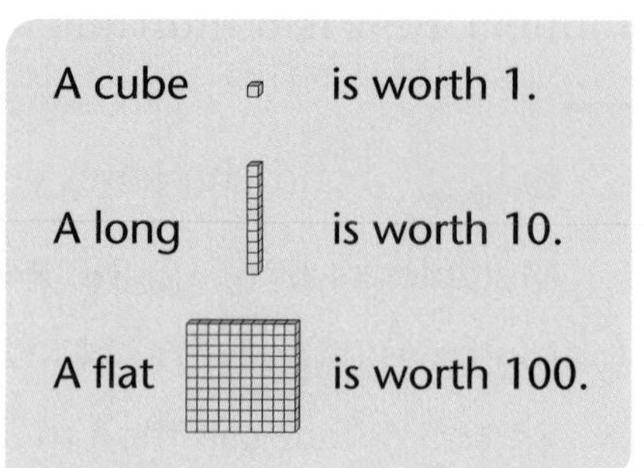

Column-Addition Method

The **column-addition method** can be used to find sums with paper and pencil, but it is not a good method for finding sums mentally.

Here is the column-addition method for adding 2-digit or 3-digit numbers.

1. Draw lines to separate the 1s, 10s, and 100s places.
2. Add the numbers in each column. Write each sum in its column.
3. If there are 2 digits in the 1s place, trade 10 ones for 1 ten.
4. If there are 2 digits in the 10s places, trade 10 tens for 1 hundred.

Example Add 248 + 187 using the column-addition method.

	100s	10s	1s
	2	**4**	**8**
	+ 1	**8**	**7**
Add the numbers in each column.	3	12	15
Two digits in the ones place. Trade 15 ones for 1 ten and 5 ones. Move the 1 ten to the tens column.	3	13	5
Two digits in the tens place. Trade 13 tens for 1 hundred and 3 tens. Move the 1 hundred to the hundreds column.	**4**	**3**	**5**

248 + 187 = 435

Larger numbers with 4 or more digits are added the same way.

Check Your Understanding

Add.

1. 327 + 252

2. 67 + 45

3. 277 + 144

4. 2,268 + 575

5. 34 + 54 + 47

6. 25 + 57 **7.** 44 + 55 **8.** 607 + 340 **9.** 1,509 + 63 **10.** 60 + 56 + 7

Check your answers on page 340.

Subtraction Methods

Trade-First Subtraction Method

The **trade-first method** is similar to the method for subtracting that most adults were taught when they were in school.

Here is the trade-first method for subtracting 2-digit or 3-digit numbers.

- Look at the digits in the 1s place. If subtracting these digits gives a negative number, trade 1 ten for 10 ones.
- Look at the digits in the 10s place. If subtracting these digits gives a negative number, trade 1 hundred for 10 tens.
- Subtract in each column.

Did You Know?

In 1498, Johann Widmann wrote the first book that used + and − symbols. Traders and shopkeepers had used both symbols long before this. They used + to show they had too much of something. They used − to show they had too little of something.

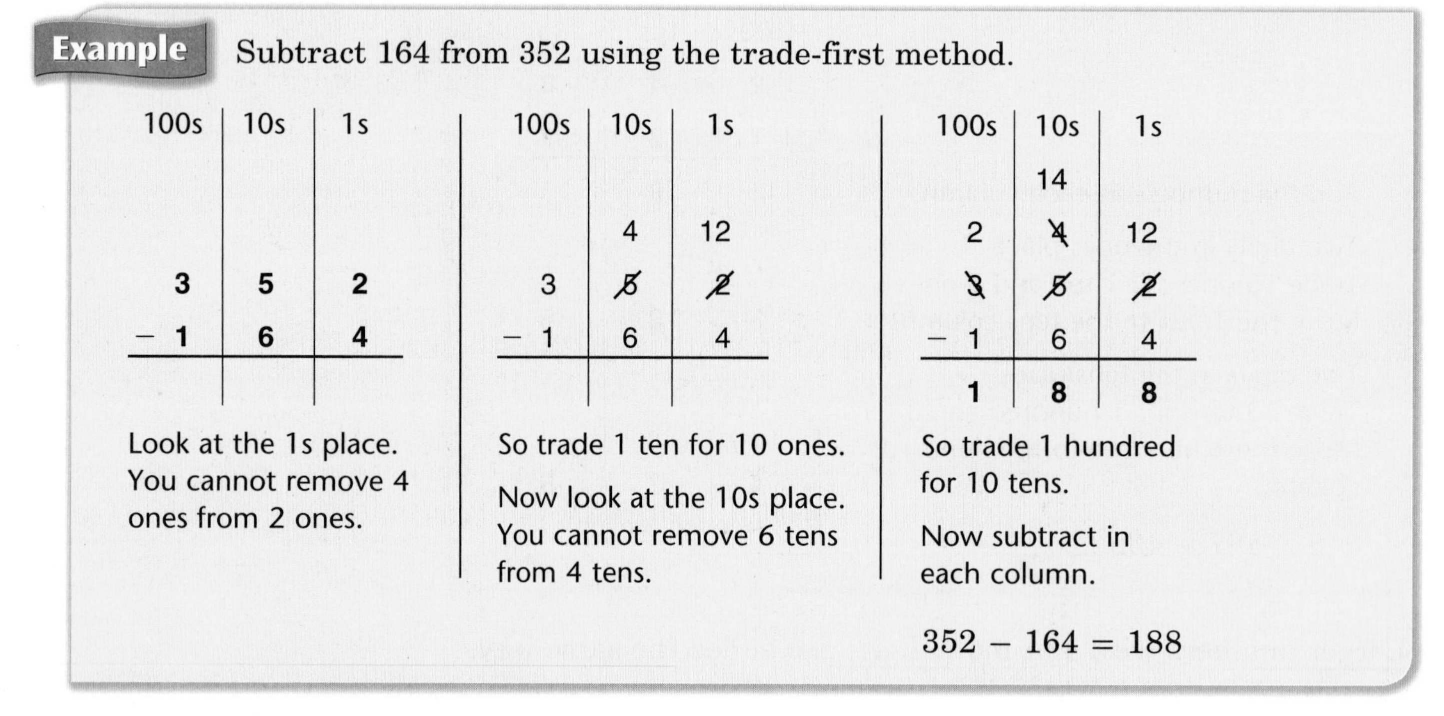

Larger numbers with 4 or more digits are subtracted the same way.

Check Your Understanding

Subtract.

1. $84 - 38$
2. $764 - 281$
3. $583 - 306$
4. $808 - 444$
5. $5,425 - 2,366$
6. $568 - 86$

Check your answers on page 340.

Base-10 blocks are useful for solving problems, but sometimes they are not available. You can draw pictures instead. The pictures sometimes used in this book to show base-10 blocks are shown in the margin.

Pictures of base-10 blocks show how the trade-first method works.

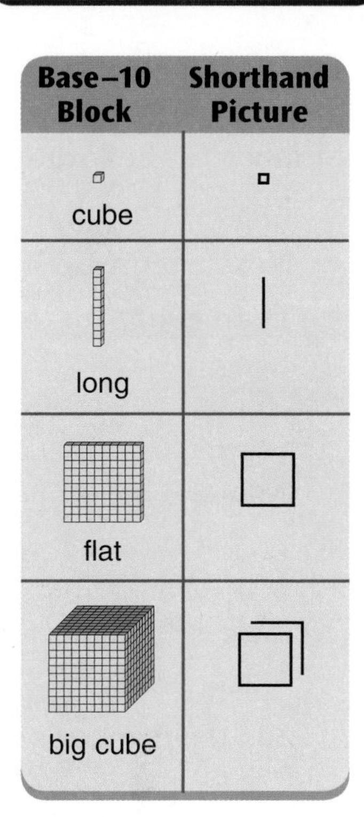

Base–10 Block	Shorthand Picture
cube	
long	
flat	
big cube	

Example $324 - 167 = ?$

Use pictures of base-10 blocks to model the larger number, 324.

Write the number to be subtracted, 167, beneath the block pictures.

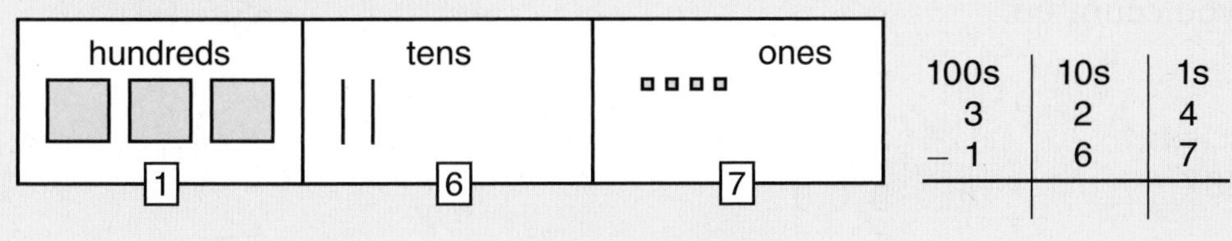

Think: Can I remove 7 cubes from 4 cubes? No
Trade 1 long for 10 cubes.

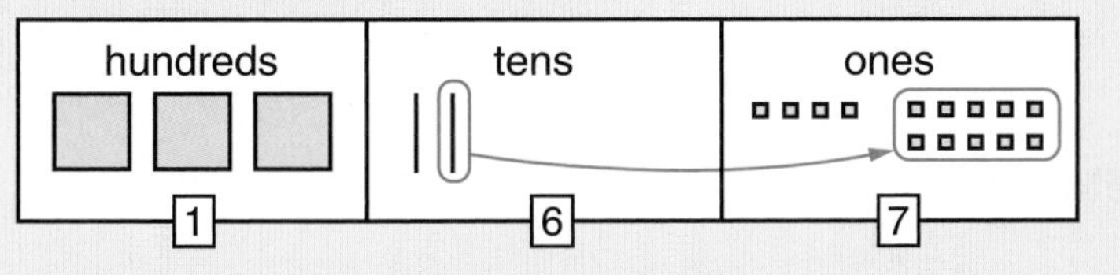

Think: Can I remove 6 longs from 1 long? No
Trade 1 flat for 10 longs.

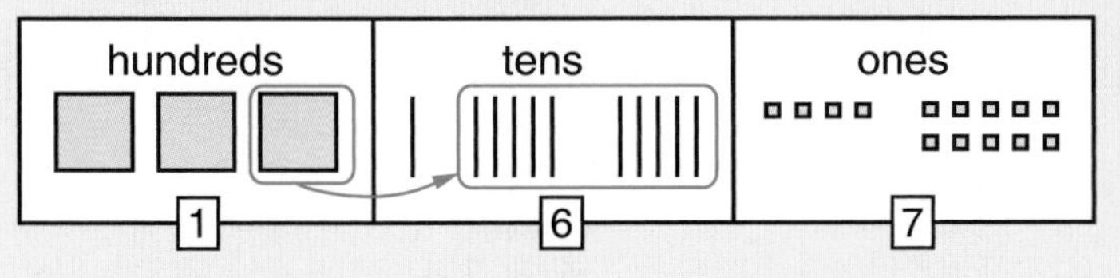

After all of the trading, the blocks look like this:

hundreds	tens	ones
1	6	7

Now subtract in each column. The difference is 157.

hundreds	tens	ones

100s	10s	1s
3	2	4
– 1	6	7
1	**5**	**7**

Counting-Up Method

You can subtract two numbers by counting up from the smaller number to the larger number.

- First count up to the nearest multiple of 10.
- Next, count up by 10s and 100s.
- Then count up to the larger number.

Example $325 - 38 = ?$

Write the smaller number, 38.

As you count from 38 up to 325, circle each number that you count up.

38	
+ (2)	Count up to the nearest 10.
40	
+ (60)	Count up to the nearest 100.
100	
+ (200)	Count up to the largest possible hundred.
300	
+ (25)	Count up to the larger number.
325	

Add the numbers you circled:
2 + 60 + 200 + 25 = 287
You counted up by 287.

$325 - 38 = 287$

Left-to-Right Subtraction Method

Starting at the left, subtract column-by-column.

Examples

	$932 - 356 = ?$	$782 - 294 = ?$
	932	**782**
Subtract the 100s.	− **300**	− **200**
	632	582
Subtract the 10s.	− **50**	− **90**
	582	492
Subtract the 1s.	− **6**	− **4**
	576	**488**

$932 - 356 = 576$

$782 - 294 = 488$

Check Your Understanding

Subtract.

1. 172 − 74 **2.** 453 − 157 **3.** 468 − 253 **4.** 915 − 56

Check your answers on page 340.

Partial-Differences Method

1. Subtract from left to right, one column at a time.
2. In some cases, the larger number is on the bottom and the smaller number is on the top. When you subtract these numbers, the difference will be a negative number.

Example 746 − 263 = ?

		7 4 6
		− 2 6 3
Subtract the 100s.	700 − 200 →	5 0 0
Subtract the 10s.	40 − 60 →	− 2 0
Subtract the 1s.	6 − 3 →	3
Find the total.	500 − 20 + 3 →	**4 8 3**

Think of 40 − 60 as first subtracting 40 and then subtracting 20 more. 40 − 40 − 20 = −20

746 − 263 = 483

Same-Change Rules

Here are the **same-change rules** for subtraction problems:

- If you add the same number to both numbers in the problem, the answer is the same.
- If you subtract the same number from both numbers in the problem, the answer is the same.

Use these rules to change the second number in the problem to a number that has 0 in the ones place. Make the *same change* to the first number. Then subtract.

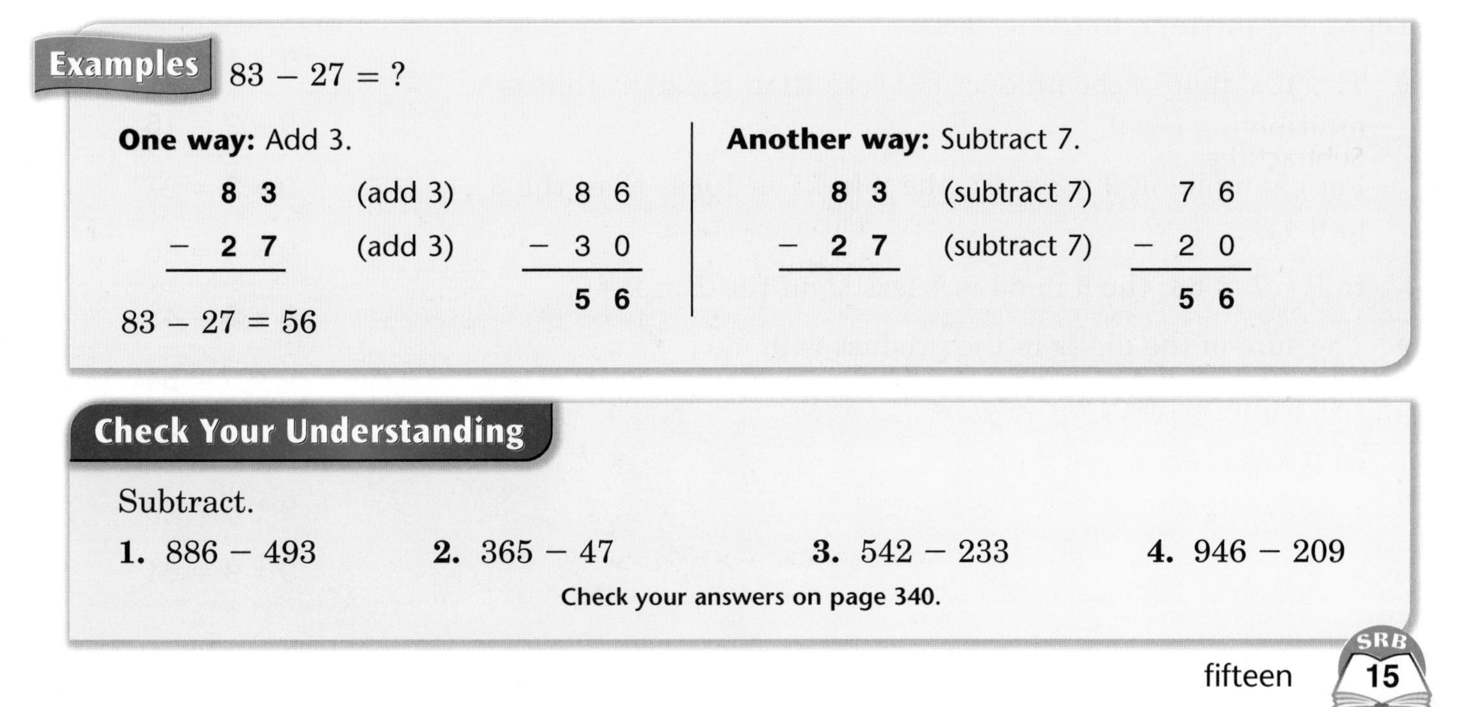

Examples 83 − 27 = ?

One way: Add 3.

8 3	(add 3)	8 6
− 2 7	(add 3)	− 3 0
		5 6

83 − 27 = 56

Another way: Subtract 7.

8 3	(subtract 7)	7 6
− 2 7	(subtract 7)	− 2 0
		5 6

Check Your Understanding

Subtract.

1. 886 − 493 **2.** 365 − 47 **3.** 542 − 233 **4.** 946 − 209

Check your answers on page 340.

Basic Multiplication Facts

The symbols × and ∗ are both used to indicate multiplication. In this book, the symbol ∗ is used most often.

A basic multiplication fact is a product of two one-digit factors.

$8 * 5 = 40$ is a basic fact. If you don't remember a basic fact, try one of the following methods:

Use Counters or Draw a Picture

To find $8 * 5$, make 8 groups of counters with 5 counters in each group, or draw a simple picture to show 8 groups of 5 objects. Then count all the objects.

×　×　×　×
××　××　××　××
××　××　××　××

×　×　×　×
××　××　××　××
××　××　××　××

Skip Count Up

To find $8 * 5$, count up by 5s, 8 times:
5, 10, 15, 20, 25, 30, 35, 40.
Use your fingers to keep track as you skip count.

Use Known Facts

The answer to a 4s fact can be found by doubling, then doubling again. For example, to find $4 * 7$, double 7 to get 14. Then double 14 to get 28.

4s Facts
Double and then double again.

The answer to an 8s fact can be found by doubling three times. For example, to find $8 * 6$, double 6 to get 12. Double again to get 24. And then double a third time to get 48.

8s Facts
Double 3 times.

The answer to a 6s fact can be found by using a related 5s fact. For example, $6 * 8$ is equal to 8 more than $5 * 8$.
$6 * 8 = 5 * 8 + 8 = 40 + 8$, or 48.

There is a **pattern** to the 9s facts:

- The 10s digit in the product is 1 less than the digit that is multiplying the 9.

  For example, in $9 * 3 = 27$, the 2 in 27 is 1 less than the 3 in $9 * 3$.

  In $9 * 7 = 63$, the 6 in 63 is 1 less than the 7 in $9 * 7$.

- The sum of the digits in the product is 9.

  For example, in $9 * 3 = 27$, $2 + 7 = 9$.

  In $9 * 7 = 63$, $6 + 3 = 9$.

9s Facts
$9 * 1 = 9$
$9 * 2 = 18$
$9 * 3 = 27$
$9 * 4 = 36$
$9 * 5 = 45$
$9 * 6 = 54$
$9 * 7 = 63$
$9 * 8 = 72$
$9 * 9 = 81$

Extended Multiplication Facts

Numbers such as 10, 100, and 1,000 are called **powers of 10.**

It is easy to multiply a whole number, *n,* by a power of 10:
To the right of the number *n,* write as many zeros as there are zeros in the power of 10.

Examples Notice that the number of zeros attached and the number of zeros in the power of 10 are the same.

10 * 57 = 570

100 * 57 = 5,700

1,000 * 57 = 57,000

10 * 40 = 400

100 * 40 = 4,000

1,000 * 40 = 40,000

100 * 320 = 32,000

10,000 * 82 = 820,000

1,000,000 * 7 = 7,000,000

If you have memorized the basic multiplication facts, you can solve problems such as 8 * 60 and 4,000 * 3 mentally.

Examples

8 * 60 = ?

Think: 8 [6s] = 48
Then 8 [60s] is 10 times as much.

8 * 60 = 10 * 48 = 480

4,000 * 3 = ?

Think: 4 [3s] = 12
Then 4,000 [3s] is 1,000 times as much.

4,000 * 3 = 1,000 * 12 = 12,000

You can use a similar method to solve problems such as 30 * 50 and 200 * 90 mentally.

Examples

30 * 50 = ?

Think: 3 [50s] = 150
Then 30 [50s] is 10 times as much.

30 * 50 = 10 * 150 = 1,500

200 * 90 = ?

Think: 2 [90s] = 180
Then 200 [90s] is 100 times as much.

200 * 90 = 100 * 180 = 18,000

Check Your Understanding

Solve these problems mentally.

1. 8 * 100 **2.** 1,000 * 41 **3.** 7 * 300 **4.** 4,000 * 9 **5.** 60 * 40 **6.** 500 * 50

Check your answers on page 340.

Multiplication Methods

Partial-Products Method

In the **partial-products method,** you must keep track of the place value of each digit. It may help to write 1s, 10s, and 100s above the columns. Each partial product is either a basic multiplication fact or an extended multiplication fact.

Example 5 * 26 = ?

Think of 26 as 20 + 6.

Multiply each part of 26 by 5.

Add the two partial products.

	100s	10s	1s	
		2	**6**	
*			**5**	
5 * 20 →	1	0	0	extended multiplication fact
5 * 6 →		3	0	basic multiplication fact
	1	**3**	**0**	

5 * 26 = 130

Example 34 * 26 = ?

Think of 26 as 20 + 6.

Think of 34 as 30 + 4.

Multiply each part of 26 by each part of 34.

Add the four partial products.

	100s	10s	1s	
		2	**6**	
*		**3**	**4**	
30 * 20 →	6	0	0	extended multiplication facts
30 * 6 →	1	8	0	extended multiplication facts
4 * 20 →		8	0	extended multiplication facts
4 * 6 →		2	4	basic multiplication fact
	8	**8**	**4**	

34 * 26 = 884

Check Your Understanding

Multiply. Write each partial product. Then add the partial products.

1. 73 * 5 **2.** 43 * 63 **3.** 40 * 27 **4.** 22 * 22 **5.** 316 * 3

Check your answers on page 340.

Lattice Method

The **lattice method** for multiplying has been used for hundreds of years. It is very easy to use if you know the basic multiplication facts.

Example 3 * 45 = ?

The box with cells and diagonals is called a **lattice.**
Write 45 above the lattice.
Write 3 on the right side of the lattice.

Multiply 3 * 5. Then multiply 3 * 4.
Write the answers in the lattice as shown.

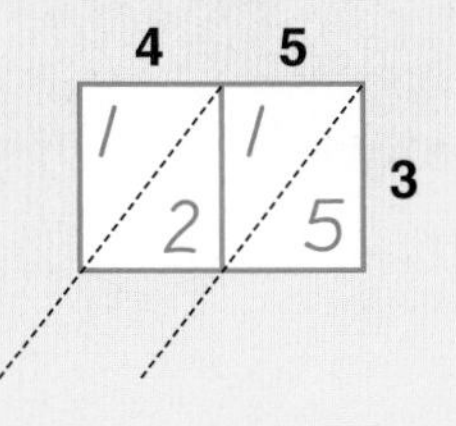

Add the numbers along each diagonal, starting at the right.

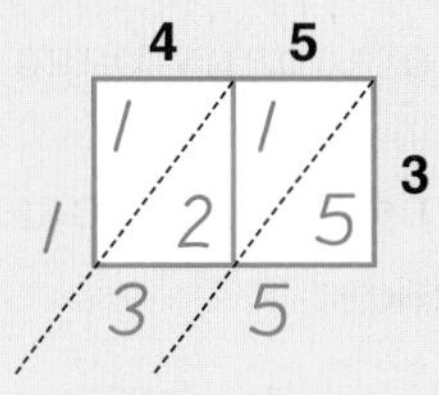

Read the answer. 3 * 45 = 135

Example 34 * 26 = ?

Write 26 above the lattice and 34 on the right side of the lattice.

Multiply 3 * 6. Then multiply 3 * 2.
Multiply 4 * 6. Then multiply 4 * 2.
Write the answers in the lattice as shown.

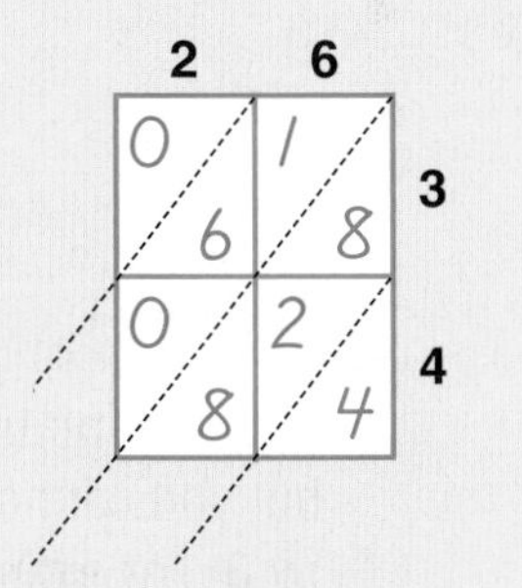

Add the numbers along each diagonal, starting at the right.
When the numbers along a diagonal add to 10 or more:

- record the ones digit, then
- add the tens digit to the sum along the diagonal above.

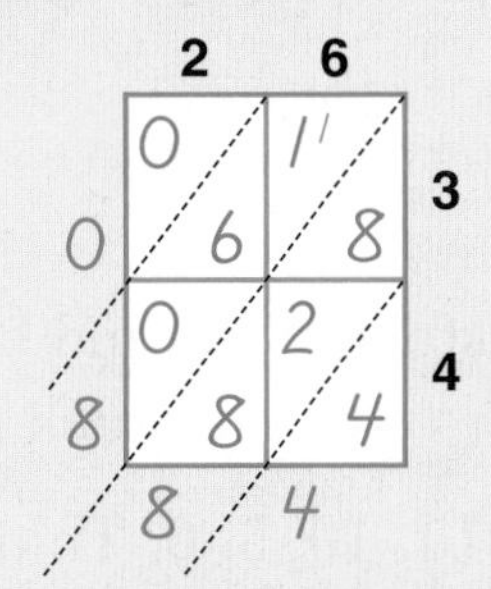

Read the answer. 34 * 26 = 884

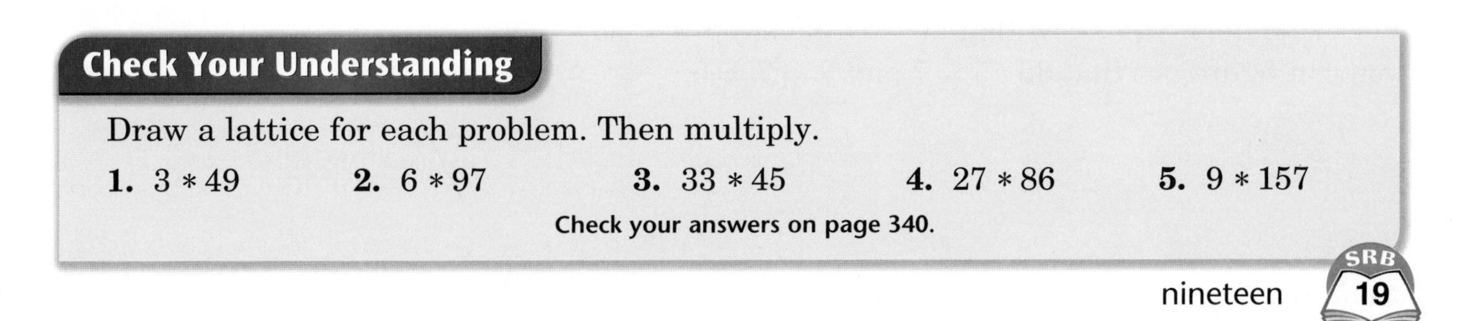

Check Your Understanding

Draw a lattice for each problem. Then multiply.

1. 3 * 49 **2.** 6 * 97 **3.** 33 * 45 **4.** 27 * 86 **5.** 9 * 157

Check your answers on page 340.

Basic Division Facts

A division fact can represent sharing equally or forming equal groups.

Sharing equally:

35 / 5 = ? 5 people share 35 pennies.
How many pennies does each person get? 7

Forming equal groups:

35 / 5 = ? There are 35 oranges in all.
5 oranges are put into each bag.
How many bags can be filled? 7

If you don't remember a basic fact, try one of the following methods:

Use Counters or Draw a Picture

To find 35 / 5, start with 35 objects.

Think: How many 5s in 35?

Make or circle groups of 5 objects each. Count the groups.

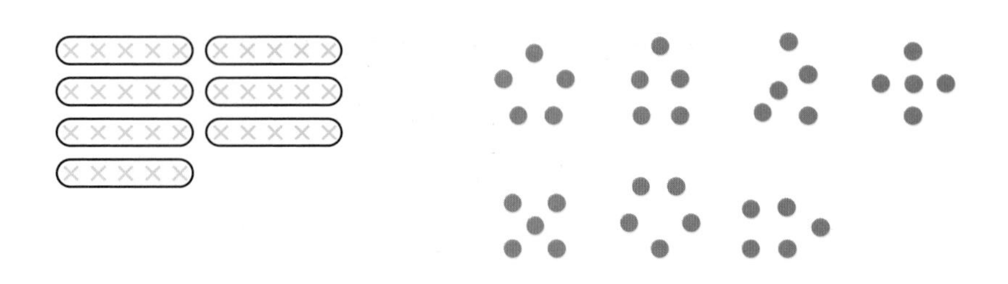

Skip Count Down

To find 35 / 5, start at 35 and count by 5s down to 0. Use your fingers to keep track as you skip count.

35, 30, 25, 20, 15, 10, 5, 0. That's 7 skips.

Use Known Multiplication Facts

Every division fact is related to a multiplication fact. For example, if you know that 5 * 7 = 35 or 7 * 5 = 35, you can figure out that 35 / 5 = 7 and 35 / 7 = 5.

Note

You can multiply by 0, but you cannot divide by 0. For example, 0 * 9 = 0 and 9 * 0 = 0, but 9 / 0 has no answer.

Extended Division Facts

Numbers such as 10, 100, and 1,000 are called **powers of 10.**

In the examples below, use the following method to divide a whole number that ends in zeros by a power of 10:

- Cross out zeros in the number, starting at the ones place.
- Cross out as many zeros as there are zeros in the power of 10.

Examples Notice that the number of zeros crossed out and the number of zeros in the power of 10 is the same.

600 / **10** = 60~~0~~

600 / **100** = 6~~00~~

9,000 / **10** = 900~~0~~

9,000 / **100** = 90~~00~~

9,000 / **1,000** = 9~~000~~

37,000 / **10** = 3700~~0~~

37,000 / **100** = 370~~00~~

37,000 / **1,000** = 37~~000~~

If you know the basic division facts, you can solve problems such as 240 / 3 and 15,000 / 5 mentally.

Examples 240 / 3 = ?

Think: 24 / 3 = 8
Then 240 / 3 is 10 times as much.

240 / 3 = 10 * 8 = 80

15,000 / 5 = ?

Think: 15 / 5 = 3
Then 15,000 / 5 is 1,000 times as much.

15,000 / 5 = 1,000 * 3 = 3,000

You can use a similar method to solve problems such as 1,800 / 30 mentally.

Example 1,800 / 30 = ?

Think: 18 / 3 = 6

First, try 6 as the answer: 6 * 30 = 180, but you want 1,800 or 10 times 180.
6 is *not* the answer.

Next, try 60 as the answer: 60 * 30 = 1,800, so 1,800 / 30 = 60.

1,800 / 30 = 60

Check Your Understanding

Solve these problems mentally.

1. 4,000 / 100 **2.** 47,000 / 1,000 **3.** 28,000 / 7 **4.** 4,800 / 10 **5.** 4,800 / 80

Check your answers on page 340.

Division Methods

Different symbols may be used to indicate division. For example, "94 divided by 6" may be written as $94 \div 6$, $6\overline{)94}$, 94 / 6, or $\frac{94}{6}$

- The number that is being divided is called the **dividend.**
- The number by which it is divided is called the **divisor.**
- The answer to a division problem is called the **quotient.**
- Some numbers cannot be divided evenly. When this happens, the answer includes a quotient and a **remainder.**

Four ways to show "123 divided by 4"

$123 \div 4$ 123 / 4

$4\overline{)123}$ $\frac{123}{4}$

123 is the dividend.
4 is the divisor.

Did You Know?

The $\div$ symbol is called an *obelus.* It was first used as a symbol for division by Johann Rahn, in 1659.

Partial-Quotients Method

In the **partial-quotients method,** it takes several steps to find the quotient. At each step, you find a partial answer (called a **partial quotient**). These partial answers are then added to find the quotient.

Study this example. To find the number of 6s in 94, first find partial quotients, then add them. Record the partial quotients in a column to the right of the original problem.

Example 94 / 6 = ?

Write partial quotients in this column.

	Partial quotients	
$6\overline{)94}$	↓	*Think:* How many [6s] are in 94? At least 10.
− 60	10	The first partial quotient is 10. 10 * 6 = 60 Subtract 60 from 94.
34		At least 5 [6s] are left in 34.
− 30	5	The second partial quotient is 5. 5 * 6 = 30 Subtract 30 from 34.
4	15	Add the partial quotients.
↑ Remainder	↑ Quotient	

The answer is 15 R4. Record the answer as $\begin{array}{r}15\text{ R4}\\6\overline{)94}\end{array}$ or write 94 / 6 → 15 R4.

The partial-quotients method works the same whether you divide by a 2-digit or a 1-digit divisor. It often helps to write down some easy facts for the divisor first.

Example 400 / 22 = ?

Some facts for 22
(to help find partial quotients):

1 * 22 = 22
2 * 22 = 44
5 * 22 = 110
10 * 22 = 220

22)400		
− 220	10	(10 [22s] in 400)
180		
− 110	5	(5 [22s] in 180)
70		
− 44	2	(2 [22s] in 70)
26		
− 22	1	(1 [22] in 26)
4	18	

Record the answer as 22)400 with 18 R4 written above, or write 400 / 22 → 18 R4.

There are different ways to find partial quotients when you use the partial-quotients method. Study the different ways in the example below. Each is correct.

Example 371 / 4 = ?

One way:

4)371	
− 200	50
171	
− 120	30
51	
− 40	10
11	
− 8	2
3	92

Another way:

4)371	
− 200	50
171	
− 160	40
11	
− 8	2
3	92

Still another way:

4)371	
− 360	90
11	
− 8	2
3	92

The answer, 92 R3, is the same for each way.

Check Your Understanding

Divide.

1. 2)75 **2.** 110 / 6 **3.** 121 ÷ 5 **4.** 3)416

Check your answers on page 340.

Here is one way you can make a table of easy facts for the divisor:

- Think about doubling the divisor, and then doubling again and again.
- Then find 10 times the divisor, and take half of this.

For example, suppose the divisor is 28.

Double this, to find	$2 * 28 = 56$.
Now double again, to find	$4 * 28 = 112$.
Double one more time, to find	$8 * 28 = 224$.
Multiply by 10, to find	$10 * 28 = 280$.
Take half of this, to find	$5 * 28 = 140$.

A table of easy facts for 28 is used in the example below. Having the easy facts written down next to the problem can make finding partial quotients a lot easier.

Example 758 / 28 = ?

$1 * 28 = 28$

$2 * 28 = 56$

$4 * 28 = 112$

$5 * 28 = 140$

$8 * 28 = 224$

$10 * 28 = 280$

$$
\begin{array}{r|l}
28\overline{)758} & \\
-\ 560 & 20 \\
\hline
198 & \\
-\ 140 & 5 \\
\hline
58 & \\
-\ 56 & 2 \\
\hline
2 & 27
\end{array}
$$

Record the answer as $28\overline{)758}$ with 27 R2 written above, or write 758 / 28 → 27 R2

Check Your Understanding

Make a table of easy facts for the divisor in each problem. Then divide.

1. 500 ÷ 15 **2.** 840 / 11 **3.** $12\overline{)300}$

Check your answers on page 340.

U.S. Traditional Addition: Whole Numbers

You can add numbers with **U.S. traditional addition.**

Did You Know?

Writing numbers above the addends in U.S. traditional addition is sometimes called "carrying." Numbers that are carried, such as the 1 in the 10s column in Step 1 below, are sometimes called "carry marks" or "carries."

Example 98,095 + 7,407 = ?

Step 1: Start with the 1s: 5 + 7 = 12.
12 = 1 ten + 2 ones
Write 2 in the 1s place below the line.
Write 1 above the numbers in the 10s place.

```
        1
  9 8 0 9 5
+   7 4 0 7
-----------
          2
```

Step 2: Add the 10s: 1 + 9 + 0 = 10.
10 tens = 1 hundred + 0 tens
Write 0 in the 10s place below the line.
Write 1 above the numbers in the 100s place.

```
      1 1
  9 8 0 9 5
+   7 4 0 7
-----------
        0 2
```

Step 3: Continue adding through the 10,000s place.

```
  1   1 1
  9 8 0 9 5
+   7 4 0 7
-----------
1 0 5 5 0 2
```

98,095 + 7,407 = 105,502

Check Your Understanding

Add.

1. 1,066 + 2,525 = ?

2. 9,649 + 803

3. 83,273,705 + 56,029,308

4. 2,099,628 + 870,975

Check your answers on page 347.

U.S. Traditional Subtraction: Whole Numbers

You can subtract numbers with **U.S. traditional subtraction.**

Example $802 - 273 = ?$

Step 1: Start with the 1s.
Since $3 > 2$, you need to regroup.
There are no tens in 802, so trade 1 hundred for 10 tens and then trade 1 ten for 10 ones:
$802 = 7$ hundreds $+ 9$ tens $+ 12$ ones.
Subtract the 1s: $12 - 3 = 9$.

$$\begin{array}{r} \scriptstyle 9 \\ \scriptstyle 7\ \cancel{10}\ 12 \\ \cancel{8}\ \cancel{0}\ \cancel{2} \\ -\ 2\ 7\ 3 \\ \hline 9 \end{array}$$

Step 2: Go to the 10s.
Subtract the 10s: $9 - 7 = 2$.

$$\begin{array}{r} \scriptstyle 9 \\ \scriptstyle 7\ \cancel{10}\ 12 \\ \cancel{8}\ \cancel{0}\ \cancel{2} \\ -\ 2\ 7\ 3 \\ \hline 2\ 9 \end{array}$$

Step 3: Go to the 100s.
Subtract the 100s: $7 - 2 = 5$.

$$\begin{array}{r} \scriptstyle 9 \\ \scriptstyle 7\ \cancel{10}\ 12 \\ \cancel{8}\ \cancel{0}\ \cancel{2} \\ -\ 2\ 7\ 3 \\ \hline 5\ 2\ 9 \end{array}$$

$802 - 273 = 529$

Check Your Understanding

Subtract.

1. $441 - 386 = ?$

2. $\begin{array}{r} 308 \\ -\ 278 \\ \hline \end{array}$

3. $\begin{array}{r} 8{,}694 \\ -\ 2{,}708 \\ \hline \end{array}$

4. $\begin{array}{r} 8{,}060 \\ -\ 6{,}409 \\ \hline \end{array}$

Check your answers on page 347.

U.S. Traditional Multiplication

You can use **U.S. traditional multiplication** to multiply.

Did You Know?

Writing 1 above the numbers in the 10s place is called "carrying" the 1. Sometimes the numbers that are carried are called "carry marks" or "carries."

Example 5 * 963 = ?

Step 1: Multiply the ones.
5 * 3 ones = 15 ones = 1 ten + 5 ones
Write 5 in the 1s place below the line.
Write 1 above the 6 in the 10s place.

```
    1
  9 6 3
*     5
-------
      5
```

Step 2: Multiply the tens.
5 * 6 tens = 30 tens
Remember the 1 ten from Step 1.
30 tens + 1 ten = 31 tens in all
31 tens = 3 hundreds + 1 ten
Write 1 in the 10s place below the line.
Write 3 above the 9 in the 100s place.

```
  3 1
  9 6 3
*     5
-------
    1 5
```

Step 3: Multiply the hundreds.
5 * 9 hundreds = 45 hundreds
Remember the 3 hundreds from Step 2.
45 hundreds + 3 hundreds = 48 hundreds in all
48 hundreds = 4 thousands + 8 hundreds
Write 8 in the 100s place below the line.
Write 4 in the 1,000s place below the line.

```
  3 1
  9 6 3
*     5
-------
4 8 1 5
```

5 * 963 = 4,815

Check Your Understanding

Multiply.

1. 7 * 49 = ?

2. 429 * 6

3. $1,808 * 7

4. 485 * 8

Check your answers on page 347.

You can use U.S. traditional multiplication to multiply by two-digit numbers.

Example 64 * 259 = ?

Step 1: Multiply 259 by the 4 in 64, as if the problem were 4 * 259.

```
    2 3
    2 5 9
*     6 4
---------
  1 0 3 6   ← The partial product
              4 * 259 = 1,036
```

Step 2: Multiply 259 by the 6 in 64, as if the problem were 6 * 259.

The 6 in 64 stands for 6 tens, so write this partial product one place to the left.

Write a 0 in the 1s place to show you are multiplying by tens.

Write the new carries above the old carries.

```
    3 5
    2 3
    2 5 9
*     6 4
---------
  1 0 3 6
1 5 5 4 0   ← 60 * 259 = 15,540
```

Step 3: Add the two partial products to get the final answer.

```
      3 5
      2 3
      2 5 9
  *     6 4
  ---------
    1 0 3 6
+ 1 5 5 4 0
-----------
  1 6 5 7 6   ← 64 * 259 = 16,576
```

64 * 259 = 16,576

Check Your Understanding

Multiply.

1. ? = 17 * 247

2. 538 * 24

3. 1,820 * 53

4. 1,907 * 75

Check your answers on page 347.

U.S. Traditional Long Division

You can use **U.S. traditional long division** to divide.

Example \$935 / 4

To begin, think about sharing \$935 among 4 people: Aimee, Brad, Carla, and Duane.

Money to be Shared

\$100 \$100 \$100 \$100 \$100
\$100 \$100 \$100 \$100
\$10 \$10 \$10
\$1 \$1 \$1 \$1 \$1

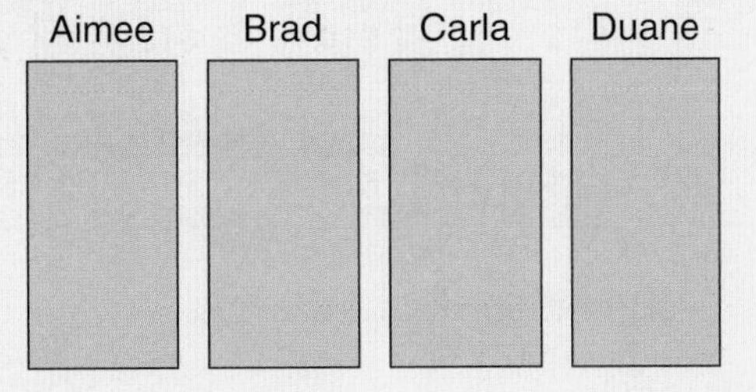

Step 1: Share the [\$100]s.

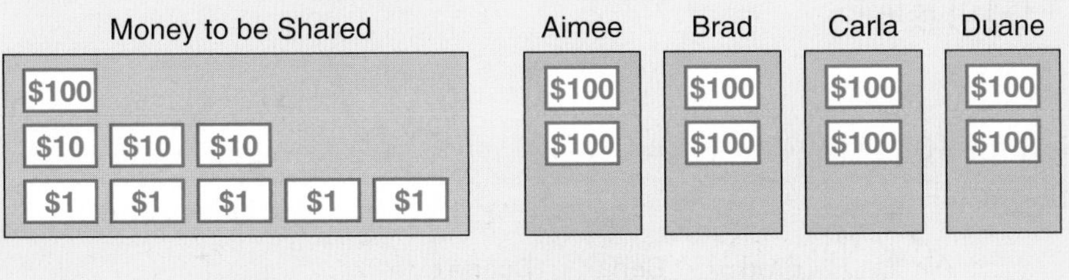

$$\begin{array}{r} 2 \\ 4\overline{)935} \\ -8 \\ \hline 1 \end{array}$$

← Each person gets 2 [\$100]s.
← 2 [\$100]s each for 4 people
← 1 [\$100] is left.

Step 2: Trade the last [\$100] for 10 [\$10]s.
That makes 13 [\$10]s in all.

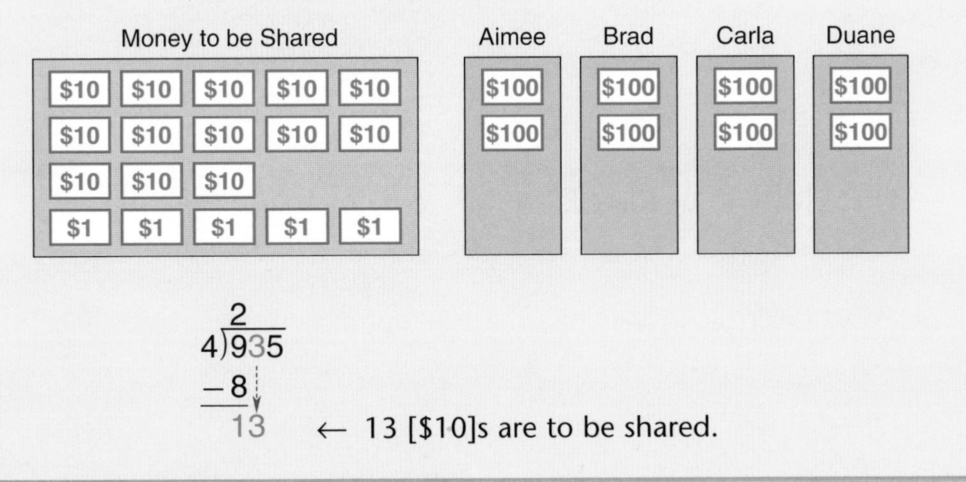

$$\begin{array}{r} 2 \\ 4\overline{)935} \\ -8 \\ \hline 13 \end{array}$$

← 13 [\$10]s are to be shared.

continued

Step 3: Share the [$10]s.

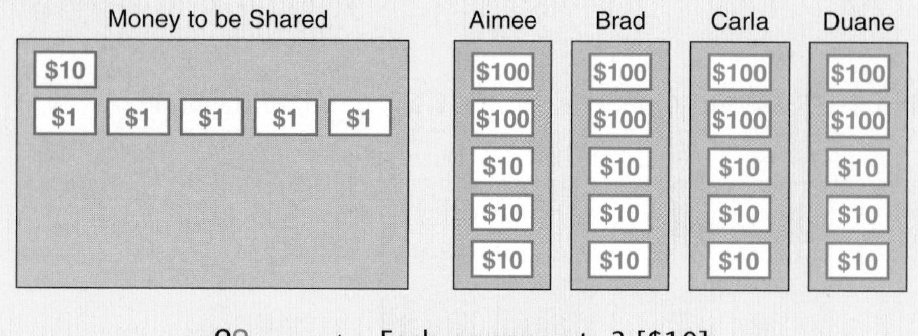

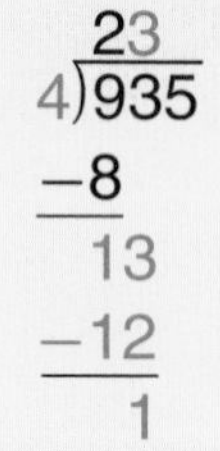

← Each person gets 3 [$10]s.

← 3 [$10]s each for 4 people

← 1 [$10] is left.

Step 4: Trade the last [$10] for 10 [$1]s.
That makes 15 [$1]s in all.

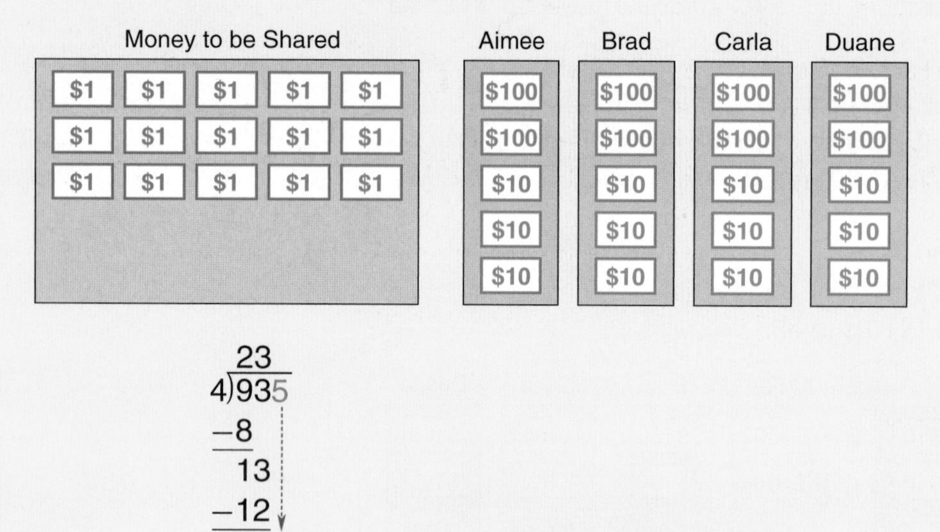

← 15 [$1]s are to be shared.

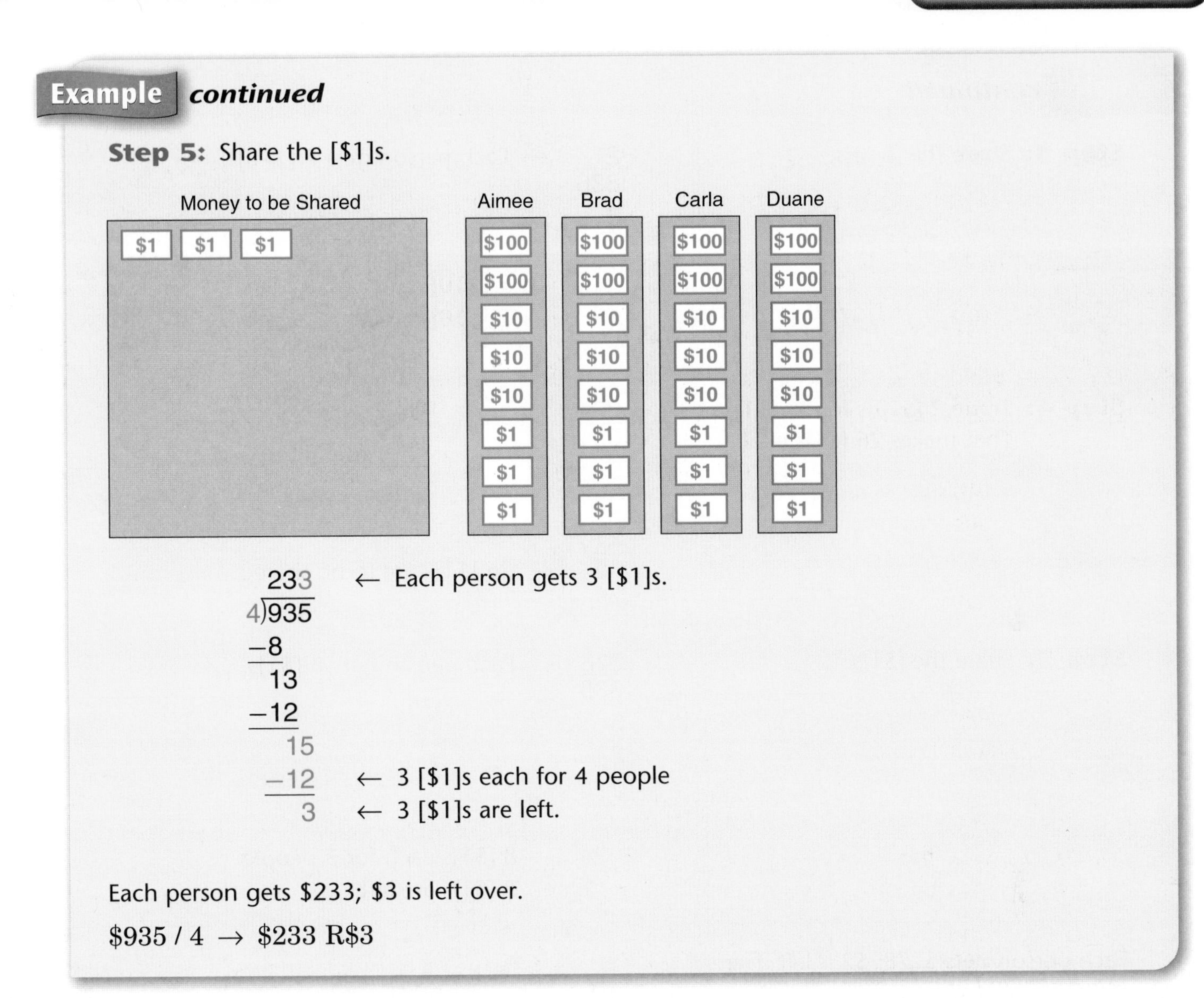

Example ***continued***

Step 5: Share the [$1]s.


```
  233    ← Each person gets 3 [$1]s.
4)935
 −8
  13
 −12
   15
  −12    ← 3 [$1]s each for 4 people
    3    ← 3 [$1]s are left.
```

Each person gets $233; $3 is left over.

$935 / 4 → $233 R$3

Use U.S. traditional long division to divide $836 among 3 people.

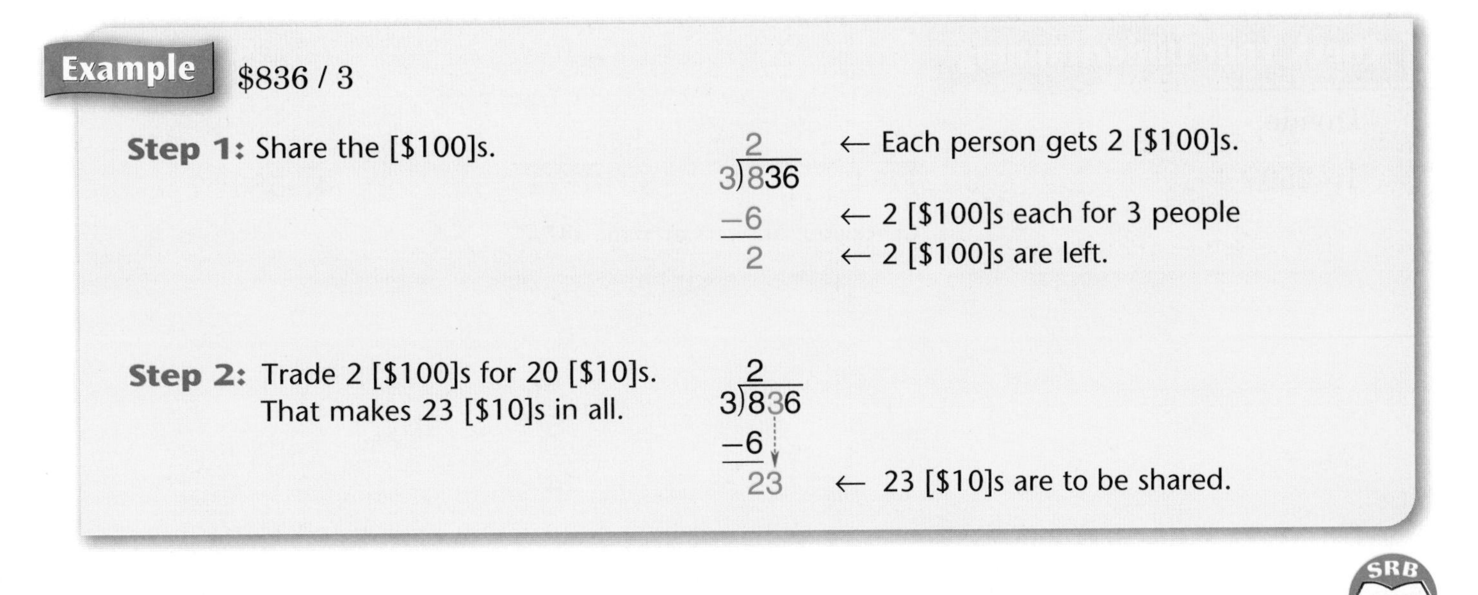

Example $836 / 3

Step 1: Share the [$100]s.

```
   2     ← Each person gets 2 [$100]s.
3)836
 −6      ← 2 [$100]s each for 3 people
  2      ← 2 [$100]s are left.
```

Step 2: Trade 2 [$100]s for 20 [$10]s. That makes 23 [$10]s in all.

```
   2
3)836
 −6↓
  23     ← 23 [$10]s are to be shared.
```

Example ***continued***

Step 3: Share the [$10]s.

$$\begin{array}{r} 27 \\ 3\overline{)836} \\ -6 \\ \hline 23 \\ -21 \\ \hline 2 \end{array}$$

← Each person gets 7 [$10]s.
← 7 [$10]s each for 3 people
← 2 [$10]s are left.

Step 4: Trade 2 [$10]s for 20 [$1]s. That makes 26 [$1]s in all.

$$\begin{array}{r} 27 \\ 3\overline{)836} \\ -6 \\ \hline 23 \\ -21 \\ \hline 26 \end{array}$$

← 26 [$1]s are to be shared.

Step 5: Share the [$1]s.

$$\begin{array}{r} 278 \\ 3\overline{)836} \\ -6 \\ \hline 23 \\ -21 \\ \hline 26 \\ -24 \\ \hline 2 \end{array}$$

← Each person gets 8 [$1]s.
← 8 [$1]s each for 3 people
← 2 [$1]s are left.

Each person gets $278; $2 is left over.
$836 / 3 → $278 R$2

Check Your Understanding

Divide.

1. 250 / 4 **2.** $2\overline{)784}$ **3.** $8\overline{)671}$ **4.** 580 / 7

Check your answers on page 347.

You can use U.S. traditional long division to divide larger numbers.

Example 8,379 / 6

Step 1: Share the thousands.

1	← Each share gets 1 thousand.
$6\overline{)8379}$	
−6	← 1 thousand ∗ 6 shares
2	← 2 thousands are left.

Step 2: Trade 2 thousands for 20 hundreds. Share the hundreds.

13	← Each share gets 3 hundreds.
$6\overline{)8379}$	
−6	
23	← 20 hundreds + 3 hundreds
−18	← 3 hundreds ∗ 6 shares
5	← 5 hundreds are left.

Step 3: Trade 5 hundreds for 50 tens. Share the tens.

139	← Each share gets 9 tens.
$6\overline{)8379}$	
−6	
23	
−18	
57	← 50 tens + 7 tens
−54	← 9 tens ∗ 6 shares
3	← 3 tens are left.

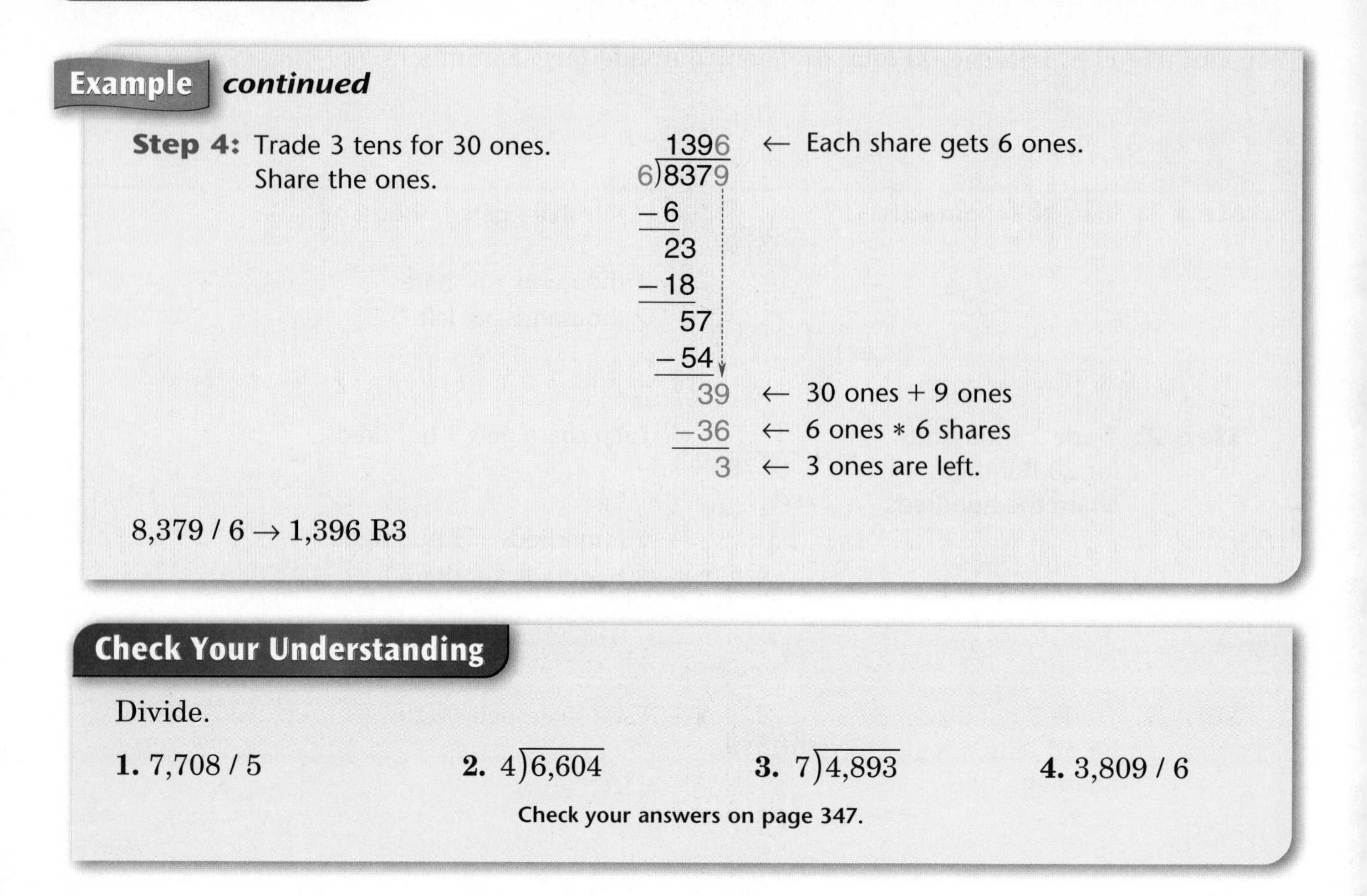

Example ***continued***

Step 4: Trade 3 tens for 30 ones. Share the ones.

```
   1396   ← Each share gets 6 ones.
 6)8379
  −6
   23
  −18
    57
   −54
     39   ← 30 ones + 9 ones
    −36   ← 6 ones * 6 shares
      3   ← 3 ones are left.
```

8,379 / 6 → 1,396 R3

Check Your Understanding

Divide.

1. 7,708 / 5 **2.** $4\overline{)6{,}604}$ **3.** $7\overline{)4{,}893}$ **4.** 3,809 / 6

Check your answers on page 347.

Decimals and Percents

Decimals

Mathematics in everyday life involves more than just **whole numbers.** We also use **decimals** and **fractions** to name numbers that are between whole numbers.

Both decimals and fractions are used to name a part of a whole thing or a part of a collection. Decimals and fractions are also used to make more precise measurements than can be made using only whole numbers.

Fractional parts of a dollar are almost always written as decimals. The receipt at the right shows that lunch cost between 25 dollars and 26 dollars. The "64" in the cost names a part of a dollar.

THE QUAY KEY WEST

0127 Table 26 #Party 2
TOM K SvrCk: 7 15:14 07/11/05

1 ICED TEA	1.25
1 CALIF BURGER	7.25
1 CHEFS SANDWICH	7.95
1 COFFEE	1.50
2 KEY LIME PIE	5.90
Sub Total:	23.85
Tax:	1.79
07/11 15:51 TOTAL:	25.64

WE HOPE YOU ENJOY YOUR STAY IN KEY WEST.

COME BACK SOON!

You probably see many other uses of decimals every day.

Weather reports give rainfall amounts in decimals. The average annual rainfall in New Orleans, Louisiana, is 66.28 inches.

Digital scales in supermarkets show the weight of fruits, vegetables, and meat with decimals.

Winners of many Olympic events are often decided by times measured to hundredths, and sometimes even thousandths, of a second. Florence Griffith-Joyner's winning time for the 100-meter run in 1988 was 10.54 seconds.

Many sports statistics use decimals. In 1993, basketball player Michael Jordan averaged 32.6 points per game. In 1901, baseball player Napoleon Lajoie had a batting average of .426.

Cars have instruments called odometers that measure distance. The word *odometer* comes from the Greek words *odos,* which means road, and *metron,* which means measure. The odometer at the right shows 12,963, which means the car has traveled at least 12,963 miles. The trip meter above it is more precise and shows tenths of a mile traveled. The trip meter at the right shows the car has traveled at least 45.6 miles since it was last reset to 0.

Decimals use the same base-ten place-value system that whole numbers use. The way you compute with decimals is very similar to the way you compute with whole numbers.

Understanding Decimals

Decimals are another way to write fractions. Many fractions have denominators of 10, 100, 1,000, and so on. It is easy to write the decimal names for fractions like these.

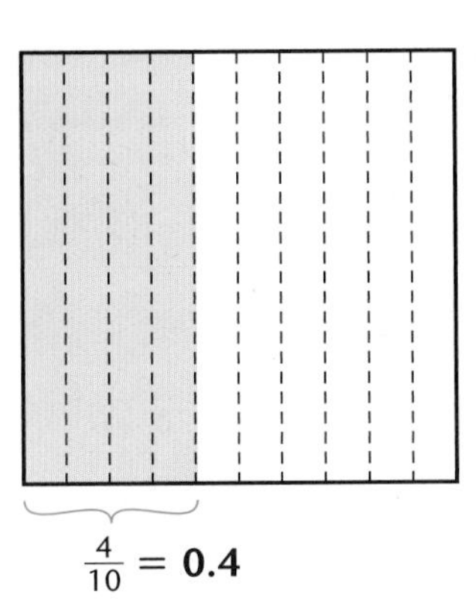

$\frac{4}{10}$ = **0.4**

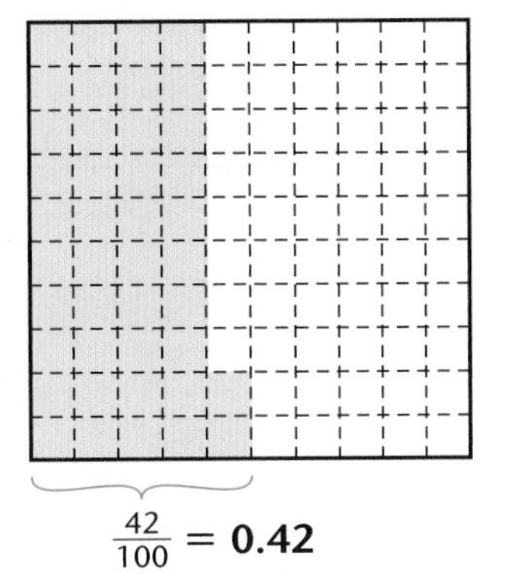

$\frac{42}{100}$ = **0.42**

This square is divided into 10 equal parts. Each part is $\frac{1}{10}$ of the square. The decimal name for $\frac{1}{10}$ is 0.1.

$\frac{4}{10}$ of the square is shaded. The decimal name for $\frac{4}{10}$ is 0.4.

This square is divided into 100 equal parts. Each part is $\frac{1}{100}$ of the square. The decimal name for $\frac{1}{100}$ is 0.01.

$\frac{42}{100}$ of the square is shaded. The decimal name for $\frac{42}{100}$ is 0.42.

Did You Know?

Decimals were invented by the Dutch scientist Simon Stevin in 1585. But there is no single worldwide form for writing decimals. Americans write the decimal 3.25, while the British write 3·25 and the Germans and French write 3,25.

In a decimal, the dot is called the **decimal point.** It separates the whole number part from the decimal part. A number with one digit after the decimal point names *tenths.* A number with two digits after the decimal point names *hundredths.* A number with three digits after the decimal point names *thousandths.*

decimal point

12.105

whole-number part | decimal part

tenths	hundredths	thousandths
$0.3 = \frac{3}{10}$	$0.23 = \frac{23}{100}$	$0.151 = \frac{151}{1,000}$
$0.7 = \frac{7}{10}$	$0.75 = \frac{75}{100}$	$0.002 = \frac{2}{1,000}$
$0.9 = \frac{9}{10}$	$0.02 = \frac{2}{100}$	$0.087 = \frac{87}{1,000}$

Check Your Understanding

1. Write each fraction as a decimal.

   a. $\frac{7}{10}$ **b.** $\frac{2}{100}$ **c.** $\frac{987}{1,000}$ **d.** $\frac{88}{100}$ **e.** $\frac{6}{1,000}$

2. Write each decimal as a fraction.

   a. 0.45 **b.** 0.6 **c.** 0.074 **d.** 0.90 **e.** 0.09

Check your answers on page 340.

Like mixed numbers, decimals are used to name numbers greater than 1.

Example

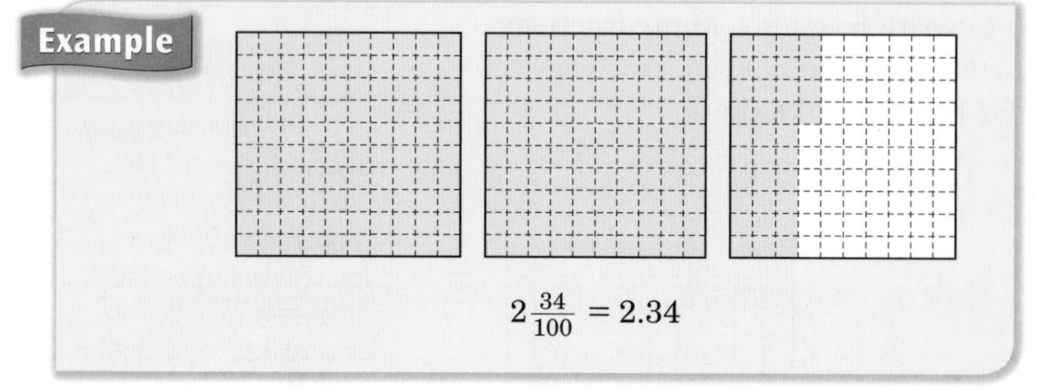

$$2\frac{34}{100} = 2.34$$

Reading Decimals

One way to read a decimal is to say it as you would a fraction or a mixed number. For example, $6.8 = 6\frac{8}{10}$, so 6.8 can be read as "six and eight-tenths." $0.001 = \frac{1}{1,000}$ and is read as "one-thousandth."

Sometimes decimals are read by first saying the whole number part, then saying "point," and finally saying the digits in the decimal part. For example, 6.8 can be read as "six point eight"; 0.15 can be read as "zero point one five." This way of reading decimals is often useful when there are many digits in the decimal.

Examples

0.12 is read as "12 hundredths" or "0 point one two." 98.6 is read as "98 and 6 tenths" or "98 point six." 2.05 is read as "2 and 5 hundredths" or "2 point zero five."

Normal body temperature is about 98.6°F.

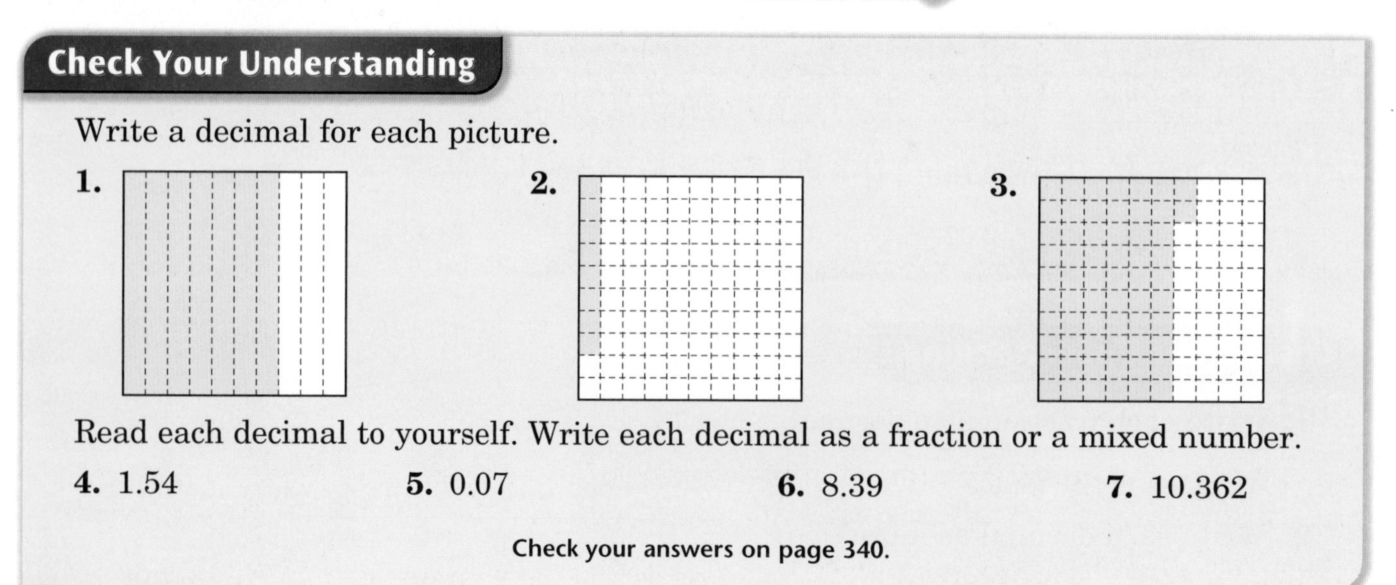

Check Your Understanding

Write a decimal for each picture.

1. **2.** **3.**

Read each decimal to yourself. Write each decimal as a fraction or a mixed number.

4. 1.54 **5.** 0.07 **6.** 8.39 **7.** 10.362

Check your answers on page 340.

Our System for Recording Numbers

The first systems for writing numbers were primitive. Ancient Egyptians used a stroke to record the number 1, a picture of an upside-down oxbow for 10, a coil of rope for 100, a lotus plant for 1,000, and a picture of a god supporting the sky for 1,000,000.

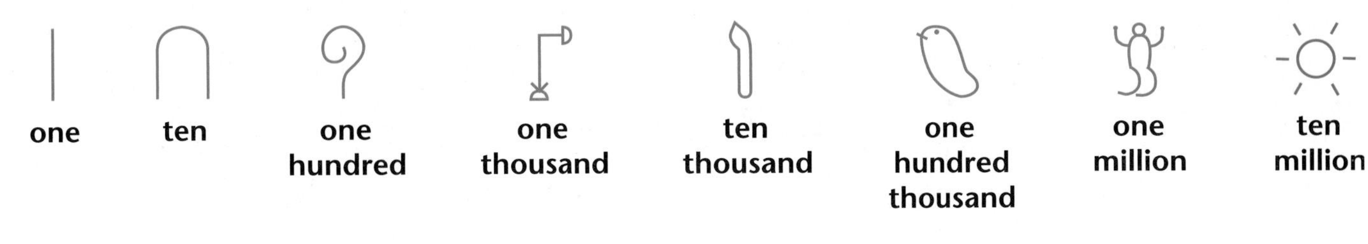

This is how an ancient Egyptian would write the number 43:

10 + 10 + 10 + 10 + 1 + 1 + 1

Our system for writing numbers was invented in India and later improved in Arabia. It is called a **base-ten** system. It uses only 10 symbols, which are called **digits:** 0, 1, 2, 3, 4, 5, 6, 7, 8, and 9. In this system, you can write any number using only these 10 digits.

For a number written in the base-ten system, each digit has a value that depends on its **place** in the number. That is why it is called a **place-value** system.

Note

It should come as no surprise that our number system uses exactly 10 symbols. People probably counted on their fingers when they first started using numbers.

1,000s thousands	**100s** hundreds	**10s** tens	**1s** ones
6	0	7	5

In the number 6,075:

6 is in the **thousands** place; its value is 6 thousands, or 6,000.

0 is in the **hundreds** place; its value is 0.

7 is in the **tens** place; its value is 7 tens, or 70.

5 is in the **ones** place; its value is 5 ones, or 5.

The 0 in 6,075 serves a very important purpose. It "holds" the hundreds place so that the 6 can be in the thousands place. When used in this way, zero is called a **placeholder.**

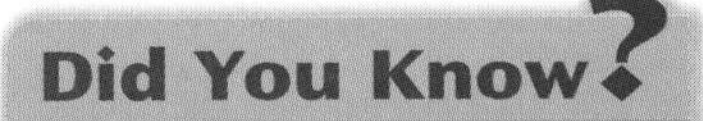

The Babylonians used a symbol for zero as early as 300 B.C. But the circle was not used as a symbol for zero until A.D. 500 in India.

The base-ten system works the same way for decimals as it does for whole numbers.

Examples

1,000s thousands	100s hundreds	10s tens	1s ones	.	0.1s tenths	0.01s hundredths	0.001s thousandths
		3	6	.	7	0	4
			3	.	2	5	0

In the number 36.704,

7 is in the **tenths** place; its value is 7 tenths, or $\frac{7}{10}$, or 0.7.
0 is in the **hundredths** place; its value is 0.
4 is in the **thousandths** place; its value is 4 thousandths, or $\frac{4}{1,000}$, or 0.004.

In the number 3.250,

2 is in the **tenths** place; its value is 2 tenths, or $\frac{2}{10}$, or 0.2.
5 is in the **hundredths** place; its value is 5 hundredths, or $\frac{5}{100}$, or 0.05.
0 is in the **thousandths** place; its value is 0.

Right to Left in the Place-Value Chart

Study this place-value chart. Look at the numbers that name the places. As you move from right to left along the chart, each number is **10 times as large as the number to its right.**

10* 10* 10* 10* 10* 10*

1,000s thousands	100s hundreds	10s tens	1s ones	.	0.1s tenths	0.01s hundredths	0.001s thousandths

one $\frac{1}{100}$ = ten $\frac{1}{1,000}$s
one $\frac{1}{10}$ = ten $\frac{1}{100}$s
one 1 = ten $\frac{1}{10}$s
one 10 = ten 1s
one 100 = ten 10s
one 1,000 = ten 100s

You use facts about the place-value chart each time you make trades using base-10 blocks.

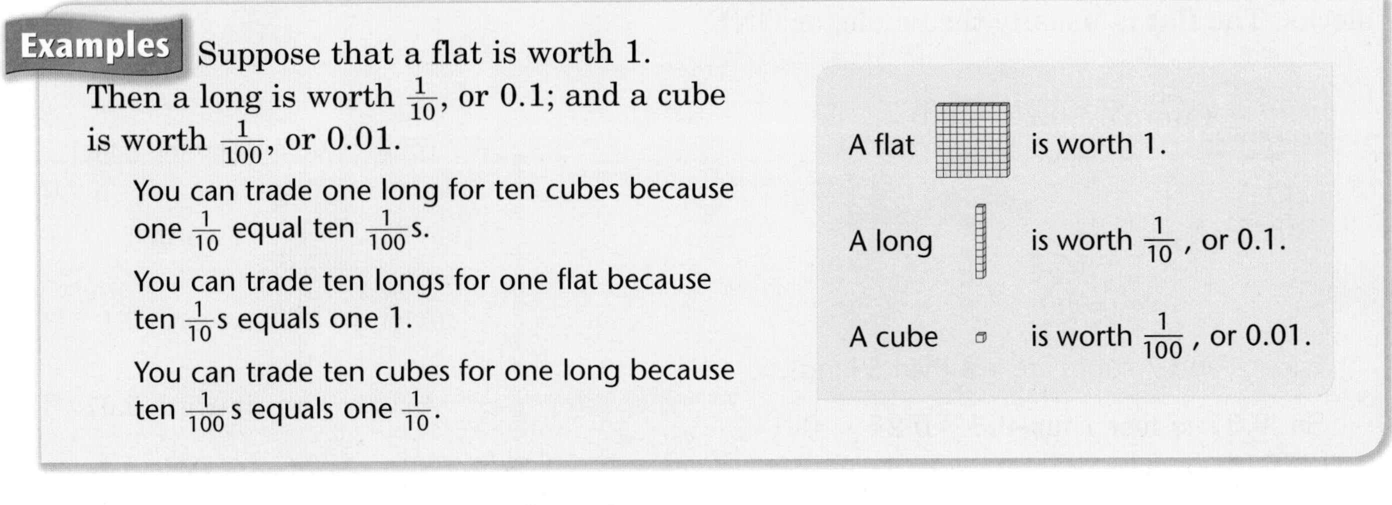

Examples Suppose that a flat is worth 1.

Then a long is worth $\frac{1}{10}$, or 0.1; and a cube is worth $\frac{1}{100}$, or 0.01.

You can trade one long for ten cubes because one $\frac{1}{10}$ equal ten $\frac{1}{100}$s.

You can trade ten longs for one flat because ten $\frac{1}{10}$s equals one 1.

You can trade ten cubes for one long because ten $\frac{1}{100}$s equals one $\frac{1}{10}$.

Left to Right in the Place-Value Chart

Study the place-value chart below. Look at the numbers that name the places. As you move from left to right along the chart, each number is **$\frac{1}{10}$ as large as the number to its left.**

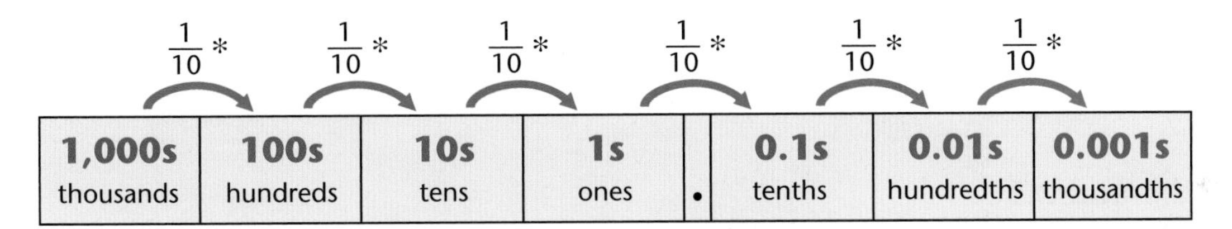

1,000s	100s	10s	1s		0.1s	0.01s	0.001s
thousands	hundreds	tens	ones	.	tenths	hundredths	thousandths

one 100 $= \frac{1}{10}$ of 1,000

one 10 $= \frac{1}{10}$ of 100

one 1 $= \frac{1}{10}$ of 10

one $\frac{1}{10} = \frac{1}{10}$ of 1

one $\frac{1}{100} = \frac{1}{10}$ of $\frac{1}{10}$

one $\frac{1}{1,000} = \frac{1}{10}$ of $\frac{1}{100}$

Check Your Understanding

1. What is the value of the digit 2 in each of these numbers?

   a. 24.7 **b.** 0.21 **c.** 31.62 **d.** 4.123 **e.** 0.002

2. Tell how much each digit in 87.654 is worth.
3. What is the smallest decimal you can write using the digits 8, 2, and 4?

Check your answers on page 340.

Comparing Decimals

One way to compare decimals is to model them with base-10 blocks. The flat is usually the whole, or ONE.

Example Compare 0.27 and 0.3.

0.27 0.3

2 longs and 7 cubes are less than 3 longs.

So, 0.27 is less than 0.3. $0.27 < 0.3$

For the examples on this page:

A flat is worth 1.

A long is worth 0.1.

A cube is worth 0.01.

Another way to compare decimals is to draw pictures of base-10 blocks.

Base-10 Blocks and Their Shorthand Pictures

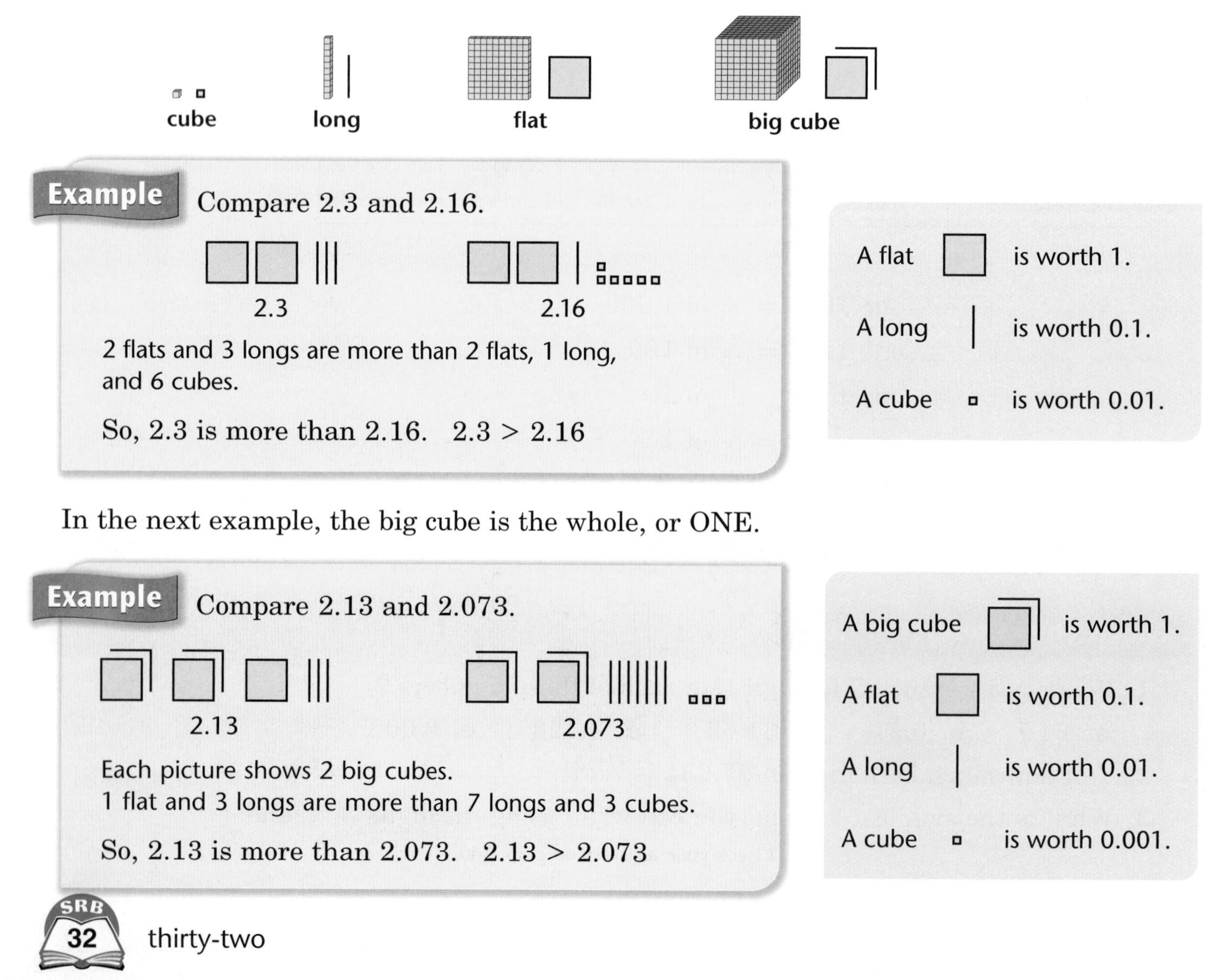

Example Compare 2.3 and 2.16.

2.3 2.16

2 flats and 3 longs are more than 2 flats, 1 long, and 6 cubes.

So, 2.3 is more than 2.16. $2.3 > 2.16$

A flat is worth 1.

A long is worth 0.1.

A cube is worth 0.01.

In the next example, the big cube is the whole, or ONE.

Example Compare 2.13 and 2.073.

2.13 2.073

Each picture shows 2 big cubes.
1 flat and 3 longs are more than 7 longs and 3 cubes.

So, 2.13 is more than 2.073. $2.13 > 2.073$

A big cube is worth 1.

A flat is worth 0.1.

A long is worth 0.01.

A cube is worth 0.001.

You can write a 0 at the end of a decimal without changing the value of the decimal: $0.7 = 0.70$. Attaching 0s is sometimes called "padding with 0s." Think of it as trading for smaller pieces.

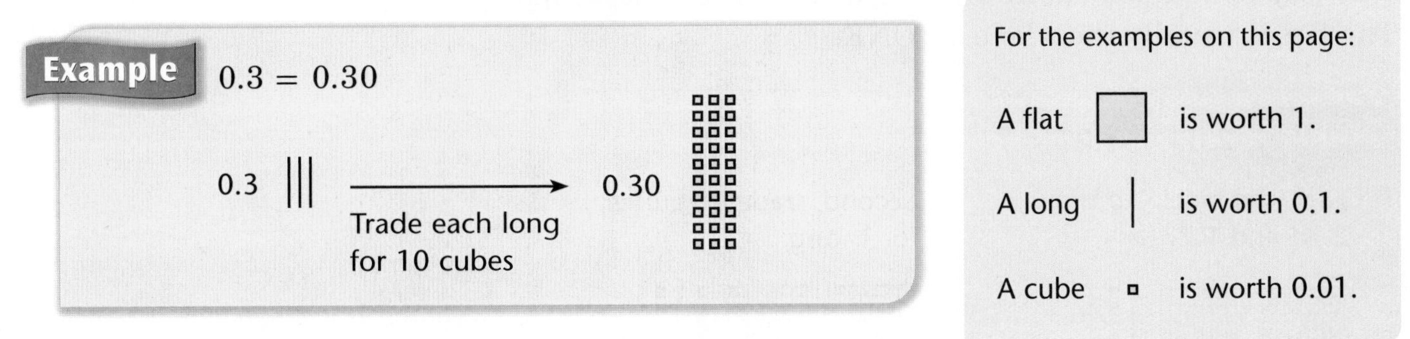

Padding with 0s makes comparing decimals easier.

Examples Compare 0.2 and 0.05.

0.2 = 0.20 (Trade 2 longs for 20 cubes.)
20 cubes are more than 5 cubes.
20 hundredths is more than 5 hundredths.

$0.20 > 0.05$, so $0.2 > 0.05$.

Compare 0.99 and 1.

1 = 1.00 (Trade 1 flat for 100 cubes.)
99 cubes are less than 100 cubes.
99 hundredths is less than 100 hundredths.

$0.99 < 1.00$, so $0.99 < 1$

A place-value chart can also be used to compare decimals.

Example Compare 3.915 and 3.972.

1s ones	.	**0.1s** tenths	**0.01s** hundredths	**0.001s** thousandths
3	.	9	1	5
3	.	9	7	2

The ones digits *are the same.* They are both worth 3.
The tenths digits *are the same.* They are both worth 9 tenths, or 0.9.

The hundredths digits are *not the same.* The 1 is worth 1 hundredth, or 0.01.
The 7 is worth 7 hundredths, or 0.07. The 7 is worth more than the 1.

So, 3.915 is less than 3.972. $3.915 < 3.972$

Check Your Understanding

Compare the numbers in each pair.

1. 0.68, 0.2 **2.** 5.39, 5.5 **3.** $\frac{1}{2}$, 0.51 **4.** 0.999, 1.1

Check your answers on page 341.

Adding and Subtracting Decimals

There are many ways to add and subtract decimals.
One way to add and subtract decimals is to use base-10 blocks.
The flat is usually the whole, or ONE.

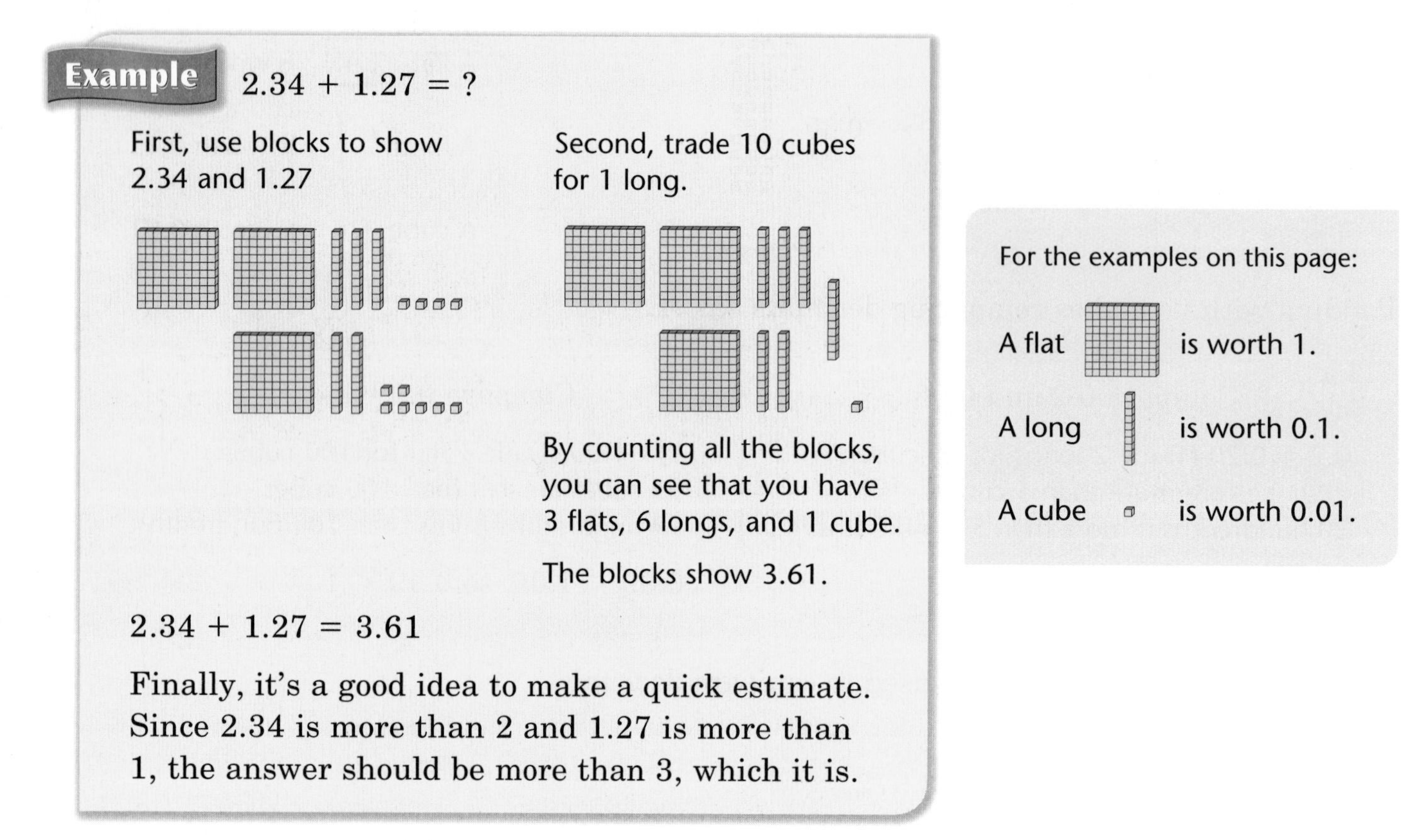

Example 2.34 + 1.27 = ?

First, use blocks to show 2.34 and 1.27

Second, trade 10 cubes for 1 long.

By counting all the blocks, you can see that you have 3 flats, 6 longs, and 1 cube.

The blocks show 3.61.

2.34 + 1.27 = 3.61

Finally, it's a good idea to make a quick estimate. Since 2.34 is more than 2 and 1.27 is more than 1, the answer should be more than 3, which it is.

For the examples on this page:

A flat is worth 1.

A long is worth 0.1.

A cube is worth 0.01.

To subtract with base-10 blocks, count out blocks for the larger number, and take away blocks for the smaller number. Then count the remaining blocks.

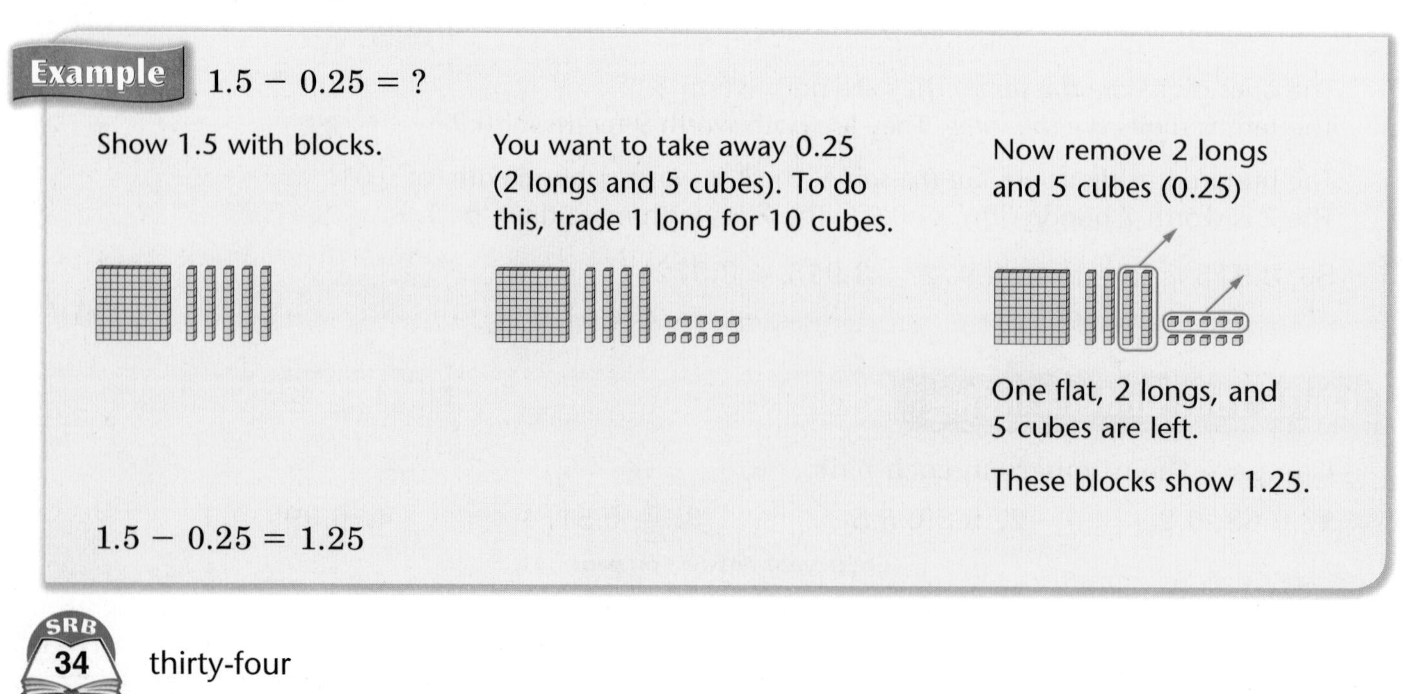

Example 1.5 − 0.25 = ?

Show 1.5 with blocks.

You want to take away 0.25 (2 longs and 5 cubes). To do this, trade 1 long for 10 cubes.

Now remove 2 longs and 5 cubes (0.25)

One flat, 2 longs, and 5 cubes are left.

These blocks show 1.25.

1.5 − 0.25 = 1.25

It is often easier to use shorthand pictures instead of real base-10 blocks.

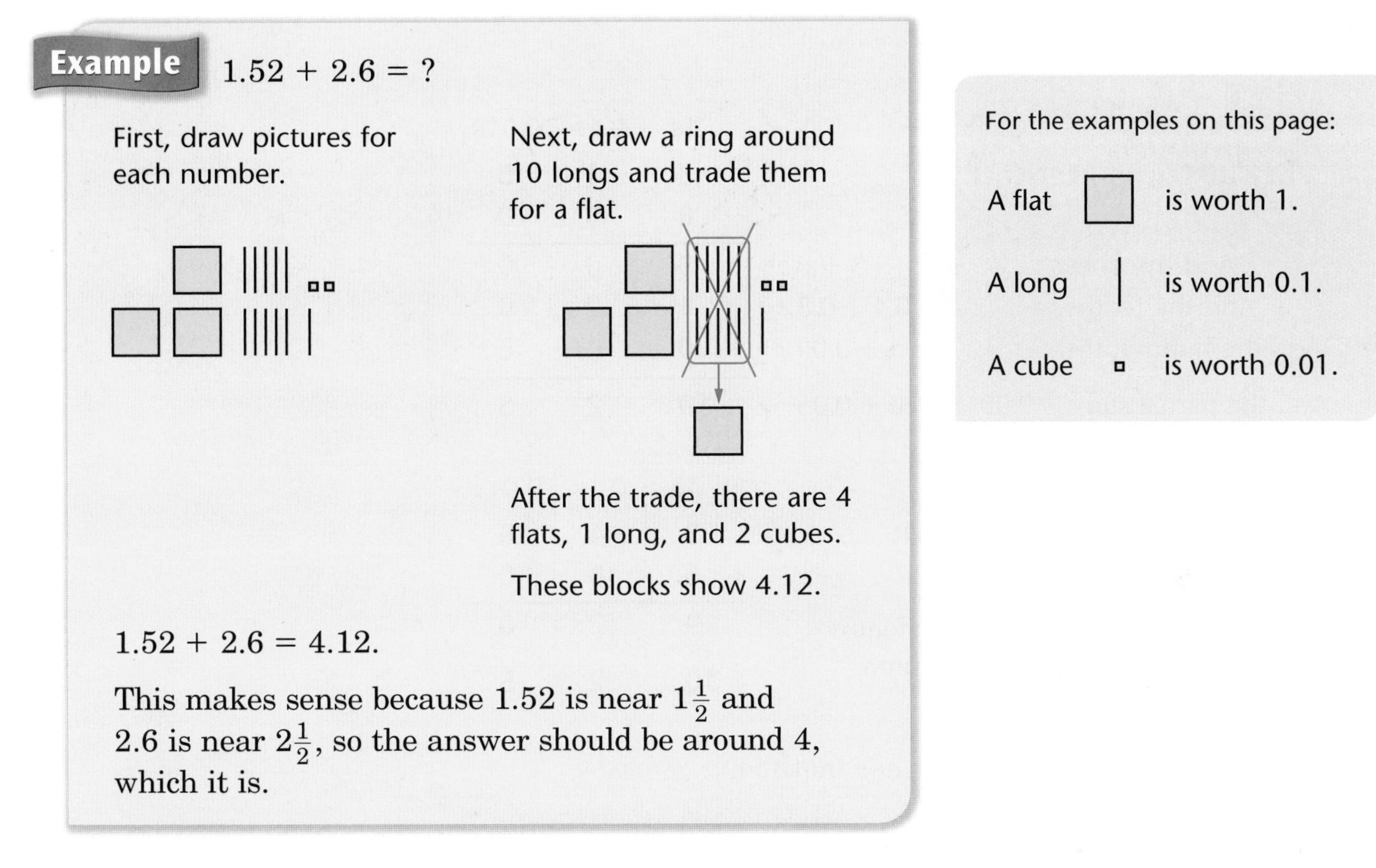

Example $1.52 + 2.6 = ?$

First, draw pictures for each number.

Next, draw a ring around 10 longs and trade them for a flat.

After the trade, there are 4 flats, 1 long, and 2 cubes.

These blocks show 4.12.

$1.52 + 2.6 = 4.12.$

This makes sense because 1.52 is near $1\frac{1}{2}$ and 2.6 is near $2\frac{1}{2}$, so the answer should be around 4, which it is.

For the examples on this page:

A flat is worth 1.

A long is worth 0.1.

A cube is worth 0.01.

Shorthand pictures of blocks are also useful for subtraction.

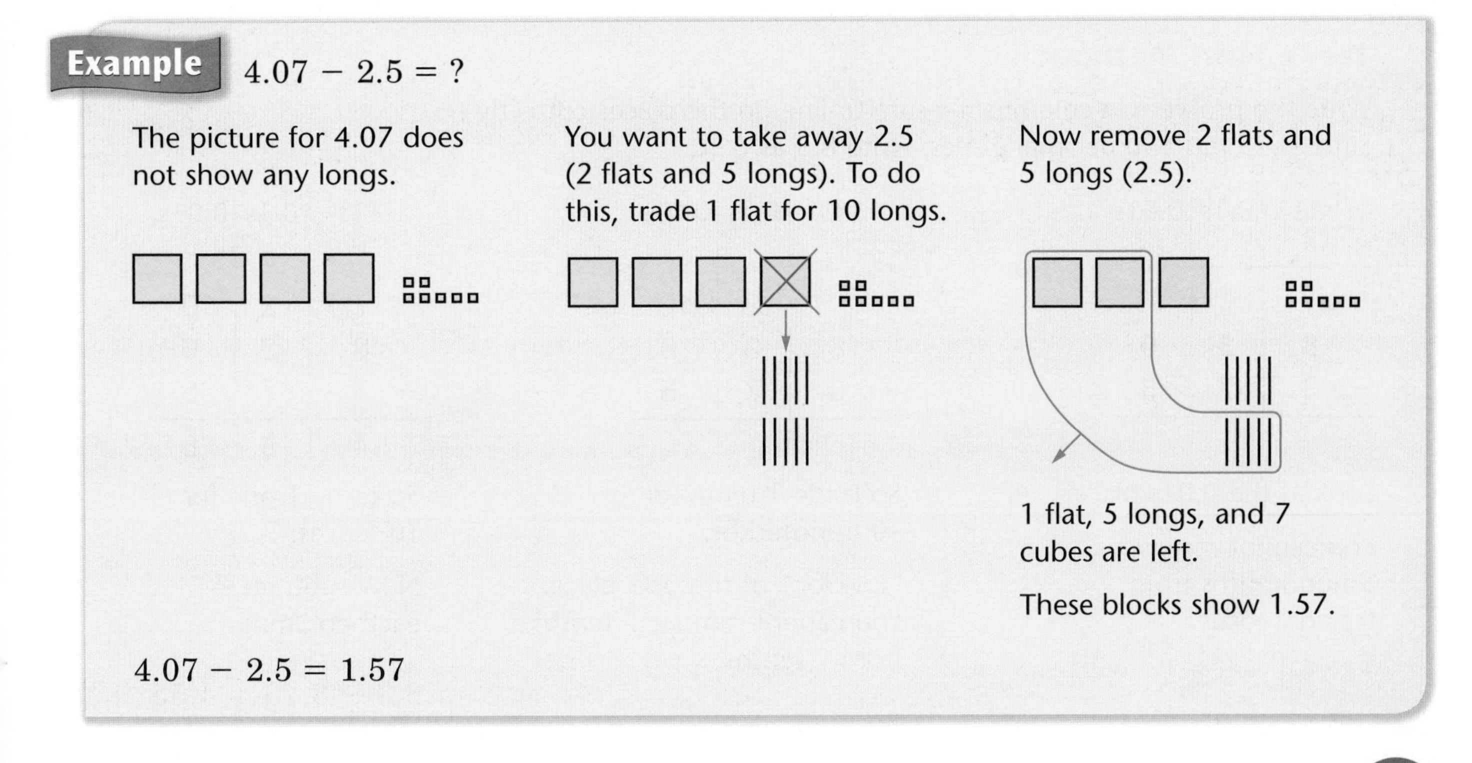

Example $4.07 - 2.5 = ?$

The picture for 4.07 does not show any longs.

You want to take away 2.5 (2 flats and 5 longs). To do this, trade 1 flat for 10 longs.

Now remove 2 flats and 5 longs (2.5).

1 flat, 5 longs, and 7 cubes are left.

These blocks show 1.57.

$4.07 - 2.5 = 1.57$

Most paper-and-pencil strategies for adding and subtracting whole numbers also work for decimals. The main difference is that you have to line up the places correctly, either by attaching 0s to the end of the numbers or by lining up the ones place.

Examples 3.45 + 6.8 = ?

Partial-Sums Method:

		1s		0.1s	0.01s
		3	.	4	5
		+ 6	.	8	0
Add the ones.	3 + 6 →	9	.	0	0
Add the tenths.	0.4 + 0.8 →	1	.	2	0
Add the hundredths.	0.05 + 0.00 →	0	.	0	5
Add the partial sums.	9.00 + 1.20 + 0.05 →	**10**	**.**	**2**	**5**

Column-Addition Method:

Add in each column.

Trade 12 tenths for 1 one and 2 tenths. Move the 1 one to the ones column.

1s	0.1s	0.01s
3.	**4**	**5**
+ 6.	**8**	**0**
9.	12	0
10.	**2**	**5**

3.45 + 6.8 = 10.25 using either method.

Example 8.3 − 3.75 = ?

Trade-First Method:

Write the problem in columns. Be sure to line up the places correctly. Since 3.75 has two decimal places, write 8.3 as 8.30.

1s	0.1s	0.01s
8.	**3**	**0**
− 3.	**7**	**5**

Look at the 0.01s place.

You cannot remove 5 hundredths from 0 hundredths.

1s	0.1s	0.01s
	2	10
8.	~~3~~	~~0~~
− 3.	7	5

So trade 1 tenth for 10 hundredths.

Now look at the 0.1s place. You cannot remove 7 tenths from 2 tenths.

1s	0.1s	0.01s
	12	
7	~~2~~	10
~~8~~.	~~3~~	~~0~~
− 3.	7	5
4.	**5**	**5**

So trade 1 one for 10 tenths.

Now subtract in each column.

8.3 − 3.75 = 4.55

Example $8.3 - 3.75 = ?$

Left-to-Right Subtraction Method:

Since 3.75 has two decimal places, write 8.3 as 8.30.

	8.30
Subtract the ones.	− **3**.00
	5.30
Subtract the tenths.	− 0.**7**0
	4.60
Subtract the hundredths.	− 0.0**5**
	4.55

$8.3 - 3.75 = 4.55$

Example $8.3 - 3.75 = ?$

Counting-Up Method:

Since 3.75 has two decimal places, write 8.3 as 8.30.

There are many ways to count up from 3.75 to 8.30. Here is one.

3.75
+ (0.25)
4.00
+ (4.00)
8.00
+ (0.30)
8.30

Add the numbers you circled and counted up by:

0.25
4.00
+ 0.30
4.55

You counted up by 4.55.

$8.3 - 3.75 = 4.55$

Check Your Understanding

Add or subtract.

1. $3.03 + 2.7$ **2.** $2.06 - 0.28$ **3.** $5.4 - 1.38$ **4.** $2.3 + 6.47$

Check your answers on page 341.

Multiplication of Decimals

You can use the same procedures for multiplying decimals that you use for whole numbers. The main difference is that with decimals you have to decide where to place the decimal point in the product.

One way to solve multiplication problems with decimals is to multiply as if both factors were whole numbers. Then adjust the product:

Step 1. Make a magnitude estimate of the product.

Step 2. Multiply as if the factors were whole numbers.

Step 3. Use the magnitude estimate to place the decimal point in the answer.

Note

A *magnitude estimate* is a very rough estimate that answers questions like: *Is the solution in the ones? Tens? Hundreds? Thousands?*

Example 15.2 * 3.6 = ?

Step 1: Make a magnitude estimate.

- Round 15.2 to 20 and 3.6 to 4.
- Since 20 * 4 = 80, the product will be in the tens. (*In the tens* means between 10 and 100.)

Step 2: Multiply as you would with whole numbers using the partial-products method. Work from left to right. Ignore the decimal points.

		152
		*** 36**
30 * 100	→	3000
30 * 50	→	1500
30 * 2	→	60
6 * 100	→	600
6 * 50	→	300
6 * 2	→	12
Add the partial products.	→	**5472**

Step 3: Place the decimal point correctly in the answer. Since the magnitude estimate is in the tens, the product must be in the tens. Place the decimal point between the 4 and the 7 in 5472.

So, 15.2 * 3.6 = 54.72.

Did You Know?

The price of a gallon of gasoline always includes an additional $\frac{9}{10}$ cent per gallon. Many people believe that this practice is deceptive.

For example, suppose that gasoline costs $2.879 per gallon. A 10-gallon purchase would cost exactly 10 * $2.879 = $28.79. A 9-gallon purchase should cost exactly 9 * $2.879 = $25.911, but the buyer is charged $25.92.

Example $3.27 * 0.8 = ?$

Step 1: Make a magnitude estimate.
- Round 3.27 to 3 and 0.8 to 1.
- Since $3 * 1 = 3$, the product will be in the ones. (*In the ones* means between 1 and 10.)

Step 2: Multiply as you would with whole numbers. Ignore the decimal points.

		327
		$*$ **8**
$8 * 300$	→	2400
$8 * 20$	→	160
$8 * 7$	→	56
$2400 + 160 + 56$	→	**2616**

Step 3: Place the decimal point correctly in the answer. Since the magnitude estimate is in the ones, the product must be in the ones. Place the decimal point between the 2 and the 6 in 2616.

So, $3.27 * 0.8 = 2.616$.

Note

Sometimes a magnitude estimate is on the "borderline" and you need to be more careful.

For example, a magnitude estimate for $18.5 * 5.2$ is $20 * 5 = 100$. The answer may be "in the 10s" or it may be "in the 100s." But the answer will be close to 100. Since $185 * 52 = 9620$, place the decimal point between the 6 and the 2: $18.5 * 5.2 = 96.20$.

There is another way to find where to place the decimal point in the product. This method is especially useful when the factors are less than 1 and have many decimal places.

Example $3.27 * 0.8 = ?$

Count the decimal places to the right of the decimal point in each factor.	2 decimal places in 3.27 1 decimal place in 0.8
Add the number of decimal places. This is how many decimal places there will be in the product.	$2 + 1 = 3$
Multiply the factors as if they were whole numbers.	$327 * 8 = 2616$
Start at the right of the product. Move the decimal point LEFT the necessary number of decimal places.	2 . 6 1 6 .

So, $3.27 * 0.8 = 2.616$.

Check Your Understanding

Multiply.

1. $1.7 * 5.7$ **2.** $2.33 * 8.4$ **3.** $0.61 * 4.04$ **4.** $0.3 * 0.021$

Check your answers on page 347.

Percents

A percent is another way to name a fraction or a decimal. **Percent** means *per hundred,* or *out of a hundred.* So 1% has the same meaning as the fraction $\frac{1}{100}$ and the decimal 0.01. And 15% is another name for the fraction $\frac{15}{100}$ and the decimal 0.15. Here are other examples.

Did You Know?

In the Latin phrase *per centum, per* means *for each,* and *centum* means *hundred.*

Examples

$35\% = \frac{35}{100} = 0.35$ $50\% = \frac{50}{100} = 0.5$ $500\% = \frac{500}{100} = 5.00$ $12.5\% = \frac{12.5}{100} = 0.125$

Example 50% of the dots in an array are red. Is that a lot of dots?

It depends on how many dots the array has. Two arrays are shown below. 50% of the dots in each array are red. The first array has 50 red dots. But the second array only has 5 red dots.

$\frac{50}{100}$ of the dots are red.
$\frac{50}{100} = 50\%$
50% of the dots are red.

$\frac{5}{10}$ of the dots are red.
But $\frac{5}{10} = \frac{50}{100}$, and $\frac{50}{100} = 50\%$.
50% of the dots are red.

Percent of a Number

Finding a percent of a number is the same as multiplying the number by the percent. This kind of problem can be solved in different ways. You can change the percent to a fraction, work with a unit percent, or change the percent to a decimal.

Many common percents are equivalent to "easy" fractions. For example, 25% is the same as $\frac{1}{4}$. Sometimes it's easier to work with the fraction than with the percent.

Example What is 25% of 36?

$25\% = \frac{25}{100} = \frac{1}{4}$, so 25% of 36 is the same as $\frac{1}{4}$ of 36.
If you divide 36 into 4 equal groups, there are 9 in each group.

So, 25% of 36 is 9.

Example What is 20% of 50?

$20\% = \frac{20}{100} = \frac{1}{5}$, so 20% of 50 is the same as $\frac{1}{5}$ of 50.
If you divide 50 into 5 equal groups, there are 10 in each group.

So, 20% of 50 is 10.

Not every percent is equal to an "easy" fraction like $\frac{1}{4}$ or $\frac{1}{5}$.
Sometimes it's easier to work with 1% instead.

Example What is 5% of 300?

$1\% = \frac{1}{100}$, so 1% of 300 is the same as $\frac{1}{100}$ of 300.
If you divide 300 into 100 equal groups, there are 3 in each group.
So, 1% of 300 is 3. Then 5% of 300 is 5 * 3, or 15.

So, 5% of 300 is 15.

Example What is 8% of 50?

1% of 50 is $\frac{1}{100}$ of 50. If you divide 50 into 100 equal groups,
there is $\frac{1}{2}$ in each group. So, 1% of 50 is $\frac{1}{2}$.

Then 8% of 50 is $8 * \frac{1}{2}$, or 4.

Another way is to change the percent to a decimal and multiply.

Example What is 35% of 45?

$35\% = \frac{35}{100} = 0.35$ 35% of 45 is the same as 0.35 * 45.
The multiplication can be done using a calculator.
Key in: .35 [×] 45 [=] Answer: 15.75

Some calculators have a [%] key, so you don't need to rename the percent as a decimal. To find 35% of 45 on such a calculator, key in 45 [×] 35 [%] [=].

35% of 45 is 15.75.

1% of 45 means 0.01 * 45.

7% of 45 means 0.07 * 45.

35% of 45 means 0.35 * 45.

The word *of* in problems like these means multiplication.

Check Your Understanding

Solve.

1. 25% of 16 **2.** 50% of 160 **3.** 75% of 80 **4.** 10% of 80 **5.** 1% of 400

Check your answers on page 341.

World of Percent

Percents are used in many places every day to give us information about our world.

Nutrition Facts

Serving Size 1 Cookie (26g/0.9 oz)
Servings Per Container 10

Amount Per Serving	
Calories 130 Calories from Fat 50	
	% Daily Value*
Total Fat 6g	**9%**
Saturated Fat 2.5g	**13%**
Trans Fat 1g	
Polyunsaturated Fat 0g	
Monounsaturated Fat 2.5g	
Cholesterol 10mg	**3%**
Sodium 35mg	**1%**
Total Carbohydrate 16g	**5%**
Dietary Fiber 2g	**8%**
Sugars 9g	
Protein 1g	

Airline Set to Cut Service 21%

SALE — 50% OFF
Everything Must Go

For Wednesday, there is a 30% chance of showers.

You can get a furnace filter that captures 94 percent of the dust, dander, pollen, and other particles that float through your house.

Voter Turnout Pegged at 55% of Registered Voters

RAISIN-LITE COOKIES
50% Less Fat
than our regular cookies

Staying in School

Education Department figures show that high school completion rates have grown over the last two decades.

Percent (0–100)
84
87
Percent completing high school by age 21–22
1980
2000

At present, computers are a leading cause of increased demand for electrical power, accounting for an estimated 5 percent of commercial demand.

Total attendance was 8% higher than in the previous year.

U.S. Traditional Addition: Decimals

You can also use **U.S. traditional addition** to add decimals.

Did You Know?

When you add decimals, make sure you line up the places properly so that tenths are added to tenths, ones to ones, and so on. *For example:*

$$\begin{array}{r} 14.5 \\ +\ 4.18 \\ \hline \end{array}$$

Example $9.87 + 2.56 = ?$

Step 1: Start with the 0.01s: $7 + 6 = 13$.
13 hundredths = 1 tenth + 3 hundredths

$$\begin{array}{r} {}^{1} \\ 9.87 \\ +\ 2.56 \\ \hline 3 \end{array}$$

Step 2: Add the 0.1s: $1 + 8 + 5 = 14$.
14 tenths = 1 whole + 4 tenths

$$\begin{array}{r} {}^{1}{}^{1} \\ 9.87 \\ +\ 2.56 \\ \hline 43 \end{array}$$

Step 3: Add the 1s: $1 + 9 + 2 = 12$.
12 ones = 1 ten + 2 ones
Remember to include the decimal point in the answer.

$$\begin{array}{r} {}^{1}{}^{1} \\ 9.87 \\ +\ 2.56 \\ \hline 12.43 \end{array}$$

$9.87 + 2.56 = 12.43$

Check Your Understanding

Add.

1. $2.78 + 3.96 = ?$

2. $\begin{array}{r} 34.29 \\ +\ 7.92 \\ \hline \end{array}$

3. $9.234 + 123.8 = ?$

4. $8.073 + 21.03 = ?$

Check your answers on page 347.

U.S. Traditional Subtraction: Decimals

You can use **U.S. traditional subtraction** to subtract decimals.

Example 7.83 – 2.89 = ?

$$\begin{array}{rccc} & & 7 & 13 \\ & 7 & . \not{8} & \not{3} \\ - & 2 & .\ 8 & 9 \\ \hline & & & 4 \end{array}$$

Step 1: Start with the 0.01s.
Since 9 > 3, you need to regroup.
Trade 1 tenth for 10 hundredths:
7.83 = 7 ones + 7 tenths + 13 hundredths.
Subtract the 0.01s: 13 – 9 = 4.

$$\begin{array}{rccc} & & 17 & \\ & 6 & \not{7} & 13 \\ & \not{7} & . \not{8} & \not{3} \\ - & 2 & .\ 8 & 9 \\ \hline & & 9 & 4 \end{array}$$

Step 2: Go to the 0.1s.
Since 8 > 7, you need to regroup.
Trade 1 one for 10 tenths:
7.83 = 6 ones + 17 tenths + 13 hundredths.
Subtract the 0.1s: 17 – 8 = 9.

$$\begin{array}{rccc} & & 17 & \\ & 6 & \not{7} & 13 \\ & \not{7} & . \not{8} & \not{3} \\ - & 2 & .\ 8 & 9 \\ \hline & 4 & .\ 9 & 4 \end{array}$$

Step 3: Go to the 1s.
You don't need to regroup.
Subtract the 1s: 6 – 2 = 4.
Remember to include the decimal point in the answer.

7.83 – 2.89 = 4.94

Check Your Understanding

Subtract.

1. 19.27 – 5.79 = ?

2. $\begin{array}{r} 4.60 \\ -\ 3.75 \\ \hline \end{array}$

3. $\begin{array}{r} 56.40 \\ -\ 6.84 \\ \hline \end{array}$

4. 52.7 – 8.75 = ?

Check your answers on page 347.

U.S. Traditional Multiplication: Decimals

You can use **U.S. traditional multiplication** to multiply money amounts.

Example 4 * $7.29 = ?

Step 1: Start with the pennies.
4 * 9 pennies = 36 pennies
36 pennies = 3 dimes + 6 pennies

```
     3
 7 . 2 9
*      4
--------
       6
```

Step 2: Multiply the dimes.
4 * 2 dimes = 8 dimes
Remember the 3 dimes from Step 1.
8 dimes + 3 dimes = 11 dimes in all
11 dimes = $1 + 1 dime

```
 1   3
 7 . 2 9
*      4
--------
     1 6
```

Step 3: Multiply the dollars.
4 * $7 = $28
Remember the $1 from Step 2.
$28 + $1 = $29 in all
Remember to include the decimal point in the answer.

```
   1   3
   7 . 2 9
*        4
----------
 2 9 . 1 6
```

4 * $7.29 = $29.16

Check Your Understanding

Multiply.

1. 6 * $445 = ?
2. ? = 3 * $5.96
3. $5,692 * 4
4. $56.25 * 8 = ?

Check your answers on page 347.

U.S. Traditional Long Division: Decimals

You can use **U.S. traditional long division** to divide money in dollars and cents notation.

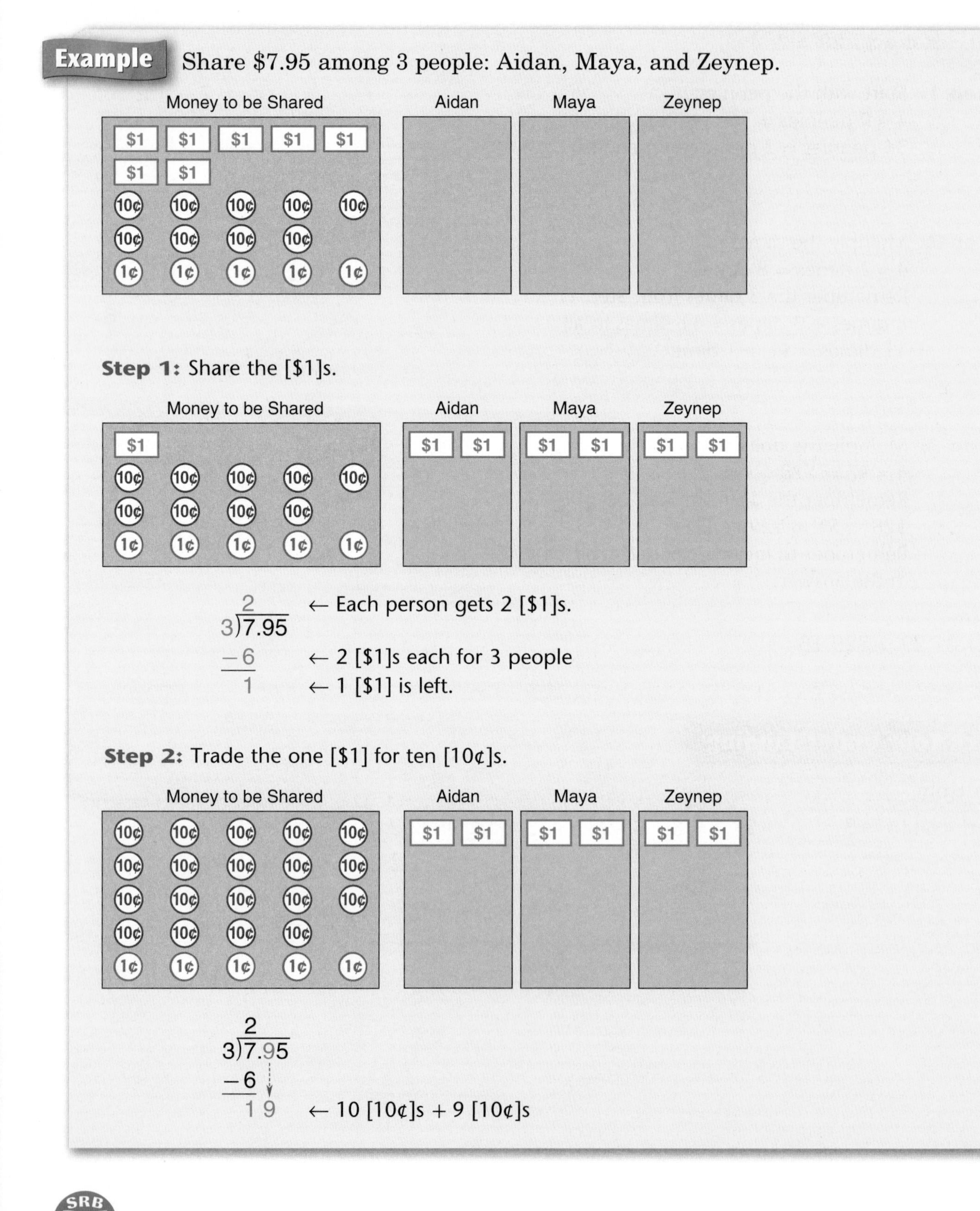

Example *continued*

Step 3: Share the [10¢]s.

Money to be Shared	Aidan	Maya	Zeynep
10¢ 1¢ 1¢ 1¢ 1¢ 1¢	\$1 \$1 10¢ 10¢ 10¢ 10¢ 10¢ 10¢	\$1 \$1 10¢ 10¢ 10¢ 10¢ 10¢ 10¢	\$1 \$1 10¢ 10¢ 10¢ 10¢ 10¢ 10¢

```
   2.6     ← Each person gets 6 [10¢]s. Write a decimal point
 3)7.95      above the line to show amounts less than $1.
  −6
   1 9
  −1 8     ← 6 [10¢]s each for 3 people
     1     ← 1 [10¢] is left.
```

Step 4: Trade the one [10¢] for ten [1¢]s.

Money to be Shared	Aidan	Maya	Zeynep
1¢ 1¢ 1¢ 1¢ 1¢ 1¢ 1¢ 1¢ 1¢ 1¢ 1¢ 1¢ 1¢ 1¢ 1¢	\$1 \$1 10¢ 10¢ 10¢ 10¢ 10¢ 10¢	\$1 \$1 10¢ 10¢ 10¢ 10¢ 10¢ 10¢	\$1 \$1 10¢ 10¢ 10¢ 10¢ 10¢ 10¢

```
   2.6
 3)7.95
  −6
   1 9
  −1 8↓
     15    ← 10 [1¢]s + 5 [1¢]s
```

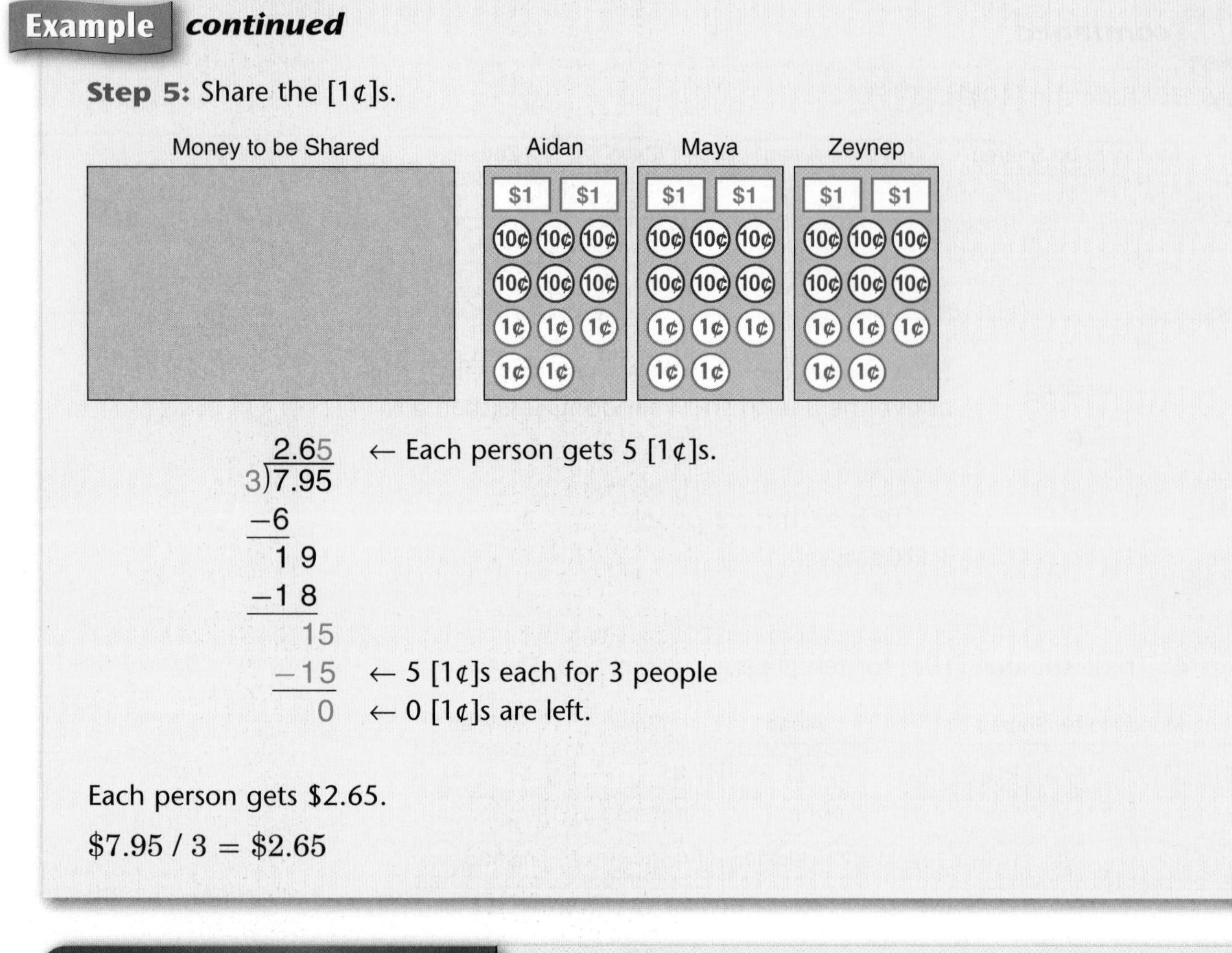

Example ***continued***

Step 5: Share the [1¢]s.


```
   2.65    ← Each person gets 5 [1¢]s.
 3)7.95
  −6
   1 9
  −1 8
     15
    −15    ← 5 [1¢]s each for 3 people
      0    ← 0 [1¢]s are left.
```

Each person gets $2.65.

$7.95 / 3 = $2.65

Check Your Understanding

Divide.

1. $6.25 / 5 **2.** $5\overline{)6.75}$ **3.** $8\overline{)4.80}$ **4.** $38.96 / 4

Check your answers on page 347.

Fractions

Uses of Fractions

The numbers $\frac{1}{2}$, $\frac{2}{3}$, $\frac{5}{4}$, $\frac{7}{1}$, and $\frac{25}{100}$ are all **fractions.** A fraction is written with two whole numbers that are separated by a fraction bar. The top number is called the **numerator.** The bottom number is called the **denominator.** The numerator of a fraction can be any whole number. The denominator can be any whole number except 0.

$\frac{1}{2}$ teaspoon

When naming fractions, name the numerator first, then name the denominator.

three-fourths $\frac{3}{4}$ ← numerator → $\frac{25}{100}$ twenty-five hundredths
← denominator →

yardstick
$\frac{3}{4}$ yard

Fractions were invented thousands of years ago to name numbers between whole numbers. People needed these in-between numbers for making careful measurements.

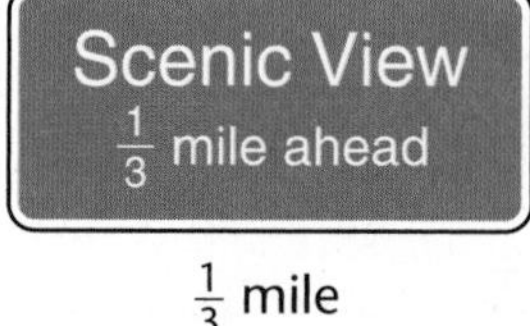

$\frac{1}{3}$ mile

Here are some examples of measurements that use fractions: $\frac{2}{3}$ cup, $\frac{3}{4}$ hour, $\frac{9}{10}$ km, and $13\frac{1}{2}$ lb.

Fractions are also used to name parts of wholes. The whole might be one single thing, like a pizza. Or, the whole might be a collection of things, like all the students in a classroom. The whole is sometimes called the ONE.

$\frac{5}{8}$ of a pizza

Examples Name the whole, or ONE, for each statement.

"Derek ate $\frac{5}{8}$ of the pizza."

The whole is the entire pizza.
The fraction, $\frac{5}{8}$, names the part of the pizza that Derek ate.

"In Mrs. Blake's classroom, $\frac{1}{2}$ of the students are girls."

The whole is the collection of all students in Mrs. Blake's classroom.
The fraction, $\frac{1}{2}$, names the part of that collection that are girls.

Mrs. Blake's Classroom
$\frac{1}{2}$ of the students are girls.

In *Everyday Mathematics,* fractions are used in other ways that may be new to you. Fractions are used in the following ways:

- to show rates (such as cost per ounce)
- to compare (such as comparing the weights of two animals)
- to name percents ($\frac{1}{2}$ is 50%)
- to show divisions (15 ÷ 3 can be written $\frac{15}{3}$)
- to show the scale of a map or a picture
- to show probabilities

Did You Know?

Arab mathematicians began to use the horizontal fraction bar around the year 1200.

They were the first to write fractions as we do today.

Here are some other examples of uses of fractions:

- Study the recipe shown at the right. Many of the amounts listed in the recipe include fractions.

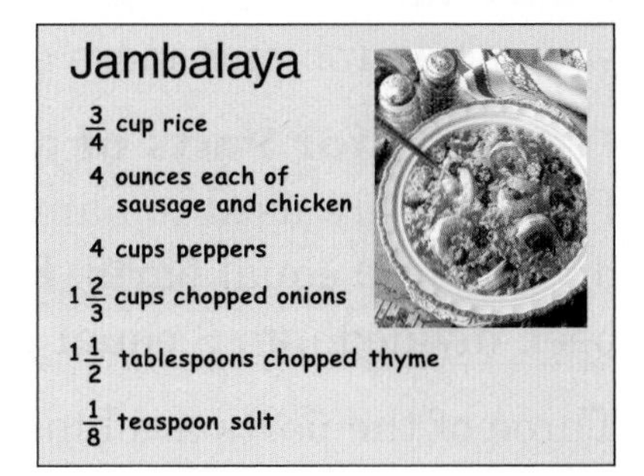

- A movie critic gave the film *Finding Nemo* a rating of $3\frac{1}{2}$ stars (on a scale of 0 to 4 stars).

Finding Nemo

- This spinner has $\frac{1}{3}$ of the circle colored red, $\frac{1}{4}$ colored blue, and $\frac{5}{12}$ colored green.

  If we spin the spinner many times, it will land on red about $\frac{1}{3}$ of the time. It will land on blue about $\frac{1}{4}$ of the time. And it will land on green about $\frac{5}{12}$ of the time.

  The probability that the spinner will land on a color that is *not* green is $\frac{7}{12}$.

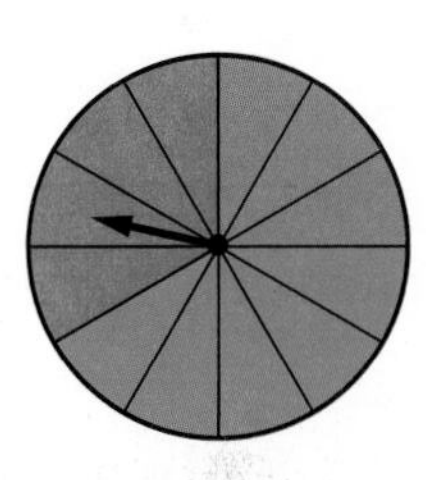

- If a map includes a **scale,** you can use the scale to estimate real-world distances. The scale on the map shown here is given as 1:10,000. This means that every distance on the map is $\frac{1}{10{,}000}$ of the real-world distance. A 1 centimeter distance on the map stands for a real-world distance of 10,000 centimeters (100 meters).

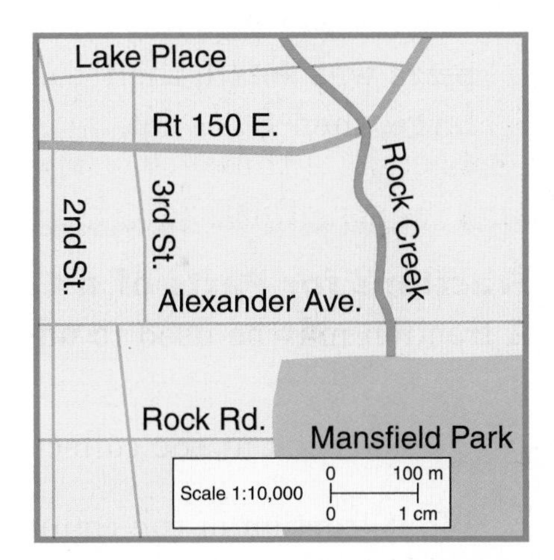

- Fractions are often used to describe clothing sizes. For example, women's shoes come in sizes 3, $3\frac{1}{2}$, 4, $4\frac{1}{2}$, and so on, up to 14.

  Part of a size chart for women's shoes is shown at the right. It gives the recommended shoe size for women whose feet are between 9 and 10 inches long.

Size Chart for Women's Shoes

Heel-to-toe length (in.)	Size
$8\frac{15}{16}$ to $9\frac{1}{16}$	6
$9\frac{2}{16}$ to $9\frac{3}{16}$	$6\frac{1}{2}$
$9\frac{4}{16}$ to $9\frac{6}{16}$	7
$9\frac{7}{16}$ to $9\frac{9}{16}$	$7\frac{1}{2}$
$9\frac{10}{16}$ to $9\frac{11}{16}$	8
$9\frac{12}{16}$ to $9\frac{14}{16}$	$8\frac{1}{2}$
$9\frac{15}{16}$ to $10\frac{1}{16}$	9

Understanding the many ways people use fractions will help you solve problems more easily.

Fractions for Parts of a Whole

Fractions are used to name a part of a whole thing that is divided into equal parts. For example, the circle at the right has been divided into 8 equal parts. Each part is $\frac{1}{8}$ of the circle.

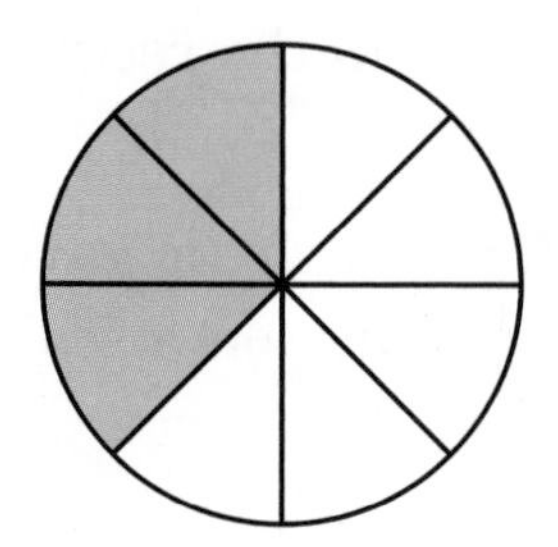

Three of the parts are blue, so $\frac{3}{8}$ (three-eighths) of the circle is blue.

$\frac{3}{8}$ ← Numerator tells the number of blue parts.
← Demominator tells the number of equal parts in the whole circle.

In *Everyday Mathematics,* the whole thing that is divided into equal parts is called the ONE. To understand a fraction used to name part of a whole, you need to know what the ONE is.

Example Sally ate half a pizza. Is that a lot?

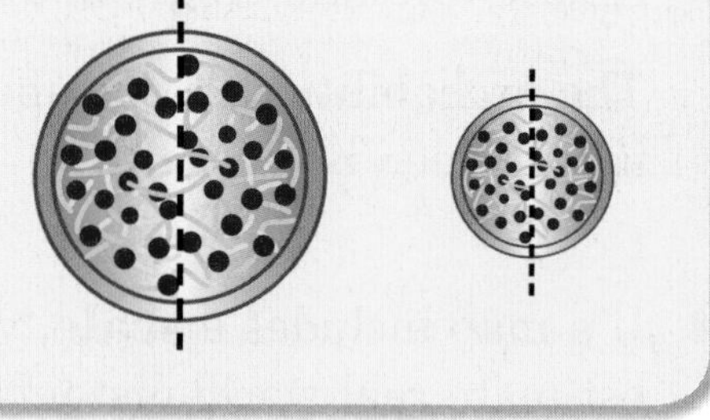

The answer depends on how big the pizza was. If the pizza was small, then $\frac{1}{2}$ *is not* a lot. If the pizza was large, then $\frac{1}{2}$ *is* a lot.

Fractions for Parts of a Collection

A fraction may be used to name part of a collection of things.

Example Look at the collection of counters.

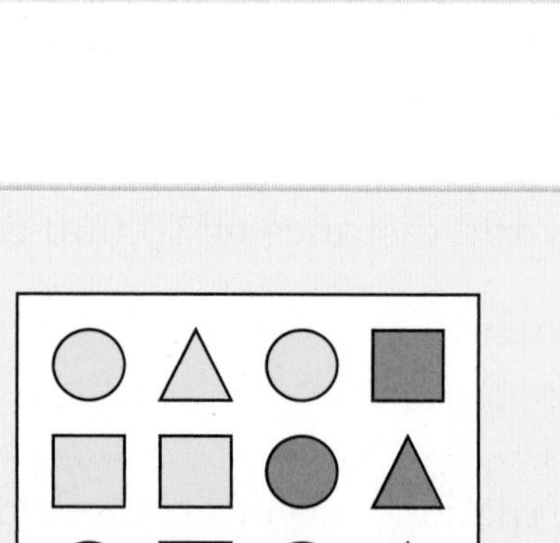

What fraction of the counters is red?

There are 12 counters in all.
Five of the counters are red.
Five out of 12 counters are red.

This fraction shows what part of the collection is red.

$\frac{5}{12}$ ← the number of red counters
← the number of counters in all

Check Your Understanding

Name the fraction of counters that are each shape in the collection above.

1. Circles **2.** Triangles **3.** Squares

Check your answers on page 341.

To understand a fraction that is used to name part of a collection, you need to know how big the whole collection is.

Example Only half of Sam's cousins can come to his party. Is that many people?

It depends on how many cousins Sam has. If Sam has only 4 cousins, then 2 cousins are coming; that's not many people. But if Sam has 24 cousins, then 12 cousins are coming. That's many people.

Fractions in Measuring

Fractions are used to make more careful measurements.

Think about the inch scale on a ruler. Suppose the spaces between the whole-inch marks are left unmarked. With a ruler like this, you can measure only to the nearest inch.

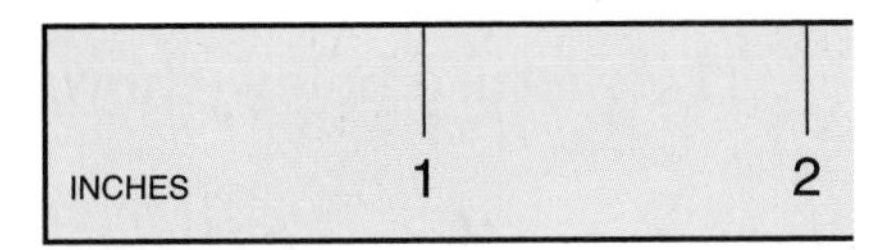

Ruler has inch marks only. You can measure to the nearest inch.

Now suppose the 1-inch spaces are divided into quarters by $\frac{1}{2}$-inch and $\frac{1}{4}$-inch marks. With this ruler, you can measure to the nearest $\frac{1}{2}$ inch or to the nearest $\frac{1}{4}$ inch.

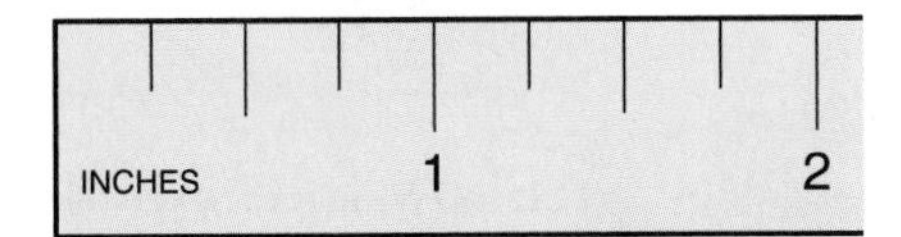

Ruler has $\frac{1}{2}$- and $\frac{1}{4}$-inch marks. You can measure to the nearest $\frac{1}{2}$ inch or to the nearest $\frac{1}{4}$ inch.

To understand a fraction used in a measurement, you need to know what the unit is. To say, "Susan lives $\frac{1}{2}$ from here" makes no sense. Susan might live half a block away or half a mile. The unit in measurement is like the ONE when fractions are used to name a part of a whole.

Fractions in Probability

A fraction may tell the chance that an event will happen. This chance, or probability, is always a number from 0 to 1. An impossible event has a probability of 0; it has no chance of happening. An event with a probability of 1 is sure to happen. An event with a probability of $\frac{1}{2}$ has an equal chance of happening or not happening.

Example When you pick a ball out of this jar without looking, the chance of getting a red ball is $\frac{5}{8}$. The chance of getting a blue ball is $\frac{3}{8}$.

$$\text{probability of picking a red ball} = \frac{\text{number of red balls}}{\text{total number of balls}} = \frac{5}{8}$$

$$\text{probability of picking a blue ball} = \frac{\text{number of blue balls}}{\text{total number of balls}} = \frac{3}{8}$$

Fractions and Division

Division problems can be written using a slash / instead of the division symbol ÷. For example, $21 \div 3$ can be written 21 / 3.

Division problems can also be written as fractions. One of the many uses of fractions is to show divisions. The example below shows that $21 \div 3$ can be written as the fraction $\frac{21}{3}$.

$21 \div 3 = 21 / 3$

and

$21 \div 3 = \frac{21}{3}$

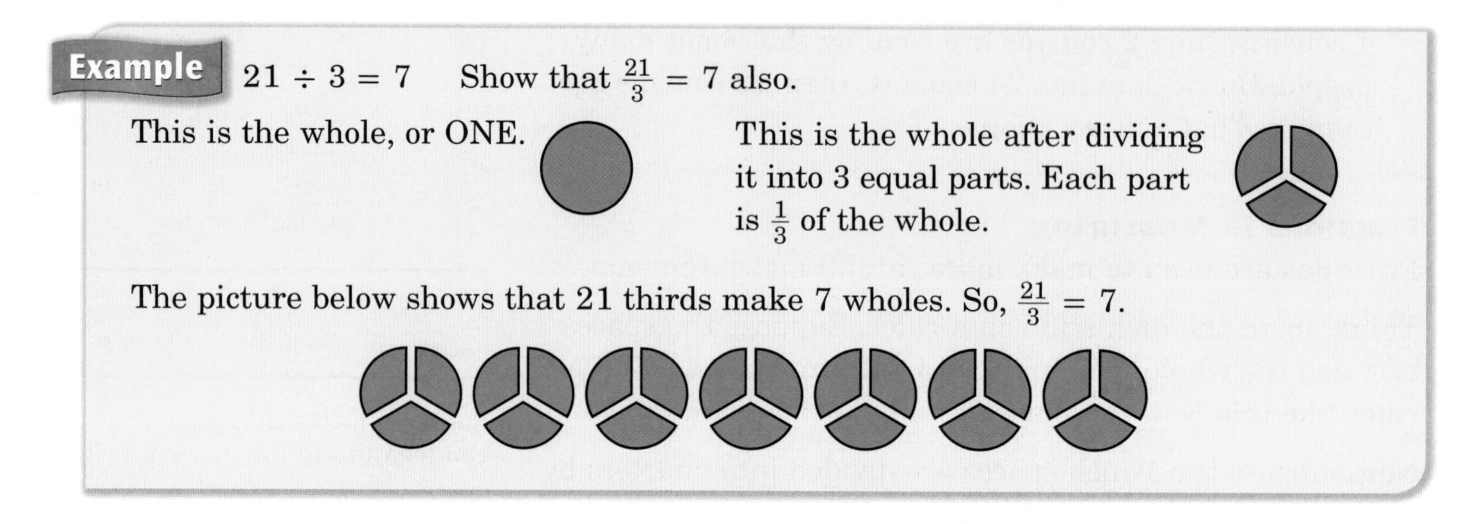

Example $21 \div 3 = 7$ Show that $\frac{21}{3} = 7$ also.

This is the whole, or ONE.

This is the whole after dividing it into 3 equal parts. Each part is $\frac{1}{3}$ of the whole.

The picture below shows that 21 thirds make 7 wholes. So, $\frac{21}{3} = 7$.

Fractions less than 1 can also be thought of as divisions.

Example Show that $3 \div 4 = \frac{3}{4}$.

Think of $3 \div 4$ as an equal-sharing problem.
Suppose 4 friends want to share 3 oranges.
They could cut or divide each orange into 4 equal parts.

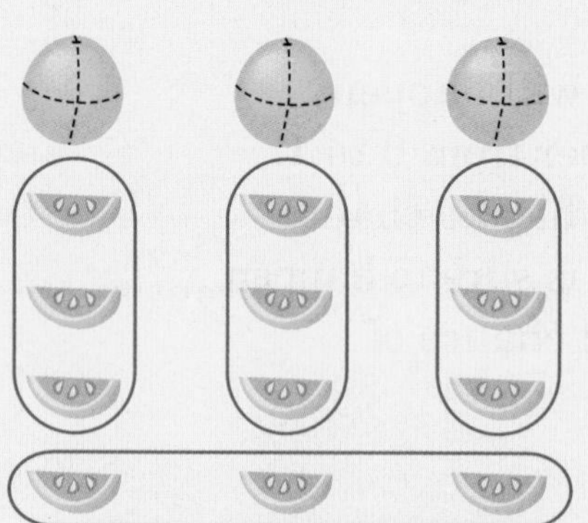

Each person gets $\frac{3}{4}$ of an orange. So, $3 \div 4 = \frac{3}{4}$.

Did You Know?

The word *fraction* is derived from the Latin word *frangere,* which means "to break." Fractions are sometimes called "broken numbers."

You can rename any fraction by dividing on your calculator. To rename $\frac{21}{3}$, think of it as a division problem and divide: Press 21 [÷] 3 [=]. The answer in the display will show 7, which is another name for $\frac{21}{3}$.

Fractions in Rates and Ratios

Fractions are often used to name rates and ratios.

A **rate** compares two numbers with *different units.* For example, 30 miles per hour is a rate that compares distance with time. It can be written as $\frac{30 \text{ miles}}{1 \text{ hour}}$.

A **ratio** is like a rate, but it compares two quantities that have the *same unit.*

Rate	Example
speed (jogging)	$\frac{\text{distance}}{\text{time}} = \frac{7 \text{ blocks}}{4 \text{ minutes}}$
price	$\frac{\text{cost}}{\text{quantity}} = \frac{99¢}{3 \text{ erasers}}$
conversion of units	$\frac{\text{distance in yards}}{\text{distance in feet}} = \frac{1 \text{ yard}}{3 \text{ feet}}$

Ratio	Example
won / lost record	$\frac{\text{games (won)}}{\text{games (lost)}} = \frac{6}{8}$
rainy days compared to total days	$\frac{\text{(rainy) days}}{\text{(total) days}} = \frac{11}{30}$

Other Uses of Fractions

Fractions are used to compare distances on maps to distances in the real world, and to describe size changes.

Example Find the real-world distance from Clay St. to S. Lake St.

Measure this distance on the map. It is 6 cm. Each distance on the map is $\frac{1}{20,000}$ of the real-world distance. So, the real-world distance equals 20,000 times the map distance. The real-world distance = 20,000 * 6 cm = 120,000 cm.

100 cm = 1 m. So, 1,200 cm = 12 m and 120,000 cm = 1,200 m.

The distance from Clay St. to S. Lake St. is 1,200 m.

Example A length of 6 centimeters on the original will be 3 centimeters on the copy.

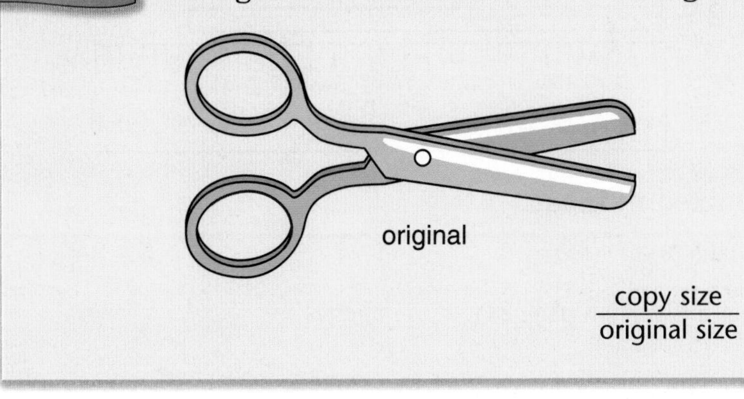

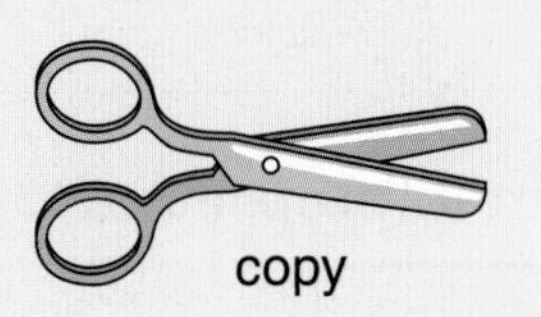

$$\frac{\text{copy size}}{\text{original size}} = \frac{1}{2}$$

Mixed Numbers

Numbers like $1\frac{1}{2}$, $2\frac{3}{5}$, and $4\frac{3}{8}$ are called **mixed numbers.** A mixed number has a whole-number part and a fraction part. In the mixed number $2\frac{3}{5}$, the whole-number part is 2 and the fraction part is $\frac{3}{5}$. A mixed number is equal to the sum of the whole-number part and the fraction part: $2\frac{3}{5} = 2 + \frac{3}{5}$. Mixed numbers are used in many of the same ways that fractions are used.

Mixed numbers can be renamed as fractions. For example, if a circle is the ONE, then $2\frac{3}{5}$ names 2 whole circles and $\frac{3}{5}$ of another circle.

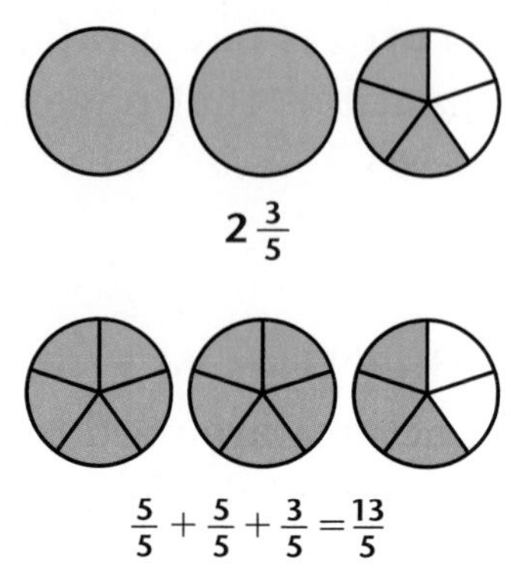

If you divide the 2 whole circles into fifths, then you can see that $2\frac{3}{5} = \frac{13}{5}$.

To rename a mixed number as a fraction, first rename 1 as a fraction with the same denominator as the fraction part. Then add all of the fractions.

For example, to rename $4\frac{3}{8}$ as a fraction, first rename 1 as $\frac{8}{8}$. Then $4\frac{3}{8} = \frac{8}{8} + \frac{8}{8} + \frac{8}{8} + \frac{8}{8} + \frac{3}{8} = \frac{35}{8}$.

Fractions like $\frac{13}{5}$ and $\frac{8}{8}$ are called **improper fractions.** An improper fraction is a fraction that is greater than or equal to 1. In an improper fraction, the numerator is greater than or equal to the denominator.

A **proper fraction** is a fraction that is less than 1. In a proper fraction, the numerator is less than the denominator.

Note

Even though they are called *improper,* there is nothing wrong about improper fractions. Do not avoid them.

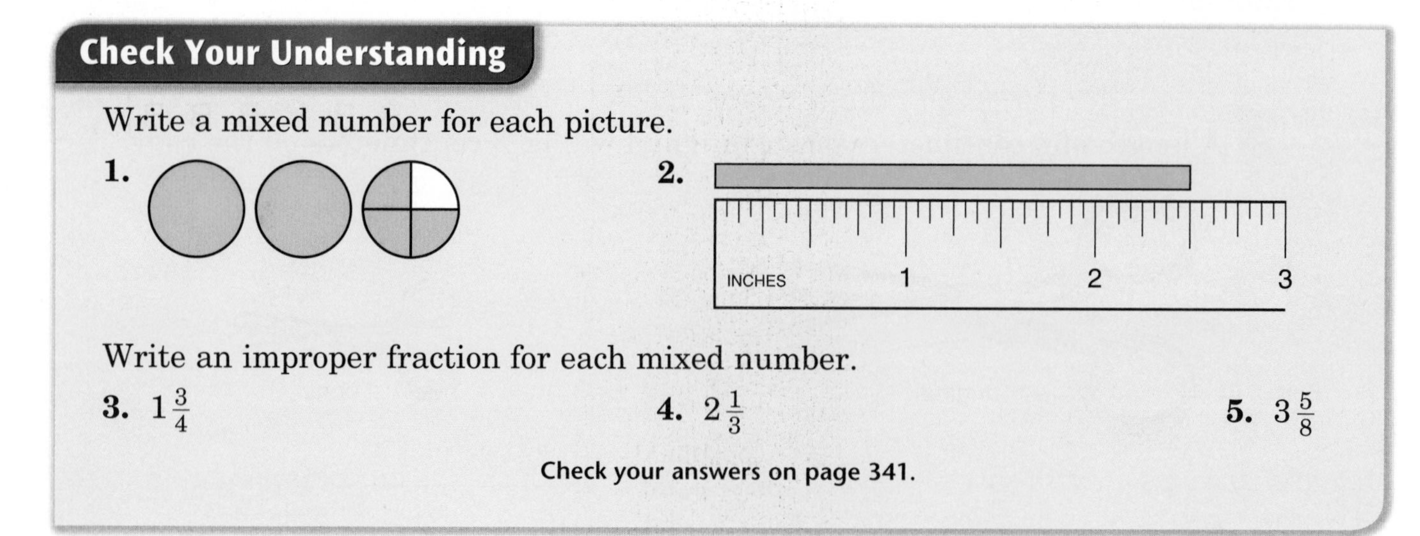

Check Your Understanding

Write a mixed number for each picture.

1.

2.

Write an improper fraction for each mixed number.

3. $1\frac{3}{4}$ **4.** $2\frac{1}{3}$ **5.** $3\frac{5}{8}$

Check your answers on page 341.

Equivalent Fractions

Two or more fractions that name the same number are called **equivalent fractions.**

Example The four circles below are the same size, but they are divided into different numbers of parts. The green areas are the same in each circle. These circles show different fractions that are equivalent to $\frac{1}{2}$.

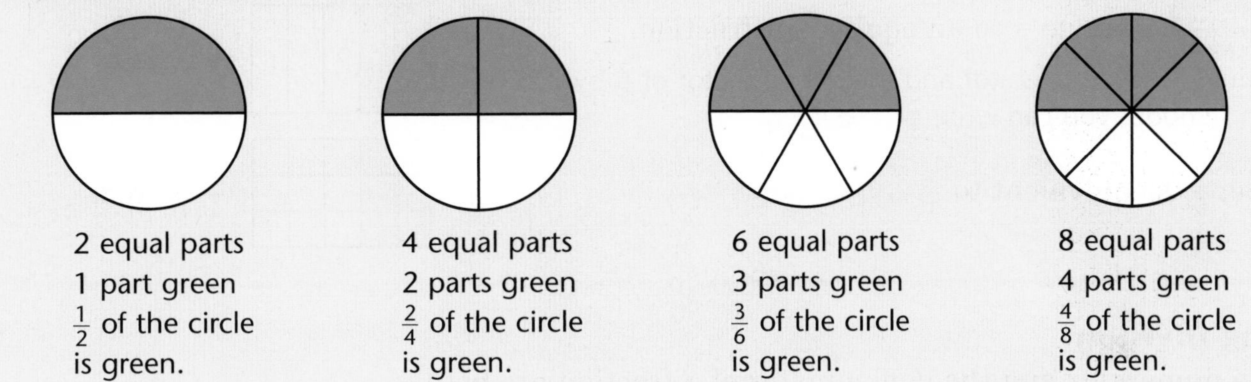

The fractions $\frac{1}{2}, \frac{2}{4}, \frac{3}{6}$ and $\frac{4}{8}$ are all equivalent. They are just different names for the part of the circle that is green.

You can write: $\frac{1}{2} = \frac{2}{4}$ $\quad \frac{1}{2} = \frac{4}{8}$ $\quad \frac{2}{4} = \frac{4}{8}$

$\frac{1}{2} = \frac{3}{6}$ $\quad \frac{2}{4} = \frac{3}{6}$ $\quad \frac{3}{6} = \frac{4}{8}$

Example On Ms. Klein's bus route, she picks up 24 students, 18 boys and 6 girls.

G	G	G	G	G	G
B	B	B	B	B	B
B	B	B	B	B	B
B	B	B	B	B	B

24 equal groups
Each group is $\frac{1}{24}$ of the total.

6 groups of girls
$\frac{6}{24}$ of the students are girls.

G G	G G	G G
B B	B B	B B
B B	B B	B B
B B	B B	B B

12 equal groups
Each group is $\frac{1}{12}$ of the total.

3 groups of girls
$\frac{3}{12}$ of the students are girls.

G G G G G G
B B B B B B
B B B B B B
B B B B B B

4 equal groups
Each group is $\frac{1}{4}$ of the total.

1 group of girls
$\frac{1}{4}$ of the students are girls.

The fractions $\frac{6}{24}$, $\frac{3}{12}$, and $\frac{1}{4}$ are all equivalent.

You can write $\frac{6}{24} = \frac{3}{12} = \frac{1}{4}$.

Rules for Finding Equivalent Fractions

Here are two shortcuts for finding equivalent fractions.

Using Multiplication

If the numerator and the denominator of a fraction are both multiplied by the same number (not 0), the result is a fraction that is equivalent to the original fraction.

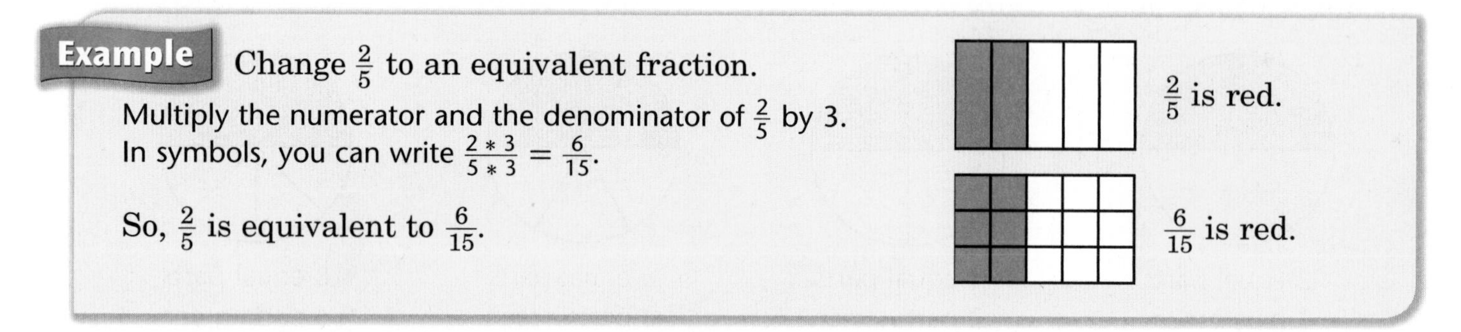

Example Change $\frac{2}{5}$ to an equivalent fraction.

Multiply the numerator and the denominator of $\frac{2}{5}$ by 3.
In symbols, you can write $\frac{2 * 3}{5 * 3} = \frac{6}{15}$.

So, $\frac{2}{5}$ is equivalent to $\frac{6}{15}$.

$\frac{2}{5}$ is red.

$\frac{6}{15}$ is red.

Using Division

If the numerator and the denominator of a fraction are both divided by the same number (not 0), the result is a fraction that is equivalent to the original fraction.

To understand why division works, use the example shown above. But start with $\frac{6}{15}$ this time and divide both numbers in the fraction by 3: $\frac{6 \div 3}{15 \div 3} = \frac{2}{5}$

The division by 3 "undoes" the multiplication by 3 that we did before. Dividing both numbers in $\frac{6}{15}$ by 3 gives an equivalent fraction, $\frac{2}{5}$.

Example Find fractions that are equivalent to $\frac{18}{24}$.

$\frac{18 \div 2}{24 \div 2} = \frac{9}{12}$ $\quad$ $\frac{18 \div 3}{24 \div 3} = \frac{6}{8}$ $\quad$ $\frac{18 \div 6}{24 \div 6} = \frac{3}{4}$

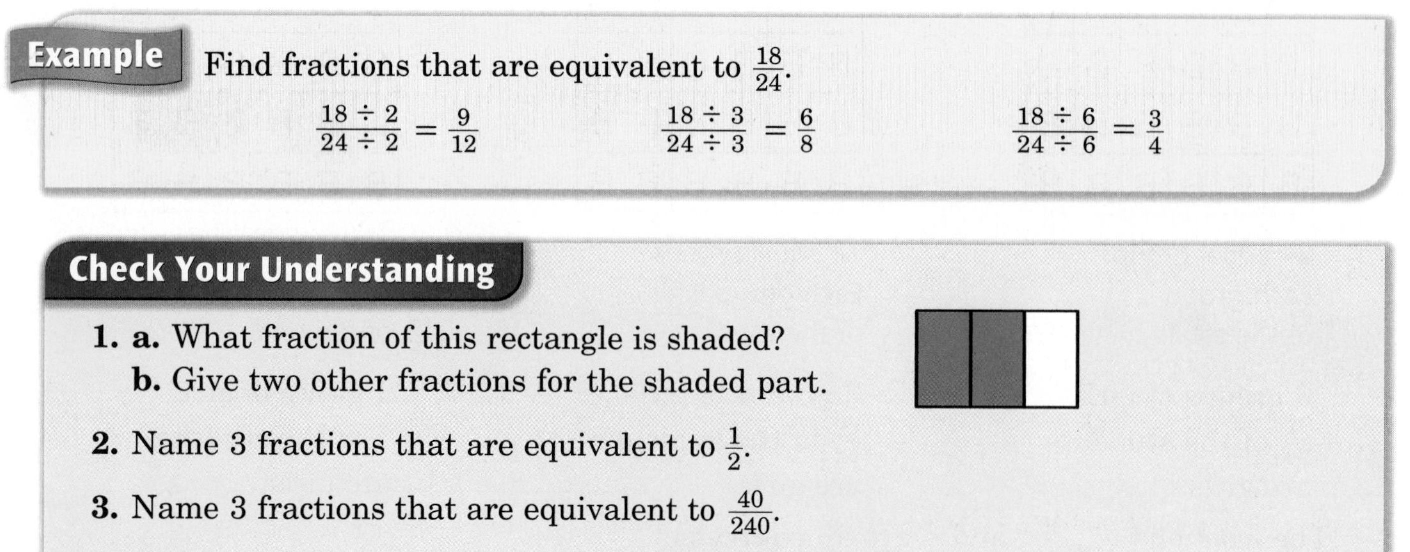

Check Your Understanding

1. **a.** What fraction of this rectangle is shaded?
   b. Give two other fractions for the shaded part.
2. Name 3 fractions that are equivalent to $\frac{1}{2}$.
3. Name 3 fractions that are equivalent to $\frac{40}{240}$.

Check your answers on page 341.

Table of Equivalent Fractions

This table lists equivalent fractions. All the fractions in a row name the same number. For example, all the fractions in the last row are names for the number $\frac{7}{8}$.

Every fraction in the first column is in simplest form. A fraction is in **simplest form** if there is no equivalent fraction with a smaller numerator and smaller denominator.

Simplest Name	Equivalent Fraction Names								
0 (zero)	$\frac{0}{1}$	$\frac{0}{2}$	$\frac{0}{3}$	$\frac{0}{4}$	$\frac{0}{5}$	$\frac{0}{6}$	$\frac{0}{7}$	$\frac{0}{8}$	$\frac{0}{9}$
1 (one)	$\frac{1}{1}$	$\frac{2}{2}$	$\frac{3}{3}$	$\frac{4}{4}$	$\frac{5}{5}$	$\frac{6}{6}$	$\frac{7}{7}$	$\frac{8}{8}$	$\frac{9}{9}$
$\frac{1}{2}$	$\frac{2}{4}$	$\frac{3}{6}$	$\frac{4}{8}$	$\frac{5}{10}$	$\frac{6}{12}$	$\frac{7}{14}$	$\frac{8}{16}$	$\frac{9}{18}$	$\frac{10}{20}$
$\frac{1}{3}$	$\frac{2}{6}$	$\frac{3}{9}$	$\frac{4}{12}$	$\frac{5}{15}$	$\frac{6}{18}$	$\frac{7}{21}$	$\frac{8}{24}$	$\frac{9}{27}$	$\frac{10}{30}$
$\frac{2}{3}$	$\frac{4}{6}$	$\frac{6}{9}$	$\frac{8}{12}$	$\frac{10}{15}$	$\frac{12}{18}$	$\frac{14}{21}$	$\frac{16}{24}$	$\frac{18}{27}$	$\frac{20}{30}$
$\frac{1}{4}$	$\frac{2}{8}$	$\frac{3}{12}$	$\frac{4}{16}$	$\frac{5}{20}$	$\frac{6}{24}$	$\frac{7}{28}$	$\frac{8}{32}$	$\frac{9}{36}$	$\frac{10}{40}$
$\frac{3}{4}$	$\frac{6}{8}$	$\frac{9}{12}$	$\frac{12}{16}$	$\frac{15}{20}$	$\frac{18}{24}$	$\frac{21}{28}$	$\frac{24}{32}$	$\frac{27}{36}$	$\frac{30}{40}$
$\frac{1}{5}$	$\frac{2}{10}$	$\frac{3}{15}$	$\frac{4}{20}$	$\frac{5}{25}$	$\frac{6}{30}$	$\frac{7}{35}$	$\frac{8}{40}$	$\frac{9}{45}$	$\frac{10}{50}$
$\frac{2}{5}$	$\frac{4}{10}$	$\frac{6}{15}$	$\frac{8}{20}$	$\frac{10}{25}$	$\frac{12}{30}$	$\frac{14}{35}$	$\frac{16}{40}$	$\frac{18}{45}$	$\frac{20}{50}$
$\frac{3}{5}$	$\frac{6}{10}$	$\frac{9}{15}$	$\frac{12}{20}$	$\frac{15}{25}$	$\frac{18}{30}$	$\frac{21}{35}$	$\frac{24}{40}$	$\frac{27}{45}$	$\frac{30}{50}$
$\frac{4}{5}$	$\frac{8}{10}$	$\frac{12}{15}$	$\frac{16}{20}$	$\frac{20}{25}$	$\frac{24}{30}$	$\frac{28}{35}$	$\frac{32}{40}$	$\frac{36}{45}$	$\frac{40}{50}$
$\frac{1}{6}$	$\frac{2}{12}$	$\frac{3}{18}$	$\frac{4}{24}$	$\frac{5}{30}$	$\frac{6}{36}$	$\frac{7}{42}$	$\frac{8}{48}$	$\frac{9}{54}$	$\frac{10}{60}$
$\frac{5}{6}$	$\frac{10}{12}$	$\frac{15}{18}$	$\frac{20}{24}$	$\frac{25}{30}$	$\frac{30}{36}$	$\frac{35}{42}$	$\frac{40}{48}$	$\frac{45}{54}$	$\frac{50}{60}$
$\frac{1}{8}$	$\frac{2}{16}$	$\frac{3}{24}$	$\frac{4}{32}$	$\frac{5}{40}$	$\frac{6}{48}$	$\frac{7}{56}$	$\frac{8}{64}$	$\frac{9}{72}$	$\frac{10}{80}$
$\frac{3}{8}$	$\frac{6}{16}$	$\frac{9}{24}$	$\frac{12}{32}$	$\frac{15}{40}$	$\frac{18}{48}$	$\frac{21}{56}$	$\frac{24}{64}$	$\frac{27}{72}$	$\frac{30}{80}$
$\frac{5}{8}$	$\frac{10}{16}$	$\frac{15}{24}$	$\frac{20}{32}$	$\frac{25}{40}$	$\frac{30}{48}$	$\frac{35}{56}$	$\frac{40}{64}$	$\frac{45}{72}$	$\frac{50}{80}$
$\frac{7}{8}$	$\frac{14}{16}$	$\frac{21}{24}$	$\frac{28}{32}$	$\frac{35}{40}$	$\frac{42}{48}$	$\frac{49}{56}$	$\frac{56}{64}$	$\frac{63}{72}$	$\frac{70}{80}$

Note

Every fraction is either in simplest form or is equivalent to a fraction in simplest form.

Lowest terms means the same as *simplest form.*

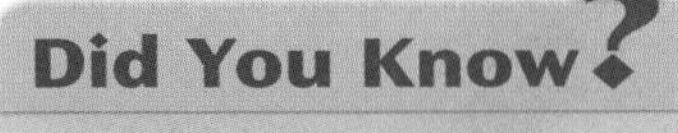

Under normal conditions, $\frac{1}{5}$ of the length of a telephone pole should be in the ground.

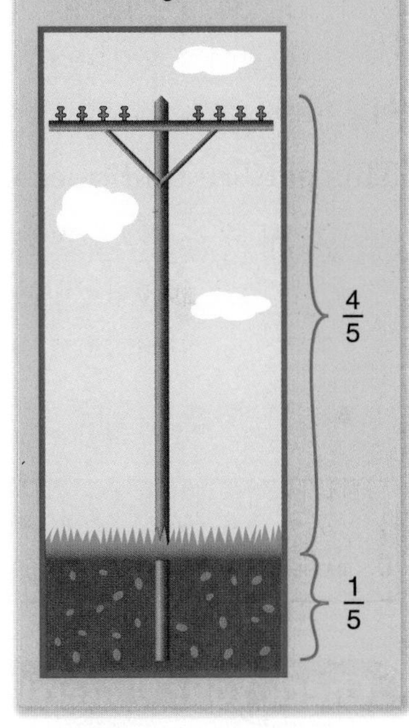

Check Your Understanding

1. True or false?

   a. $\frac{1}{2} = \frac{9}{18}$ **b.** $\frac{5}{8} = \frac{20}{32}$ **c.** $\frac{2}{5} = \frac{12}{40}$ **d.** $\frac{0}{2} = \frac{0}{120}$

2. **a.** Use the table to find 3 other fractions that are equivalent to $\frac{1}{5}$.

   b. Add 2 more equivalent fractions that are not in the table.

Check your answers on page 341.

Equivalent Fractions on a Ruler

Rulers marked in inches usually have tick marks of different lengths. The longest tick marks on the ruler below show the whole inches. The marks used to show half inches, quarter inches, and eighths of an inch become shorter and shorter. The shortest marks show the sixteenths of an inch.

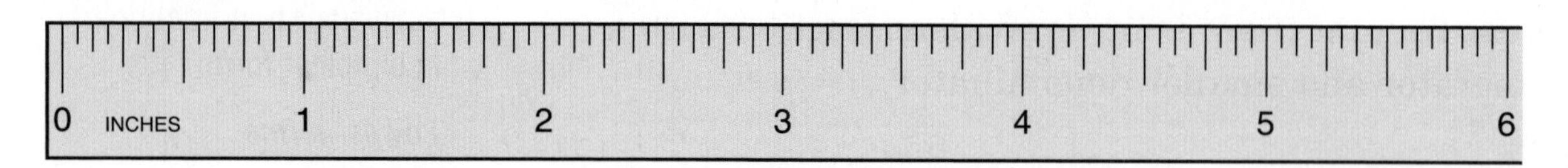

Every tick mark on this ruler can be named by a number of sixteenths. Some tick marks can also be named by eighths, fourths, halves, and ones. The picture below shows the pattern of fraction names for a part of the ruler.

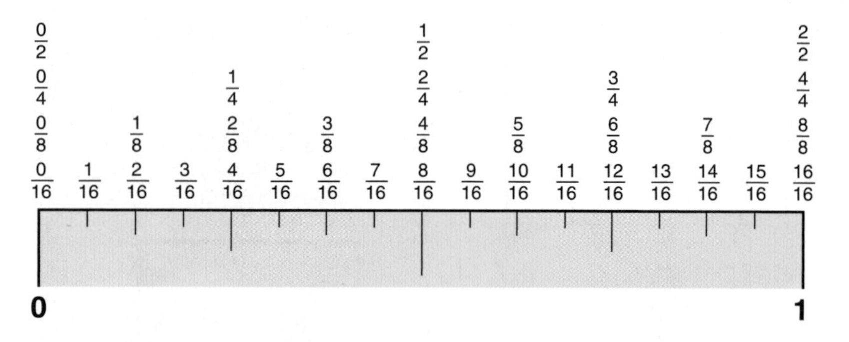

This pattern continues past 1 inch, with mixed numbers naming the tick marks.

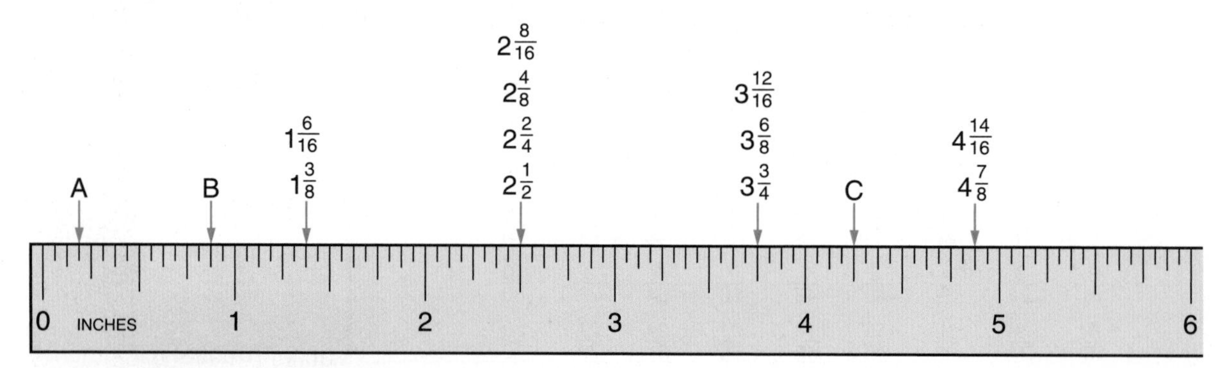

Check Your Understanding

1. Name a fraction or mixed number for each mark labeled A, B, and C on the ruler above.
2. What is the length of this nail?

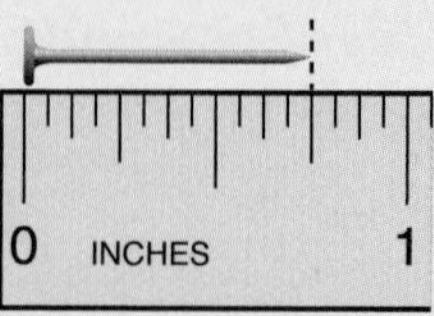

   a. in quarter inches **b.** in eighths of an inch

   c. in sixteenths of an inch

Check your answers on page 341.

Comparing Fractions

When you compare fractions, you have to pay attention to both the numerator and the denominator.

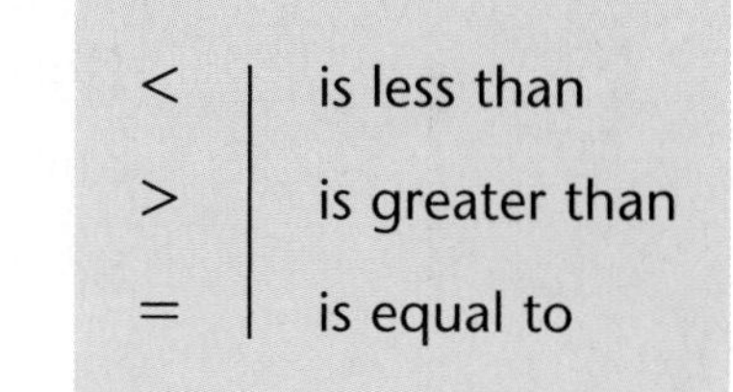

$<$	is less than
$>$	is greater than
$=$	is equal to

Like Denominators

Fractions are easy to compare when they have the same denominator. For example, to decide which is larger, $\frac{7}{8}$ or $\frac{5}{8}$, think of them as 7 eighths and 5 eighths. Just as 7 bananas is more than 5 bananas, and 7 dollars is more than 5 dollars, 7 eighths is more than 5 eighths.

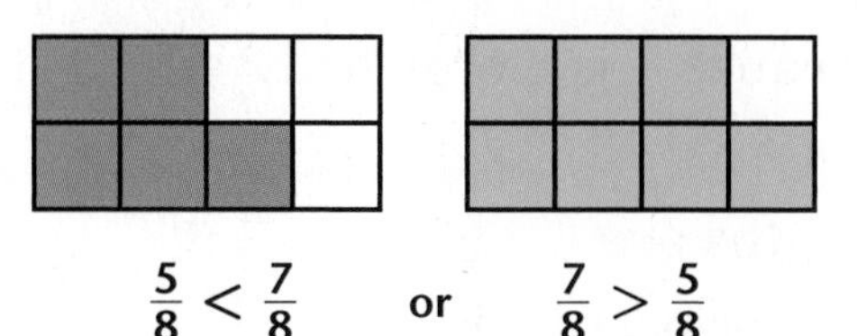

$\frac{5}{8} < \frac{7}{8}$ or $\frac{7}{8} > \frac{5}{8}$

To compare fractions that have the same denominators, just look at the numerators. The fraction with the larger numerator is larger.

Examples $\frac{4}{5} > \frac{3}{5}$ because $4 > 3$. $\frac{2}{9} < \frac{7}{9}$ because $2 < 7$.

Note

Fractions with **like denominators** have the same denominator.

$\frac{1}{4}$ and $\frac{3}{4}$ have like denominators.

Fractions with **like numerators** have the same numerator.

$\frac{2}{3}$ and $\frac{2}{5}$ have like numerators.

Like Numerators

If the numerators of two fractions are the same, then the fraction with the smaller denominator is larger. Remember, a smaller denominator means the ONE has fewer parts and each part is bigger. For example, $\frac{3}{5} > \frac{3}{8}$ because fifths are bigger than eighths, so 3 fifths is more than 3 eighths.

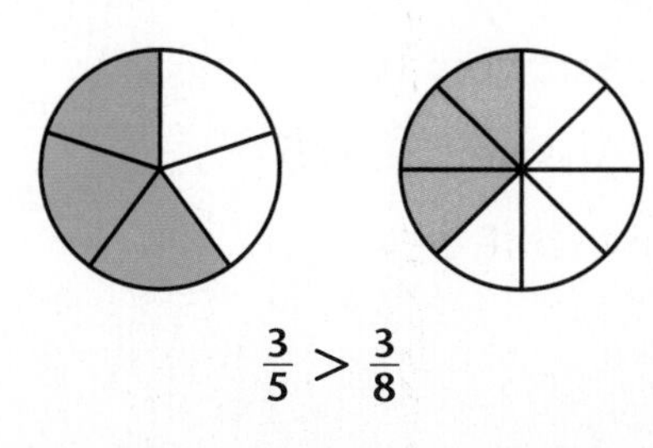

$\frac{3}{5} > \frac{3}{8}$

Did You Know?

Use of the symbol = for "equal to" dates back to 1571. Use of the symbols > and < for "greater than" and "less than" dates back to 1631.

Examples $\frac{1}{2} > \frac{1}{3}$ because halves are bigger than thirds.
$\frac{3}{8} < \frac{3}{4}$ because eighths are smaller than fourths.

Unlike Numerators and Unlike Denominators

Several strategies can help you compare fractions when *both* the numerators and the denominators are different.

Comparing to $\frac{1}{2}$

Compare $\frac{3}{7}$ and $\frac{5}{8}$.
Notice that $\frac{5}{8}$ is more than $\frac{1}{2}$ and $\frac{3}{7}$ is less than $\frac{1}{2}$.
So, $\frac{3}{7} < \frac{5}{8}$.

0 $\frac{3}{7}$ $\frac{1}{2}$ $\frac{5}{8}$ 1

Comparing to 0 or 1

Comparing fractions to 0 or 1 can also be helpful. For example, $\frac{7}{8} > \frac{3}{4}$ because $\frac{7}{8}$ is closer to 1. ($\frac{7}{8}$ is $\frac{1}{8}$ away from 1 but $\frac{3}{4}$ is $\frac{1}{4}$ away from 1. Since eighths are smaller than fourths, $\frac{7}{8}$ is closer to 1.)

0 $\frac{1}{2}$ $\frac{3}{4}$ $\frac{7}{8}$ 1

Using Equivalent Fractions

One way to compare fractions that *always* works is to find equivalent fractions that have the same denominator. For example, to compare $\frac{5}{8}$ and $\frac{3}{5}$, look at the table of equivalent fractions on page 51. The table shows that both fifths and eighths can be written as 40ths: $\frac{5}{8} = \frac{25}{40}$ and $\frac{3}{5} = \frac{24}{40}$. Since $\frac{25}{40} > \frac{24}{40}$, you know that $\frac{5}{8} > \frac{3}{5}$.

Using Decimal Equivalents

Using decimal equivalents is another way to compare fractions that *always* works. For example, to compare $\frac{2}{5}$ and $\frac{3}{8}$, use a calculator to change both fractions to decimals:

$\frac{2}{5}$: Key in: 2 [÷] 5 [=] Answer: 0.4
$\frac{3}{8}$: Key in: 3 [÷] 8 [=] Answer: 0.375

Since 0.4 > 0.375, you know that $\frac{2}{5} > \frac{3}{8}$.

Note

Remember that fractions can be used to show division problems.

$\frac{a}{b} = a \div b$

Check Your Understanding

Compare. Write <, >, or = in each box.

1. $\frac{3}{5} \square \frac{3}{7}$
2. $\frac{2}{3} \square \frac{4}{9}$
3. $\frac{3}{8} \square \frac{5}{8}$
4. $\frac{2}{6} \square \frac{2}{5}$

Check your answers on page 341.

Adding and Subtracting Fractions

Like Denominators

Adding or subtracting fractions that have the same denominator is easy: Just add or subtract the numerators, and keep the same denominator.

$\frac{2}{7} + \frac{3}{7} = \frac{5}{7}$

You can use division to put the answer in simplest form.

Examples $\frac{3}{8} + \frac{1}{8} = \frac{4}{8} = \frac{4 \div 4}{8 \div 4} = \frac{1}{2}$

$\frac{7}{10} - \frac{3}{10} = \frac{4}{10} = \frac{4 \div 2}{10 \div 2} = \frac{2}{5}$

Unlike Denominators

When you are adding and subtracting fractions that have unlike denominators, you must be especially careful. One way is to model the problem with pattern blocks. Remember that different denominators mean the ONE is divided into different numbers (and different sizes) of parts.

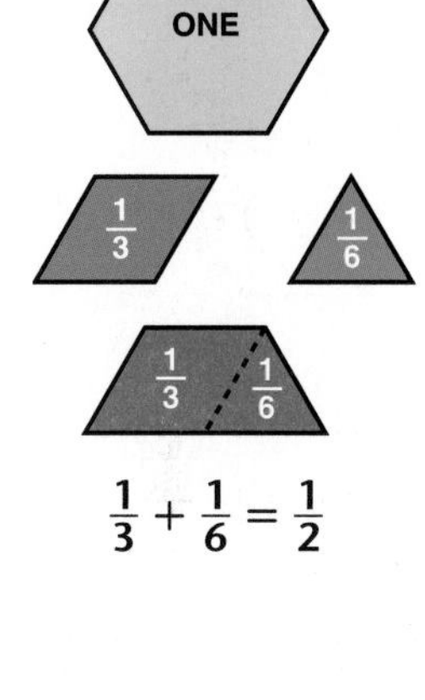

Example $\frac{1}{3} + \frac{1}{6} = ?$

If the hexagon is ONE, then the rhombus is $\frac{1}{3}$ and the triangle is $\frac{1}{6}$.

When you put one rhombus and one triangle together, you will find that they make a trapezoid. If the hexagon is ONE, then the trapezoid is $\frac{1}{2}$.

So, $\frac{1}{3} + \frac{1}{6} = \frac{1}{2}$.

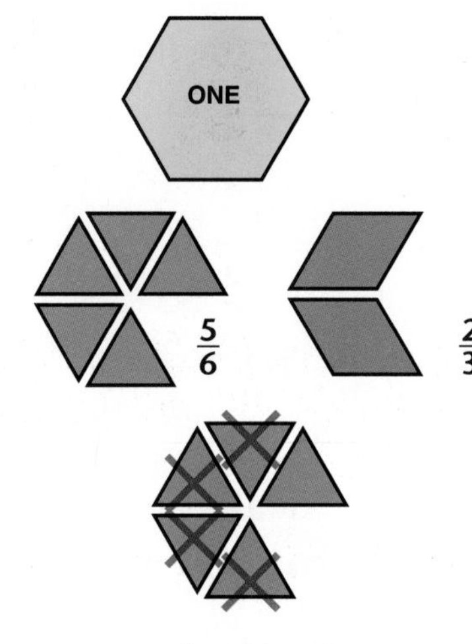

$\frac{5}{6} - \frac{2}{3} = \frac{1}{6}$

Example $\frac{5}{6} - \frac{2}{3} = ?$

If the hexagon is ONE, then $\frac{5}{6}$ is 5 triangles and $\frac{2}{3}$ is 2 rhombuses.

To take away $\frac{2}{3}$ (2 rhombuses) from $\frac{5}{6}$ (5 triangles), you would need to take away 4 triangles.

Then there would be 1 triangle or $\frac{1}{6}$ left.

So, $\frac{5}{6} - \frac{2}{3} = \frac{1}{6}$.

Adding and Subtracting Mixed Numbers

Like Denominators

One way to add or subtract mixed numbers is to add or subtract the whole numbers and fractions separately. If the fractions in the mixed numbers have the same denominators, just add or subtract the numerators. Then use division, if needed, to write the answer in simplest form.

Example $1\frac{5}{8} + 2\frac{1}{8} = ?$

Add the fractions.

$$\begin{array}{r} 1\mathbf{\frac{5}{8}} \\ +\ 2\mathbf{\frac{1}{8}} \\ \hline \mathbf{\frac{6}{8}} \end{array}$$

$\frac{5}{8}$

$+\ \frac{1}{8}$

$\frac{6}{8}$

Add the whole numbers.

$$\begin{array}{r} \mathbf{1}\frac{5}{8} \\ +\ \mathbf{2}\frac{1}{8} \\ \hline \mathbf{3}\frac{6}{8} \end{array}$$

1   1 $\frac{5}{8}$

+ 2   1   1 $+\ \frac{1}{8}$

3   1   1   1 $\frac{6}{8}$

$1\frac{5}{8} + 2\frac{1}{8} = 3\frac{6}{8} = 3\frac{3}{4}$

Example $7\frac{3}{5} - 6\frac{1}{5} = ?$

Subtract the fractions.

$$\begin{array}{r} 7\mathbf{\frac{3}{5}} \\ -\ 6\mathbf{\frac{1}{5}} \\ \hline \mathbf{\frac{2}{5}} \end{array}$$

Subtract the whole numbers.

$$\begin{array}{r} \mathbf{7}\frac{3}{5} \\ -\ \mathbf{6}\frac{1}{5} \\ \hline \mathbf{1}\frac{2}{5} \end{array}$$

$7\frac{3}{5} - 6\frac{1}{5} = 1\frac{2}{5}$

Some calculators have special keys for entering mixed numbers so you can add and subtract them. See pages 199 and 200.

Unlike Denominators

You can also add or subtract the whole numbers and fractions separately if two mixed numbers have fractions with unlike denominators. In these examples, you may use pattern blocks to model the fractions.

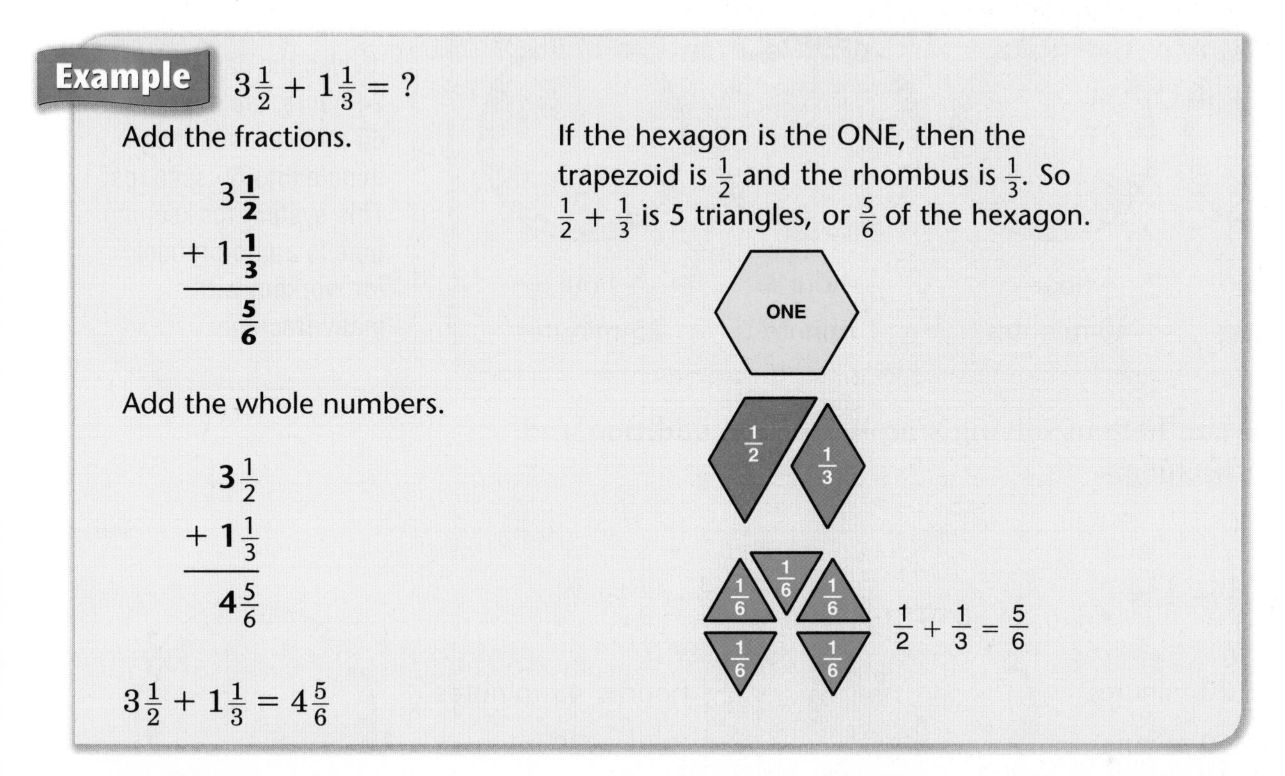

Example $3\frac{1}{2} + 1\frac{1}{3} = ?$

Add the fractions.

$$\begin{array}{r} 3\mathbf{\frac{1}{2}} \\ +\ 1\mathbf{\frac{1}{3}} \\ \hline \mathbf{\frac{5}{6}} \end{array}$$

Add the whole numbers.

$$\begin{array}{r} \mathbf{3}\frac{1}{2} \\ +\ \mathbf{1}\frac{1}{3} \\ \hline \mathbf{4}\frac{5}{6} \end{array}$$

$3\frac{1}{2} + 1\frac{1}{3} = 4\frac{5}{6}$

If the hexagon is the ONE, then the trapezoid is $\frac{1}{2}$ and the rhombus is $\frac{1}{3}$. So $\frac{1}{2} + \frac{1}{3}$ is 5 triangles, or $\frac{5}{6}$ of the hexagon.

Example $6\frac{2}{3} - 4\frac{1}{6} = ?$

Subtract the fractions.

$$\begin{array}{r} 6\mathbf{\frac{2}{3}} \\ -\ 4\mathbf{\frac{1}{6}} \\ \hline \mathbf{\frac{1}{2}} \end{array}$$

Subtract the whole numbers.

$$\begin{array}{r} \mathbf{6}\frac{2}{3} \\ -\ \mathbf{4}\frac{1}{6} \\ \hline \mathbf{2}\frac{1}{2} \end{array}$$

$6\frac{2}{3} - 4\frac{1}{6} = 2\frac{1}{2}$

If the hexagon is the ONE, then $\frac{2}{3}$ is 2 rhombuses and $\frac{1}{6}$ is 1 triangle. The 2 rhombuses are the same as 4 triangles. So $\frac{2}{3}$ (4 triangles) – $\frac{1}{6}$ (1 triangle) leaves 3 triangles, or $\frac{1}{2}$ the hexagon.

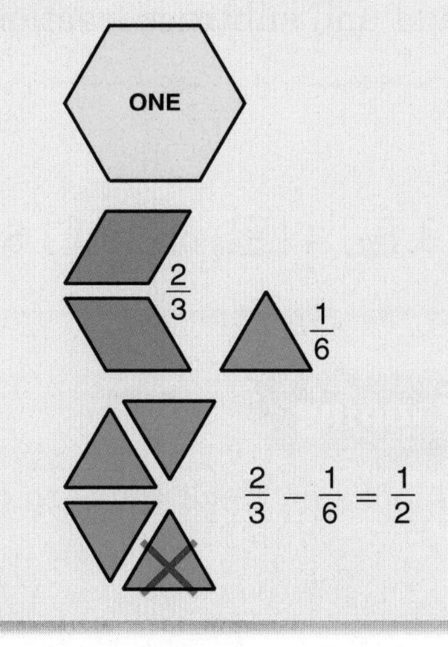

Clock Fractions

A clock face can be used to model fractions with 2, 3, 4, 5, 6, 10, 12, 15, 20, 30, or 60 in the denominator.

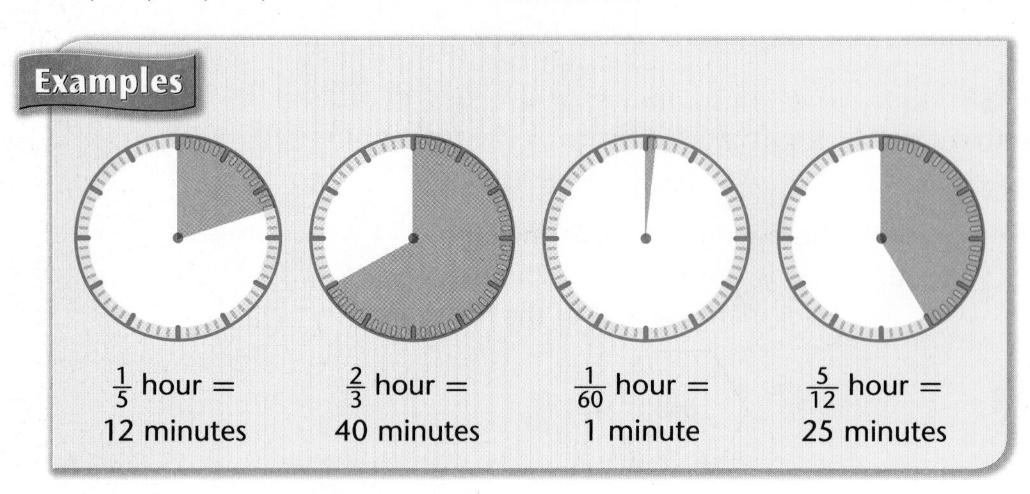

$\frac{1}{5}$ hour = 12 minutes

$\frac{2}{3}$ hour = 40 minutes

$\frac{1}{60}$ hour = 1 minute

$\frac{5}{12}$ hour = 25 minutes

Note

Thousands of years ago, the ancient Babylonians divided the day into 24 hours, the hour into 60 minutes, and the minute into 60 seconds. This system for keeping time is a good model for working with many fractions.

A clock face can help in solving simple fraction addition and subtraction problems.

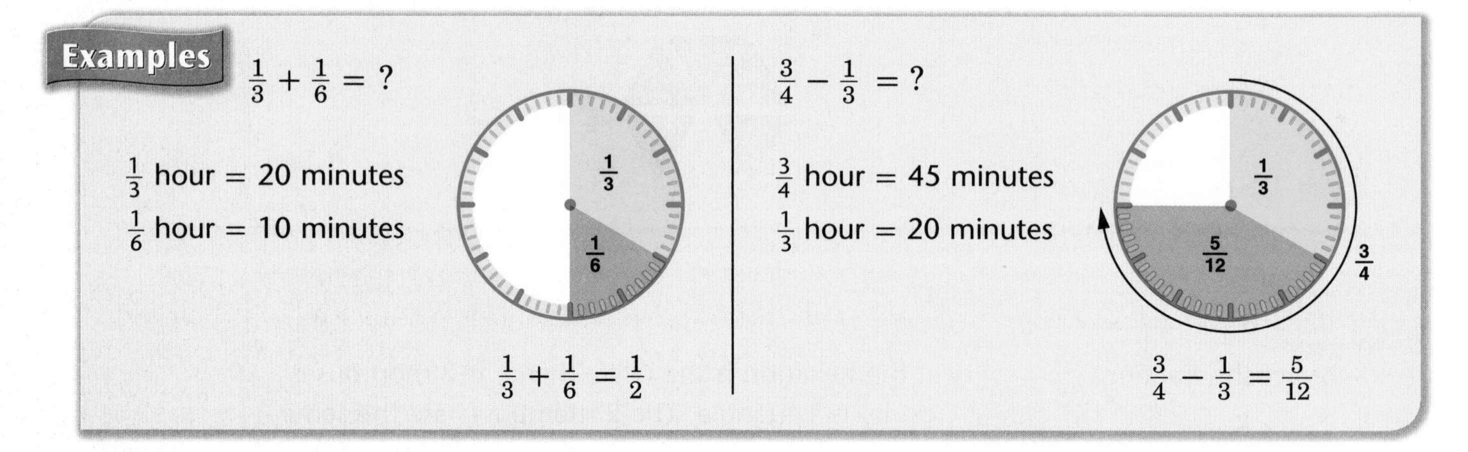

$\frac{1}{3} + \frac{1}{6} = ?$

$\frac{1}{3}$ hour = 20 minutes

$\frac{1}{6}$ hour = 10 minutes

$\frac{1}{3} + \frac{1}{6} = \frac{1}{2}$

$\frac{3}{4} - \frac{1}{3} = ?$

$\frac{3}{4}$ hour = 45 minutes

$\frac{1}{3}$ hour = 20 minutes

$\frac{3}{4} - \frac{1}{3} = \frac{5}{12}$

Using a Calculator

Some calculators can add and subtract fractions.

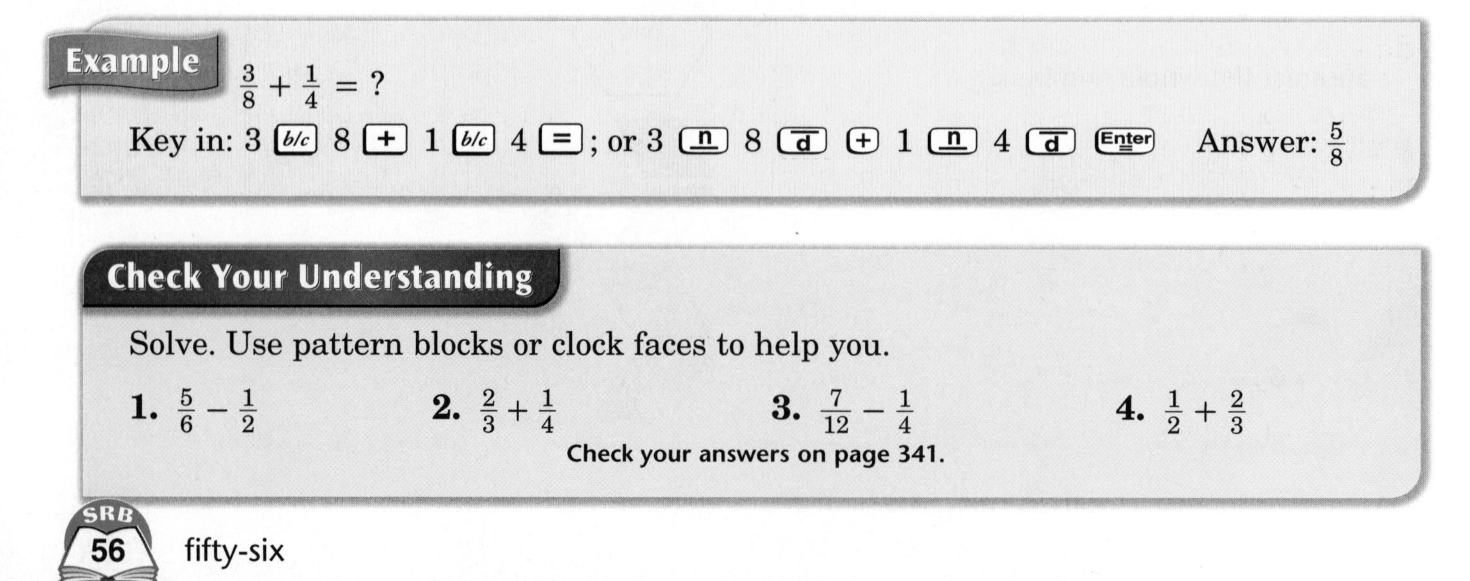

Example $\frac{3}{8} + \frac{1}{4} = ?$

Key in: 3 [b/c] 8 [+] 1 [b/c] 4 [=]; or 3 [n] 8 [d] [+] 1 [n] 4 [d] [Enter] Answer: $\frac{5}{8}$

Check Your Understanding

Solve. Use pattern blocks or clock faces to help you.

1. $\frac{5}{6} - \frac{1}{2}$ **2.** $\frac{2}{3} + \frac{1}{4}$ **3.** $\frac{7}{12} - \frac{1}{4}$ **4.** $\frac{1}{2} + \frac{2}{3}$

Check your answers on page 341.

Sometimes tools like pattern blocks or clock faces are not helpful for solving a fraction addition or subtraction problem. Here is a method that always works.

Using a Like Denominator

To add or subtract fractions that have different denominators, first rename them as fractions with a like denominator. A quick like denominator to use is the product of the denominators.

Example $\frac{1}{4} + \frac{2}{3} = ?$

A quick way to find a like denominator for these fractions is to multiply the denominators: $4 * 3 = 12$.

Rename $\frac{1}{4}$ and $\frac{2}{3}$ as 12ths:

$\frac{1}{4} = \frac{1 * 3}{4 * 3} = \frac{3}{12}$

$\frac{2}{3} = \frac{2 * 4}{3 * 4} = \frac{8}{12}$

So, $\frac{1}{4} + \frac{2}{3} = \frac{3}{12} + \frac{8}{12} = \frac{11}{12}$.

Example $\frac{3}{4} - \frac{2}{5} = ?$

A like denominator for these fractions is $4 * 5 = 20$.

Rename $\frac{3}{4}$ and $\frac{2}{5}$ as 20ths:

$\frac{3}{4} = \frac{3 * 5}{4 * 5} = \frac{15}{20}$

$\frac{2}{5} = \frac{2 * 4}{5 * 4} = \frac{8}{20}$

So, $\frac{3}{4} - \frac{2}{5} = \frac{15}{20} - \frac{8}{20} = \frac{7}{20}$.

Did You Know?

A *furlong* is a unit of distance, equal to $\frac{1}{8}$ mile. It is often used to measure distances in horse and dog races.

To add a distance in furlongs (eighths of a mile) and a distance given in tenths of a mile, you could rename the fractions using $8 * 10 = 80$ as a like denominator.

If two fractions are renamed so that they have the same denominator, that denominator is called a **common denominator.**

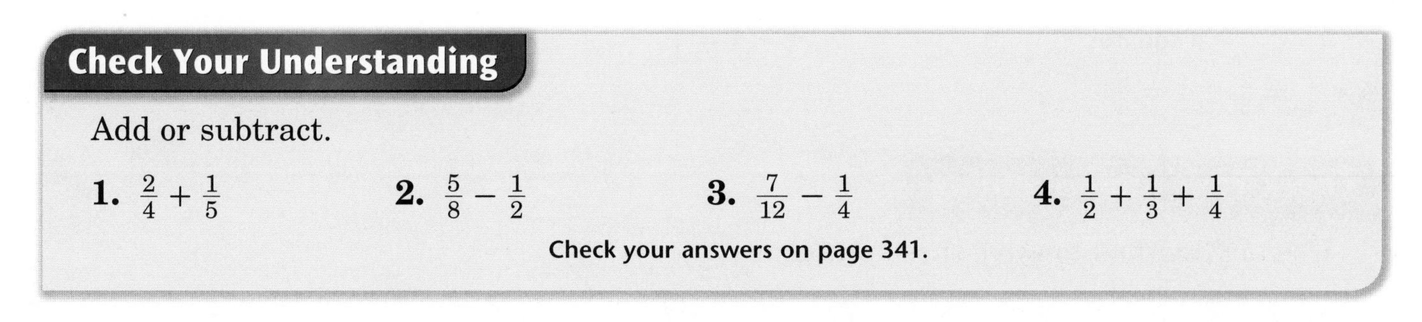

Check Your Understanding

Add or subtract.

1. $\frac{2}{4} + \frac{1}{5}$ **2.** $\frac{5}{8} - \frac{1}{2}$ **3.** $\frac{7}{12} - \frac{1}{4}$ **4.** $\frac{1}{2} + \frac{1}{3} + \frac{1}{4}$

Check your answers on page 341.

Multiplying Fractions and Whole Numbers

There are several ways to think about multiplying a whole number and a fraction.

Using a Number Line

One way to multiply a whole number and a fraction is to think about "hops" on a number line. The whole number tells how many hops to make, and the fraction tells how long each hop should be. For example, to solve $4 * \frac{2}{3}$, imagine taking 4 hops on a number line, each $\frac{2}{3}$ unit long.

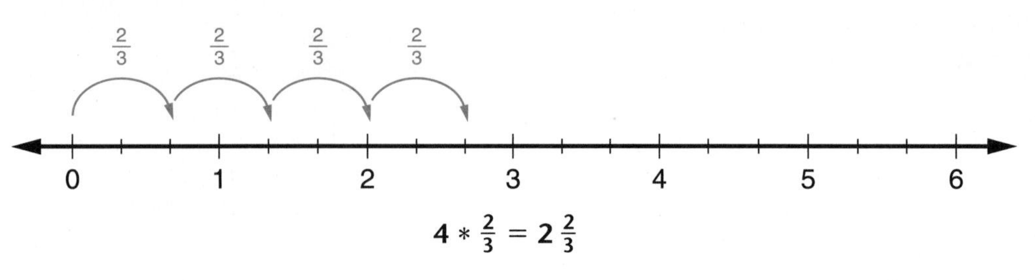

$4 * \frac{2}{3} = 2\frac{2}{3}$

Using Addition

You can use addition to multiply a fraction and a whole number. For example, to find $4 * \frac{2}{3}$, draw 4 models of $\frac{2}{3}$. Then add up all of the fractions.

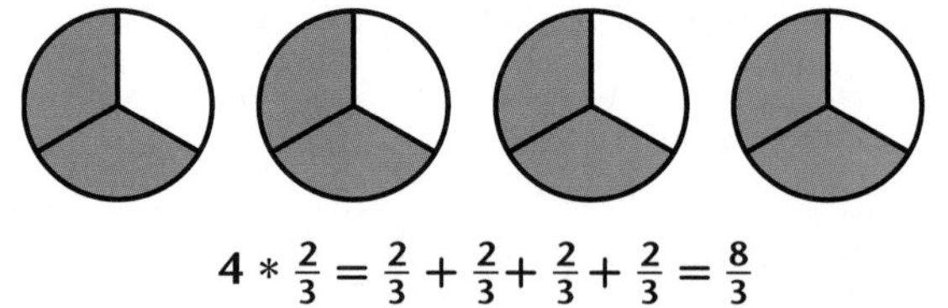

$4 * \frac{2}{3} = \frac{2}{3} + \frac{2}{3} + \frac{2}{3} + \frac{2}{3} = \frac{8}{3}$

Using Fraction of an Area

You can think of multiplying with a fraction as finding the fraction of an area. For example, to solve $4 * \frac{2}{3}$ (which is the same as $\frac{2}{3} * 4$), find $\frac{2}{3}$ of an area that is 4 square units.

The rectangle on the left has an area of 4 square units. The shaded area of the rectangle on the right has an area of $\frac{8}{3}$ square units (8 small rectangles, each with an area of $\frac{1}{3}$.)

So, $\frac{2}{3}$ of the rectangle area = the shaded area = $\frac{8}{3}$.

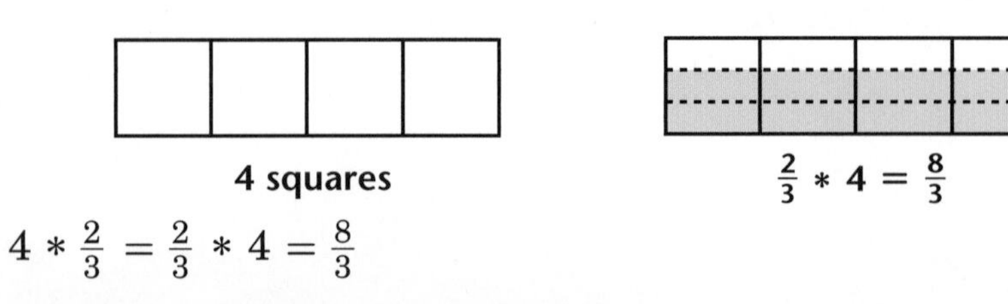

4 squares

$\frac{2}{3} * 4 = \frac{8}{3}$

$4 * \frac{2}{3} = \frac{2}{3} * 4 = \frac{8}{3}$

$\frac{3}{5}$ of 20 means $\frac{3}{5} * 20$.

$\frac{1}{6}$ of 18 means $\frac{1}{6} * 18$.

$\frac{3}{4}$ of 24 means $\frac{3}{4} * 24$.

The word "of" in problems like these means multiplication.

Check Your Understanding

Use any method to solve these problems.

1. $6 * \frac{2}{3}$ **2.** $3 * \frac{4}{5}$ **3.** $\frac{3}{4} * 6$ **4.** $4 * \frac{3}{4}$ **5.** $\frac{4}{5} * 5$

Check your answers on page 341.

Finding a Fraction of a Set

You can think of multiplication with fractions as finding a fraction of a set. For example, think of the problem $\frac{2}{5} * 30$ as "What is $\frac{2}{5}$ of 30¢?" One way to solve this problem is first to find $\frac{1}{5}$ of 30, and then use that answer to find $\frac{2}{5}$ of 30.

$\frac{2}{5}$ of 30 means $\frac{2}{5} * 30$.

Example $\frac{2}{5} * 30 = ?$ Think of the problem as "What is $\frac{2}{5}$ of 30?"

Step 1: Find $\frac{1}{5}$ of 30.

To do this, divide the 30 pennies into 5 equal groups. Then count the number of pennies in one group.

$30 \div 5 = 6$, so $\frac{1}{5}$ of 30 is 6.

Step 2: Next find $\frac{2}{5}$ of 30.

Since $\frac{1}{5}$ of 30 is 6, $\frac{2}{5}$ of 30 is $2 * 6 = 12$.

$\frac{2}{5} * 30 = \frac{2}{5}$ of $30 = 12$

5 equal groups, with 6 in each group

Example $\frac{2}{3} * 15 = ?$ Think of the problem as "What is $\frac{2}{3}$ of 15?"

Step 1: Find $\frac{1}{3}$ of 15.

Divide 15 pennies into 3 equal groups.
$15 \div 3 = 5$, so $\frac{1}{3}$ of 15 is 5.

Step 2: Next find $\frac{2}{3}$ of 15.
Since $\frac{1}{3}$ of 15 is 5, $\frac{2}{3}$ of 15 is $2 * 5 = 10$.

$\frac{2}{3} * 15 = \frac{2}{3}$ of $15 = 10$

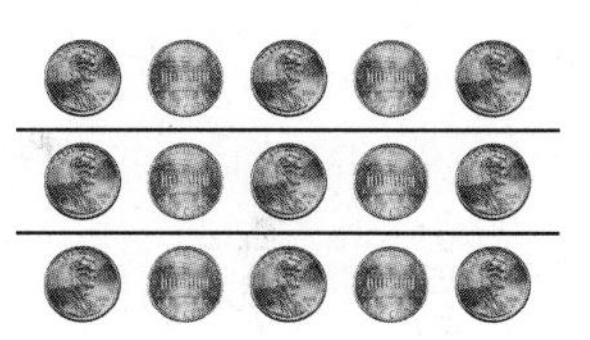

3 equal groups, with 5 in each group

Check Your Understanding

Find each answer.

1. $\frac{1}{4} * 28 = ?$
2. $\frac{3}{5}$ of $20 = ?$
3. $16 * \frac{5}{8} = ?$
4. Rita and Hunter earned $12 raking lawns. Since Rita did most of the work, she got $\frac{2}{3}$ of the money. How much did each person get?

Check your answers on page 341.

Negative Numbers and Rational Numbers

People have used **counting numbers** (1, 2, 3, and so on) for thousands of years. Long ago people found that the counting numbers did not meet all of their needs. They needed numbers for in-between measures such as $2\frac{1}{2}$ inches and $6\frac{5}{6}$ hours.

Fractions were invented to meet these needs. Fractions can also be renamed as decimals and percents. Most of the numbers you have seen are fractions or can be renamed as fractions.

Examples Rename as fractions: 0, 12, 15.3, 3.75, and 25%.

$0 = \frac{0}{1}$ $12 = \frac{12}{1}$ $15.3 = \frac{153}{10}$ $3.75 = \frac{375}{100}$ $25\% = \frac{25}{100}$

Note

Every **whole number** (0, 1, 2, and so on) can be renamed as a fraction. For example, 0 can be written as $\frac{0}{1}$. And 8 can be written as $\frac{8}{1}$.

However, even fractions did not meet every need. For example, problems such as $5 - 7$ and $2\frac{3}{4} - 5\frac{1}{4}$ have answers that are less than 0 and cannot be named as fractions. (Fractions, by the way they are defined, can never be less than 0.) This led to the invention of **negative numbers.** Negative numbers are numbers that are less than 0. The numbers $-\frac{1}{2}$, -2.75, and -100 are negative numbers. The number -2 is read "negative 2."

Note

Numbers like -2.75 and -100 may not look like negative fractions, but they can be renamed as negative fractions.

$-2.75 = -\frac{11}{4}$, and

$-100 = -\frac{100}{1}$

Negative numbers serve several purposes:

- To express locations such as temperatures below zero on a thermometer and depths below sea level
- To show changes such as yards lost in a football game
- To extend the number line to the left of zero
- To calculate answers to many subtraction problems

The **opposite** of every positive number is a negative number, and the opposite of every negative number is a positive number. The number 0 is neither positive nor negative; 0 is also its own opposite.

The diagram at the right shows this relationship.

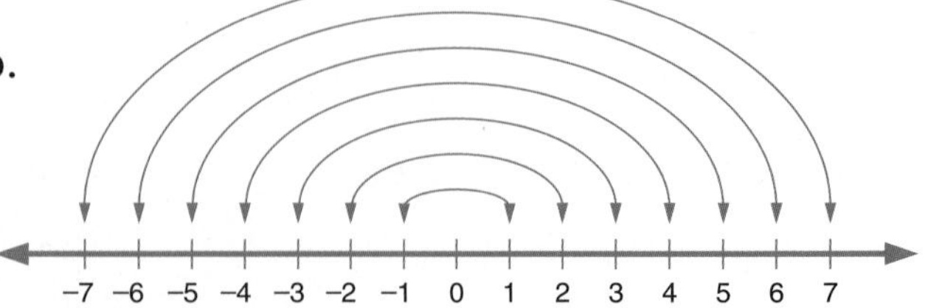

The **rational numbers** are all the numbers that can be written or renamed as fractions or as negative fractions.

Fractions, Decimals, and Percents

Fractions, decimals, and percents are different ways to write numbers. Sometimes it is easier to work with a fraction instead of a decimal or a percent. Other times it is easier to work with a decimal or a percent.

Did You Know?

The U.S. Constitution did not take effect until 9 of the 13 original states had approved it.

$\frac{9}{13} = 0.69$ and

$\frac{9}{13} = 69\%$

(decimal and percent rounded to 2 digits)

Renaming a Fraction as a Decimal

You can rename a fraction as a decimal if you can find an equivalent fraction with a denominator of 10, 100, or 1,000. This only works for certain fractions.

Examples

$\frac{1}{2} = \frac{1 * 50}{2 * 50} = \frac{50}{100} = 0.50$

$\frac{4}{5} = \frac{4 * 2}{5 * 2} = \frac{8}{10} = 0.8$

Note

Remember that $\frac{a}{b} = a \div b$ is true for any fraction $\frac{a}{b}$.

Another way to rename a fraction as a decimal is to divide the numerator by the denominator. You can use a calculator for this division.

Examples

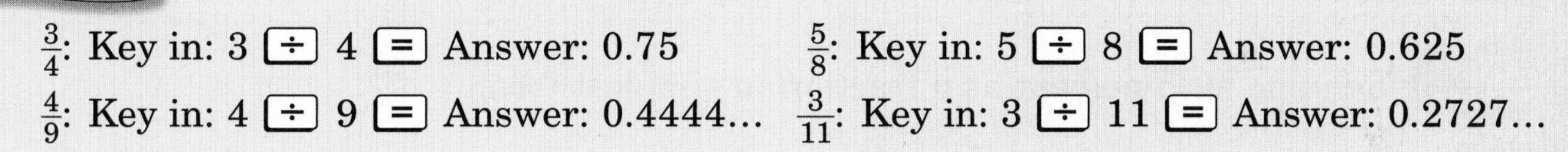

$\frac{3}{4}$: Key in: 3 ÷ 4 = Answer: 0.75

$\frac{5}{8}$: Key in: 5 ÷ 8 = Answer: 0.625

$\frac{4}{9}$: Key in: 4 ÷ 9 = Answer: 0.4444...

$\frac{3}{11}$: Key in: 3 ÷ 11 = Answer: 0.2727...

Renaming a Decimal as a Fraction

To change a decimal to a fraction, write the decimal as a fraction with a denominator of 10, 100, or 1,000. Then you can rename the fraction in simplest form.

Examples Write each decimal as a fraction.

For 0.5, the rightmost digit is 5, which is in the 10ths place. So, $0.5 = \frac{5}{10}$, or $\frac{1}{2}$.

For 0.307, the rightmost digit is 7, which is in the 1,000ths place. So, $0.307 = \frac{307}{1{,}000}$.

For 4.75, the rightmost digit is 5, which is in the 100ths place. So, $4.75 = \frac{475}{100}$ (a fraction) or $4\frac{75}{100}$ or $4\frac{3}{4}$ (mixed numbers).

Note

This method will work for most of the decimal numbers you see. But it will *not* work for every decimal number. For example, 0.4444... *cannot* be written as a fraction with a denominator of 10, 100, 1,000, or any other power of 10.

Renaming a Decimal as a Percent

To rename a decimal as a percent, try to write the decimal as a fraction with a denominator of 100. Then use the meaning of percent (number of hundredths) to rename the fraction as a percent.

Examples Rename each decimal as a percent.

$0.5 = 0.50 = \frac{50}{100} = 50\%$ $\quad$ $0.01 = \frac{1}{100} = 1\%$ $\quad$ $1.2 = 1.20 = \frac{120}{100} = 120\%$

Renaming a Percent as a Decimal

To rename a percent as a decimal, try to rename it as a fraction with a denominator of 100. Then rename the fraction as a decimal.

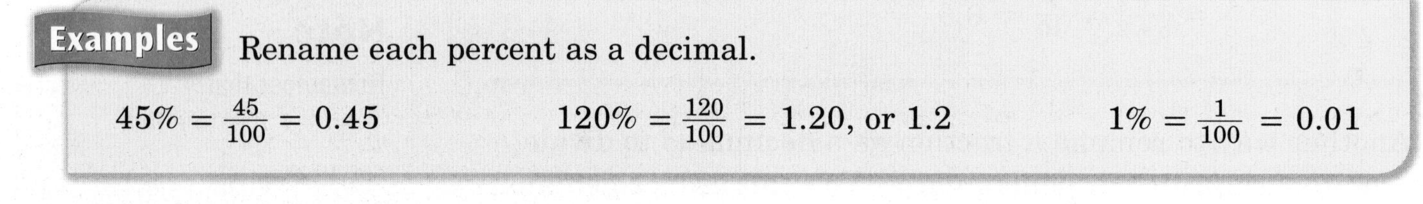

Examples Rename each percent as a decimal.

$45\% = \frac{45}{100} = 0.45$ $\quad$ $120\% = \frac{120}{100} = 1.20$, or 1.2 $\quad$ $1\% = \frac{1}{100} = 0.01$

Renaming a Percent as a Fraction

To rename a percent as a fraction, try to write it as a fraction with a denominator of 100.

Examples Rename each percent as a fraction in simplest form.

$50\% = \frac{50}{100} = \frac{1}{2}$ $\quad$ $75\% = \frac{75}{100} = \frac{3}{4}$ $\quad$ $1\% = \frac{1}{100}$ $\quad$ $200\% = \frac{200}{100} = 2$

Renaming a Fraction as a Percent

To rename a fraction as a percent, try to rename it as a fraction with a denominator of 100. Then rename the fraction as a percent.

Examples Rename each fraction as a percent.

$\frac{1}{2} = 0.50 = \frac{50}{100} = 50\%$ $\quad$ $\frac{3}{5} = 0.60 = \frac{60}{100} = 60\%$ $\quad$ $\frac{3}{8} = 0.375 = \frac{37.5}{100} = 37.5\%$

Check Your Understanding

Write each number as a fraction, a decimal, and a percent.

1. $\frac{1}{2}$ $\quad$ **2.** 0.75 $\quad$ **3.** 10% $\quad$ **4.** $\frac{4}{5}$

Check your answers on page 341.

Sound, Music, and Mathematics

Musicians make patterns of sound to create music. Mathematics can help us understand how both sound and music are created.

Sound

Every sound you hear begins with a vibration—a back and forth motion. For musical instruments to produce sound, something must be set in motion.

The sound of a drum starts when a person beats the drum head. When the drum head stops vibrating, the sound stops.

The sound of a guitar starts when a person plucks or strums the strings. Each vibrating string moves back and forth at the same rate until it stops moving. When the strings stop vibrating, the sound stops.

The sound of a flute starts when a person blows across the mouthpiece. A column of air moves back and forth inside the flute. When the player stops blowing, the column of air stops vibrating and the sound of the flute stops.

The rate at which a string, a drum head, or a column of air vibrates is called the **frequency.** Higher frequency vibrations produce higher-pitched notes. Frequency is measured in Hertz (Hz), or "vibrations per second." The human ear can hear vibrations from about 15 Hz to 20,000 Hz.

Instrument Length and Pitch

Many instruments rely on a vibrating column of air to make sound. A longer column of air vibrates at a lower frequency and makes a deeper- or lower-pitched note. Shorter vibrating air columns make higher-pitched notes.

Here are some instruments you may have heard, along with the frequency of the lowest note that can be played on the instrument. What happens to the frequency as the instruments get shorter?

bassoon, 58 Hz

clarinet, 139 Hz

oboe, 233 Hz

piccolo, 587 Hz

Mathematics ... Every Day

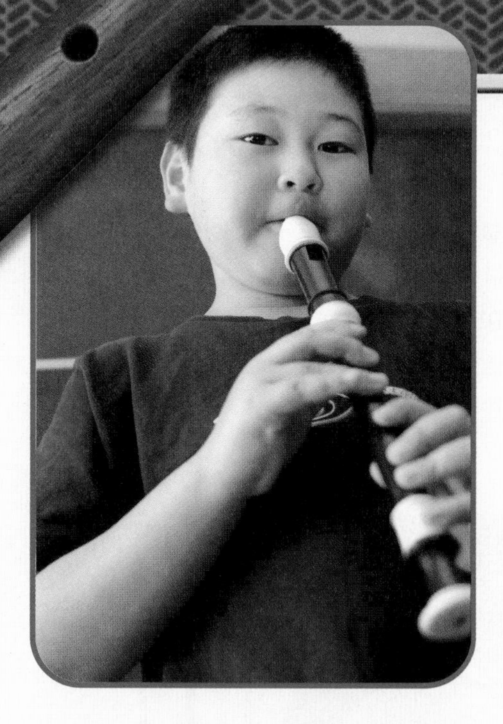

A recorder can play a range of pitches. By covering all of the finger holes on a recorder, the musician creates the longest possible column of air, and the lowest-pitched note. With all holes uncovered, a high note is produced.

The piccolo has a very short column of air within it, so it produces high-pitched notes. Piccolos produce notes in the range of about 600 to 4,000 Hz, which humans can hear easily.

This pan flute, from Peru, is played by blowing across the edges of hollow tubes of different lengths. Short tubes produce high-pitched notes, and long tubes produce low-pitched notes. The player slides the instrument from side to side to change notes.

Because the alto saxophone is much longer than the piccolo, its sound is lower-pitched. Saxophones use a reed, which is a carefully-shaped piece of cane. The musician blows into the mouthpiece, which causes the reed to vibrate. This starts the vibration of the column of air.

Percussion Instruments

Drums are percussion instruments. The size of the instrument affects the pitch it can play. The size and tightness of the drum head and the materials that the drum head is made from also affect the pitch.

In a trap set, the largest drum—the base or "kick" drum—produces the lowest-pitched notes. Each drum can be tuned up or down by tightening or loosening the heads.

A drummer holds a West African talking drum, or *donno*, between the upper arm and the body. Squeezing the strings with the upper arm tightens the drumhead and raises the pitch of the drum. Releasing the strings loosens the drumhead and lowers the pitch of the drum.

The steel drum, from the Caribbean island of Trinidad, is made by cutting off the top of a steel oil barrel. Each small rounded section of the drum head is shaped to play a different pitched note. The pitch of the instrument can be very high because the small metal sections vibrate rapidly.

Mathematics ... Every Day

Stringed Instruments

The pitch of the notes that a stringed instrument can play is related to the length, diameter, and tension of the strings.

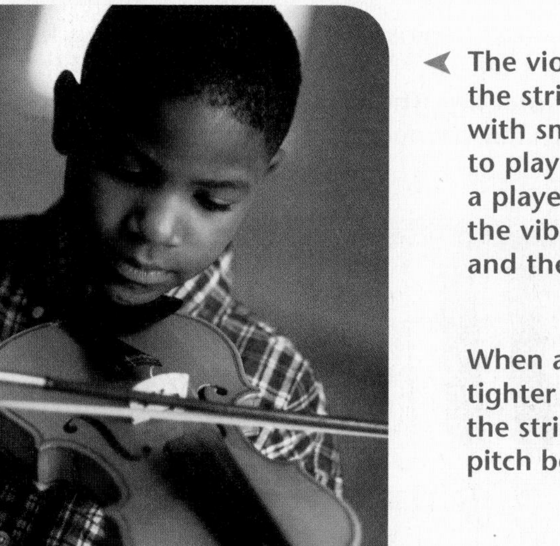

The violin, the smallest member of the string family, has short strings with small diameters. It is designed to play high-pitched notes. When a player presses down on a string, the vibrating part is shortened and the pitch becomes higher.

When a musician winds a string tighter around its tuning peg, the string is tightened and the pitch becomes higher.

This man is tuning his stringed instrument. Tightening a string raises the pitch. Loosening a string lowers the pitch.

Compared to the violin, the cello has longer strings of greater diameters. It is designed to play low-pitched notes.

The Piano

Looking closely at the way a piano works can help you see some of the mathematical relationships in music.

A piano's sound begins when a player presses a key. This causes a felt-covered wooden hammer to hit the strings for that key. The strings then vibrate to produce sound. Each key produces a note with a different pitch.

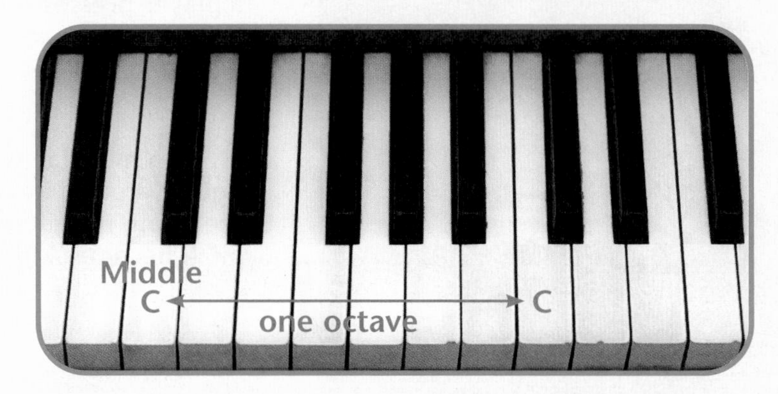

An octave begins and ends on a note with the same name. For example, the keys between "Middle C" and the C to the right of it represent 1 octave. There are 8 octaves on most pianos. The names of the white and black keys in an octave repeat eight times.

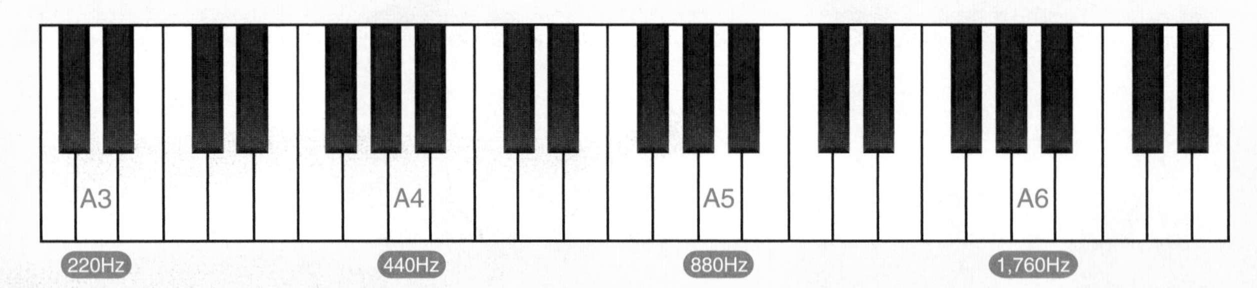

As you move to the right on the piano keyboard, the frequencies get higher. What patterns do you see in the frequencies?

This tuning fork vibrates 440 times per second. A piano tuner tightens or loosens the A4 string until its pitch exactly matches the pitch of the vibrating tuning fork. Then all other strings are tightened or loosened based on that note.

What patterns can you find in music? How have you seen mathematics used in music?

Data and Probability

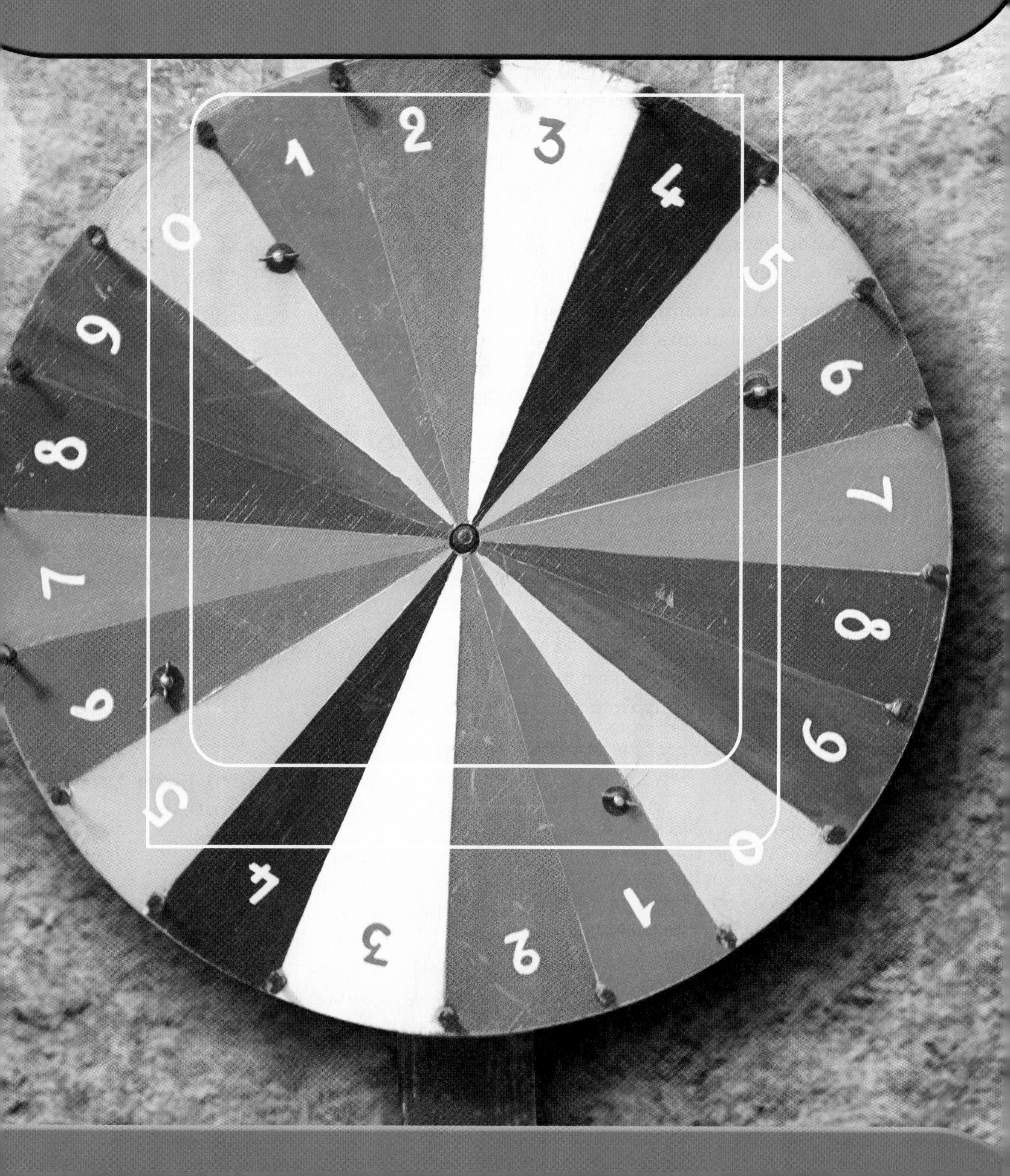

Collecting Data

There are different ways to collect information about something. You can count, measure, ask questions, or observe and describe what you see. The information you collect is called **data.**

Surveys

A **survey** is a study that collects data. Much of the information used to make decisions comes from surveys. Many surveys collect data about people. Stores survey customers to find out what products they should carry. Television stations survey viewers to learn what programs are popular. The survey data are collected in several ways, including face-to-face interviews, telephone interviews, and written questions that are answered and returned by mail.

Not all surveys gather information about people. For example, there are surveys about cars, buildings, and animal groups.

Example

A bird survey is conducted each year in the Chicago area during December and January. Bird watchers list the different bird species they see, and they count the number of each species seen. The lists are combined to create a final data set.

From a recent Chicago bird survey:

Species	Number of Birds Seen
blue heron	10
canada goose	3,768
house sparrow	2,446
robin	213

Samples

A **sample** is a smaller group chosen to represent the whole group. Data are collected only from this smaller group.

Example

A survey of teenagers collects data on people aged 13 to 19. There are about 28 million teenagers in the United States. It is not possible to collect data from every one of them. Data are collected from a sample of teenagers instead.

Average Time Spent Each Day
(in hours:minutes)

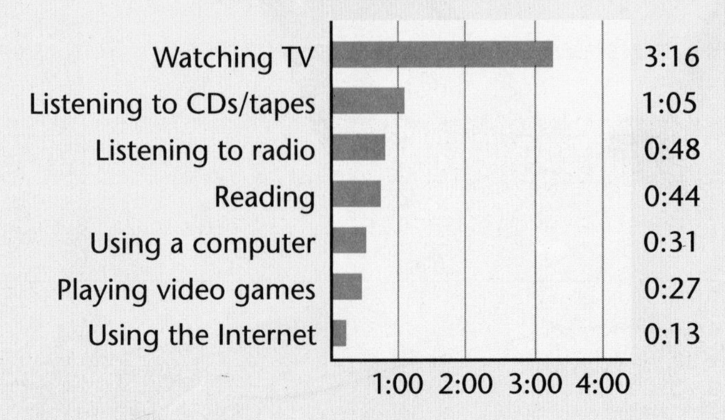

Organizing Data

Once the data have been collected, it helps to organize them to make them easier to understand. **Line plots** and **tally charts** are two methods of organizing data.

Example Mr. Jackson's class got the following scores on a five-word spelling test. Make a line plot and a tally chart to show the data below.

5 3 5 0 4 4 5 4 4 4 2 3 4 5 3 5 4 3 4 4

Scores on a 5-Word Spelling Test

Number of Students

				X	
				X	
				X	
				X	
				X	X
			X	X	X
			X	X	X
			X	X	X
X		X	X	X	X
0	1	2	3	4	5

Number Correct

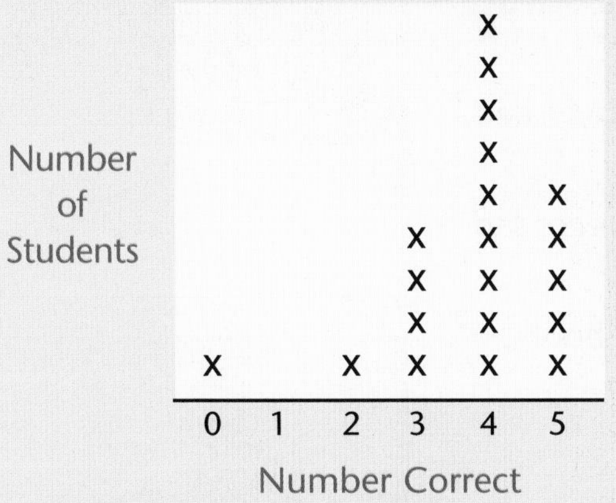

In this **line plot,** there are 4 Xs above the number 3.

Four students got a score of 3 on the test.

Scores on a 5-Word Spelling Test

Number Correct	Number of Students
0	/
1	
2	/
3	////
4	𝍸 ////
5	𝍸

In this **tally chart,** there are 4 tallies to the right of 3.

Four students got a score of 3 on the test.

Both the line plot and the tally chart help to organize the data. They make it easier to describe the data. For example,

- Five students had 5 words correct.
- 4 correct is the score that came up most often.
- 0 correct and 2 correct are scores that came up least often.
- No student got exactly 1 correct.

Check Your Understanding

Here are the number of hits made by 14 players in a baseball game.

4 1 0 2 1 3 2 1 0 2 0 2 0 3

Organize the data. **1.** Make a tally chart. **2.** Make a line plot.

Check your answers on page 341.

Sometimes the data are spread out over a wide range of numbers. This makes a tally chart and a line plot difficult to draw. In such cases, you can make a tally chart in which the results are grouped.

Example Ms. Beck asked her students to make a count of the number of books they read over the summer. These were the results:

9 13 3 10 16 0 12 5 16 4 10 11 13 5 20

5 15 1 12 24 7 13 0 38 2 11 14 18 6 12

The table below sorts the data into **intervals** of 5. This is called a tally chart of **grouped data.**

- If a student read from 0 to 4 books, one tally mark is recorded for the "0–4" interval.
- If a student read from 5 to 9 books, one tally mark is recorded for the "5–9" interval, and so on.

The chart shows that most students read fewer than 20 books. The most frequent number of books read was from 10 to 14.

Books Read by Students

Number of Books	Number of Students
0–4	~~////~~ /
5–9	~~////~~ /
10–14	~~////~~ ~~////~~ /
15–19	////
20–24	//
25 or more	/

Check Your Understanding

Michael Jordan played in 12 games of the 1996 NBA Playoffs. He scored the following numbers of points:

35 29 26 44 28 46 27 35 21 35 17 45

Copy and complete the tally chart of grouped data.

Number of Points	Number of Games
10–19	
20–29	
30–39	
40–49	

Check your answer on page 341.

Statistical Landmarks

The **landmarks** for a set of data are used to describe the data.

- The **minimum** is the smallest value.
- The **maximum** is the largest value.
- The **range** is the difference between the maximum and the minimum.
- The **mode** is the value or values that occur most often.
- The **median** is the middle value.

The dividing area between opposite lanes of traffic on some highways is called the *median* or the *median strip.*

Example Here is a record of children's absences for one week at Medgar Evers School.

Monday	Tuesday	Wednesday	Thursday	Friday
27	19	12	16	16

Find the landmarks for the data.

Minimum (smallest) number: 12
Range of numbers: 27 − 12 = 15

Maximum (largest) number: 27
Mode (most frequent number): 16

To find the median (middle value):

- List the numbers in order from smallest to largest or largest to smallest. — 12 16 16 19 27
- Cross out one number from each end of the list. — ~~12~~ 16 16 19 ~~27~~
- Continue crossing out one number from each end of the list. — ~~12~~ ~~16~~ [16] ~~19~~ ~~27~~
- The median is the number that remains after all others have been crossed out. — median

There may not be landmarks for some sets of data. For example, if you collect data about hair color, there is no "largest color" or "middle value color." But, you can still find the mode. The mode is the hair color that occurs most often.

Check Your Understanding

Here are math quiz scores (number correct) for 11 students: 3 1 2 2 1 1 2 0 0 4 2

Find these landmarks for the data.

1. minimum **2.** maximum **3.** range **4.** mode **5.** median

Check your answers on page 341.

Example The **line plot** shows students' scores on a 20-word spelling test. Find the landmarks for the data.

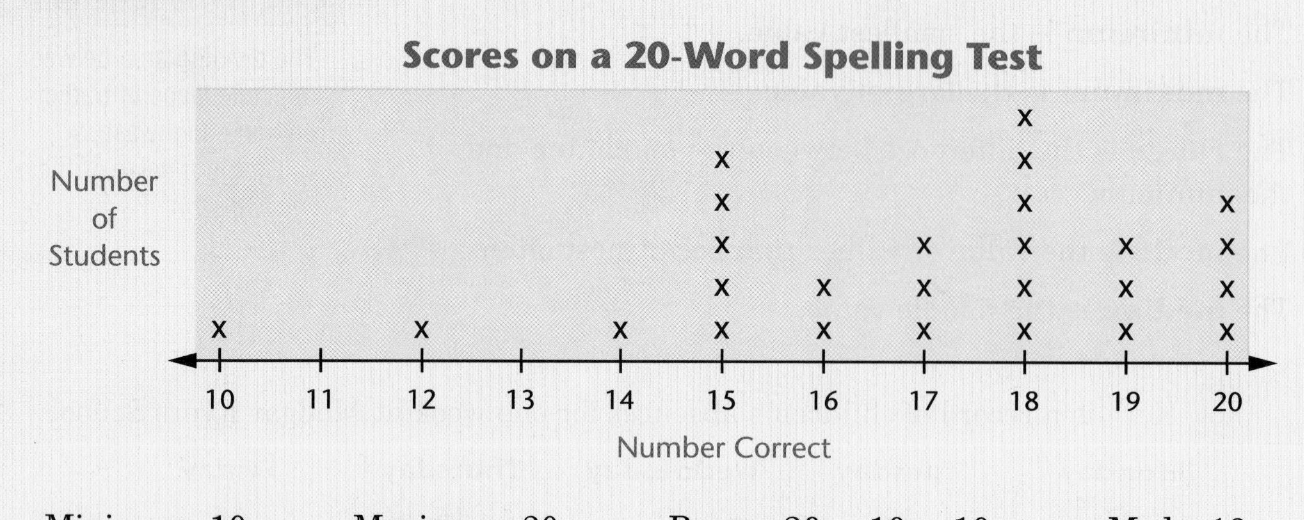

Minimum: 10 Maximum: 20 Range: 20 − 10 = 10 Mode: 18

To find the median (middle value), first list the numbers in order:

10 12 14 15 15 15 15 15 16 16 17 17 17 18 18 18 18 18 18 19 19 19 20 20 20 20

Cross out one number from each end of this list.
Continue to cross out one number from each end until there are only two numbers left.
The two numbers remaining are the middle scores.

~~10~~ ~~12~~ ~~14~~ ~~15~~ ~~15~~ ~~15~~ ~~15~~ ~~15~~ ~~16~~ ~~16~~ ~~17~~ ~~17~~ 17 18 ~~18~~ ~~18~~ ~~18~~ ~~18~~ ~~18~~ ~~19~~ ~~19~~ ~~19~~ ~~20~~ ~~20~~ ~~20~~ ~~20~~

middle scores

There are two middle scores, 17 and 18.

The median is 17.5, which is the number halfway between 17 and 18.

Check Your Understanding

1. Here are math quiz scores (number correct) for 12 students:

   1 3 2 2 3 2 1 4 4 0 4 3

   Find the minimum, maximum, range, mode, and median for this set of data.

2. Find the median for this set of numbers: 23 9 12 16 18 7 23 7 12 31

Check your answers on page 341.

The Mean (or Average)

The **mean** of a set of numbers is often called the *average*.
To find the mean:

Step 1: Add the numbers.

Step 2: Divide the sum by the number of addends.

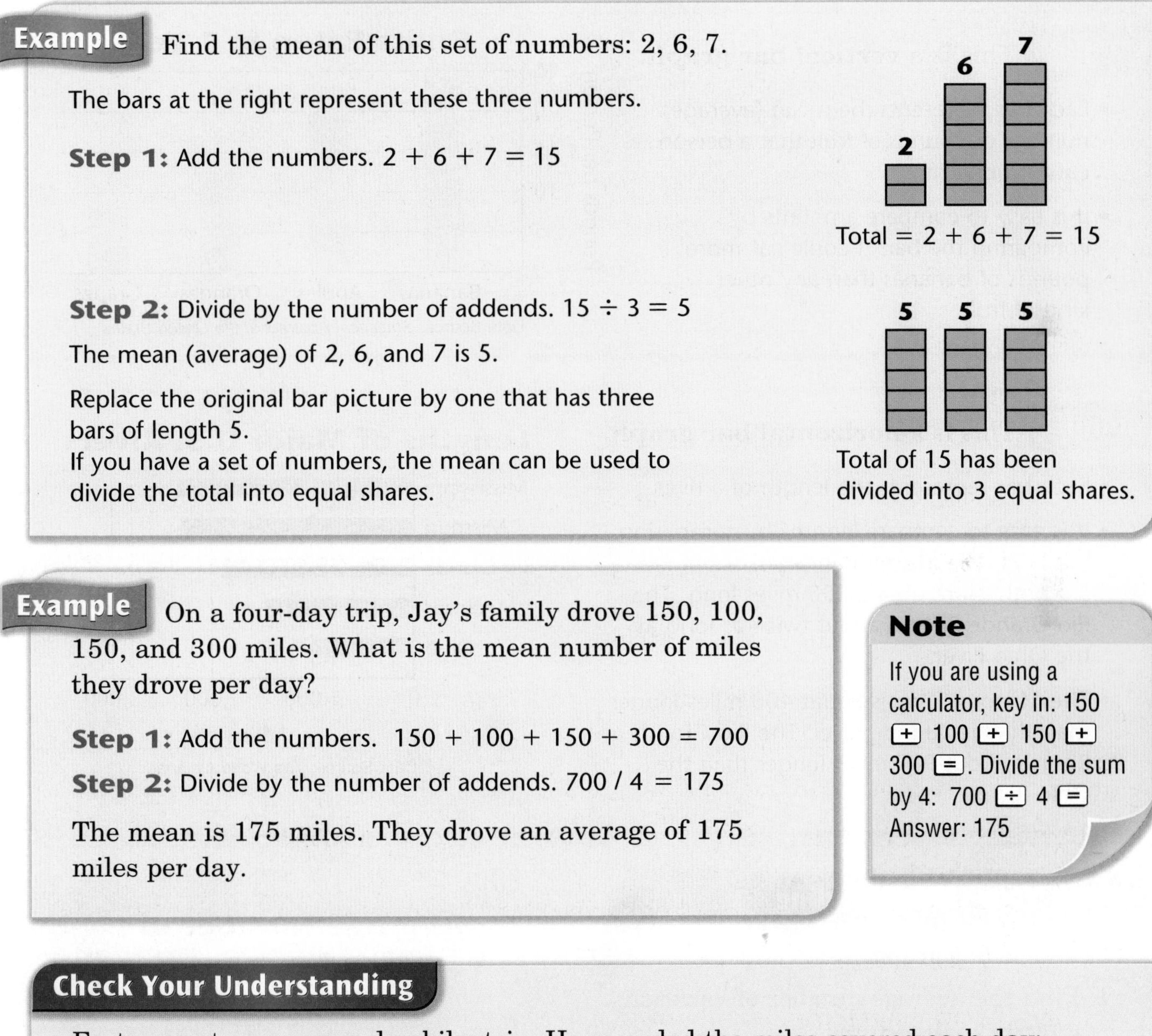

Example Find the mean of this set of numbers: 2, 6, 7.

The bars at the right represent these three numbers.

Step 1: Add the numbers. $2 + 6 + 7 = 15$

Total $= 2 + 6 + 7 = 15$

Step 2: Divide by the number of addends. $15 \div 3 = 5$

The mean (average) of 2, 6, and 7 is 5.

Replace the original bar picture by one that has three bars of length 5.
If you have a set of numbers, the mean can be used to divide the total into equal shares.

Total of 15 has been divided into 3 equal shares.

Example On a four-day trip, Jay's family drove 150, 100, 150, and 300 miles. What is the mean number of miles they drove per day?

Step 1: Add the numbers. $150 + 100 + 150 + 300 = 700$

Step 2: Divide by the number of addends. $700 / 4 = 175$

The mean is 175 miles. They drove an average of 175 miles per day.

Note

If you are using a calculator, key in: 150 [+] 100 [+] 150 [+] 300 [=]. Divide the sum by 4: 700 [÷] 4 [=]
Answer: 175

Check Your Understanding

Foster went on a seven-day bike trip. He recorded the miles covered each day:

Mon: 18 Tue: 20 Wed: 17 Thurs: 23 Fri: 25 Sat: 26 Sun: 25

1. Find the total distance. **2.** Find the mean (average) distance covered per day.

Check your answers on page 341.

Bar Graphs

A **bar graph** is a drawing that uses bars to represent numbers. Bar graphs display information in a way that makes it easy to show comparisons. A bar graph has a title that describes the information in the graph. Each bar has a label. Units are given to show how something was counted or measured. When possible, the graph gives the source of the information.

Example This is a **vertical bar graph.**

- Each bar represents the mean (average) number of pounds of fruit that a person eats in one year.
- It is easy to compare amounts by comparing the bars. People eat more pounds of bananas than any other kind of fruit.

Example This is a **horizontal bar graph.**

- Each bar represents the length of a river.
- It is easy to compare lengths by comparing the bars. The Missouri and Mississippi Rivers are both more than 2,000 miles long. The Rio Grande River is about twice as long as the Ohio River.
- The Missouri River is about 400 miles longer than the Rio Grande River. The Rio Grande River is about 900 miles longer than the Ohio River.

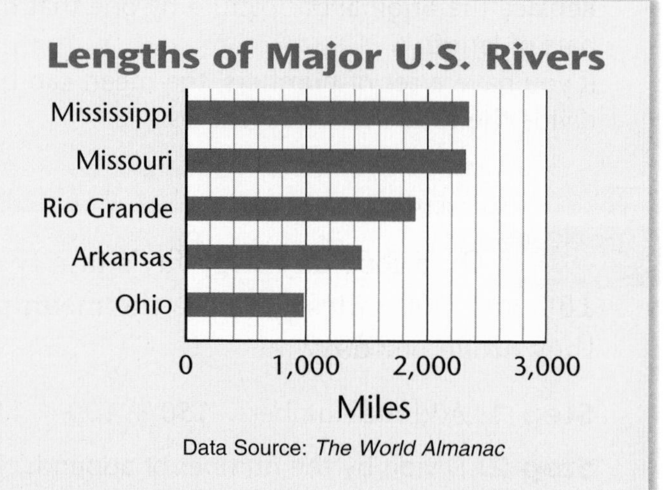

Check Your Understanding

Read the bar graph.

1. Find the average number of vacation days in each country.
2. Compare the number of vacation days in Italy and the United States.
3. Compare the number of vacation days in Canada and the United States.

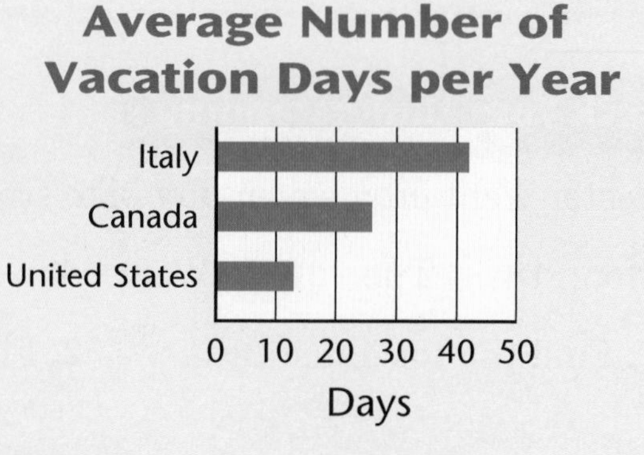

Check your answers on page 342.

Pictographs

A **pictograph** uses picture symbols to show numbers. Pictographs are similar to tally charts, with each pictograph symbol replacing one or more tally marks.

Example The pictograph below shows how many children chose certain foods as their favorite foods.

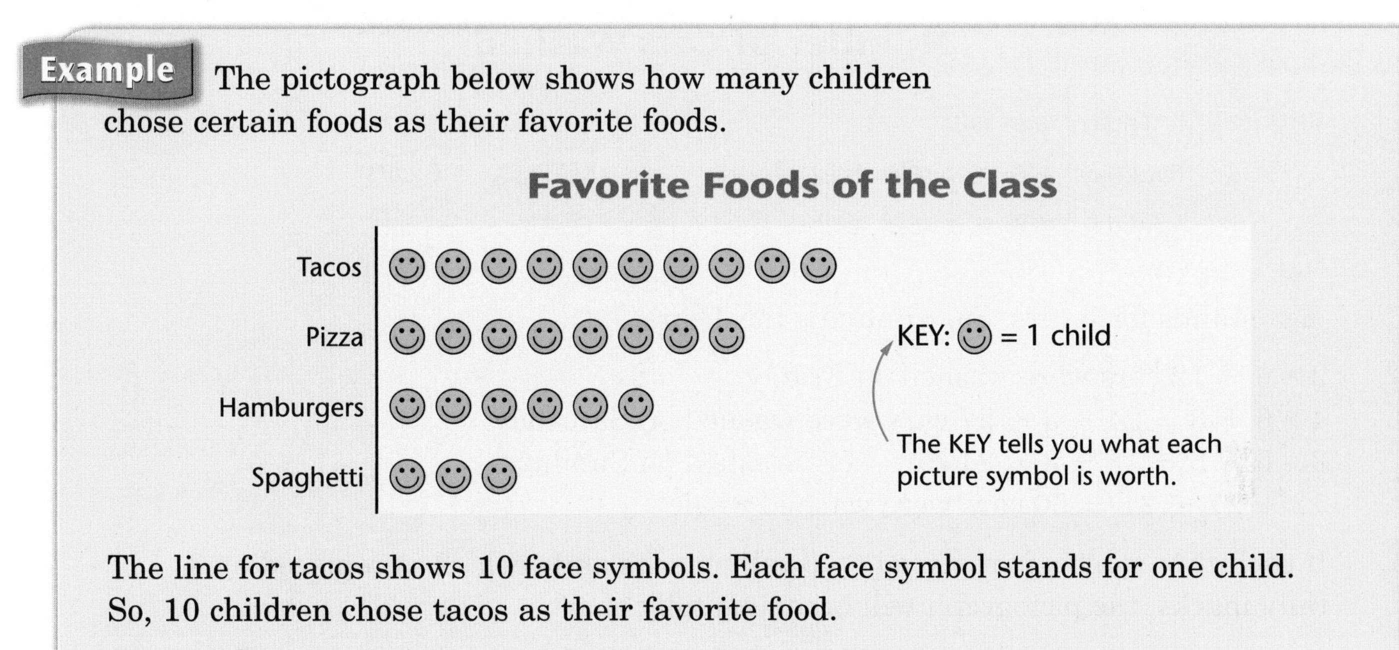

The line for tacos shows 10 face symbols. Each face symbol stands for one child. So, 10 children chose tacos as their favorite food.

If each ☺ symbol is replaced by one tally mark, the pictograph will become a tally chart.

When you use a pictograph, always check the KEY first.

Example This pictograph shows how many children in Lincoln School are in each grade.

Number of Children in Each Grade

3rd grade	☺☺☺☺☺☺☺☺☺	KEY: ☺ = 10 children
4th grade	☺☺☺☺☺☺☺	
5th grade	☺☺☺☺☺☺☺☺	

The line for 3rd grade shows 9 face symbols.
Each face symbol stands for 10 children.
So, there are $9 * 10 = 90$ children in the 3rd grade of Lincoln School.

If each ☺ symbol is replaced by 10 tally marks 𝍸𝍸, the pictograph will become a tally chart.

In some pictographs, you may see only part of a picture symbol.
Use the KEY to decide how much this part of the symbol is worth.

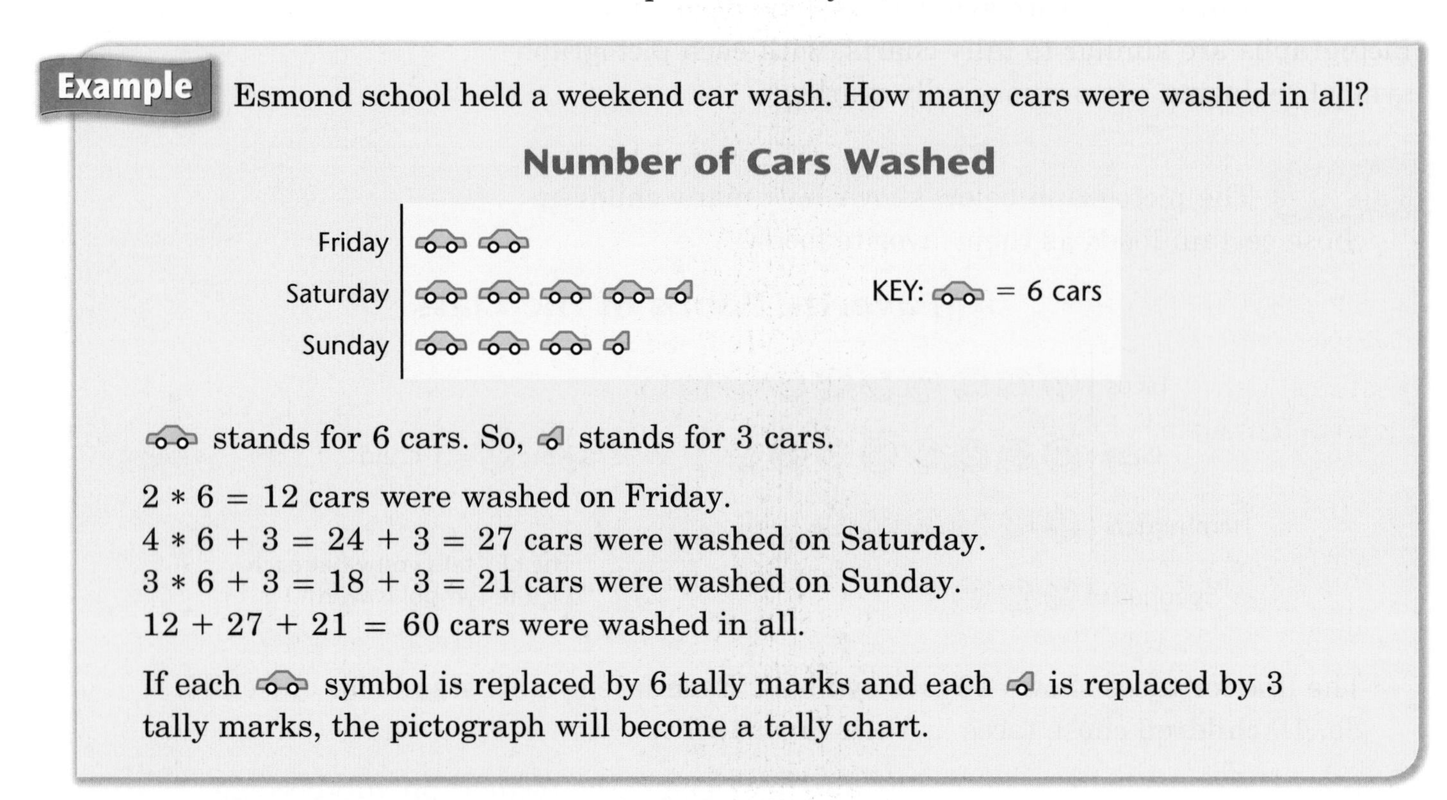

Example Esmond school held a weekend car wash. How many cars were washed in all?

stands for 6 cars. So, stands for 3 cars.

$2 * 6 = 12$ cars were washed on Friday.
$4 * 6 + 3 = 24 + 3 = 27$ cars were washed on Saturday.
$3 * 6 + 3 = 18 + 3 = 21$ cars were washed on Sunday.
$12 + 27 + 21 = 60$ cars were washed in all.

If each symbol is replaced by 6 tally marks and each is replaced by 3 tally marks, the pictograph will become a tally chart.

Check Your Understanding

Use the pictograph to answer the questions.

Number of Children Who Ride Bicycles to School

3rd Grade
4th Grade
5th Grade

KEY: = 4 children

1. How many children in 4th grade ride a bicycle to school?
2. What is the total number of children in grades 3 through 5 who ride a bicycle to school?
3. Draw a tally chart that shows the same information as the pictograph.

Check your answers on page 342.

Line Graphs

Line graphs are used to display information that shows trends. They often show how something has changed over a period of time.

Line graphs are sometimes called **broken-line graphs.** Line segments connect the points on the graph. Joined end to end, the segments look like a broken line.

Line graphs have both a horizontal and a vertical scale. Each of these scales is called an **axis** (plural: **axes**). Each axis is labeled to show what is being measured or counted and what the unit of measure or count unit is.

Broken-Line Graph

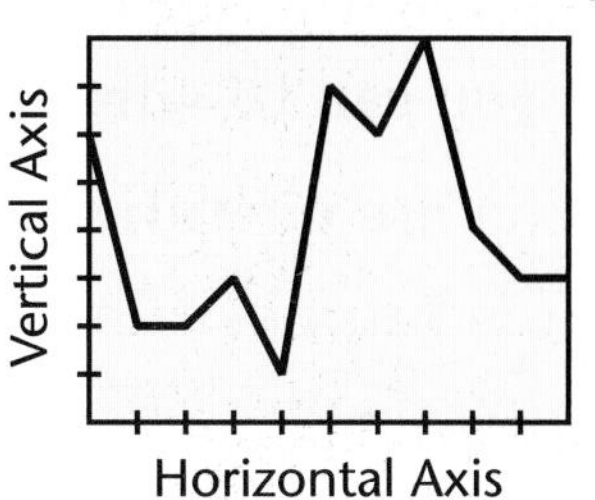

The segments joined end-to-end look like a broken line.

Example The broken-line graph at the right shows the number of farms in the United States from 1950 to 2000.

The horizontal axis is divided into 10-year intervals. The vertical axis shows the number of farms in the United States, in millions.

By studying the graph, you can see the trend or direction in the number of farms over the last 50 years.

- The number has decreased in each 10-year interval from 1950 to 1990.
- From 1990 to 2000, the number of farms stayed about the same.

Number of U.S. Farms

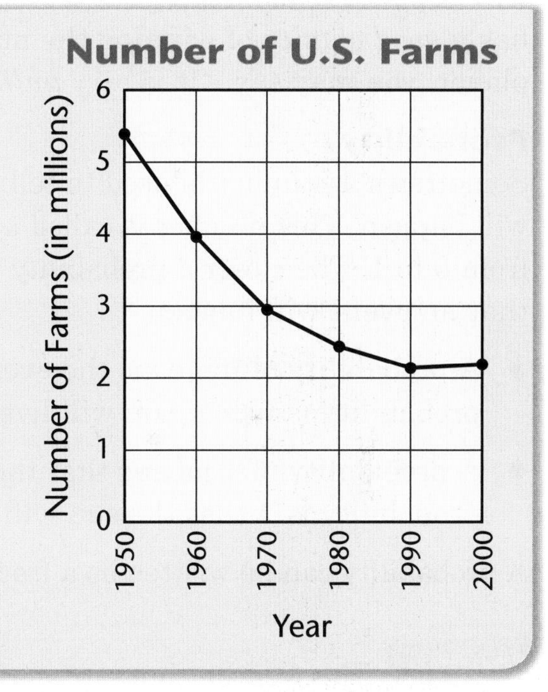

Check Your Understanding

The table at the right shows the number of cars in a school parking lot at different times. Make a line graph to show this information.

Cars in a Parking Lot

Time	Number of Cars
6 A.M.	3
7 A.M.	13
8 A.M.	35
9 A.M.	50
10 A.M.	45

Check your answer on page 342.

Chance and Probability

Chance

Things that happen are called **events.** There are many events that you can be sure about.

- You are **certain** that the sun will rise tomorrow.
- It is **impossible** for you to grow to be 10 feet tall.

There are also many events that you *cannot* be sure about.

- You cannot be sure that you will get a letter tomorrow.
- You cannot be sure whether it will be sunny next Tuesday.

Sometimes you might talk about the **chance** that something will happen. If Joan is a good tennis player, you may say, "Joan has a *good chance* of winning the match." If Joan is a poor player, you may say, "It is *very unlikely* that Joan will win."

Probability

Sometimes a number is used to tell the chance that something will happen. This number is called a **probability.** It is a number from 0 to 1. The closer a probability is to 1, the more likely it is that an event will happen.

- A probability of 0 means the event is *impossible.* The probability is 0 that you will live to the age of 200.
- A probability of 1 means that the event is *certain*. The probability is 1 that the sun will rise tomorrow.

A probability can be written as a fraction, a decimal, or a percent.

Did You Know?

In the 18th century, the mathematician Pierre Laplace calculated the probability that the sun would rise as 99.9999452%. Laplace himself did not actually believe that this probability was correct.

Example The weather bureau predicts that there is an 80% chance of rain today and a 3 in 4 chance of rain tomorrow. On which day is it more likely to rain?

A probability of 80% can also be written as 0.8, or $\frac{80}{100}$.
A 3 in 4 chance means that the probability is $\frac{3}{4}$, 0.75, or 75%.

Since 0.8 is greater than 0.75, it is more likely to rain today.

A 9 in 10 chance means the probability is $\frac{9}{10}$.

A 55 in 100 chance means the probability is $\frac{55}{100}$.

A 1 in 1,000 chance means the probability is $\frac{1}{1,000}$.

Check Your Understanding

The weather bureau predicts a 15% chance of snow on Sunday and a 1 in 6 chance of snow on Monday. On which day is it more likely to snow?

Check your answer on page 342.

Calculating a Probability

The examples below show four common ways for finding probabilities.

Make a Guess

John guesses that he has an 80% chance (an 8 in 10 chance) of returning home by 9 o'clock.

Conduct an Experiment

Kathleen dropped 100 tacks: 60 landed point up and 40 landed point down. The fraction of tacks that landed point up is $\frac{60}{100}$. Kathleen estimates the probability that the next tack she drops will land point up is $\frac{60}{100}$, or 60%.

Use a Data Table

Art got 48 hits in his last 100 times at bat.

He estimates the probability that he will get a hit the next time at bat is $\frac{48}{100}$, or 48%.

Hits	Walks	Outs	Total
48	11	41	100

Assume That All Possible Results Have the Same Chance

A standard die has 6 faces and is shaped like a cube. You can assume that each face has the same $\frac{1}{6}$ chance of coming up.

An 8-sided die has 8 faces and is shaped like a regular octahedron. You can assume that each face has the same $\frac{1}{8}$ chance of coming up.

A standard deck of playing cards has 52 cards. Suppose the cards are shuffled and one card is drawn. You can assume that each card has the same $\frac{1}{52}$ chance of being drawn.

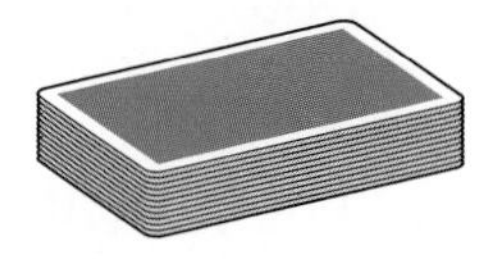

Suppose that a spinner is divided into 12 equal sections. When you spin the spinner, you can assume that each section has the same $\frac{1}{12}$ chance of being landed on.

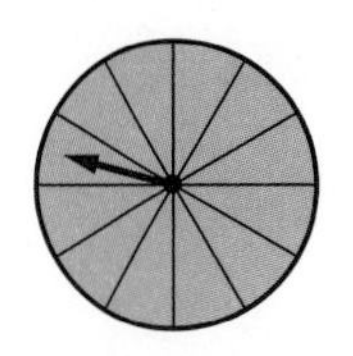

Equally Likely Outcomes

It is often useful to list all of the possible results for a situation. Each possible result is called an **outcome.** If all of the possible outcomes have the same probability, they are called **equally likely outcomes.**

Examples If you roll a 6-sided die, the number of dots that turns up may be 1, 2, 3, 4, 5, or 6. There are six possible outcomes. You can assume that each face of the die has the same $\frac{1}{6}$ chance of coming up. The outcomes are equally likely.

The spinner at the right is divided into 10 equal parts. If you spin the spinner, it will land on one of the numbers 1 through 10. There are 10 possible outcomes. You can assume that each number has a $\frac{1}{10}$ chance of being landed on, because the sections are equal parts of the spinner. The outcomes are equally likely.

The spinner at the right is divided into two sections. If you spin the spinner, it will land on red or blue. Red and blue are the two possible outcomes. However, the outcomes are *not* equally likely, because there is a greater chance of landing on red than on blue.

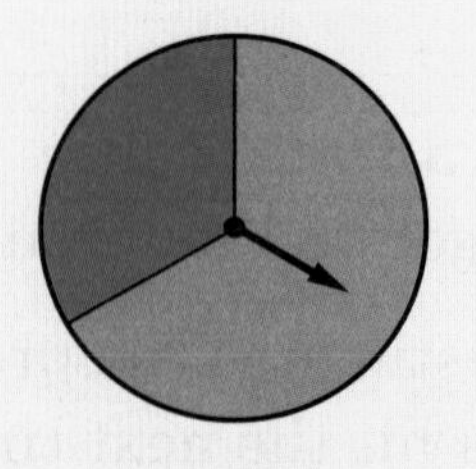

A Probability Formula

Finding the probability of an event is easy if all of the outcomes are equally likely. Follow these steps:

1. List all the possible outcomes.
2. Look for any outcomes that will make the event happen. These outcomes are called **favorable outcomes.** Circle each favorable outcome.
3. Count the number of possible outcomes. Count the number of favorable outcomes.

   The probability of the event is $\frac{\text{number of favorable outcomes}}{\text{number of possible outcomes}}$.

Example Amy, Beth, Carol, Dave, Edgar, Frank, and George are on a camping trip. They decide to choose a leader. Each child writes his or her name on an index card. The cards are put into a paper bag and mixed. One card will be drawn, and the child whose name is drawn will become leader. Find the probability that a girl will be selected.

What is the *event* you want to find the probability of? Draw a girl's name.

How many *possible outcomes* are there? Seven. Any of the 7 names might be drawn.

Are the outcomes *equally likely?* Yes. Names were written on identical cards and the cards were mixed in the bag. Each name has the same $\frac{1}{7}$ chance of being drawn.

Which of the possible outcomes are *favorable outcomes?* Amy, Beth, and Carol. Drawing any one of these 3 names will make the event happen.

List the possible outcomes and circle the favorable outcomes.

(Amy) **(Beth)** **(Carol)** **Dave** **Edgar** **Frank** **George**

The probability of drawing a girl's name equals $\frac{\text{number of favorable outcomes}}{\text{number of possible outcomes}} = \frac{3}{7}$.

Example You have one each of the number cards 1, 4, 6, 8, and 10. Draw one card without looking. What is the probability that the number is between 5 and 9?

Event: Get a number between 5 and 9.

Possible outcomes: 1, 4, 6, 8, and 10
The card is drawn without looking, so the outcomes are equally likely.

Favorable outcomes: 6 and 8 **1** **4** **(6)** **(8)** **10**

The probability of getting a number between 5 and 9 equals

$$\frac{\text{number of favorable outcomes}}{\text{number of possible outcomes}} = \frac{2}{5}, \text{ or } 0.4, \text{ or } 40\%.$$

Listing all the possible outcomes is sometimes confusing. Study the example below.

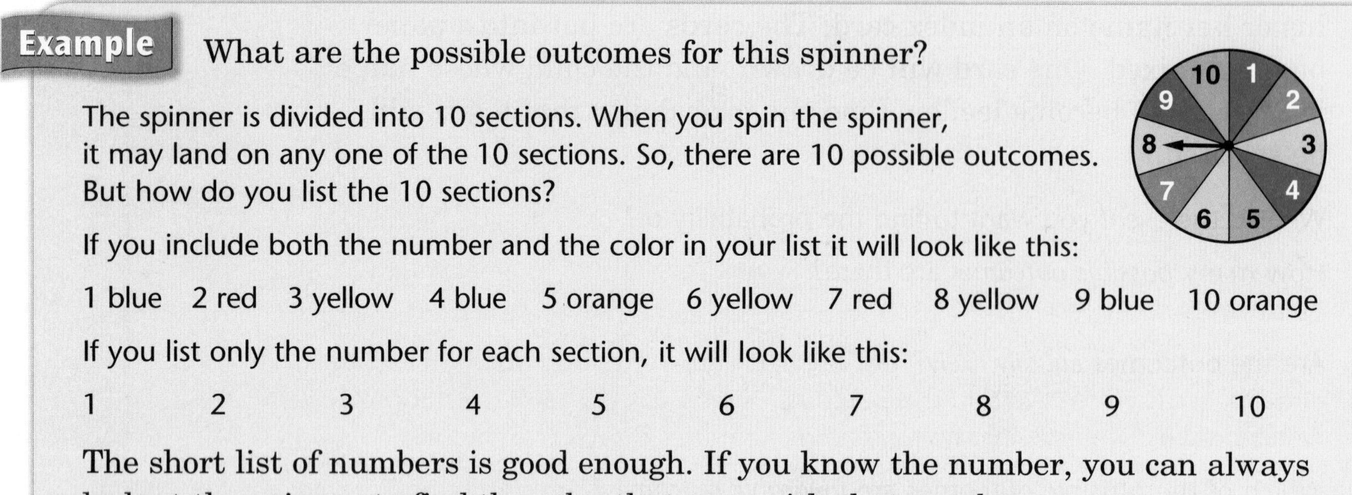

Example What are the possible outcomes for this spinner?

The spinner is divided into 10 sections. When you spin the spinner, it may land on any one of the 10 sections. So, there are 10 possible outcomes. But how do you list the 10 sections?

If you include both the number and the color in your list it will look like this:

1 blue 2 red 3 yellow 4 blue 5 orange 6 yellow 7 red 8 yellow 9 blue 10 orange

If you list only the number for each section, it will look like this:

1 2 3 4 5 6 7 8 9 10

The short list of numbers is good enough. If you know the number, you can always look at the spinner to find the color that goes with that number.

Example What is the probability that the above spinner will land on a prime number?

Event: Land on a prime number.

Possible outcomes: 1, 2, 3, 4, 5, 6, 7, 8, 9, 10

The sections are equal parts of the spinner, so the outcomes are equally likely.

Favorable outcomes: the prime numbers 2, 3, 5, and 7

1 ② ③ 4 ⑤ 6 ⑦ 8 9 10

The probability of landing on a prime number equals

$$\frac{\text{number of favorable outcomes}}{\text{number of possible outcomes}} = \frac{4}{10}, 0.4, \text{ or } 40\%.$$

Example What is the probability that the above spinner will land on a blue or yellow section that has an even number?

The possible outcomes are 1, 2, 3, 4, 5, 6, 7, 8, 9, and 10.

The favorable outcomes are the sections that have an even number *and* have the color blue or yellow. Only three sections meet these conditions: the sections numbered 4, 6, and 8.

1 2 3 ④ 5 ⑥ 7 ⑧ 9 10

The probability of landing on a section that has an even number and is blue or yellow equals

$$\frac{\text{number of favorable outcomes}}{\text{number of possible outcomes}} = \frac{3}{10}, 0.3, \text{ or } 30\%.$$

In some problems, there may be several outcomes that look exactly the same.

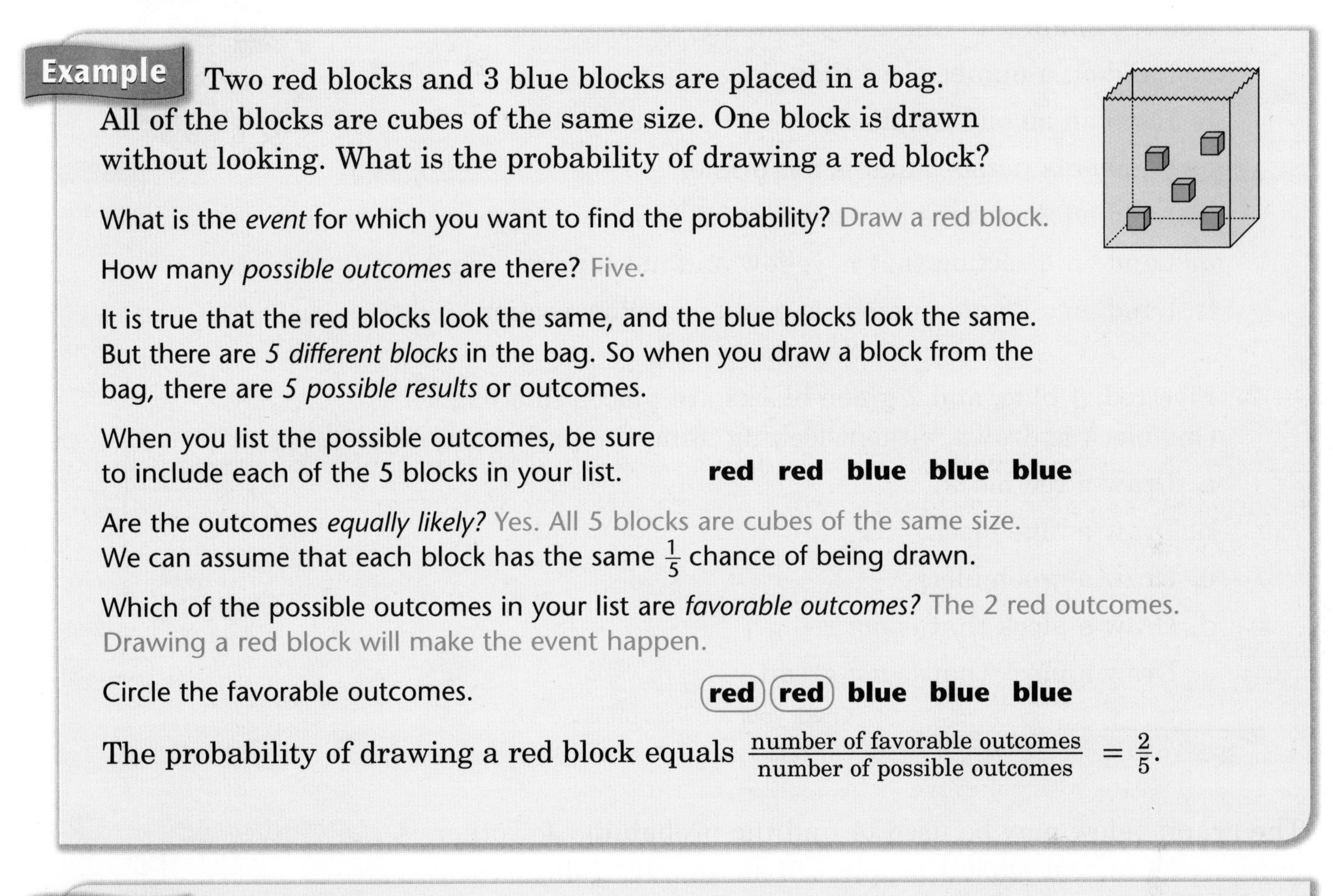

Example Two red blocks and 3 blue blocks are placed in a bag. All of the blocks are cubes of the same size. One block is drawn without looking. What is the probability of drawing a red block?

What is the *event* for which you want to find the probability? Draw a red block.

How many *possible outcomes* are there? Five.

It is true that the red blocks look the same, and the blue blocks look the same. But there are *5 different blocks* in the bag. So when you draw a block from the bag, there are *5 possible results* or outcomes.

When you list the possible outcomes, be sure to include each of the 5 blocks in your list. **red red blue blue blue**

Are the outcomes *equally likely?* Yes. All 5 blocks are cubes of the same size. We can assume that each block has the same $\frac{1}{5}$ chance of being drawn.

Which of the possible outcomes in your list are *favorable outcomes?* The 2 red outcomes. Drawing a red block will make the event happen.

Circle the favorable outcomes. **(red) (red) blue blue blue**

The probability of drawing a red block equals $\frac{\text{number of favorable outcomes}}{\text{number of possible outcomes}} = \frac{2}{5}$.

Example A bag contains 1 green, 2 blue, and 3 red counters. The counters are the same, except for color. One counter is drawn. What is the probability of drawing a blue counter?

Event: Draw a blue counter.

Possible outcomes: green, blue, blue, red, red, red
The counters are the same, except for color.
So the 6 outcomes are equally likely.

Favorable outcomes: blue, blue **green (blue) (blue) red red red**

The probability of drawing a blue counter equals $\frac{\text{number of favorable outcomes}}{\text{number of possible outcomes}} = \frac{2}{6}$, or $\frac{1}{3}$.

Check Your Understanding

1. Use the spinner to find the probability of each event.
   a. Land on a number less than 7.
   b. Land on an odd number.
   c. Land on a number that is not prime.
   d. Land on an orange or a red section.
   e. Land on a section that is yellow and not a prime number.
   f. Land on a prime number in a red or yellow section.

2. Five red, 4 blue, and 2 green blocks are placed in a bag. One block is drawn without looking. Find the probability of each event:
   a. Draw a red block.
   b. Draw a blue block.
   c. Draw a green block.
   d. Draw a block that is *not* red.
   e. Draw a block that is *not* green.

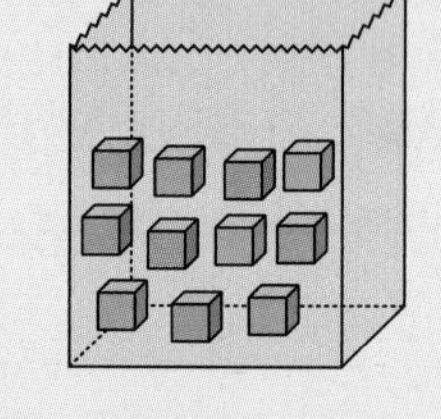

Check your answers on page 342.

The graph below may be used to find the probability for other group sizes that at least 2 people will share a birthday.

Birthday Probabilities

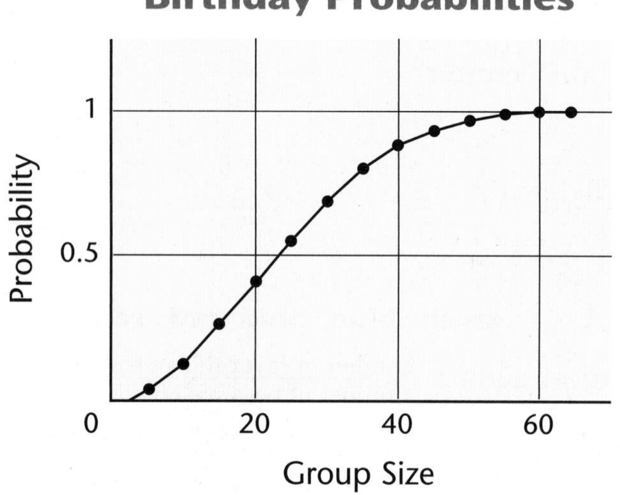

Did You Know?

If there are 23 people in a room, the chance that at least 2 of them have the same birthday is 50.7%.

If there are at least 60 people in a room, you are almost certain to find that at least 2 of them have the same birthday.

If these are at least 367 people in a room, you are certain to find that at least 2 of them the same birthday. Why?

Geometry and Constructions

Geometry in Our World

The world is filled with geometry. There are angles, segments, lines, and curves everywhere you look. There are 2-dimensional and 3-dimensional shapes of every type.

Many wonderful geometric patterns can be seen in nature. You can find patterns in flowers, spider webs, leaves, seashells, even your own face and body.

The ideas of geometry are also found in the things people create. Think of the games you play. Checkers is played with round pieces. The gameboard is covered with squares. Basketball and tennis are played with spheres. They are played on rectangular courts that are painted with straight and curved lines. The next time you play or watch a game, notice how geometry is important to the way the game is played.

The places we live in are built from plans that use geometry. Buildings almost always have rectangular rooms. Outside walls and roofs often include sections that have triangular shapes. Archways are curved and are often shaped like semicircles (half circles). Staircases may be straight or spiral. Buildings and rooms are often decorated with beautiful patterns. You see these decorations on doors and windows; on walls, floors, and ceilings; and on railings of staircases.

The clothes people wear are often decorated with geometric shapes. So are the things they use every day. Everywhere in the world, people create things using geometric patterns. Examples include quilts, pottery, baskets, and tiles. Some patterns are shown here. Which are your favorites?

Make a practice of noticing geometric shapes around you. Pay attention to bridges, buildings, and other structures. Look at the ways in which simple shapes such as triangles, rectangles, and circles are combined. Notice interesting designs. Share these with your classmates and your teacher.

In this section, you will study geometric shapes and learn how to construct them. As you learn, try to create your own beautiful designs.

Points and Line Segments

A **point** is a location in space. You often make a dot with a pencil to show where a point is.

Letters are used to name points. The letter names make it easy to talk about the points. For example, in the illustration at the right, point A is closer to point B than it is to point P. Point P is closer to point B than it is to point A.

A B

P

A **line segment** is made up of 2 points and the straight path between them. You can use any tool with a straight edge to draw the path between 2 points.

- The two points are called the **endpoints** of the line segment.
- The line segment is the shortest path between the endpoints.

Did You Know?

The ancient Greeks used letters to name points and lines. This use of letters has been traced back to Hippocrates of Chios (about 440 B.C.).

The line segment below is called *line segment AB* or *line segment BA*.

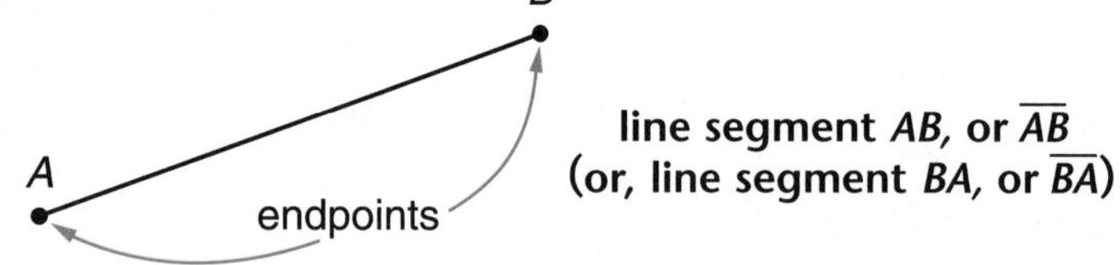

The symbol for a line segment is a raised bar. The bar is written above the letters that name the endpoints for the segment. The name of the line segment above can be written $\overline{AB}$ or $\overline{BA}$.

Straightedge and Ruler

A **straightedge** is a strip of wood, plastic, or metal that may be used to draw a straight path. A **ruler** is a straightedge that is marked so that it may be used to measure lengths.

- Every ruler is a straightedge.
- However, every straightedge is not a ruler.

ruler

CM 1 2 3 4 5 6 7 8 9 10 11

straightedge

Rays and Lines

A **ray** is a straight path that has a starting point and goes on forever in *one* direction.

To draw a ray, draw a line segment and extend the path beyond one endpoint. Then add an arrowhead to show that the path goes on forever.

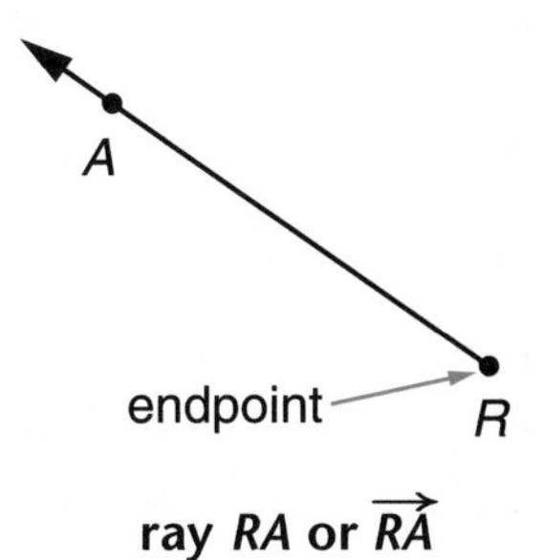

ray *RA* or $\overrightarrow{RA}$

The ray at the right is called ray *RA*.

Point *R* is the endpoint of ray *RA*. The endpoint is always the first letter in the name of a ray. The second letter can be any other point on the ray.

The symbol for a ray is a raised arrow, pointing to the right. For example, ray *RA* can be written $\overrightarrow{RA}$.

A **line** is a straight path that goes on forever in *both* directions.

To draw a line, draw a line segment and extend the path beyond each endpoint. Then add an arrowhead at each end.

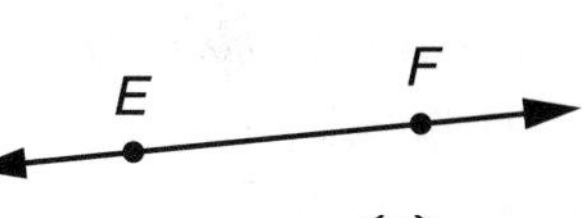

line *EF*, or $\overleftrightarrow{EF}$
(or, line *FE*, or $\overleftrightarrow{FE}$)

The symbol for a line is a raised bar with two arrowheads $\longleftrightarrow$. The name of the line at the right can be written either as $\overleftrightarrow{FE}$ or as $\overleftrightarrow{EF}$. You can name a line by listing any two points on the line, in any order.

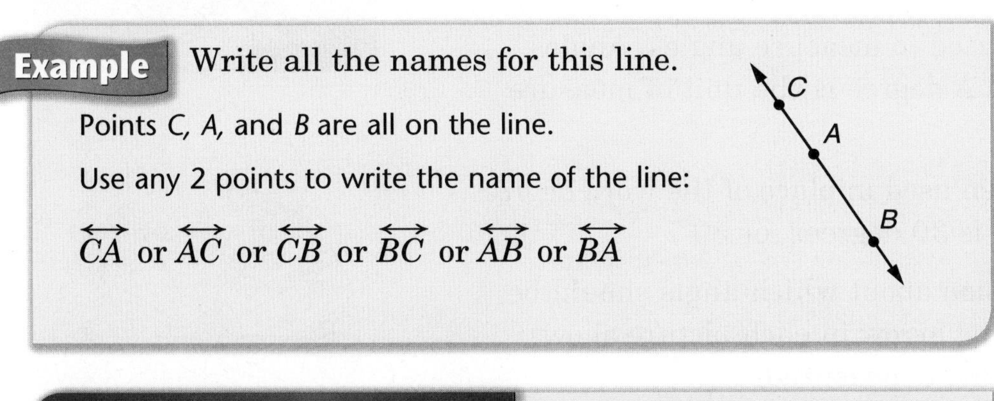

Example Write all the names for this line.

Points *C*, *A*, and *B* are all on the line.

Use any 2 points to write the name of the line:

$\overleftrightarrow{CA}$ or $\overleftrightarrow{AC}$ or $\overleftrightarrow{CB}$ or $\overleftrightarrow{BC}$ or $\overleftrightarrow{AB}$ or $\overleftrightarrow{BA}$

Check Your Understanding

1. Draw and label $\overleftrightarrow{GH}$.
2. Draw and label a point *T* that is not on $\overleftrightarrow{GH}$.
3. Draw and label $\overrightarrow{TG}$ and $\overrightarrow{TH}$.

Check your answers on page 342.

Angles

An **angle** is formed by 2 rays or 2 line segments that share the same endpoint.

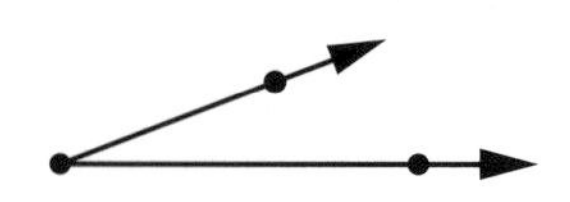
angle formed by 2 rays

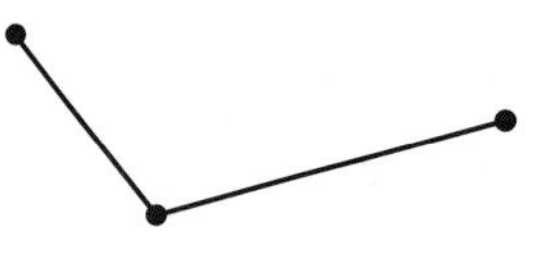
angle formed by 2 segments

The endpoint where the rays or segments meet is called the **vertex** of the angle. The rays or segments are called the **sides** of the angle.

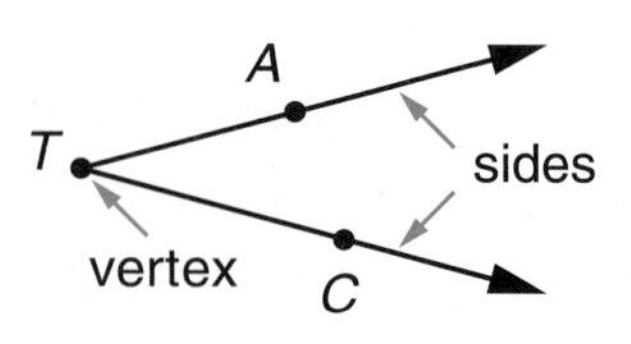

Naming Angles

The symbol for an angle is ∠. An angle can be named in two ways:

1. Name the vertex. The angle shown above is angle *T*. Write this as $\angle T$.
2. Name 3 points: the vertex and one point on each side of the angle. The angle above can be named angle *ATC* ($\angle ATC$) or angle *CTA* ($\angle CTA$). The vertex must always be listed in the middle, between the points on the sides.

Measuring Angles

The **protractor** is a tool used to measure angles. Angles are measured in **degrees.** A degree is the unit of measure for the size of an angle.

The degree symbol ° is often used in place of the word *degrees*. The measure of $\angle T$ above is 30 degrees, or 30°.

Sometimes there is confusion about which angle should be measured. The small curved arrow in each picture shows which angle opening should be measured.

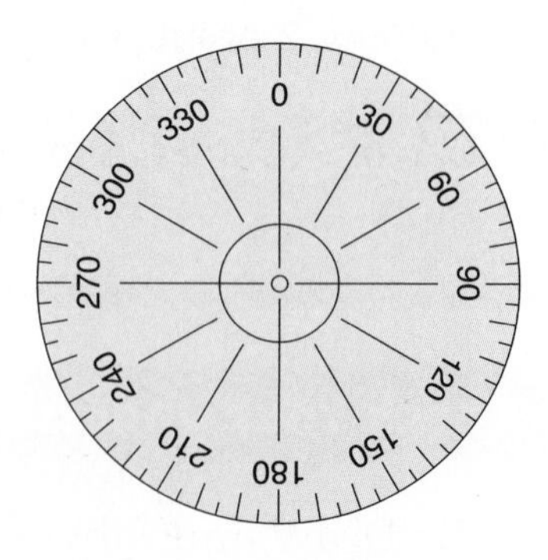
full-circle protractor

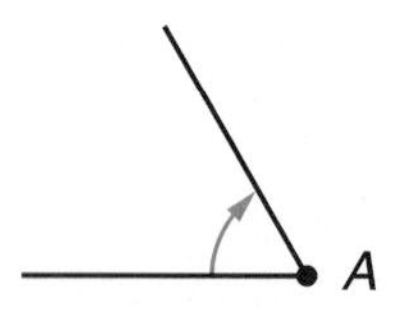

Measure of ∠A is 60°

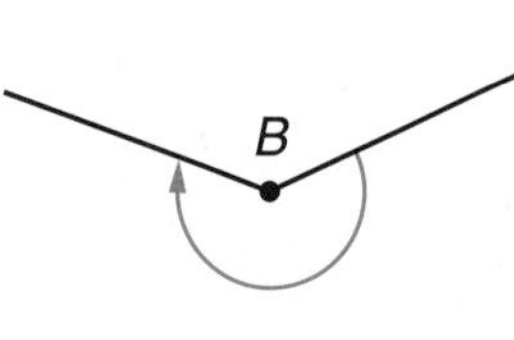

Measure of ∠B is 225°

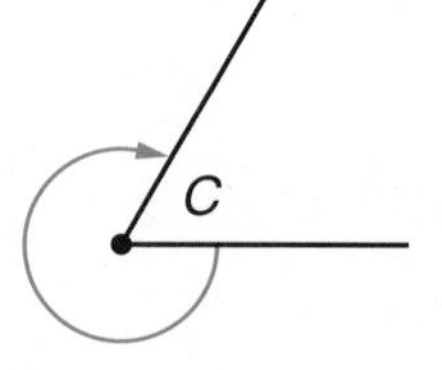

Measure of ∠C is 300°

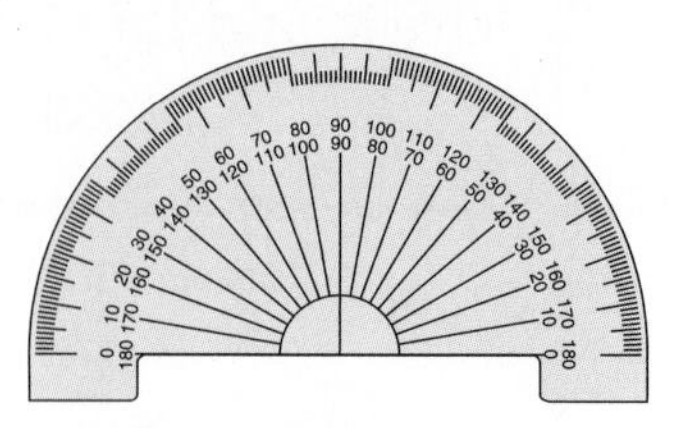
half-circle protractor

Classifying Angles

Angles may be classified according to size.

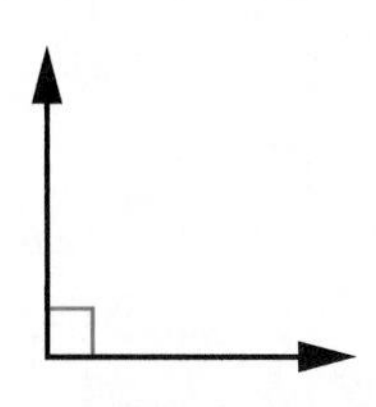

A **right angle** measures 90°.

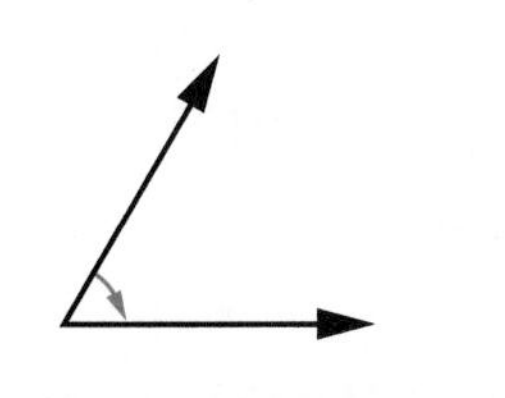

An **acute angle** measures between 0° and 90°.

An **obtuse angle** measures between 90° and 180°.

A **straight angle** measures 180°.

A **reflex angle** measures between 180° and 360°.

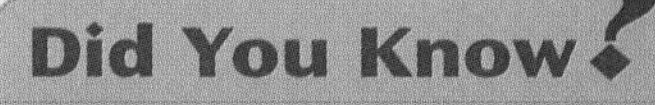

The degree symbol ° was first used in 1558, in a book on astronomy. The angle symbol ∠ was first used in a 1634 mathematics book. The less than symbol < was also used in that book to represent an angle.

A **right angle** is an angle whose sides form a square corner. You may draw a small corner symbol inside an angle to show that it is a right angle.

A **straight angle** is an angle whose sides form one straight path.

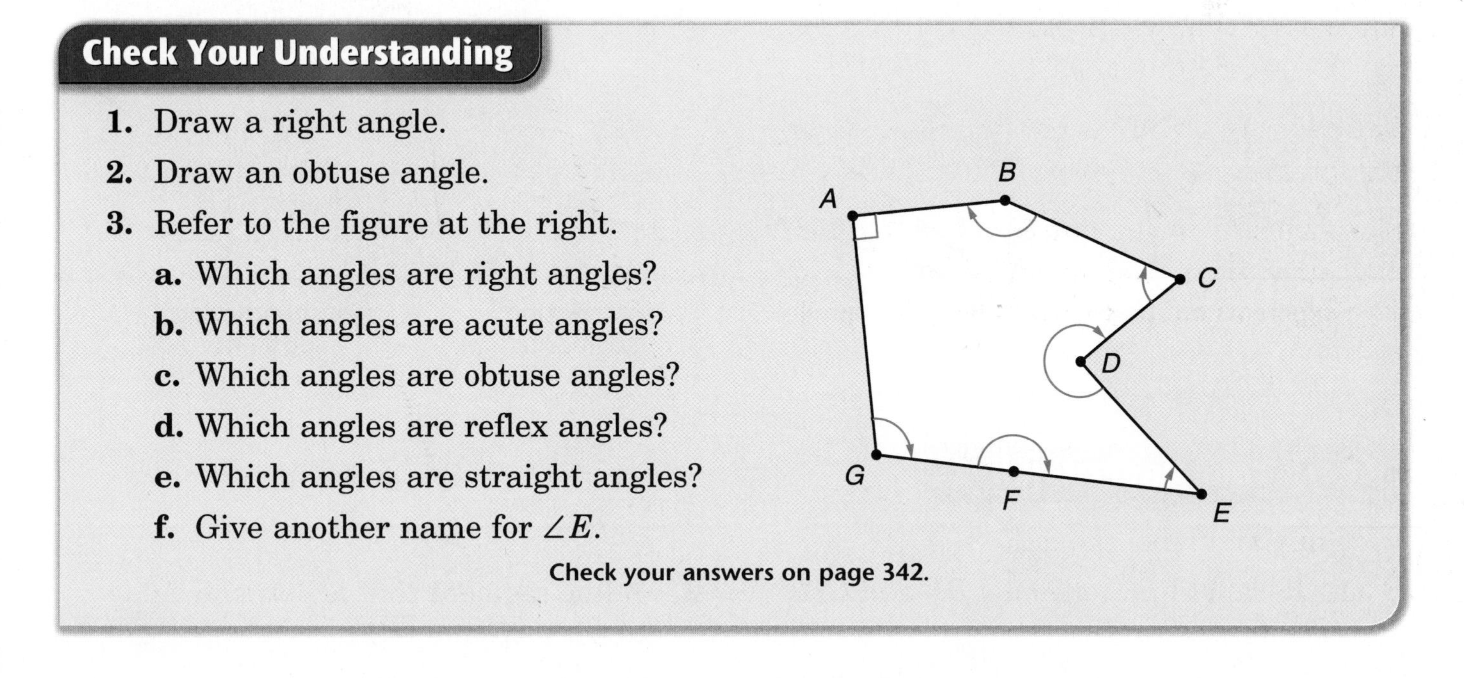

Check Your Understanding

1. Draw a right angle.
2. Draw an obtuse angle.
3. Refer to the figure at the right.
   a. Which angles are right angles?
   b. Which angles are acute angles?
   c. Which angles are obtuse angles?
   d. Which angles are reflex angles?
   e. Which angles are straight angles?
   f. Give another name for $\angle E$.

Check your answers on page 342.

Parallel Lines and Segments

Parallel lines are lines on a flat surface that never cross or meet. Think of a railroad track that goes on forever. The two rails are parallel lines. The rails never meet or cross and they are always the same distance apart.

Parallel line segments are parts of lines that are parallel. The top and bottom edges of this page are parallel. If each edge were extended forever in both directions, the lines would be parallel. The symbol for parallel is a pair of vertical lines ||. If $\overline{BF}$ and $\overline{TG}$ are parallel, write $\overline{BF} \parallel \overline{TG}$.

If lines or segments cross or meet each other, they **intersect**. Lines or segments that intersect and form right angles are called **perpendicular** lines or segments.

The symbol for perpendicular is $\perp$, which looks like an upside-down letter T. If $\overleftrightarrow{RS}$ and $\overleftrightarrow{XY}$ are perpendicular, write $\overleftrightarrow{RS} \perp \overleftrightarrow{XY}$.

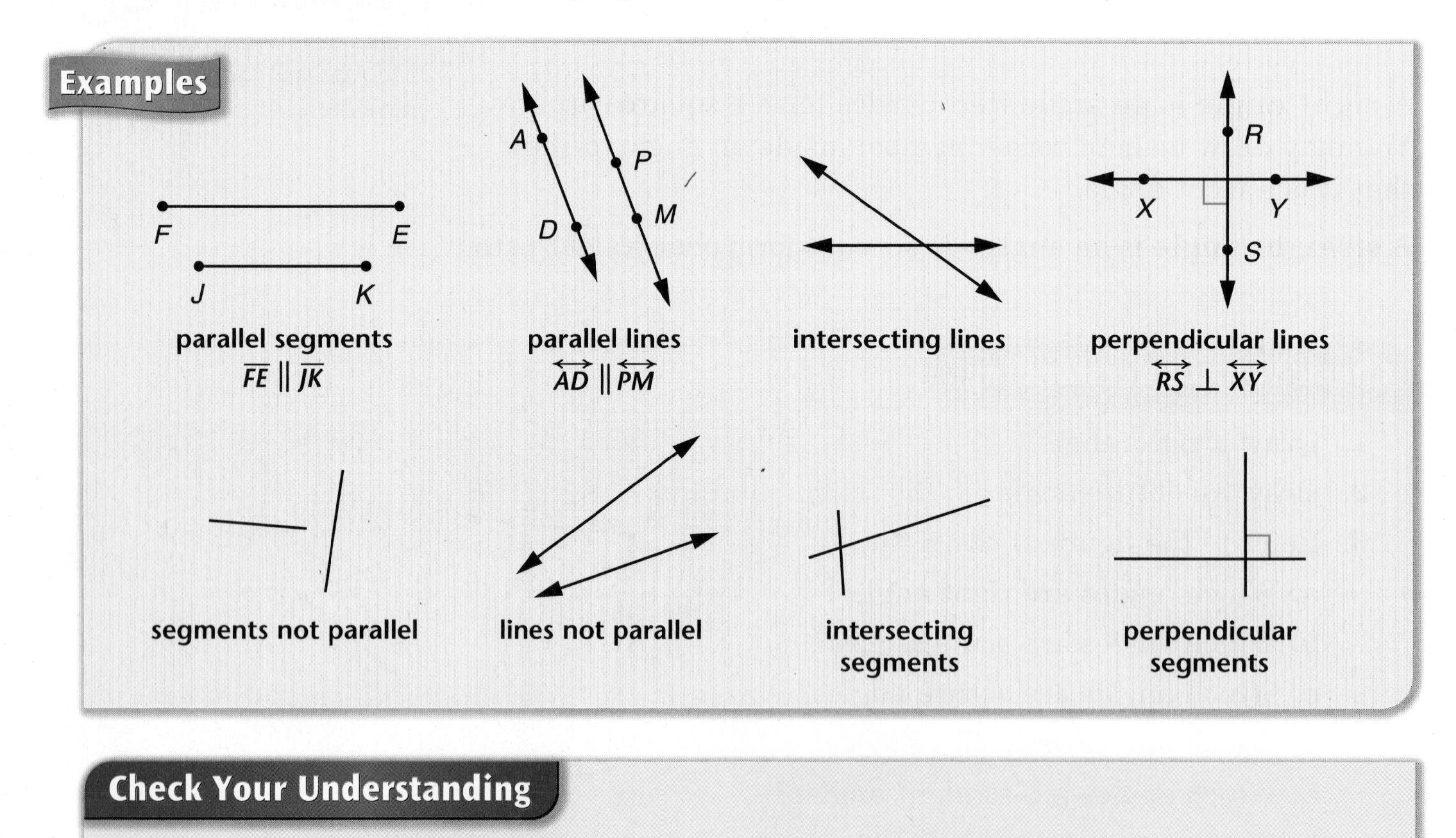

Check Your Understanding

Draw and label the following.

1. Parallel line segments EF and GH
2. A line segment that is perpendicular to both $\overline{EF}$ and $\overline{GH}$

Check your answers on page 342.

Line Segments, Rays, Lines, and Angles

Figure	Symbol	Name and Description
A	A	**point:** A location in space
E, F, endpoints	$\overline{EF}$ or $\overline{FE}$	**line segment:** A straight path between 2 points called its endpoints
N, endpoint, M	$\overrightarrow{MN}$	**ray:** A straight path that goes on forever in one direction from an endpoint
P, R	$\overleftrightarrow{PR}$ or $\overleftrightarrow{RP}$	**line:** A straight path that goes on forever in both directions
vertex, S, T, P	$\angle T$ or $\angle STP$ or $\angle PTS$	**angle:** Two rays or line segments with a common endpoint, called the vertex
B, A, S, R	$\overleftrightarrow{AB} \parallel \overleftrightarrow{RS}$	**parallel lines:** Lines that never cross or meet and are everywhere the same distance apart
	$\overline{AB} \parallel \overline{RS}$	**parallel line segments:** Segments that are parts of lines that are parallel
R, E, D, S	none	**intersecting lines:** Lines that cross or meet
	none	**intersecting line segments:** Segments that cross or meet
B, R, S, C	$\overleftrightarrow{BC} \perp \overleftrightarrow{RS}$	**perpendicular lines:** Lines that intersect at right angles
	$\overline{BC} \perp \overline{RS}$	**perpendicular line segments:** Segments that intersect at right angles

Check Your Understanding

Draw and label each of the following.

1. point M **2.** $\overleftrightarrow{RT}$ **3.** $\angle TRY$

4. $\overline{XY}$ **5.** $\overline{DE} \parallel \overline{KL}$ **6.** $\overrightarrow{FG}$

Check your answers on page 342.

Polygons

A **polygon** is a flat, 2-dimensional figure made up of line segments called **sides**. A polygon can have any number of sides, as long as it has at least three sides.

- The sides of a polygon are connected end to end and make one closed path.
- The sides of a polygon do not cross.

Each endpoint where sides meet is called a **vertex**. The plural of the word *vertex* is **vertices**.

Figures That Are Polygons

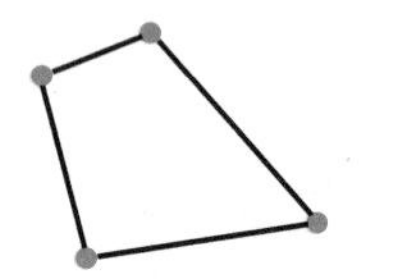

4 sides, 4 vertices

3 sides, 3 vertices

7 sides, 7 vertices

Figures That Are NOT Polygons

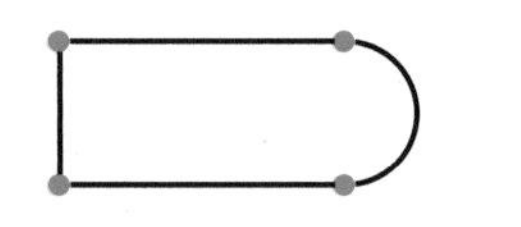

All sides of a polygon must be line segments. Curved lines are not line segments.

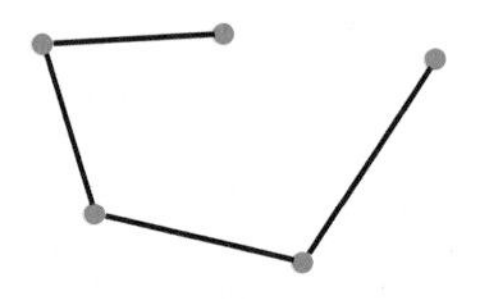

The sides of a polygon must form a closed path.

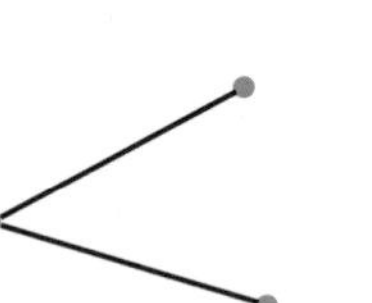

A polygon must have at least 3 sides.

The sides of a polygon must not cross.

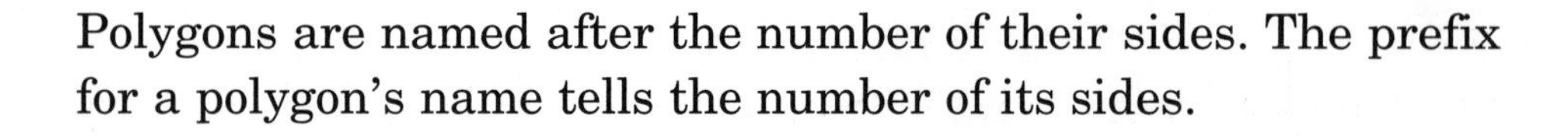

Polygons are named after the number of their sides. The prefix for a polygon's name tells the number of its sides.

Prefixes	
tri-	3
quad-	4
penta-	5
hexa-	6
hepta-	7
octa-	8
nona-	9
deca-	10
dodeca-	12

Convex Polygons

A **convex** polygon is a polygon in which all the sides are pushed outward. The polygons below are convex.

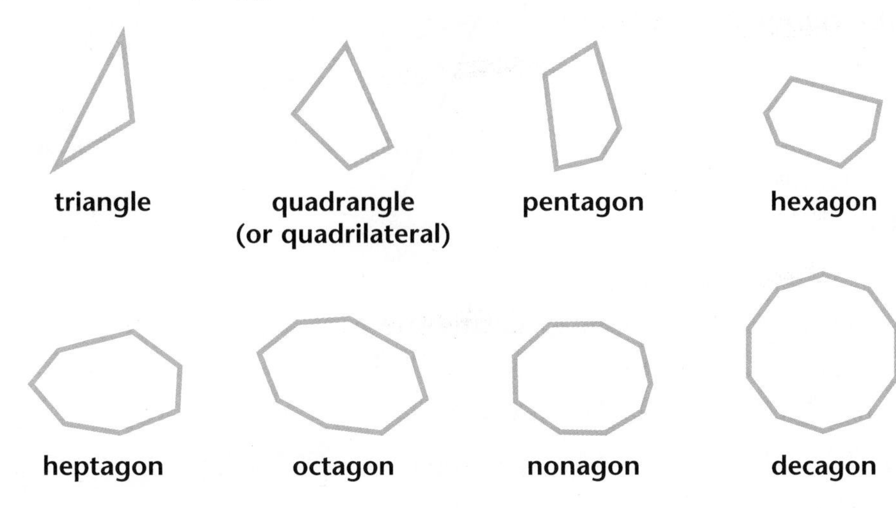

triangle · quadrangle (or quadrilateral) · pentagon · hexagon

heptagon · octagon · nonagon · decagon

Nonconvex (Concave) Polygons

A **nonconvex,** or **concave**, polygon is a polygon in which at least two sides are pushed in. The polygons at the right are nonconvex.

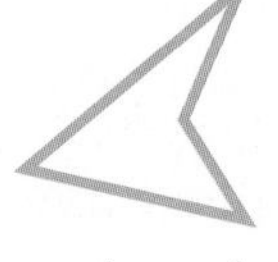

quadrangle
(or quadrilateral)

pentagon

hexagon

octagon

Regular Polygons

A polygon is a **regular polygon** if (1) all the sides have the same length; and (2) all the angles inside the figure are the same size. A regular polygon is always convex. The polygons below are regular.

equilateral triangle

square

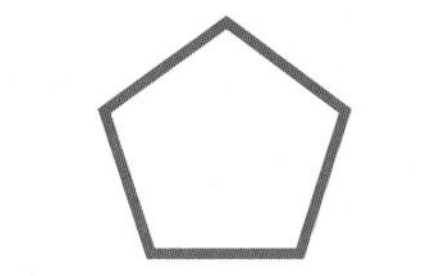

regular pentagon

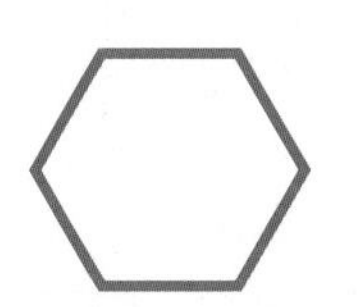

regular hexagon

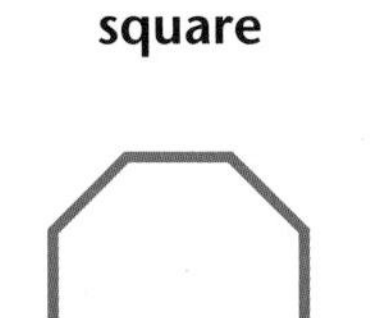

regular octagon

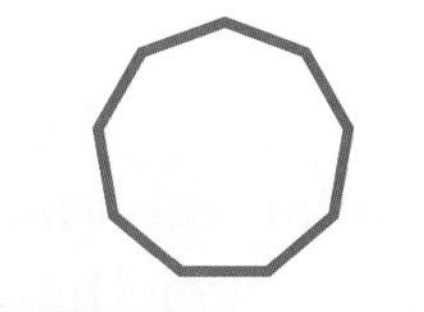

regular nonagon

Check Your Understanding

1. What is the name of a polygon having **a.** 4 sides? **b.** 6 sides? **c.** 8 sides?
2. **a.** Draw a convex hexagon. **b.** Draw a concave octagon.
3. Explain why the cover of this book is not a regular polygon.

Check your answers on page 342.

Triangles

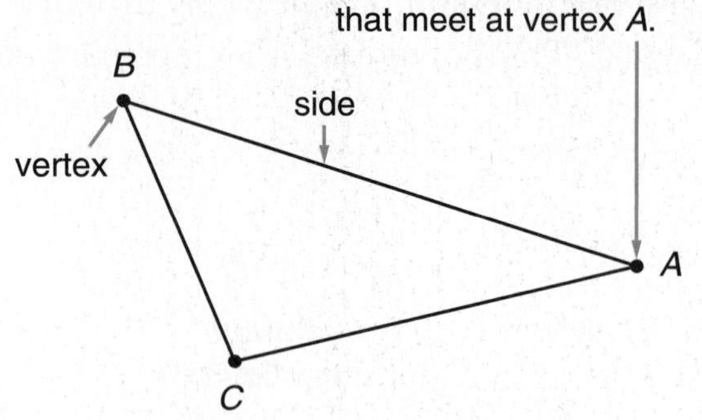

Triangles are the simplest type of polygon. The prefix *tri-* means *three*. All triangles have 3 vertices, 3 sides, and 3 angles.

For the triangle shown here:

- The vertices are the points B, C, and A.
- The sides are $\overline{BC}$, $\overline{BA}$, and $\overline{CA}$.
- The angles are $\angle B$, $\angle C$, and $\angle A$.

Triangles have 3-letter names. You name a triangle by listing the letter names for the vertices, in order. The triangle above has 6 possible names: triangle *BCA, BAC, CAB, CBA, ABC*, and *ACB*.

Triangles have many different sizes and shapes. You will work with two types of triangles that have been given special names.

An **equilateral triangle** is a triangle whose three sides all have the same length. Equilateral triangles have many different sizes, but all equilateral triangles have the same shape.

Did You Know?

In about 150 A.D., the Greek mathematician Heron was the first to use $\triangle$ as a symbol for triangle.

A **right triangle** is a triangle with one right angle (square corner). Right triangles have many different shapes and sizes.

A right triangle cannot be an equilateral triangle because the side opposite the right angle is always longer than each of the other sides.

Check Your Understanding

1. **a.** Draw and label an equilateral triangle named *SAC*.
   b. Write the five other possible names for this triangle.
2. Draw a right triangle with two sides that are the same length.

Check your answers on page 343.

Quadrangles

A **quadrangle** is a polygon that has 4 sides. Another name for quadrangle is **quadrilateral**. The prefix *quad-* means *four*. All quadrangles have 4 vertices, 4 sides, and 4 angles.

For the quadrangle shown here:

- The sides are $\overline{RS}$, $\overline{ST}$, $\overline{TU}$, and $\overline{UR}$.
- The vertices are R, S, T, and U.
- The angles are $\angle R$, $\angle S$, $\angle T$, and $\angle U$.

A quadrangle is named by listing in order the letter names for the vertices. The quadrangle above has 8 possible names:

RSTU, RUTS, STUR, SRUT, TURS, TSRU, URST, UTSR

Some quadrangles have two pairs of parallel sides. These quadrangles are called **parallelograms**.

Reminder: Two sides are parallel if they never meet, no matter how far they are extended.

Did You Know?

In 150 A.D., Heron used the symbol = to indicate parallel lines and segments. The parallel symbol || written vertically was first used by William Oughtred, in 1677.

Figures That Are Parallelograms

Opposite sides are parallel in each figure.

Figures That Are NOT Parallelograms

No parallel sides

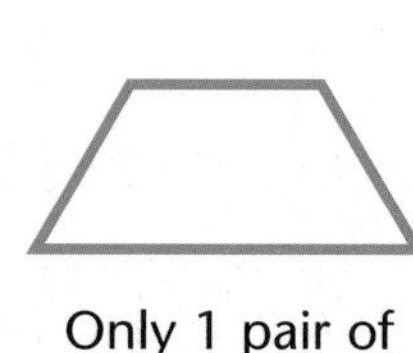

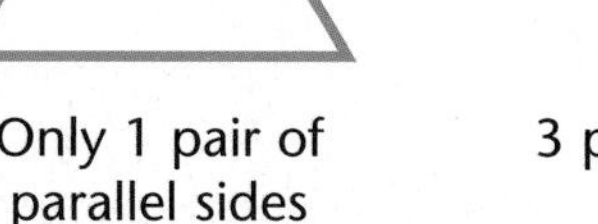

Only 1 pair of parallel sides

3 pairs of parallel sides

A parallelogram must have exactly 2 pairs of parallel sides.

Many special types of quadrangles have been given names. Some of these are parallelograms, but others are not parallelograms. See the table on the next page for examples of each type.

This tree diagram shows how the different types of quadrangles are related. For example, quadrangles are divided into two major groups—*parallelograms* and *not parallelograms.*

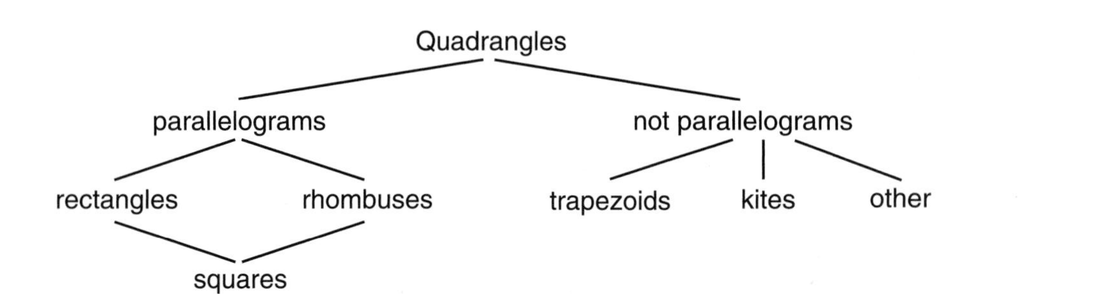

Quadrangles That Are Parallelograms		
rectangle		**Rectangles** are parallelograms. A rectangle has 4 right angles (square corners). The sides do not all have to be the same length.
rhombus		**Rhombuses** are parallelograms. A rhombus has 4 sides that are all the same length. The angles of a rhombus are usually not right angles, but they may be.
square		**Squares** are parallelograms. A square has 4 right angles (square corners). Its 4 sides are all the same length. All squares are rectangles. All squares are also rhombuses.

Quadrangles That Are NOT Parallelograms		
trapezoid		**Trapezoids** have exactly 1 pair of parallel sides. The 4 sides of a trapezoid can all have different lengths.
kite		A **kite** is a quadrangle with 2 pairs of equal sides. The equal sides are next to each other. The 4 sides cannot all have the same length. (A rhombus is not a kite.)
other		Any polygon with 4 sides that is not a parallelogram, a trapezoid, or a kite.

Check Your Understanding

What is the difference between the quadrangles in each pair below?

1. a square and a rectangle

2. a kite and a rhombus

3. a trapezoid and a parallelogram

Check your answers on page 343.

Geometric Solids

Polygons and circles are flat, **2-dimensional** figures. The surfaces they enclose take up a certain amount of area, but they do not have any thickness and do not take up any volume. **Three-dimensional** shapes have length, width, *and* thickness. They take up volume. Boxes, chairs, and balls are all examples.

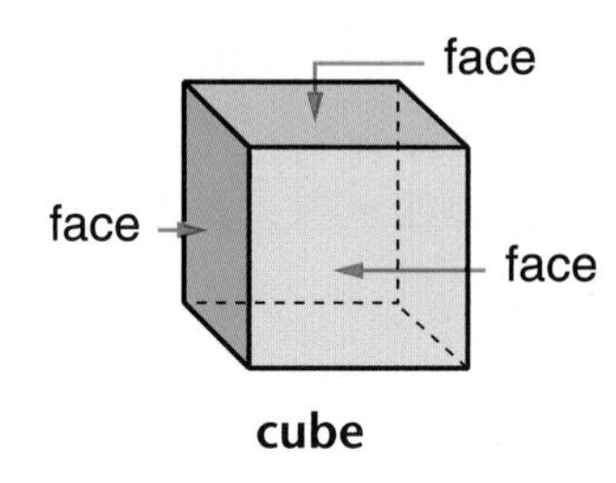

cube

A **geometric solid** is the surface or surfaces that surround a 3-dimensional shape. The surfaces of a geometric solid may be flat or curved or both. A **flat surface** of a solid is called a **face.** A **curved surface** of a solid does not have any special name.

A **cube** has 6 square faces that are the same size. Three of the cube's faces cannot be seen in the figure at the right.

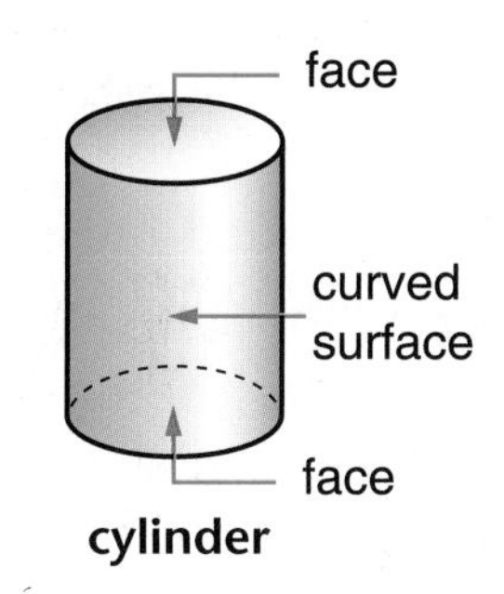

cylinder

A **cylinder** has 3 surfaces. The flat top and flat bottom are faces that are formed by circles. A curved surface connects the top and bottom faces. A food can is a good model of a cylinder.

A **cone** has 2 surfaces. The flat bottom is a face that is formed by a circle. A curved surface is connected to the bottom face and comes to a point. An ice cream cone is a good model of a cone. However, keep in mind that a cone is closed; it has a "lid."

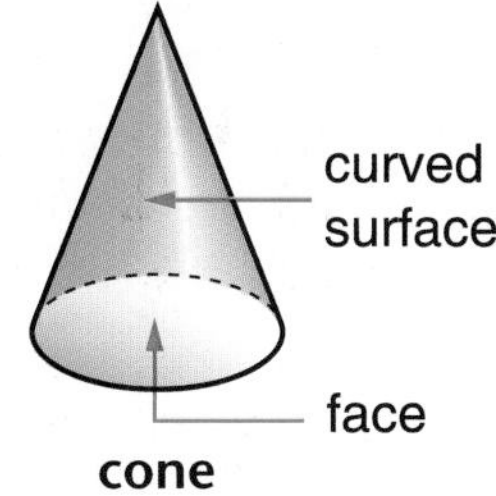

cone

The **edges** of a geometric solid are the line segments or curves where surfaces meet. A corner of a geometric solid is called a **vertex** (plural *vertices*). A vertex is usually a point at which edges meet, but the vertex of a cone is an isolated corner. It is completely separated from the edge of the cone.

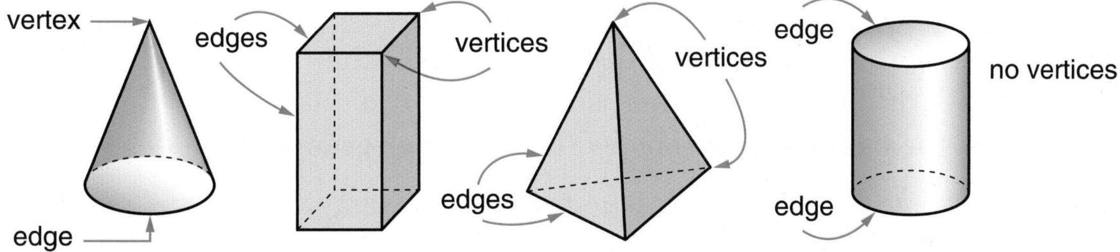

sphere
no edges, no vertices

A **sphere** has one curved surface but no edges and no vertices. A basketball or globe is a good model of a sphere.

Check Your Understanding

1. **a.** How are cylinders and cones alike? **b.** How do they differ?
2. **a.** How are spheres and cones alike? **b.** How do they differ?

Check your answers on page 343.

Polyhedrons

A **polyhedron** is a geometric solid whose surfaces are all formed by polygons. These surfaces are the faces of the polyhedron. A polyhedron does not have any curved surfaces.

Pyramids and **prisms** are two important kinds of polyhedrons.

Did You Know?

A rhombicuboctahedron has 26 faces. Eighteen of them are squares and 8 are triangles.

Polyhedrons That Are Pyramids

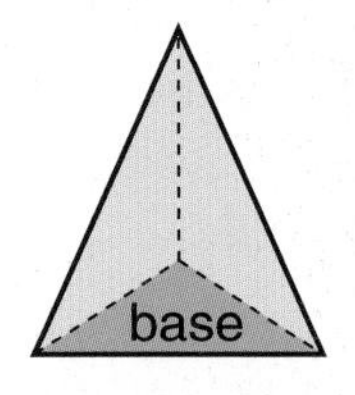

triangular pyramid

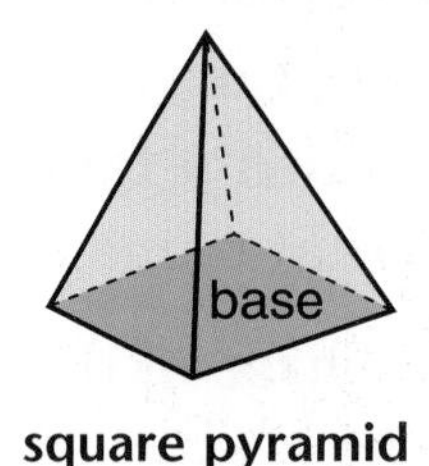

square pyramid

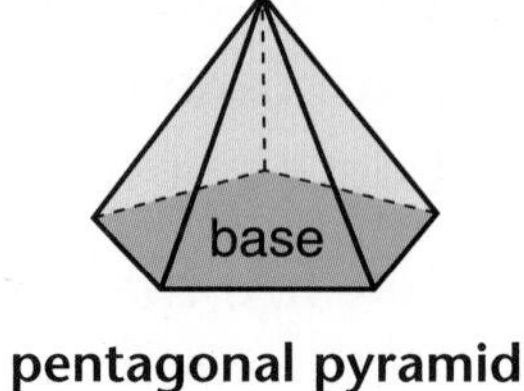

pentagonal pyramid

hexagonal pyramid

The shaded face of each pyramid above is called the **base** of the pyramid. The shape of the base is used to name the pyramid. For example, the base of a square pyramid has a square shape. The faces of a pyramid that are not the base are all shaped like triangles and meet at the same vertex.

Polyhedrons That Are Prisms

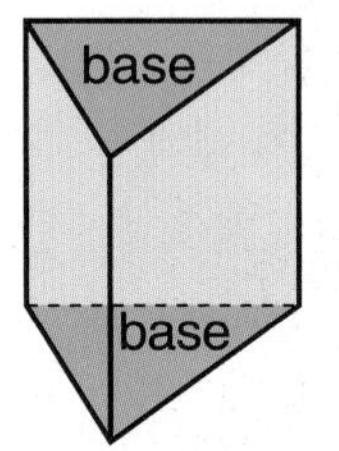
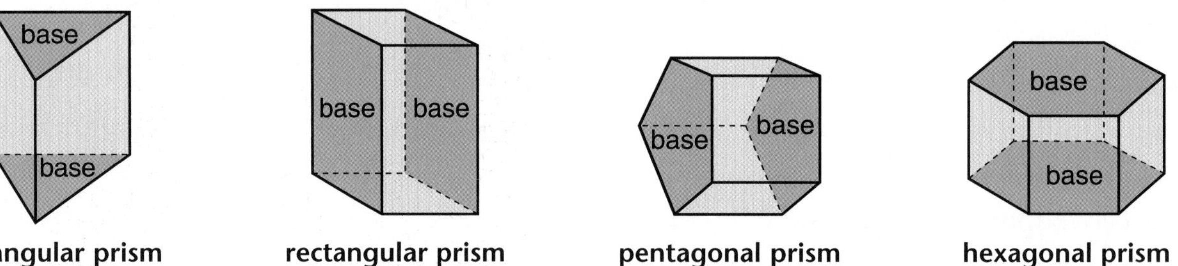

triangular prism

rectangular prism

pentagonal prism

hexagonal prism

The two shaded faces of each prism above are called the **bases** of the prism. The bases of a prism are the same size and shape. They are parallel. All other faces join the bases and are shaped like parallelograms.

The shape of the bases of a prism is used to name the prism. For example, the bases of a pentagonal prism have the shape of a pentagon.

Many polyhedrons are not pyramids or prisms. Some are illustrated below.

Polyhedrons That Are NOT Pyramids or Prisms

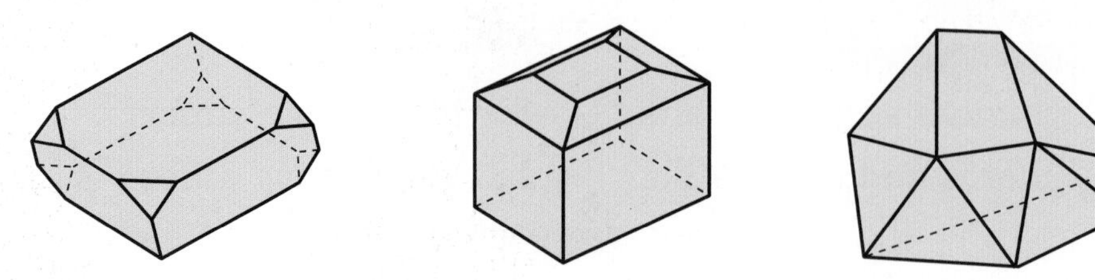

Regular Polyhedrons

A polyhedron is **regular** if:

- Each face is formed by a regular polygon.
- The faces all have the same size and shape.
- All of the vertices look exactly the same.

There are only five kinds of regular polyhedrons.

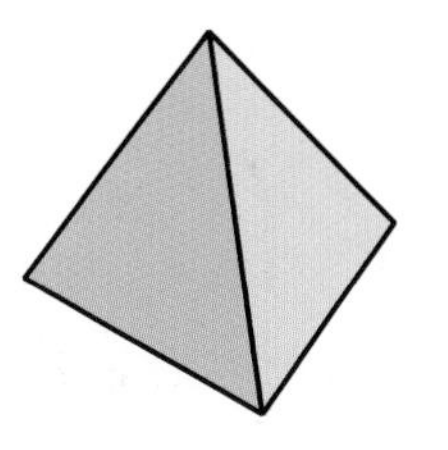
regular tetrahedron

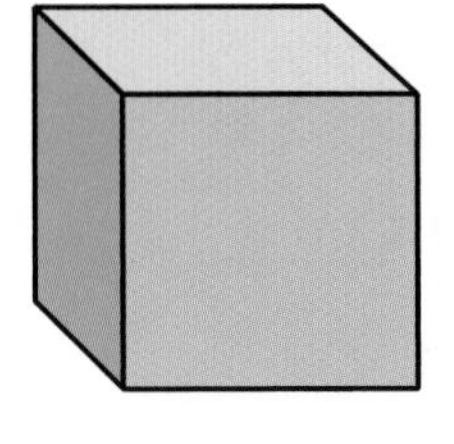
cube

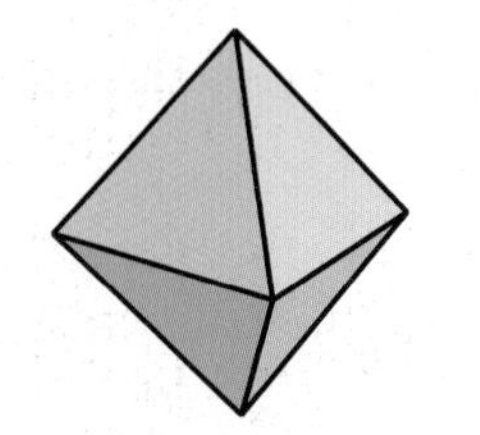
regular octahedron

regular dodecahedron

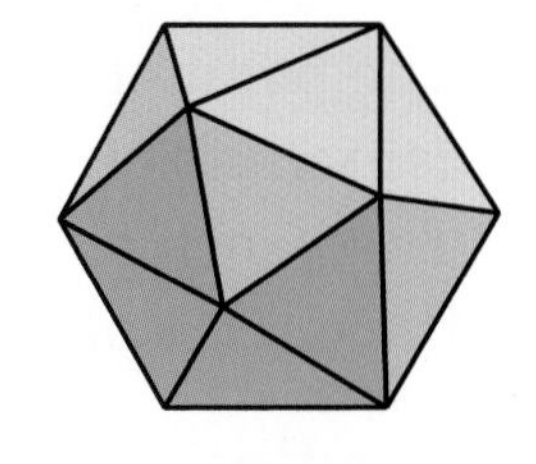
regular icosahedron

Name	Shape of Face	Number of Faces
regular tetrahedron	equilateral triangle	4
cube	square	6
regular octahedron	equilateral triangle	8
regular dodecahedron	regular pentagon	12
regular icosahedron	equilateral triangle	20

Check Your Understanding

1. How many faces does a rectangular pyramid have?
2. How many faces of a rectangular pyramid have a rectangular shape?
3. How many faces does a rectangular prism have?
4. How many faces of a rectangular prism have the shape of a parallelogram?
5. Which solid has more faces: a triangular pyramid or a triangular prism?
6. Which regular polyhedrons have faces that are formed by equilateral triangles?
7. How many edges does a regular octahedron have?
8. How many vertices does a regular octahedron have?
9. **a.** How are regular tetrahedrons and regular octahedrons alike?
   b. How are they different?

Check your answers on page 343.

Circles and Spheres

A **circle** is a curved line that forms one closed path on a flat surface. All of the points on a circle are the same distance from the **center of the circle.**

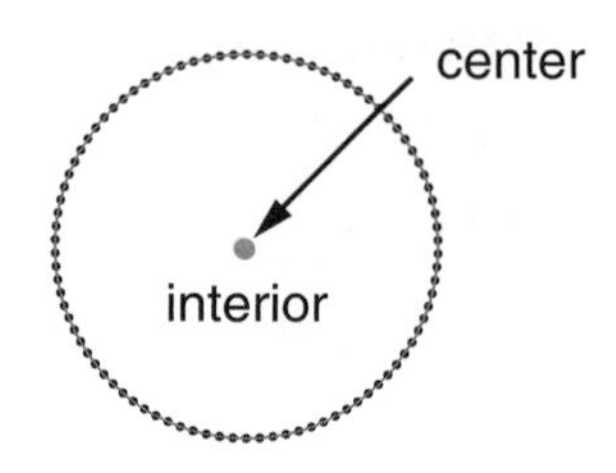

The center is not part of the circle because the interior is not part of the circle.

The **compass** is a tool used to draw circles.

- ♦ The point of a compass, called the **anchor,** is placed at the center of the circle.
- ♦ The pencil in a compass traces out a circle. Every point on the circle will be the same distance from the anchor.

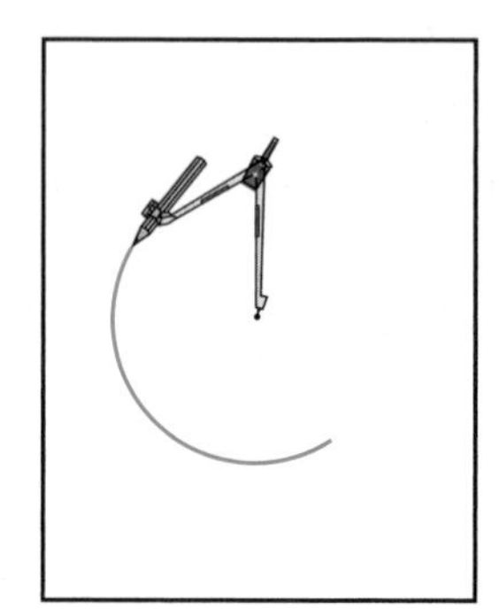

The **radius** of a circle is any line segment that connects the center of the circle with any point on the circle. The word *radius* can also refer to the length of this line segment.

The **diameter** of a circle is any line segment that passes through the center of the circle and has both of its endpoints on the circle. The word *diameter* can also refer to the length of this line segment.

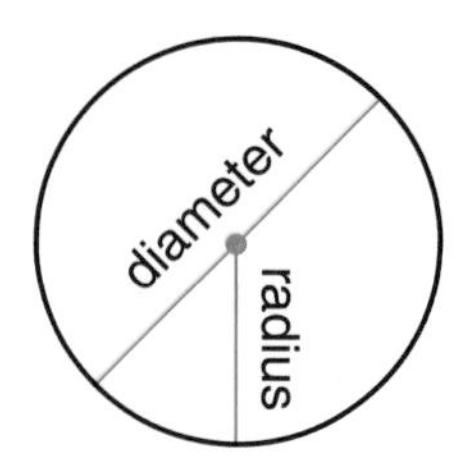

A **sphere** is a geometric solid that has a single curved surface. Spheres are shaped like basketballs or globes. All of the points on the surface of the sphere are the same distance from the **center of the sphere.**

All spheres have the same shape, but spheres do not all have the same size. The size of a sphere is the distance across its center.

- ♦ The line segment *RS* passes through the center of the sphere. This line segment is called a **diameter of the sphere.**
- ♦ The length of the line segment *RS* is also called the diameter of the sphere.

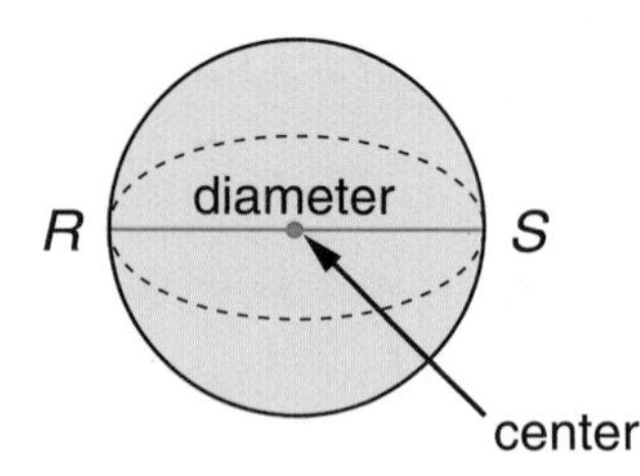

Globes and basketballs are examples of spheres that are hollow. Their interiors are empty. The hollow interior is not part of the sphere. The sphere includes only the points on its curved surface.

Marbles and baseballs are examples of spheres that have solid interiors. In cases like these, think of the solid interior as part of the sphere.

Ganymede, Jupiter's largest moon, is the largest moon in the Solar System. Its diameter is about 3,260 miles.

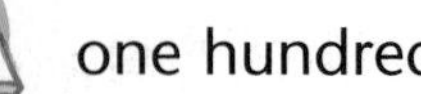

Congruent Figures

Sometimes figures have the same shape and size. These figures are **congruent.** Figures are congruent if they match exactly when one figure is placed on top of the other.

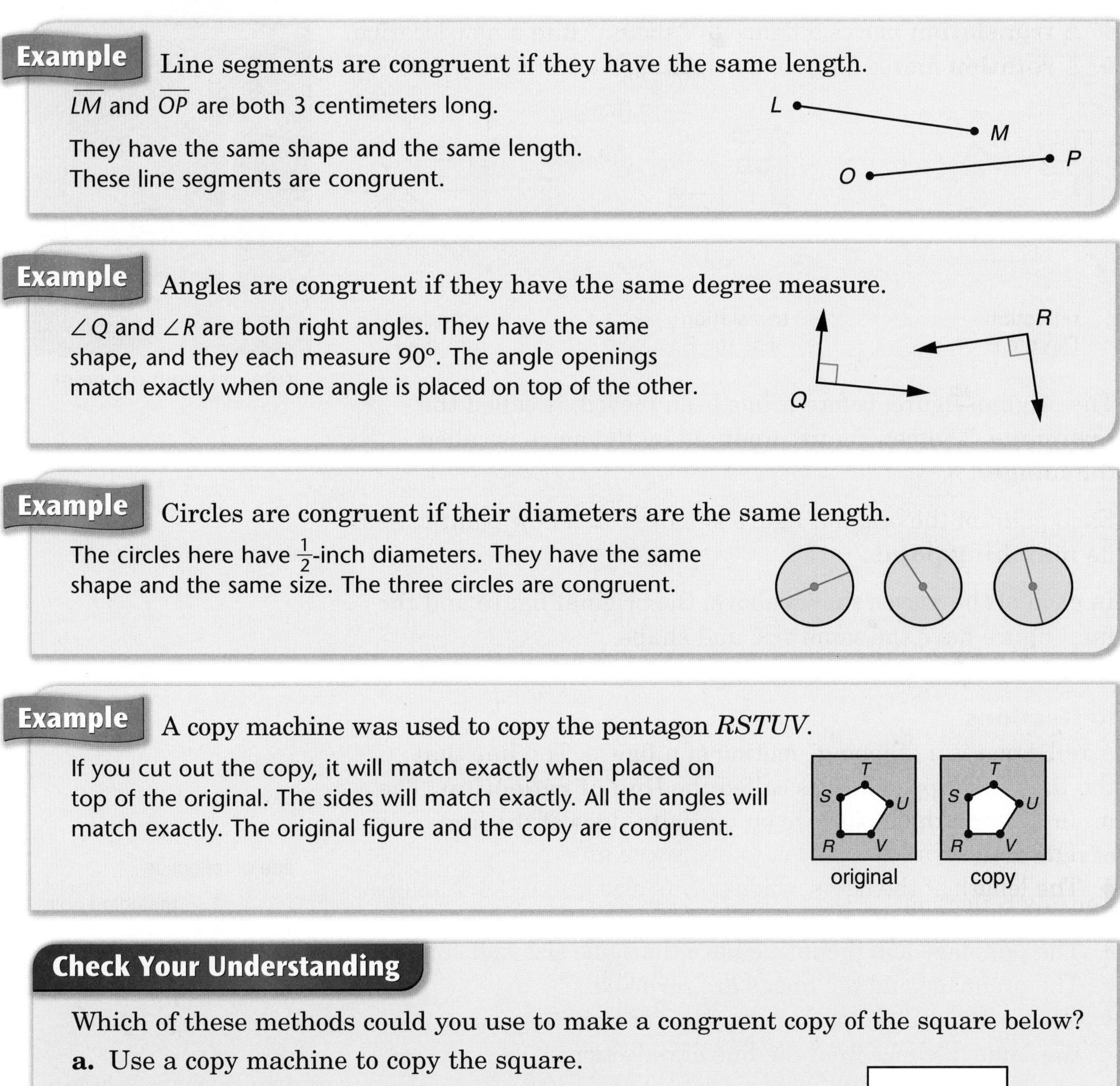

Example Line segments are congruent if they have the same length.

$\overline{LM}$ and $\overline{OP}$ are both 3 centimeters long.

They have the same shape and the same length. These line segments are congruent.

Example Angles are congruent if they have the same degree measure.

$\angle Q$ and $\angle R$ are both right angles. They have the same shape, and they each measure 90°. The angle openings match exactly when one angle is placed on top of the other.

Example Circles are congruent if their diameters are the same length.

The circles here have $\frac{1}{2}$-inch diameters. They have the same shape and the same size. The three circles are congruent.

Example A copy machine was used to copy the pentagon *RSTUV*.

If you cut out the copy, it will match exactly when placed on top of the original. The sides will match exactly. All the angles will match exactly. The original figure and the copy are congruent.

Check Your Understanding

Which of these methods could you use to make a congruent copy of the square below?

a. Use a copy machine to copy the square.

b. Use tracing paper and trace over the square.

c. Cut out the square and trace around it.

d. Measure the sides with a ruler, then draw the sides at right angles to each other using a protractor

Check your answers on page 343.

Reflections, Translations, and Rotations

A geometric figure can be moved from one place to another in three different ways.

- A **reflection** moves a figure by "flipping" it over a line.
- A **translation** moves a figure by "sliding" it to a new location.
- A **rotation** moves a figure by "turning" it around a point.

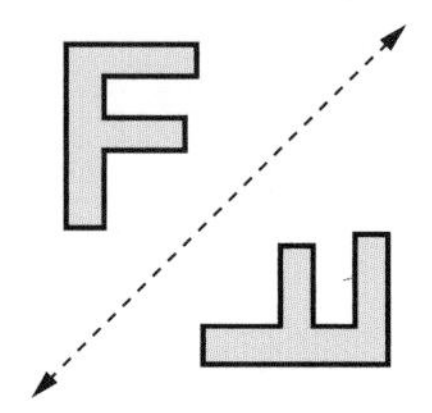

reflection
Flip the **F.**

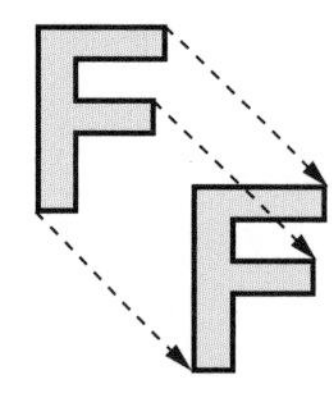

translation
Slide the **F.**

rotation
Turn the **F.**

rotation around a point

The original figure, before it has been moved, is called the **preimage.** The new figure produced by the move is called the **image.**

Each point of the original figure is moved to a new point called its **matching point.**

In each of the moves shown above, the original figure and the final figure have the same size and shape.

Reflections

A reflection is a "flipping" motion of a figure. The line that the figure is flipped over is called the **line of reflection.** The preimage and the image are on opposite sides of the line of reflection.

For any reflection:

- The preimage and the image have the same size and shape.
- The preimage and the image are reversed.
- Each point and the point it flips to (its matching point) are the same distance from the line of reflection.

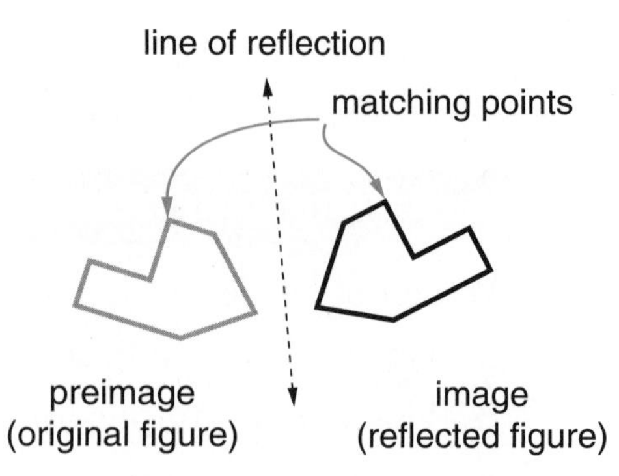

Translations

A translation is a "sliding" motion of a figure. Each point of the figure slides the same distance in the same direction. Imagine the letter T drawn on grid paper.

- If each point of the letter T slides 6 grid squares to the right, the result is a *horizontal translation*.
- If each point of the letter T slides 8 grid squares upward, the result is a *vertical translation*.
- Suppose that each point of the letter T slides 6 grid squares to the right, then 8 grid squares upward. The result is the same as a *diagonal translation*.

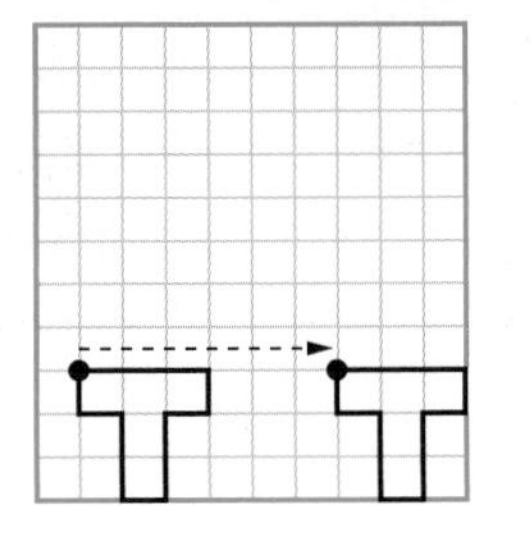

horizontal translation

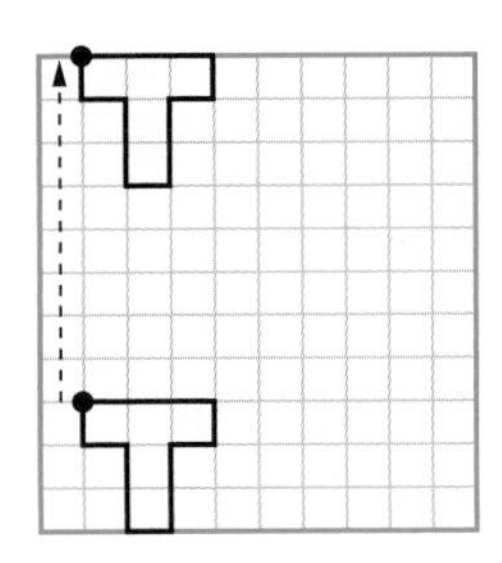

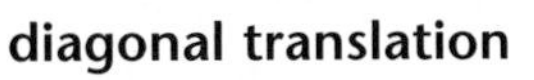

vertical translation

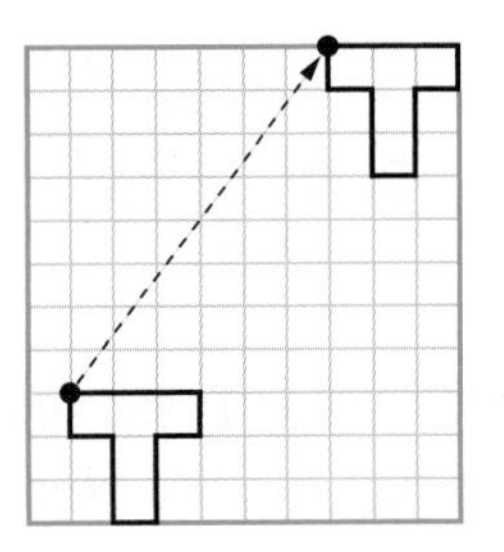

diagonal translation

Rotations

When a figure is rotated, it is turned a certain number of degrees around a particular point.

A figure can be rotated *clockwise* (the direction that clock hands move).

A figure can also be rotated *counterclockwise* (the opposite direction of the way clock hands move).

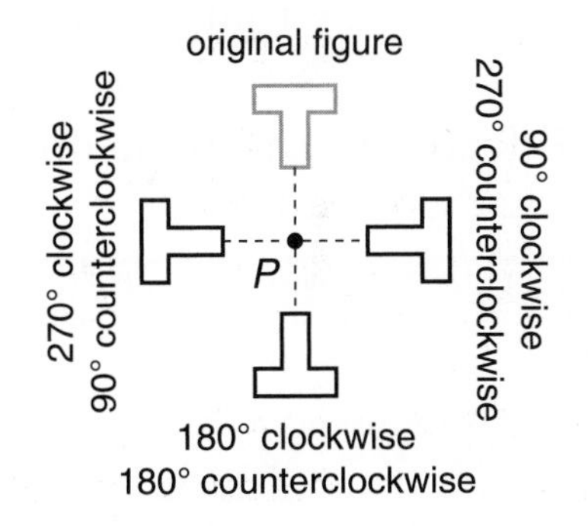

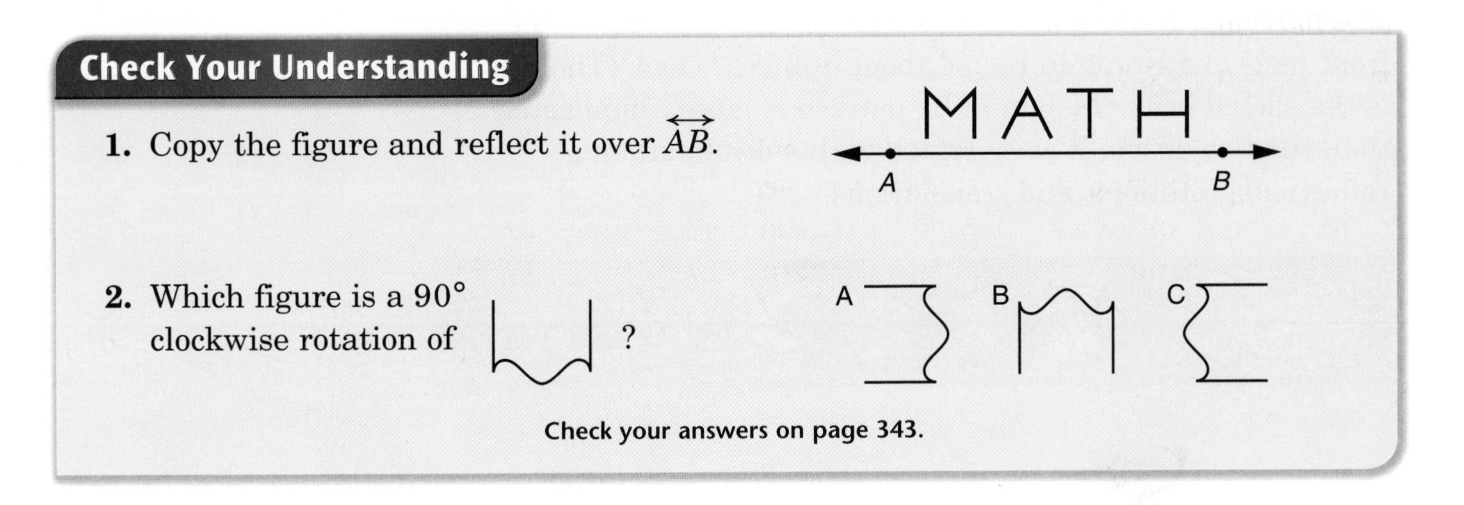

Check Your Understanding

1. Copy the figure and reflect it over $\overleftrightarrow{AB}$.

2. Which figure is a 90° clockwise rotation of [figure] ?

Check your answers on page 343.

Frieze Patterns

A frieze pattern is a design made of shapes that are in line. Frieze patterns are often found on the walls of buildings, on the borders of rugs and tiled floors, and on clothing.

In many frieze patterns, the same design is reflected over and over. For example, the following frieze pattern was used to decorate a sash worn by a Mazahua woman from San Felipe Santiago pueblo in the state of New Mexico. The strange-looking animals reflected in the frieze are probably meant to be horses.

Some frieze patterns are made by repeating (translating) the same design instead of reflecting it. These patterns look as if they were made by sliding the design along the strip. An example of such a frieze pattern is the elephant and horse design below that was found on a woman's sarong from Sumba, Indonesia. The elephants and horses repeated in the frieze are all facing in the same direction.

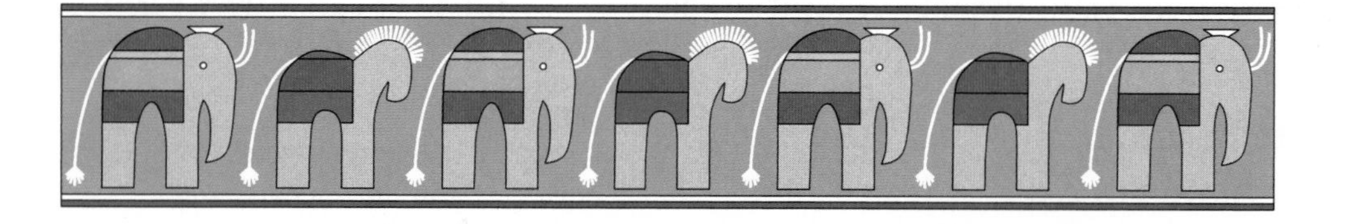

The following frieze pattern is similar to one painted on the front page of a Koran in Egypt about 600 years ago. (The Koran is the sacred book of Islam.) The pattern is more complicated than the two above. It was created with a combination of reflections, rotations, and translations.

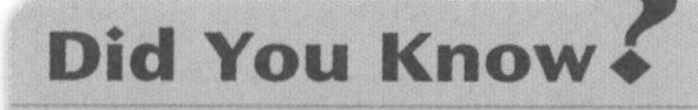

The Mayan people decorated their buildings with painted sculpture, wood carvings, and stone mosaics. The decorations were usually arranged in wide **friezes.**

Line Symmetry

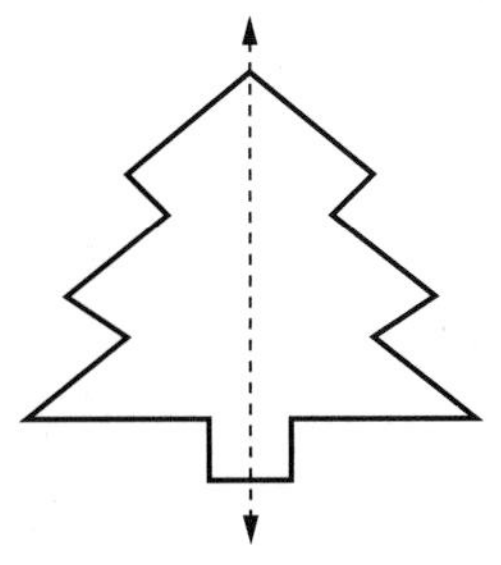
line of symmetry

A dashed line is drawn through the figure at the right. The line divides the figure into two parts. Both parts look alike but are facing in opposite directions.

The figure is **symmetric about a line.** The dashed line is called a **line of symmetry** for the figure.

You can use a reflection to get the figure shown at the right.

- Think of the line of symmetry as a line of reflection.
- Reflect the left side of the figure over the line.
- Together, the left side and its reflection (the right side) form the figure.

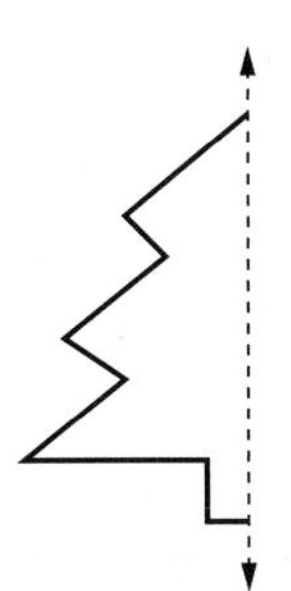
Reflect the left side to get the figure above.

An easy way to check whether a figure has *line symmetry* is to fold it in half. If the two halves match exactly, the figure is symmetric. The fold line is the line of symmetry.

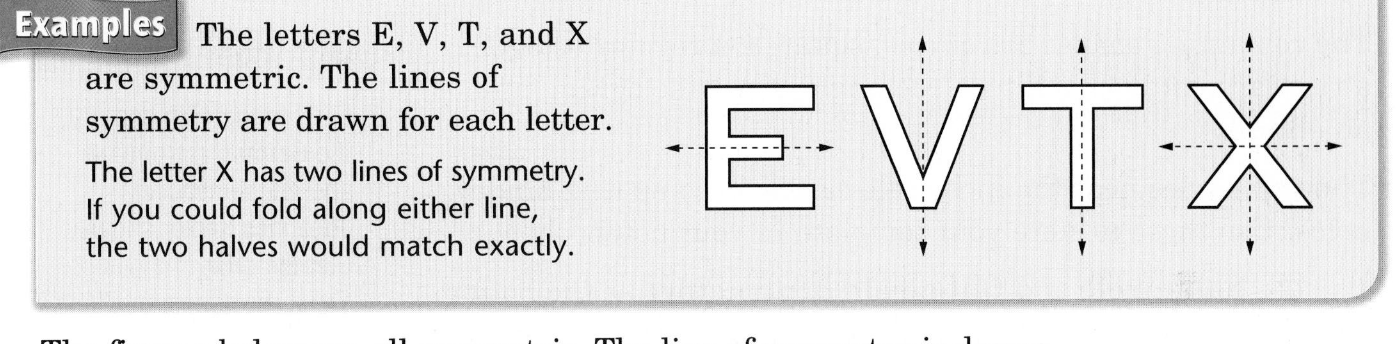

Examples The letters E, V, T, and X are symmetric. The lines of symmetry are drawn for each letter.

The letter X has two lines of symmetry. If you could fold along either line, the two halves would match exactly.

The figures below are all symmetric. The line of symmetry is drawn for each figure. If there is more than one line of symmetry, they are all drawn.

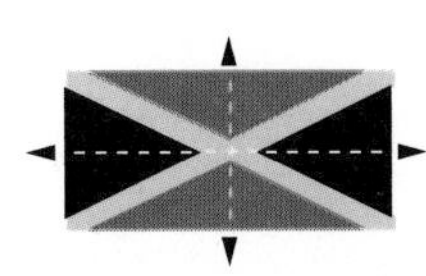
flag of Jamaica

butterfly

human body

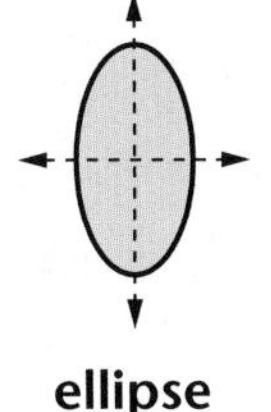
ellipse

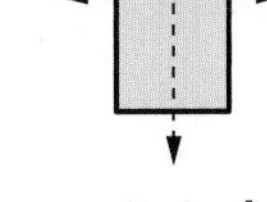
rectangle

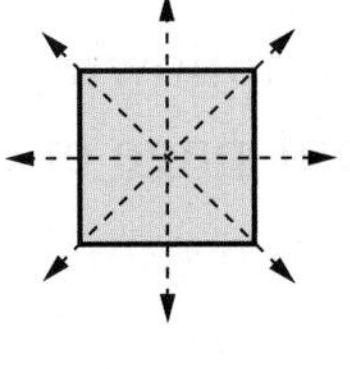
square

Check Your Understanding

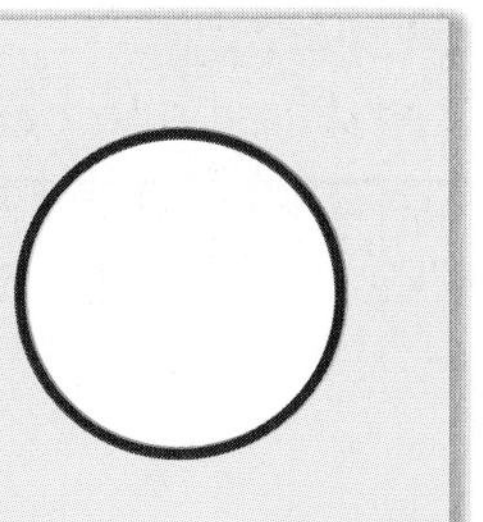

1. Trace each pattern-block (PB) shape on your Geometry Template onto a sheet of paper. Draw the lines of symmetry for each shape.
2. How many lines of symmetry does a circle have?

Check your answers on page 343.

The Geometry Template

The **Geometry Template** has many uses.

The template has two rulers. The inch scale measures in inches and fractions of an inch. The centimeter scale measures in centimeters and millimeters. Use either side of the template as a straightedge for drawing line segments.

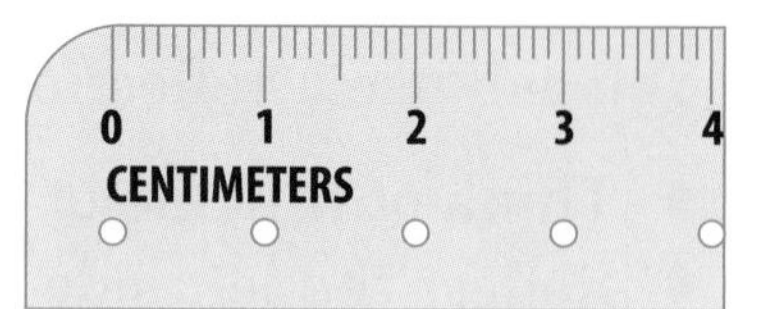

There are 17 different geometric figures on the template. The figures labeled "PB" are **pattern-block shapes.** These are half the size of real pattern blocks. There is a hexagon, a trapezoid, two different rhombuses, an equilateral triangle, and a square. These will come in handy for some of the activities you will do this year.

Each triangle on the template is labeled with the letter T and a number. Triangle "T1" is an equilateral triangle whose sides all have the same length. Triangles "T2" and "T5" are right triangles. Triangle "T3" has sides that all have different lengths. Triangle "T4" has two sides of the same length.

The remaining shapes are circles, squares, a regular octagon, a regular pentagon, a kite, a rectangle, a parallelogram, and an ellipse.

Early in the 17th century, the German astronomer and mathematician Johannes Kepler showed that the orbit of a planet about the sun is an ellipse.

The two circles near the inch scale can be used as ring-binder holes. Use these to store your template in your notebook.

Use the **half-circle** and **full-circle protractors** at the bottom of the template to measure and draw angles. You will construct and measure circle graphs in *Fifth Grade Everyday Mathematics* with the **Percent Circle** (at the top of the template).

Notice the tiny holes near the 0-, $\frac{1}{4}$-, $\frac{2}{4}$-, and $\frac{3}{4}$-inch marks of the inch scale and at each inch mark from 1 to 7. On the centimeter side, the holes are placed at each centimeter mark from 0 to 10. These holes can be used to draw circles.

Example Draw a circle with a 3-inch radius.

Place one pencil point in the hole at 0. Place another pencil point in the hole at 3 inches. Hold the pencil at 0 inches steady while rotating the pencil at 3 inches (along with the template) to draw the circle.

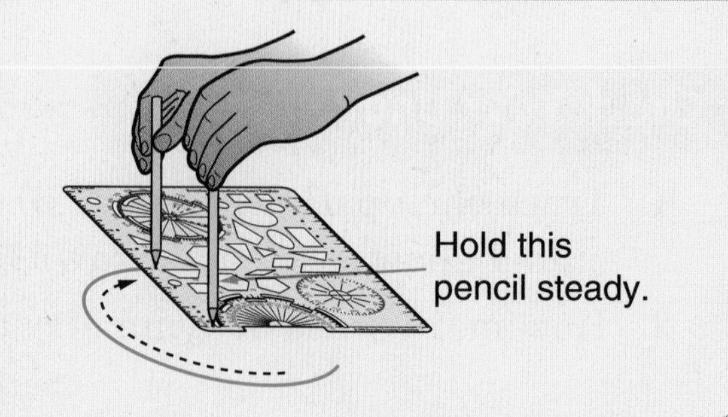

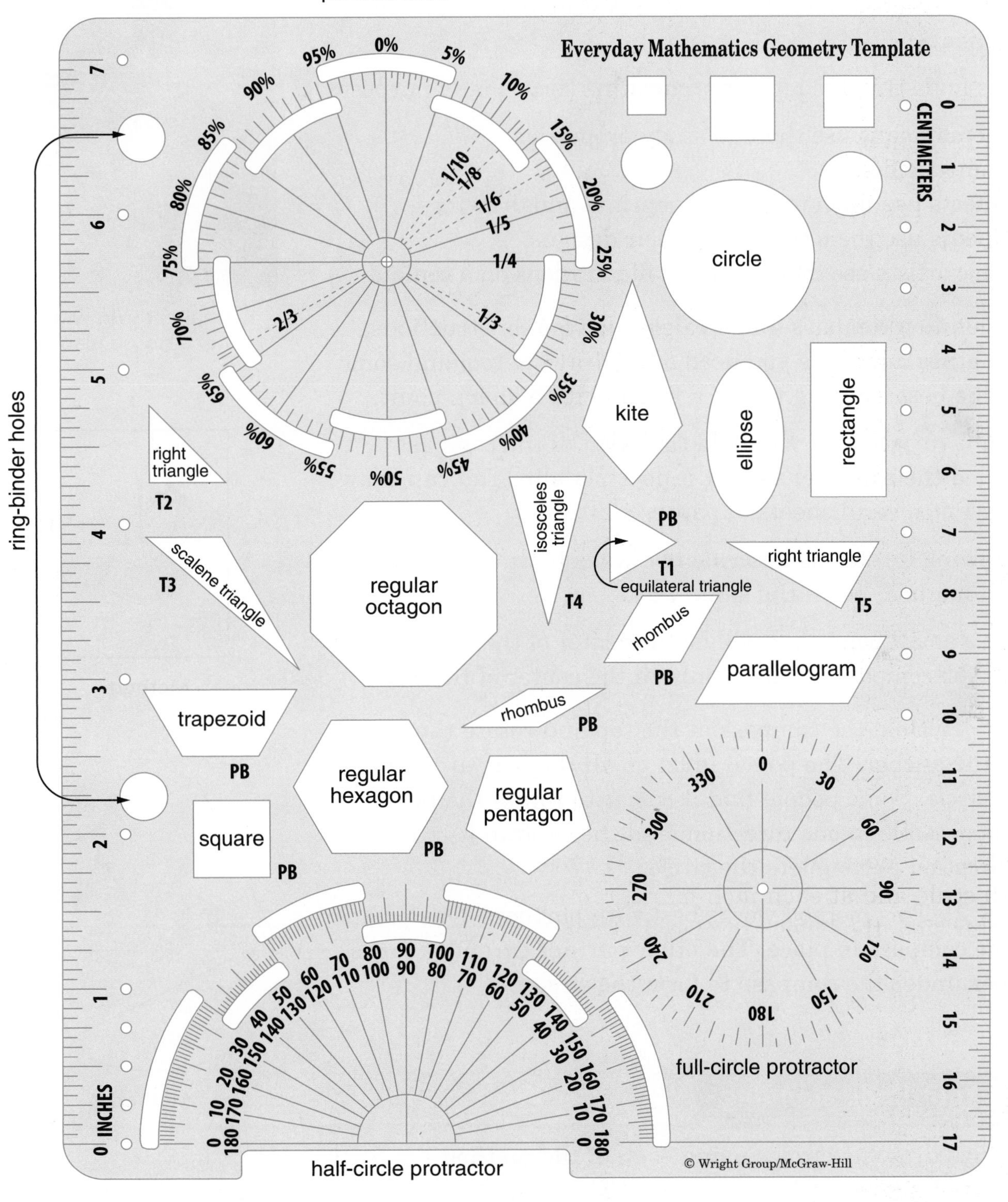
percent circle
Everyday Mathematics Geometry Template
0%
5%
10%
15%
20%
25%
30%
35%
40%
45%
50%
55%
60%
65%
70%
75%
80%
85%
90%
95%
1/10
1/8
1/6
1/5
1/4
1/3
2/3
CENTIMETERS
circle
kite
ellipse
rectangle
right triangle
T2
scalene triangle
T3
regular octagon
isosceles triangle
T4
PB
T1
equilateral triangle
right triangle
T5
rhombus
parallelogram
trapezoid
rhombus
regular hexagon
regular pentagon
square
ring-binder holes
full-circle protractor
half-circle protractor
INCHES
© Wright Group/McGraw-Hill

SRB
111

Compass-and-Straightedge Constructions

Many geometric figures can be drawn with only a compass and a straightedge. The compass is used to draw circles and to mark off lengths. The straightedge is used to draw straight line segments.

Architect's drawing of a house plan

Compass-and-straightedge **constructions** serve many purposes.

- Mathematicians use them to study properties of geometric figures.
- Architects use them to make blueprints and drawings.
- Engineers use them to develop their designs.
- Graphic artists use them to create illustrations on a computer.

In addition to a compass and straightedge for constructions, the only other materials you need are a drawing tool and some paper. The best drawing tool is a pencil with a sharp point.

Draw on a surface that will hold the point of the compass (also called the **anchor**) so that it does not slip. You can draw on a stack of several sheets of paper.

The following directions describe two ways to draw circles. For each method, begin the same way.

- Draw a small point that will be the center of the circle.
- Press the compass anchor firmly on the center of the circle.

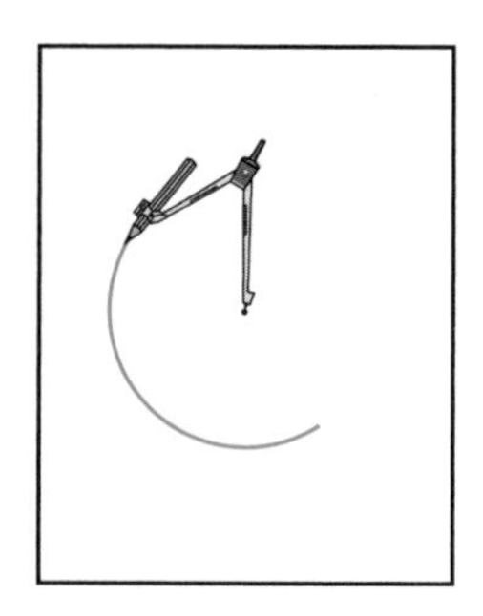

Method 1

Method 1: Hold the compass at the top and rotate the pencil around the anchor. The pencil must go all the way around to make a circle. Some people find it easier to rotate the pencil as far as possible in one direction, and then rotate it in the other direction to complete the circle.

Method 2: This method works best with partners. One partner holds the compass in place. The other partner carefully turns the paper under the compass to form the circle.

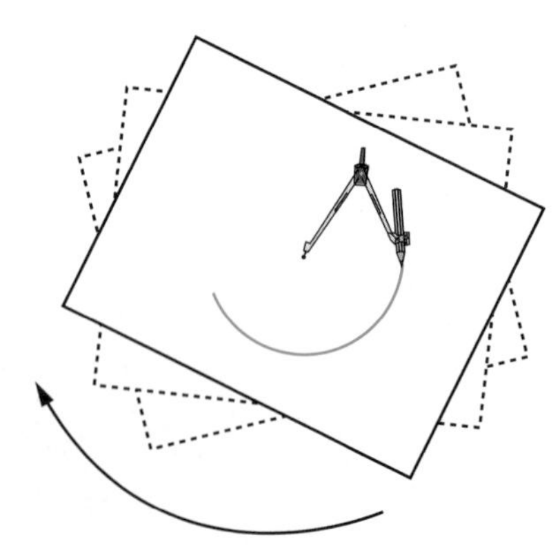

Method 2

Check Your Understanding

Practice drawing circles using each of the methods described above.

Copying a Line Segment

Follow each step carefully. Use a clean sheet of paper.

Step 1: Draw line segment *AB*.

Step 2: Draw a second line segment. It should be longer than segment *AB*. Label one of its endpoints *C*.

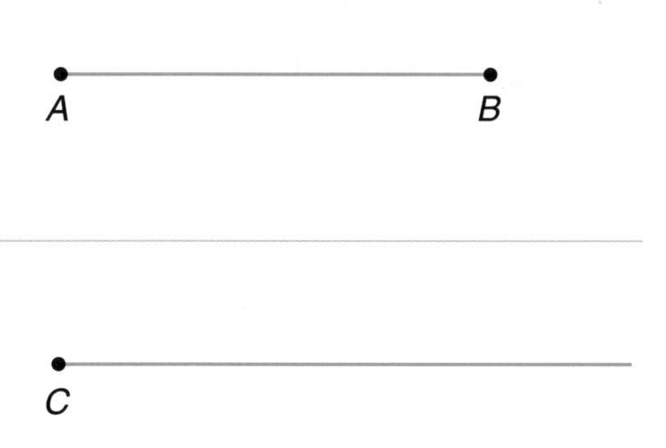

Step 3: Place the compass anchor at *A* and the pencil point at *B*.

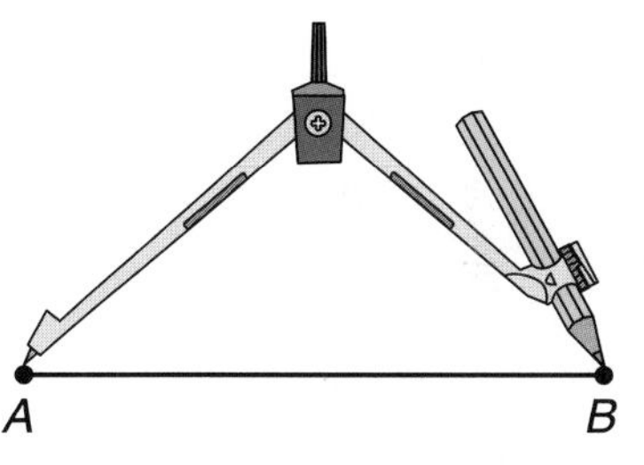

Step 4: Without changing your compass opening, place the compass anchor on *C* and draw a small arc that crosses the line segment. Label the point where the arc crosses the line segment as *D*.

Line segment *CD* should be the same length as line segment *AB*.

Line segment *CD* is **congruent** to line segment *AB*.

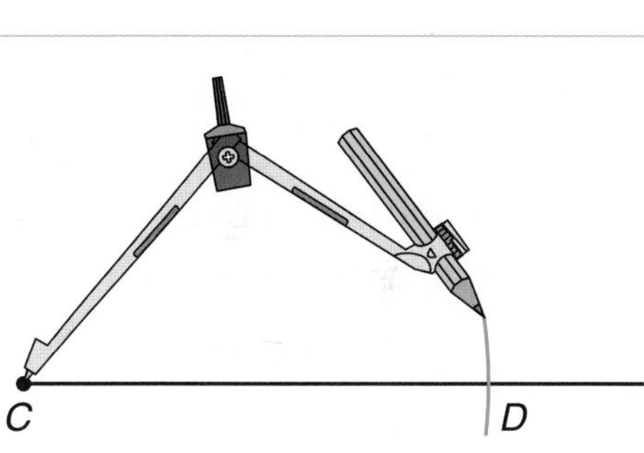

Check Your Understanding

1. Draw a line segment. Using a compass and straightedge only, copy the line segment.
2. After you make your copy, measure the segments with a ruler to see how accurately you copied the original line segment.

Constructing a Parallelogram

Follow each step carefully. Use a clean sheet of paper.

Step 1: Draw an angle and label it *ABC*.

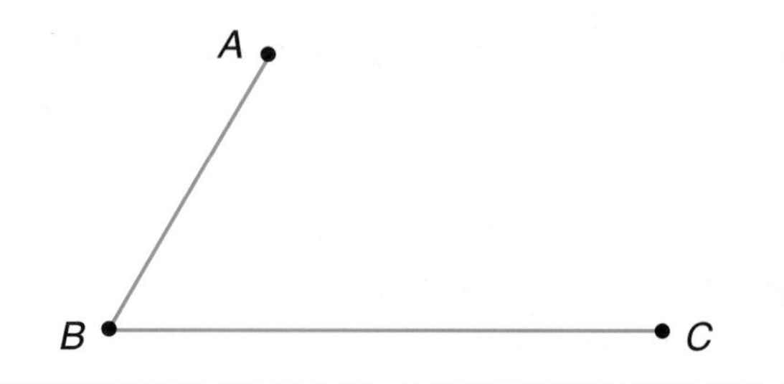

Step 2: Place the compass anchor at *B* and the pencil point at *C*. Without changing your compass opening, place the compass anchor on *A* and draw an arc.

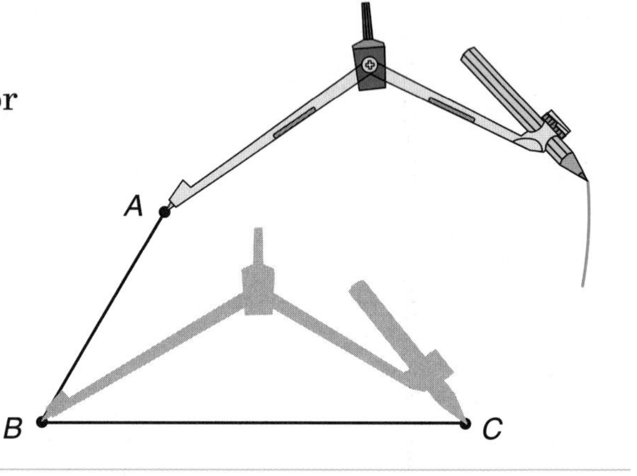

Step 3: Place the compass anchor at *B* and the pencil point at *A*. Without changing your compass opening, place the compass anchor on *C* and draw another arc that crosses the first arc. Label the point where the two arcs cross as *D*.

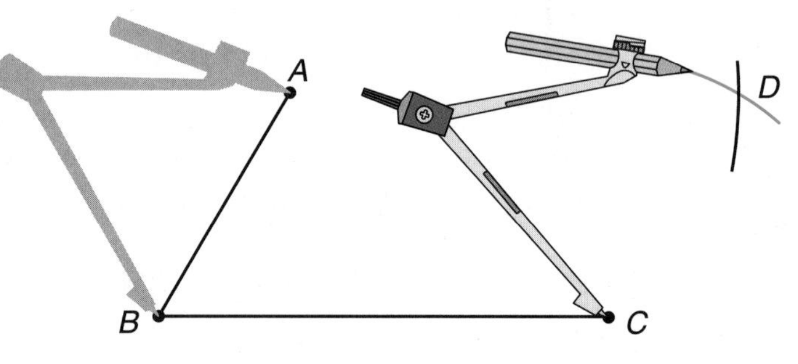

Step 4: Draw line segments *AD* and *CD*.

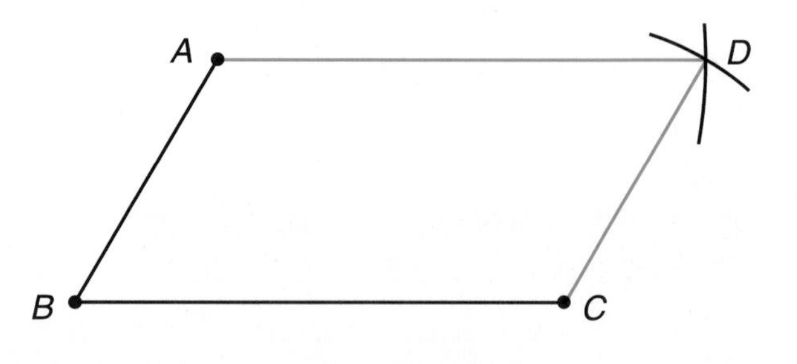

Check Your Understanding

Use a compass and straightedge to construct a parallelogram.

Constructing a Regular Inscribed Hexagon

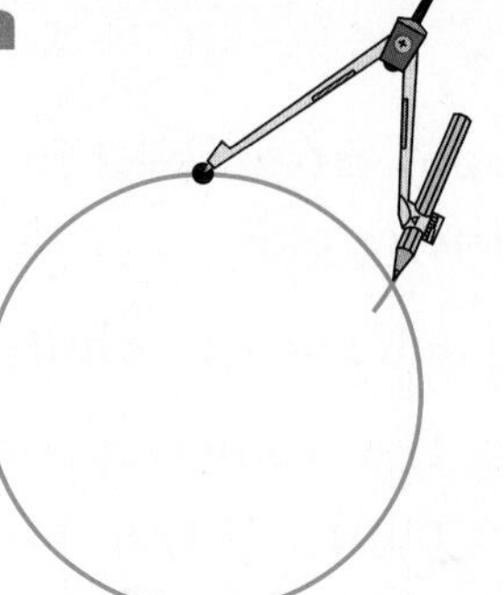

Follow each step carefully. Use a clean sheet of paper.

Step 1: Draw a circle and keep the same compass opening. Make a dot on the circle. Place the compass anchor on the dot and make a mark with the pencil on the circle. Keep the same compass opening for Steps 2 and 3.

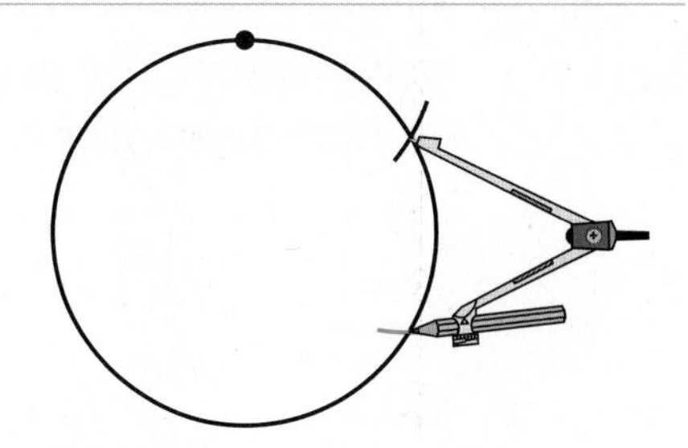

Step 2: Place the compass anchor on the mark you just made. Make another mark with the pencil on the circle.

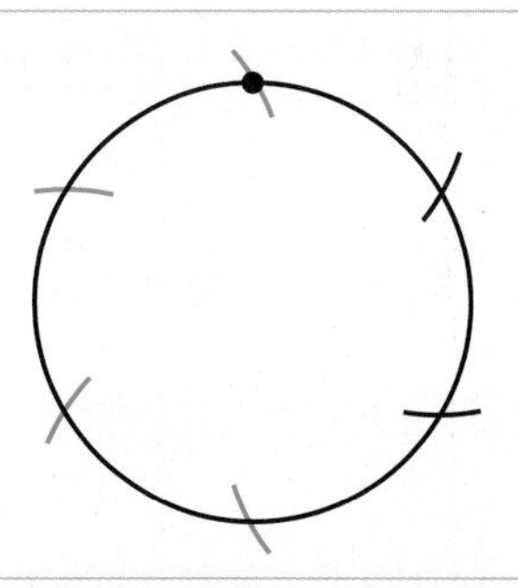

Step 3: Do this four more times to divide the circle into 6 equal parts. The sixth mark should be on the dot you started with or very close to it.

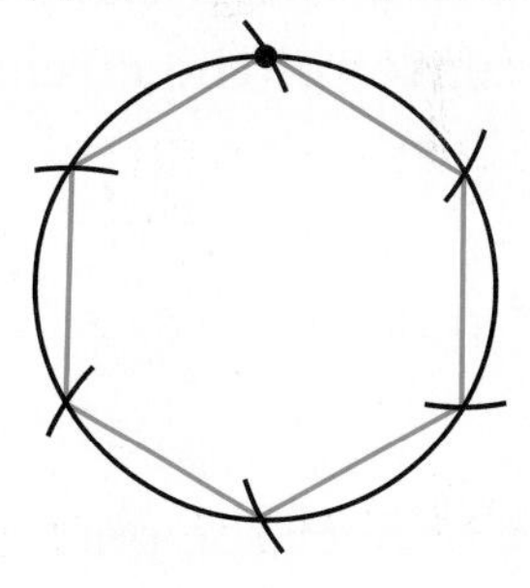

Step 4: With your straightedge, connect the 6 marks on the circle to form a regular hexagon.

Use your compass to check that the sides of the hexagon are all the same length.

The hexagon is **inscribed** in the circle because each vertex of the hexagon is on the circle.

Check Your Understanding

1. Draw a circle. Using a compass and straightedge, construct a regular hexagon that is inscribed in the circle.
2. Draw a line segment from the center of the circle to each vertex of the hexagon to form 6 triangles. Use your compass to check that the sides of each triangle are the same length.

Constructing an Inscribed Square

A square is **inscribed** in a circle if all the vertices of the square are on the circle.

Follow each step carefully. Use a clean sheet of paper.

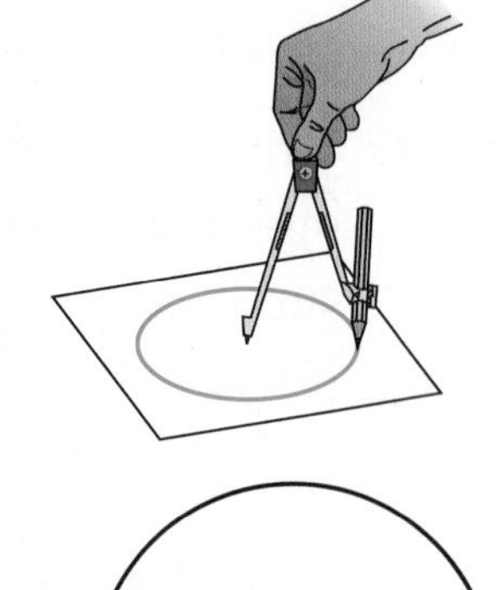

Step 1: Use your compass to draw a circle on a sheet of paper. Cut it out.

With your pencil, make a dot in the center of the circle where the compass anchor left a hole. Mark on both the front and the back of the circle.

Step 2: Fold the circle in half. Make sure the edges match and the fold line passes through the center.

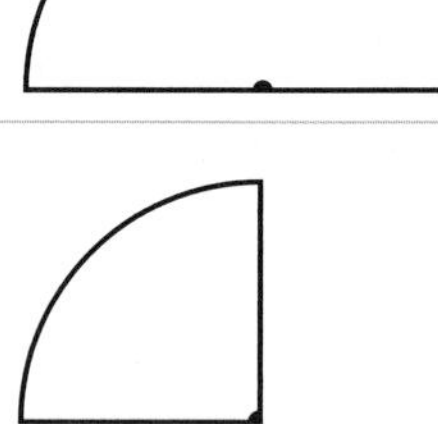

Step 3: Fold the circle in half again so that the edges match.

Step 4: Unfold your circle. The folds should pass through the center of the circle and form four right angles.

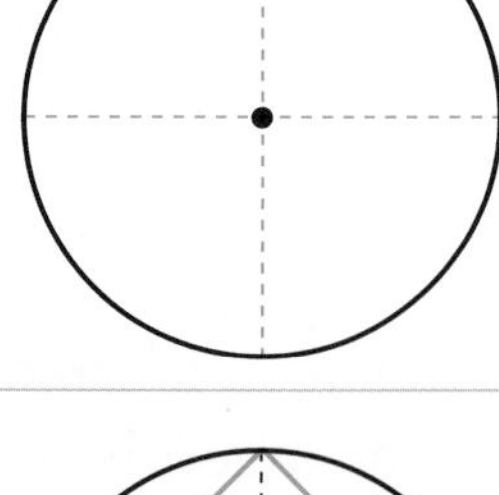

Step 5: Connect the endpoints of the folds with a straightedge to make a square that is inscribed in the circle.

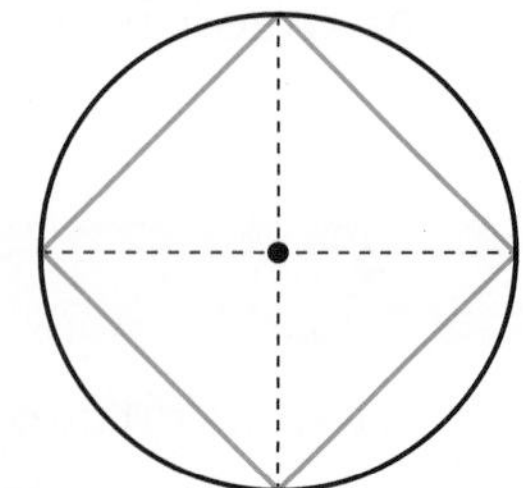

Check Your Understanding

Construct a square inscribed in a circle.

Constructing a Perpendicular Line Segment (Part 1)

Let P be a point on line segment AB. You can construct a line segment that is perpendicular to line segment AB at point P.

Follow each step carefully. Use a clean sheet of paper.

Step 1: Draw line segment AB. Make a dot on $\overline{AB}$, and label it as P.

A P B

Step 2: Place the compass anchor on P, and draw an arc that crosses $\overline{AB}$. Label the point where the arc crosses the segment as C.

Keeping the compass anchor on point P and keeping the same compass opening, draw another arc that crosses $\overline{AB}$. Label the point where the arc crosses the segment as D.

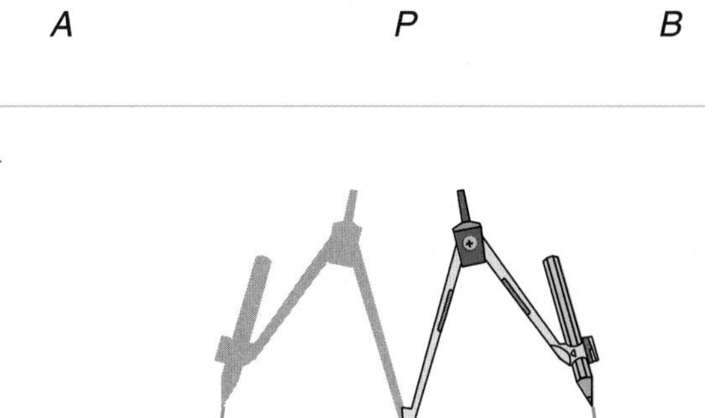

Step 3: Make sure the compass opening is greater than the length of $\overline{CP}$. Place the compass anchor on C and draw an arc above $\overline{AB}$.

Keeping the same compass opening, place the compass anchor on D and draw another arc above $\overline{AB}$ that crosses the first arc.

Label the point where the two arcs cross as Q.

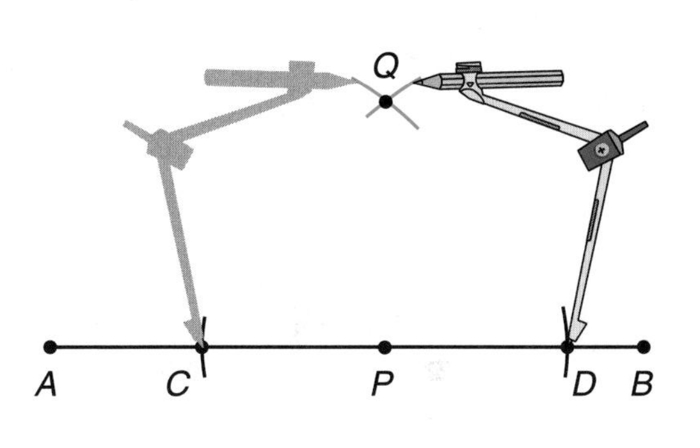

Step 4: Draw $\overline{QP}$.

$\overline{QP}$ is **perpendicular** to $\overline{AB}$.

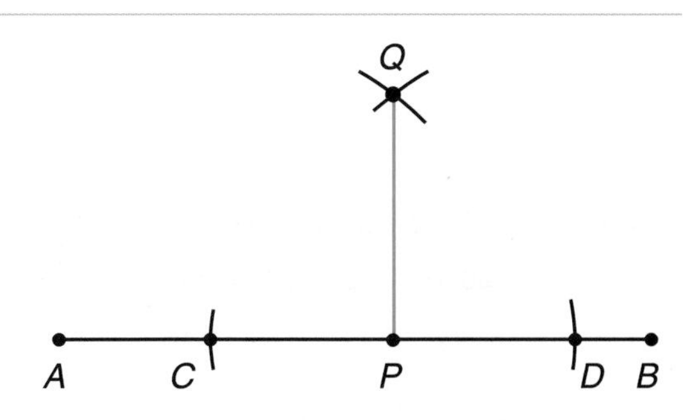

Check Your Understanding

Draw a line segment. Draw a point on the segment and label it as R.

Use a compass and straightedge. Construct a line segment through point R that is perpendicular to the segment you drew.

R

Use a protractor to check that the segments are perpendicular.

Constructing a Perpendicular Line Segment (Part 2)

Let M be a point that is *not* on line segment PQ. You can construct a line segment with one endpoint at M that is perpendicular to line segment PQ.

Follow each step carefully. Use a clean sheet of paper.

Step 1: Draw line segment PQ.

Draw a point M not on $\overline{PQ}$.

Step 2: Place the compass anchor on M and draw an arc that crosses $\overline{PQ}$ at two points.

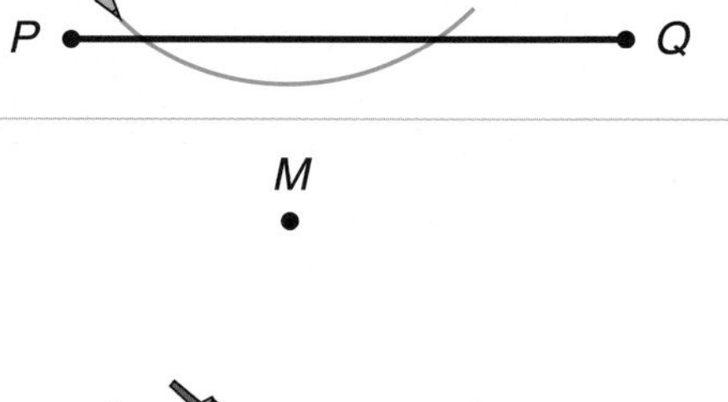

Step 3: Place the compass anchor on one of the points and draw an arc below $\overline{PQ}$.

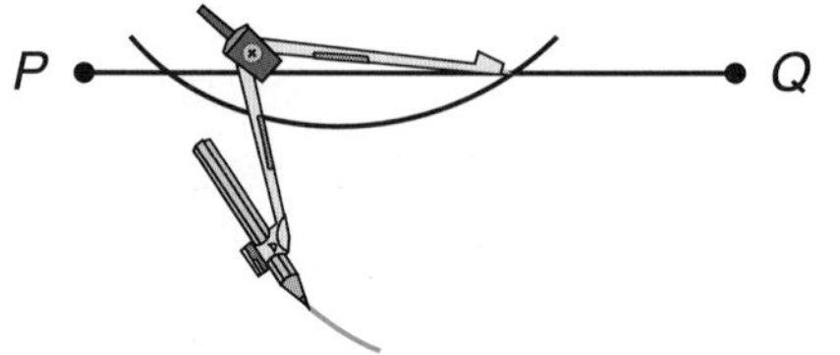

Step 4: Keeping the same compass opening, place the compass anchor on the other point and draw another arc that crosses the first arc.

Label the point where the two arcs cross as N. Then draw the line segment MN.

$\overline{MN}$ is **perpendicular** to $\overline{PQ}$.

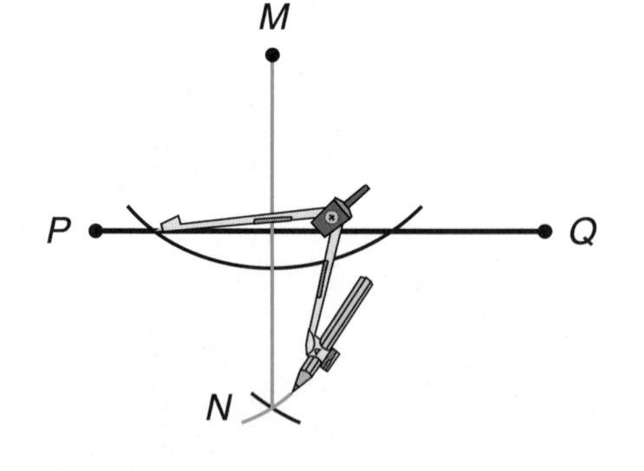

Check Your Understanding

Draw a line segment HI and a point G above the line segment. Using a compass and straightedge, construct a line segment from point G that is perpendicular to $\overline{HI}$.

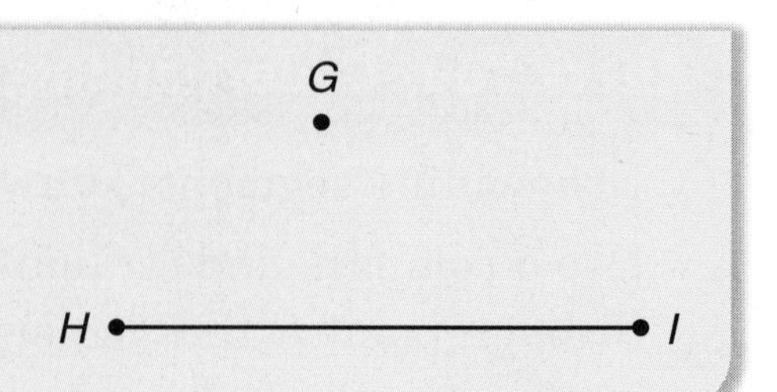

Mathematics and Architecture

Architecture is the design and creation of structures for living, working, worshipping, playing, and other activities. The geometric shapes in a structure and the way they work together give a building its form and strength.

In many parts of the world, an architect draws plans for a structure, and a builder carries them out. ➤

When a building is complete, all of its geometric shapes fit together like pieces in a puzzle. What geometric shapes can you find in this photograph?

Homes

People everywhere need places to live, but their homes can be quite different. The shape and structure of homes are influenced by the materials available and the way people live.

Nomadic people need homes that are sturdy, yet easy to move. Some people in Mongolia live in cylindrical homes called *gers.* ▼

▲ The walls of a ger are made with criss-crossing wooden slats. Large pieces of felt or animal skin are wrapped around the walls and roof and fastened with rope.

The Indians of New Mexico used mud bricks to build villages of connected homes. Why do you think the rooms are shaped like cubes? ▼

Pieces of straw mixed with the mud make the bricks durable. ➤

Homes with rectangular patterns on the walls can be seen throughout Northern Europe. The steep angle of the straw roof lets rain and snow run off easily. ➤

Mathematics ... Every Day

Homes for Powerful People

Homes for a country's leaders or royalty are designed to be awe-inspiring and beautiful. Many important buildings use symmetry to create a feeling of balance and order.

Bodiam castle in Sussex, England was built in 1385 for a royal councilor to the King of England. Notice the symmetry to the left and right sides of the bridge.

Matsumoto castle in Nagano, Japan was built in the 16th century. There are several ornate rectangular floors stacked one on top of the other.

The White House is home to the president of the United States. Notice the number of columns and windows on either side of the building.

The Grand Palace in Bangkok, Thailand was built in 1782. Note the many different geometrical shapes and patterns in this complex structure.

Ceremonial Buildings

In many societies, the most fantastic architecture can be found in places of worship or monuments for tombs.

In ancient Egypt's old kingdom, the Pharaohs chose the pyramid as the shape for their tombs.

The shapes and symmetry of the Byodo-In temple are a common feature of Japanese architecture.

The Taj Mahal is a tomb that was built in India in the 17th century.

This photograph shows some of the detailed patterns on an outside wall of the Taj Mahal.

Mathematics ... Every Day

Modern Buildings

Frank Lloyd Wright and I.M. Pei are two of the many architects who brought new ideas to architecture in the 20th century.

Frank Lloyd Wright, a native of Wisconsin, designed a number of buildings during the first half of the 20th century. One of his well-known projects is the Guggenheim Museum, which was built in New York City in the 1950s. ➤

▲ This is the ceiling of the Guggenheim, viewed from inside.

I. M. Pei was born in China, but later emigrated to the United States. He designed the Bank of China Building in Hong Kong. ▼

Pei often combines simple geometric forms in his designs. This is the Rock-and-Roll Hall of Fame Museum in Cleveland, Ohio. ▼

Other Structures

How many different shapes can you find in these structures? Which shapes do you think give strength to the structures? Which ones do you think add beauty? Which ones do you think do both? Neither?

Eiffel Tower, Paris, France

Sydney Opera House
Sydney, Australia

Tip of the
Chrysler Building,
New York City

Azadhi Tower, Tehran, Iran

Measurement

Natural Measures and Standard Units

Systems of weights and measures have been used in many parts of the world since ancient times. People measured lengths and weights long before they had rulers and scales.

Ancient Measures of Weight

Shells and grains, such as wheat or rice, were often used as units of weight. For example, a small item might be said to weigh 300 grains of rice. Large weights were often compared to the load that could be carried by a man or a pack animal.

Ancient Measures of Length

People used **natural measures** based on the human body to measure length and distance. Some of these units are shown below.

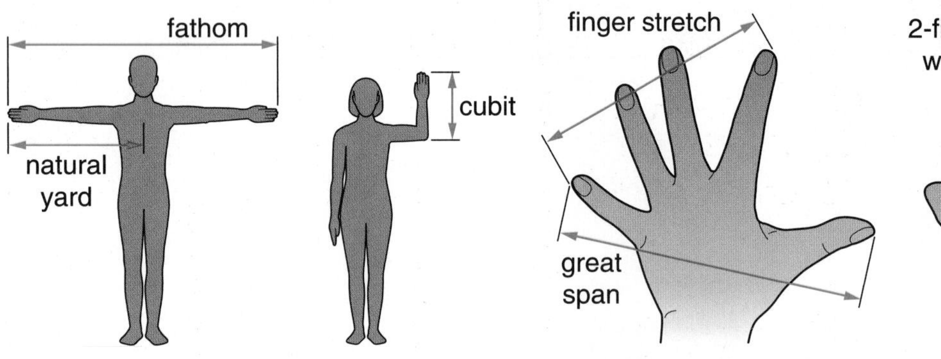

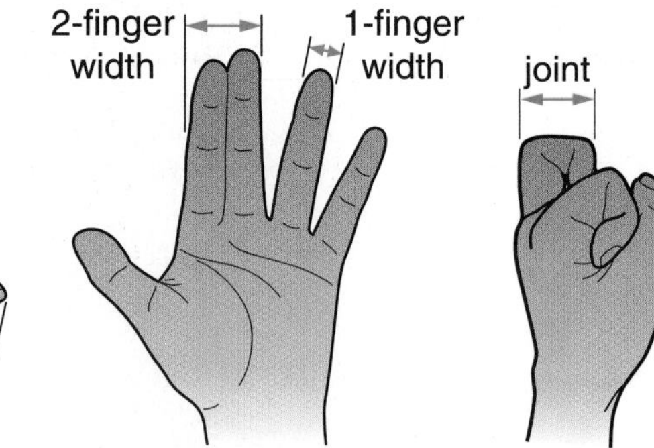

Standard Units of Length and Weight

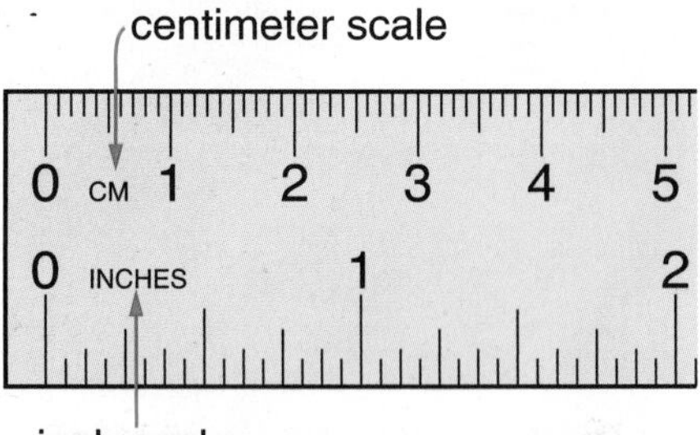

Using shells and grains to measure weight is not exact. Even if the shells and grains are of the same type, they vary in size and weight.

Using body lengths to measure length is not exact either. Body measures depend upon the person who is doing the measuring. The problem is that different persons have hands and arms of different lengths.

One way to solve this problem is to make **standard units** of length and weight. Most rulers are marked off using inches and centimeters as standard units. Bath scales are marked off using pounds and kilograms as standard units. Standard units never change and are the same for everyone. If two people measure the same object using standard units, their measurements will be the same or almost the same.

Did You Know?

The first metric standards were adopted in France in 1799. They were a standard meter and a kilogram bar.

The Metric System and the U.S. Customary System

About 200 years ago, a system of weights and measures called the **metric system** was developed. The metric system uses standard units for length, weight, and temperature. In the metric system:

- The **meter** is the standard unit for length. The symbol for a meter is **m.** A meter is about the width of a front door.
- The **gram** is the standard unit for weight. The symbol for a gram is **g.** A paper clip weighs about $\frac{1}{2}$ gram.
- The **Celsius degree** or **°C** is the standard unit for temperature. Water freezes at 0°**C** and boils at 100°**C**. Room temperature is about 20°**C**.

Scientists almost always use the metric system for measurement. It is easy to use because it is a base-ten system. Larger and smaller units are defined by multiplying or dividing the standard units (given above) by powers of ten: 10, 100, 1,000, and so on.

Example All metric units of length are based on the meter.

Each unit is defined by multiplying or dividing the meter by a power of 10.

Units of Length Based on the Meter	Prefix	Meaning
1 decimeter (dm) = $\frac{1}{10}$ meter	deci-	$\frac{1}{10}$
1 centimeter (cm) = $\frac{1}{100}$ meter	centi-	$\frac{1}{100}$
1 millimeter (mm) = $\frac{1}{1{,}000}$ meter	milli-	$\frac{1}{1{,}000}$
1 kilometer (km) = 1,000 meters	kilo-	1,000

Note

The U.S. customary system is not based on powers of 10. This makes it more difficult to use than the metric system. For example, to change inches to yards, you must know that 36 inches equals 1 yard.

The metric system is used in most countries around the world. In the United States, however, the **U.S. customary system** is used for everyday purposes. The U.S. customary system uses standard units such as the **inch, foot, yard, mile, ounce, pound,** and **ton.**

Check Your Understanding

1. Which units below are in the metric system?

   foot millimeter pound inch gram meter centimeter yard

2. What does the prefix "milli-" mean?
3. 5 meters = ? millimeters

Check your answers on page 343.

Length

Length is the measure of the distance between two points. Length is usually measured with a ruler. The edges of the Geometry Template are rulers. Tape measures, yardsticks, and metersticks are rulers for measuring longer distances.

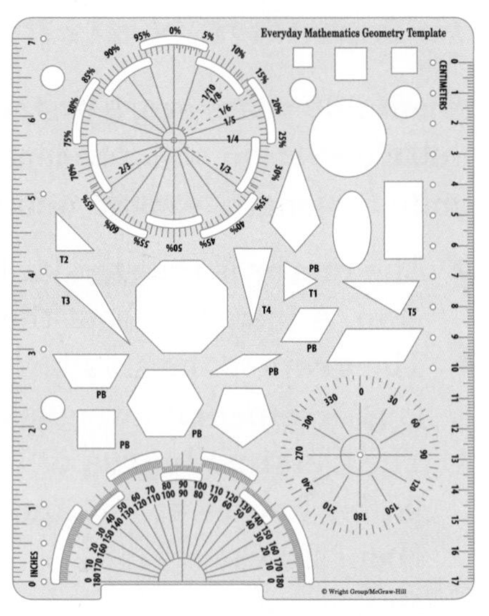

Part of a meterstick is shown here. The meter has been divided into smaller units—centimeters and millimeters.

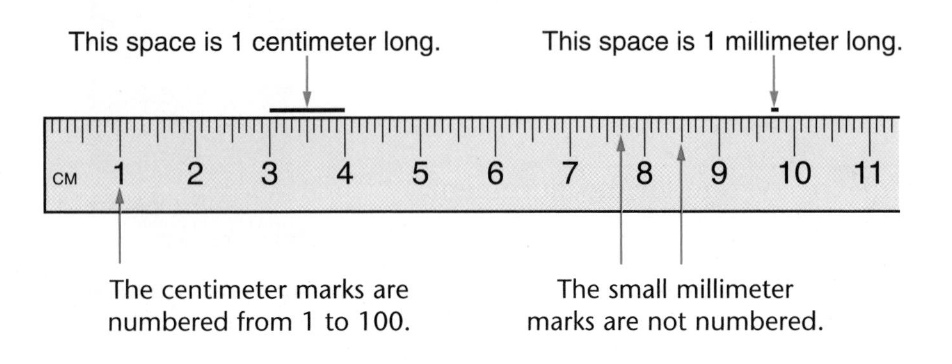

Did You Know?

Grant's Rhinoceros Beetle, also known as the "Western Hercules Beetle," is about $3\frac{1}{2}$ in. long.

On rulers, inches are usually divided into halves, quarters, eighths, and sixteenths. There are usually different-size marks to show different fractions of an inch.

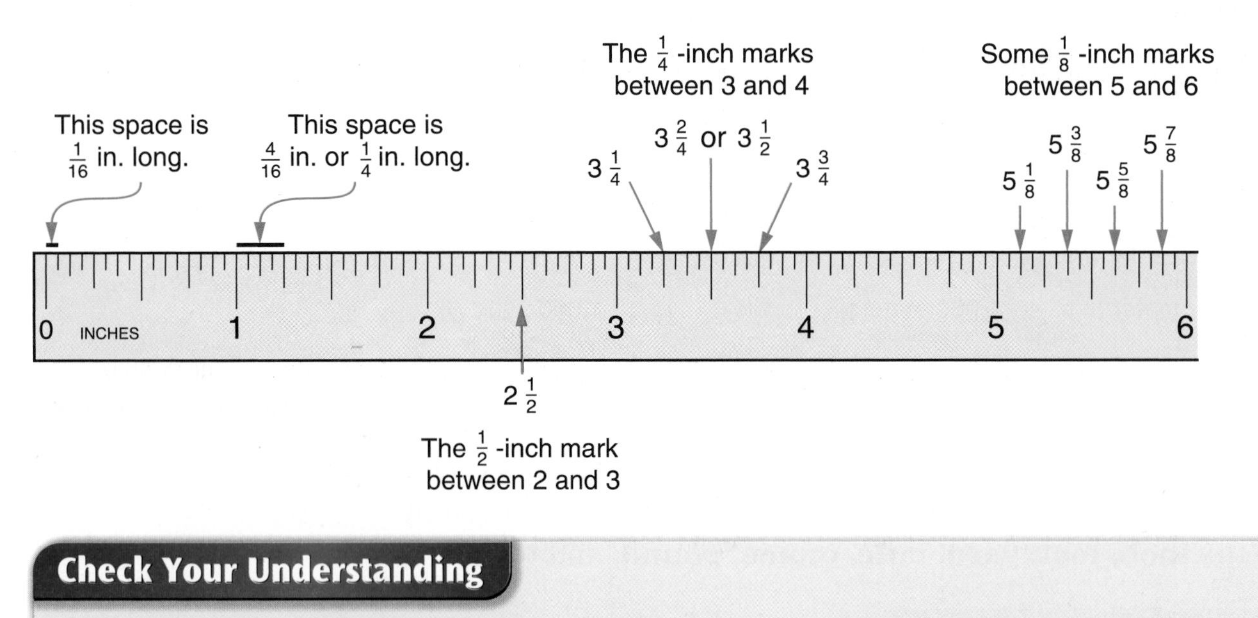

Check Your Understanding

Measure these line segments. Record each length in centimeters (cm) and millimeters (mm).

1. ____________ **2.** ____________ **3.** ____________

4. Measure each line segment above to the nearest quarter-inch.

Check your answers on page 343.

Converting Units of Length

The table below shows how different units of length in the metric system compare. You can use this table to rewrite a length using a different unit.

Did You Know?

The Roe River in Montana is only 201 ft, or 67 yd long.

Comparing Metric Units of Length				Symbols for Units of Length	
1 cm = 10 mm	1 m = 1,000 mm	1 m = 100 cm	1 km = 1,000 m	mm = millimeter	cm = centimeter
1 mm = $\frac{1}{10}$ cm	1 mm = $\frac{1}{1,000}$ m	1 cm = $\frac{1}{100}$ m	1 m = $\frac{1}{1,000}$ km	m = meter	km = kilometer

Examples Use the above table to rewrite each length using a different unit.

Replace the unit given with an equal length that uses the new unit.

Problem	Solution
27 centimeters = ? millimeters	27 cm = 27 * 10 mm = 270 mm
27 centimeters = ? meters	27 cm = 27 * $\frac{1}{100}$ m = $\frac{27}{100}$ m = 0.27 m
6.4 kilometers = ? meters	6.4 km = 6.4 * 1,000 m = 6,400 m
7.3 meters = ? centimeters	7.3 m = 7.3 * 100 cm = 730 cm

The table below shows how different units of length in the U.S. customary system compare. You can use this table to rewrite a length using a different unit.

Comparing U.S. Customary Units of Length				Symbols for Units of Length	
1 ft = 12 in.	1 yd = 36 in.	1 yd = 3 ft	1 mi = 5,280 ft	in. = inch	ft = foot
1 in. = $\frac{1}{12}$ ft	1 in. = $\frac{1}{36}$ yd	1 ft = $\frac{1}{3}$ yd	1 ft = $\frac{1}{5,280}$ mi	yd = yard	mi = mile

Examples Use the above table to rewrite each length using a different unit.

Replace the unit given with an equal length that uses the new unit.

Problem	Solution
9 feet = ? inches	9 ft = 9 * 12 in. = 108 in.
9 feet = ? yards	9 ft = 9 * $\frac{1}{3}$yd = $\frac{9}{3}$yd = 3 yd
4 miles = ? feet	4 mi = 4 * 5,280 ft = 21,120 ft
144 inches = ? yards	144 in. = 144 * $\frac{1}{36}$ yd = $\frac{144}{36}$ yd = 4 yd

Personal References for Units of Length

Sometimes it is hard to remember just how long a centimeter or a yard is, or how a kilometer and a mile compare. You may not have a ruler, yardstick, or tape measure handy. When this happens, you can estimate lengths by using the lengths of common objects and distances that you know.

Some examples of personal references for length are given below. A good personal reference is something that you often see or use, so you don't forget it. A good personal reference also does not change size. For example, a wooden pencil is not a good personal reference for length, because it gets shorter as it is sharpened.

The diameter of a quarter is about 1 in.

The thickness of pattern blocks is about 1 cm.

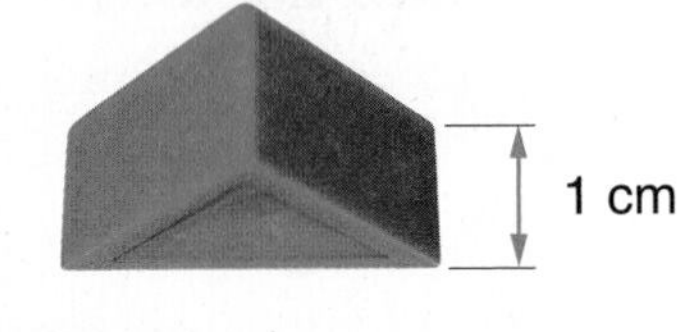

Personal References for Metric Units of Length

About 1 millimeter	About 1 centimeter
Thickness of a dime Thickness of a thumbtack point Thickness of a paper match (the thin edge)	Thickness of a crayon Width of the head of a thumbtack Thickness of a pattern block
About 1 meter	**About 1 kilometer**
One big step (for an adult) Width of a front door Tip of the nose to tip of the thumb, with arm extended (for an adult)	1,000 big steps (for an adult) Length of 10 football fields (including the end zones)

Personal References for U.S. Customary Units of Length

About 1 inch	About 1 foot
Length of a paper clip Width (diameter) of a quarter Width of a man's thumb	A man's shoe length Length of a license plate Length of this book
About 1 yard	**About 1 mile**
One big step (for an adult) Width of a front door Tip of the nose to tip of the thumb, with arm extended (for an adult)	2,000 average-size steps (for an adult) Length of 15 football fields (including the end zones)

Note

The personal references for 1 meter can also be used for 1 yard. 1 yard equals 36 inches; 1 meter is about 39.37 inches. One meter is often called a "fat yard," which means one yard plus one hand width.

Did You Know?

Recently, the tallest man in the world was measured at 7 ft 8.9 in. (2.359 m) in Tunisia.

Perimeter

Sometimes we want to know the **distance around** a shape. The distance around a shape is called the **perimeter** of the shape. To measure perimeter, we use units of length such as inches or meters or miles.

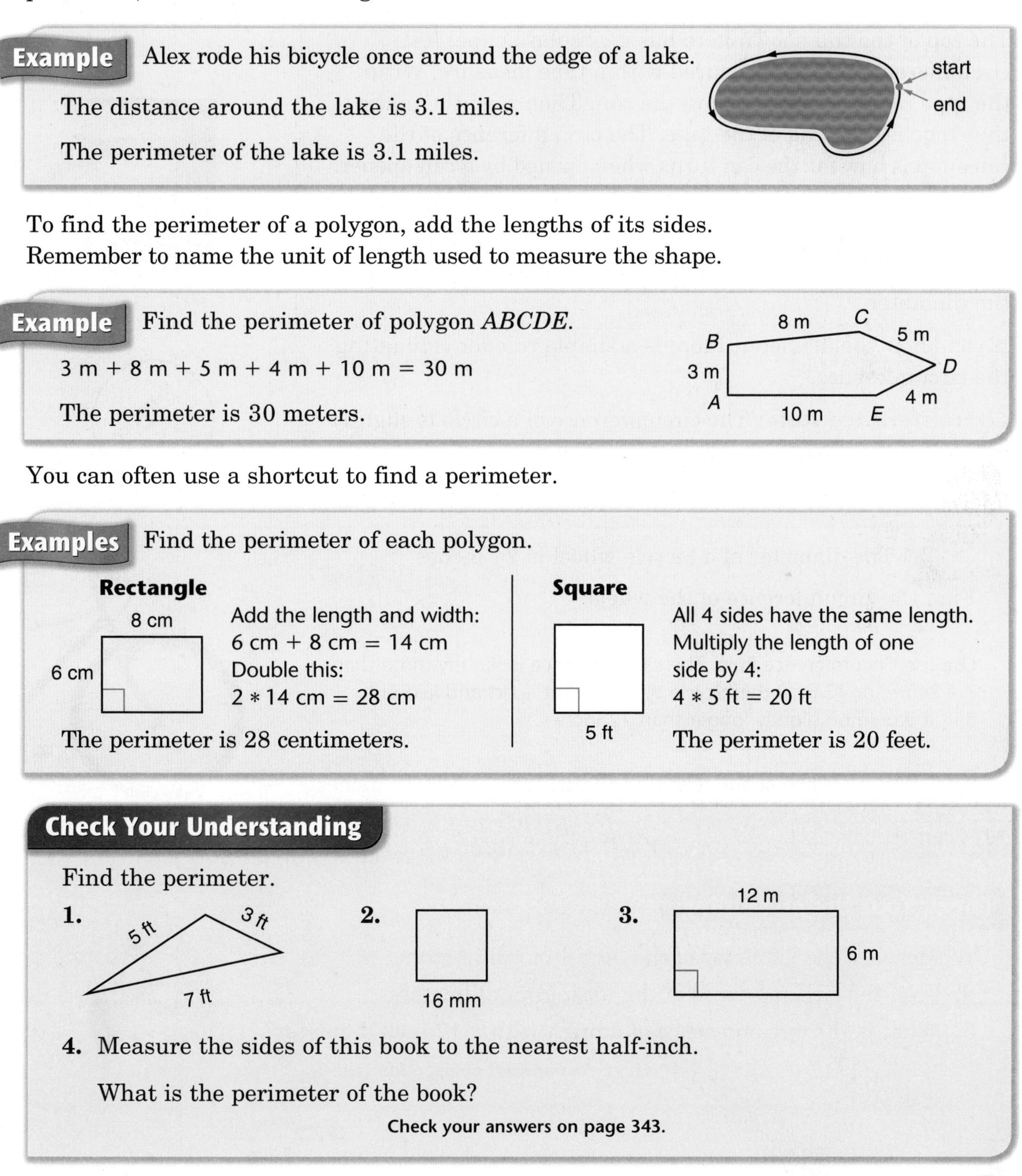

Example Alex rode his bicycle once around the edge of a lake.

The distance around the lake is 3.1 miles.

The perimeter of the lake is 3.1 miles.

To find the perimeter of a polygon, add the lengths of its sides. Remember to name the unit of length used to measure the shape.

Example Find the perimeter of polygon *ABCDE*.

3 m + 8 m + 5 m + 4 m + 10 m = 30 m

The perimeter is 30 meters.

You can often use a shortcut to find a perimeter.

Examples Find the perimeter of each polygon.

Rectangle

Add the length and width:
6 cm + 8 cm = 14 cm
Double this:
2 * 14 cm = 28 cm

The perimeter is 28 centimeters.

Square

All 4 sides have the same length.
Multiply the length of one side by 4:
4 * 5 ft = 20 ft

The perimeter is 20 feet.

Check Your Understanding

Find the perimeter.

1.

2.

3.

4. Measure the sides of this book to the nearest half-inch.

What is the perimeter of the book?

Check your answers on page 343.

Circumference

The perimeter of a circle is the **distance around** the circle.

The perimeter of a circle has a special name. It is called the **circumference** of the circle.

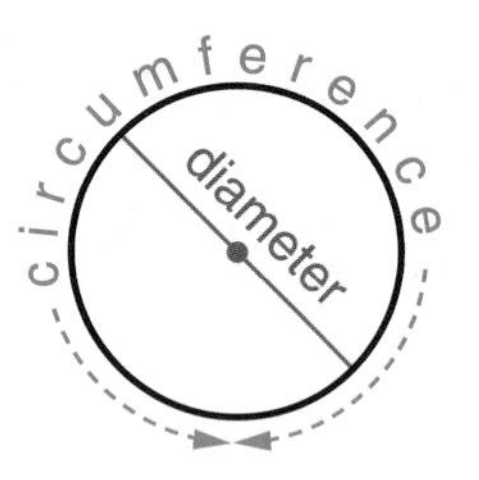

The top of the can shown here has a circular shape. Its circumference can be measured with a tape measure. Wrap the tape measure once around the can. Then read the mark that touches the end of the tape. The circumference of the can's top is how far the can turns when opened by a can opener.

The **diameter** of a circle is any line segment that passes through the center of the circle and has both endpoints on the circle. The length of a diameter segment is also called the diameter.

If you know the diameter, there is a simple rule for estimating the circumference.

circumference
diameter

Circumference Rule: The circumference of a circle is slightly more than three times the diameter of the circle.

Example The diameter of a bicycle wheel is 24 inches. Find the circumference of the wheel.

Use the Circumference Rule. The circumference is slightly more than 3 * 24 in., or 72 in. If the bicycle tire were cut apart and laid out flat, it would be slightly longer than 72 inches.

The circumference of the wheel is slightly more than 72 inches.

Check Your Understanding

1. Measure the diameter of the nickel in millimeters.
2. Find the circumference of the nickel in millimeters.
3. What is the circumference of a pizza with a 12-inch diameter?

Check your answers on page 343.

Area

Sometimes you want to know the amount of **surface inside** a shape. The amount of surface inside a shape is called its **area.** You can find the area of a shape by counting the number of squares of a certain size that cover the inside of the shape. The squares must cover the entire inside of the shape and must not overlap, have any gaps, or cover any surface outside of the shape. Sometimes a shape cannot be covered by an exact number of squares. If this is so, count the number of whole squares and the fractions of squares that cover the shape.

Example What is the area of the rectangle?

The rectangle at the right is covered by squares that are 1 centimeter on each side. Each square is called a **square centimeter (cm^2).**

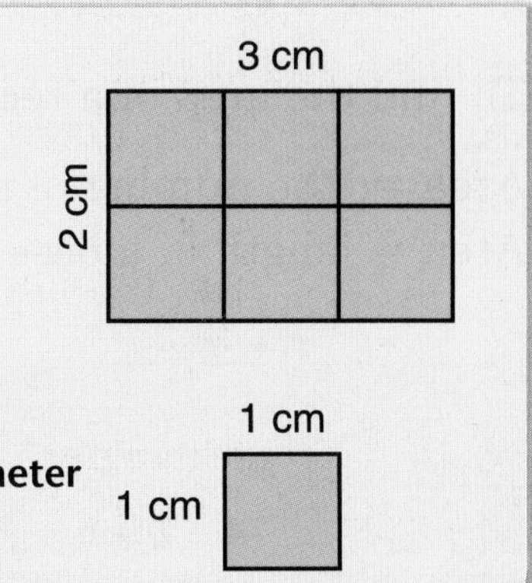

Six of the squares cover the rectangle.
The area of the rectangle is 6 square centimeters.
This is written as 6 sq. cm, or 6 cm^2.

1 square centimeter (actual size)

Reminder: Be careful not to confuse the **area** of a shape with its **perimeter.** The **area** is the amount of surface *inside* the shape. The **perimeter** is the distance *around* the shape. Area is measured in units such as square inches, square feet, square centimeters, square meters, and square miles. Perimeter is measured in units such as inches, feet, centimeters, meters, and miles.

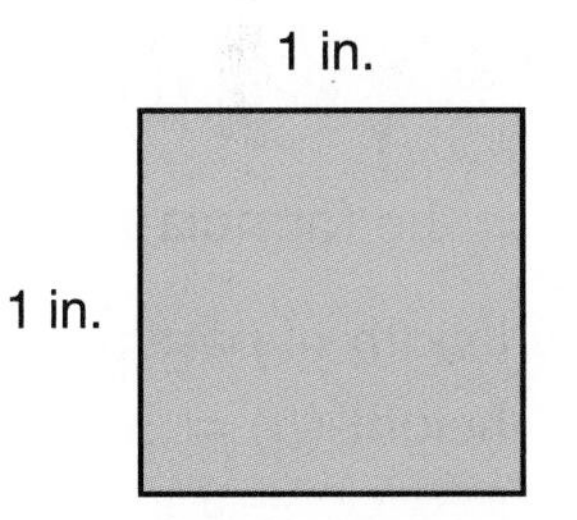

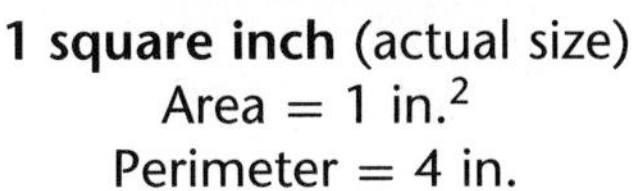
1 square inch (actual size)
Area = 1 $in.^2$
Perimeter = 4 in.

There are many situations where it is important to know the area.

- You may want to install carpeting in your living room. You will need to find the area of the floor to figure out how much carpeting to buy. For units, you would use **square yards (yd^2)** in the United States and **square meters (m^2)** in the rest of the world.
- You may want to paint the walls and ceilings of the rooms in your home. You will need to find the total area of all the surfaces to be painted to figure out how many gallons of paint to buy. Labels on cans of paint usually tell about how many **square feet (ft^2)** of surface can be painted with the paint in that can.
- In the "World Tour" section of this book, the area of each country you visit is given in **square miles.** This information is important when comparing the sizes of different countries.

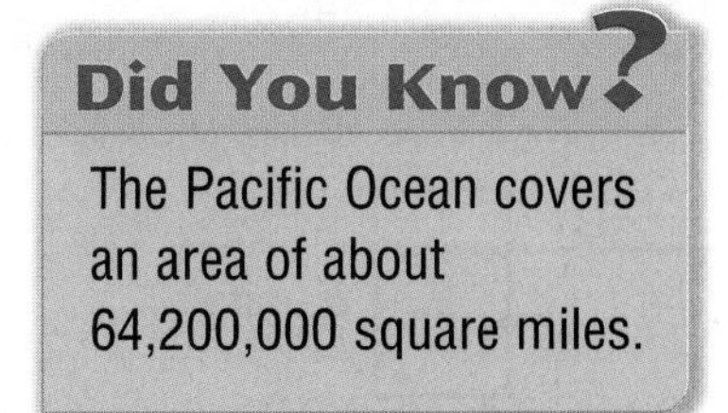
Did You Know?

The Pacific Ocean covers an area of about 64,200,000 square miles.

Area of a Rectangle

When you cover a rectangular shape with unit squares, the squares can be arranged into rows. Each row will contain the same number of squares and fractions of squares.

Example Find the area of the rectangle.

3 rows with 5 squares in each row for a total of 15 squares

Area = 15 square units

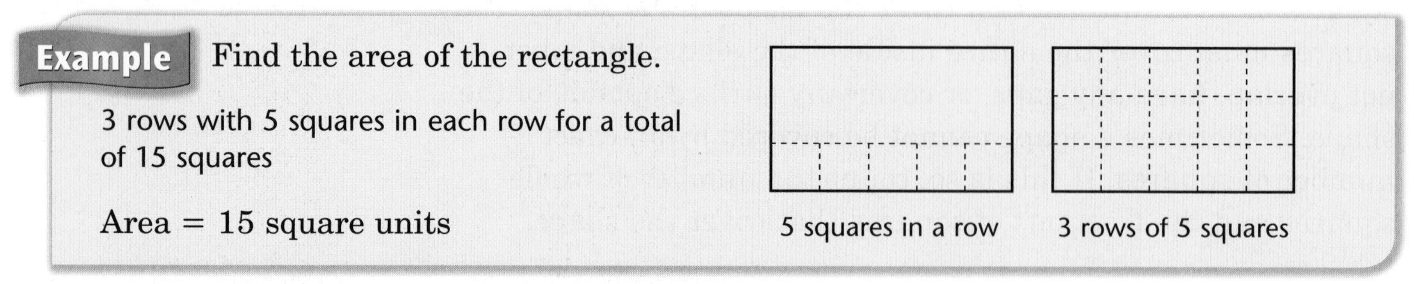

To find the area of a rectangle, use either formula below:

Area = (the number of squares in 1 row) * (the number of rows)
Area = length of a base * height

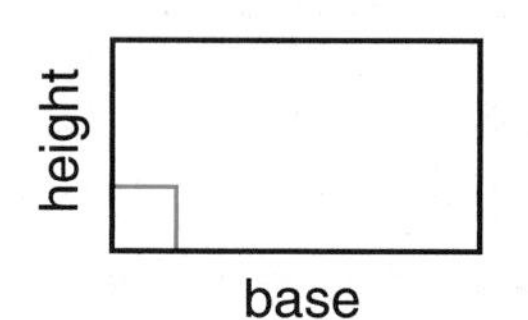

Area Formulas	
Rectangle	**Square**
$A = b * h$ A is the area, b is the length of a base, h is the height of the rectangle.	$A = s^2$ A is the area and s is the length of a side of the square.

Either pair of parallel sides in a rectangle can be chosen as its **bases.** The **height** of a rectangle is the shortest distance between its bases.

Examples Find the area of the rectangle.

Use the formula $A = b * h$.

- length of base (b) = 4 in.
- height (h) = 3 in.
- area (A) = 4 in. * 3 in. = 12 in.^{2}

The area of the rectangle is 12 in.^{2}

3 in.
4 in.

Find the area of the square.

Use the formula $A = s^2$.

- length of a side (s) = 5 ft
- area (A) = 5 ft * 5 ft = 25 ft^2

The area of the square is 25 ft^2.

5 ft

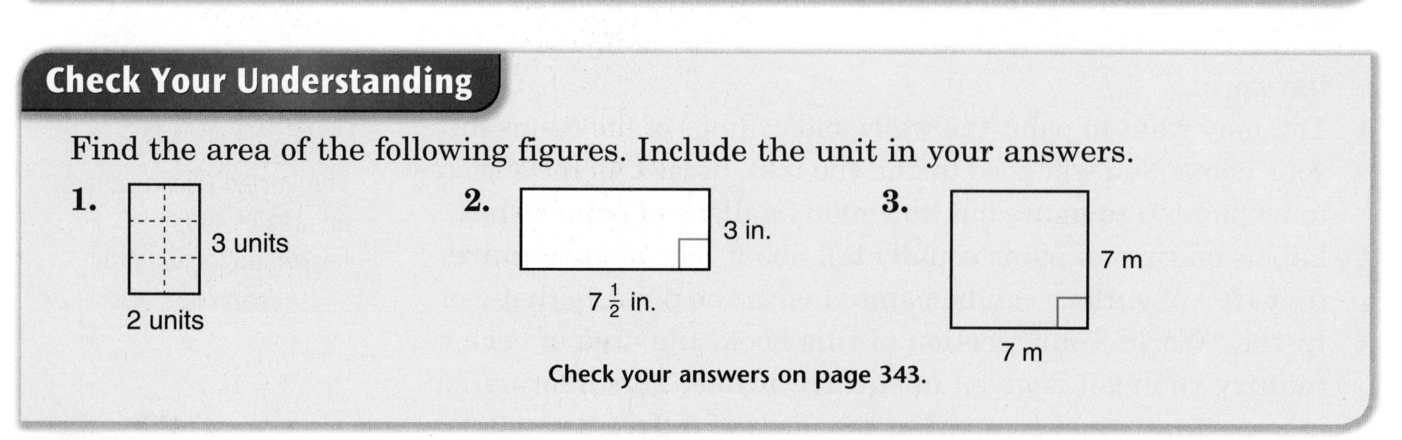

Check your answers on page 343.

Area of a Parallelogram

In a parallelogram, either pair of opposite sides can be chosen as its **bases.** The **height** of the parallelogram is the shortest distance between the two bases.

In the parallelograms at the right, the height is shown by a dashed line that is **perpendicular** (at a right angle) to the base. In the second parallelogram, the base has been extended and the dashed height line falls outside the parallelogram.

Any parallelogram can be cut into two pieces that will form a rectangle. This rectangle will have the same base length and height as the parallelogram. It will also have the same area as the parallelogram.

You can find the area of the parallelogram in the same way you find the area of the rectangle—by multiplying the length of the base by the height.

height
base

height
base

height
base

height
base

Formula for the Area of a Parallelogram

$A = b * h$

A is the area, b is the length of the base, and h is the height of the parallelogram.

Example Find the area of the parallelogram.

Use the formula $A = b * h$.

- length of base $(b) = 6$ cm
- height $(h) = 3$ cm
- area $(A) = 6 \text{ cm} * 3 \text{ cm} = 18 \text{ cm}^2$

The area of the parallelogram is 18 cm^2.

3 cm
6 cm

Check Your Understanding

Find the area of each parallelogram. Include the unit in your answers.

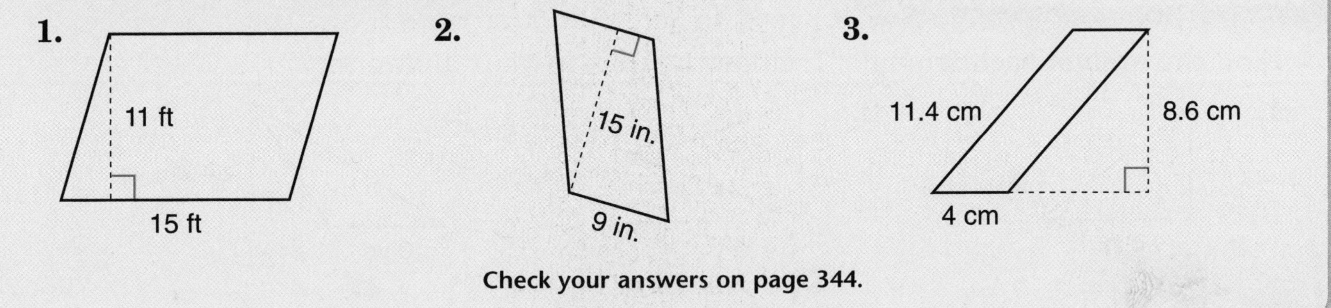

Check your answers on page 344.

Area of a Triangle

Any of the sides of a triangle can be chosen as its **base.** The **height** of the triangle is the shortest distance between the chosen base and the **vertex** opposite the base.

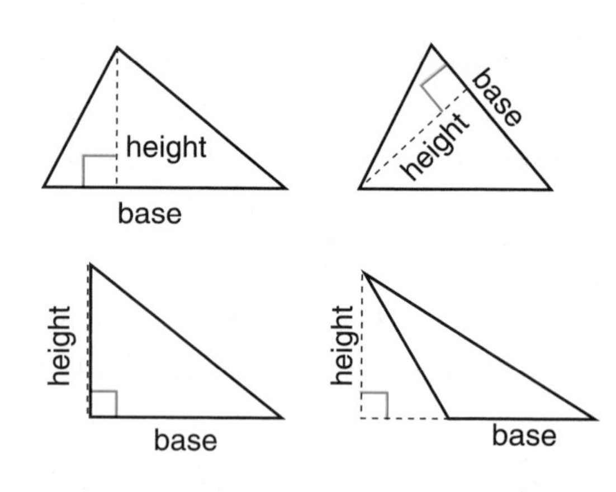

In the triangles at the right, the height is shown by a dashed line that is **perpendicular** (at a right angle) to the base. In one of the triangles, the base has been extended and the dashed height line falls outside the triangle. In the right triangle shown, the height line is one of the sides of the triangle.

Any triangle can be combined with a second triangle of the same size and shape to form a parallelogram. Each triangle at the right has the same size base and height as the parallelogram. The area of each triangle is half the area of the parallelogram. Therefore, the area of a triangle is half the product of the base length multiplied by the height.

Area Formulas	
Parallelograms	**Triangles**
$A = b * h$	$A = \frac{1}{2} * (b * h)$
A is the area, *b* is the length of a base, *h* is the height.	*A* is the area, *b* is the length of a base, *h* is the height.

Example Find the area of the triangle.

Use the formula $A = \frac{1}{2} * (b * h)$.

4 in.

7 in.

- length of base $(b) = 7$ in.
- height $(h) = 4$ in.
- area $(A) = \frac{1}{2} * (7 \text{ in.} * 4 \text{ in.}) = \frac{1}{2} * 28 \text{ in.}^2 = \frac{28}{2} \text{ in.}^2 = 14 \text{ in.}^2$

The area of the triangle is 14 in.^{2}.

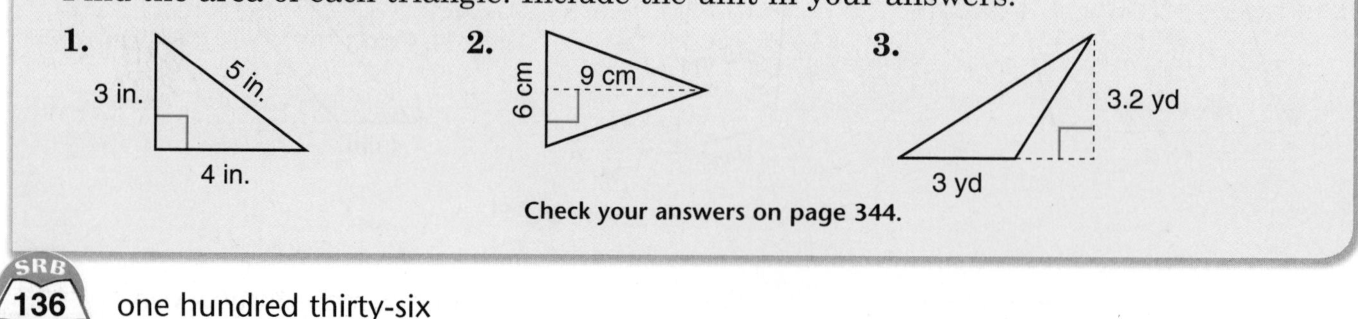

Volume and Capacity

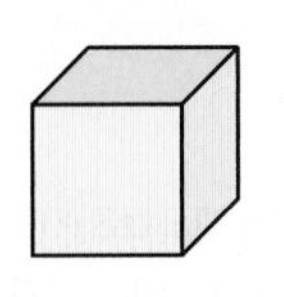

1 cubic centimeter
(actual size)

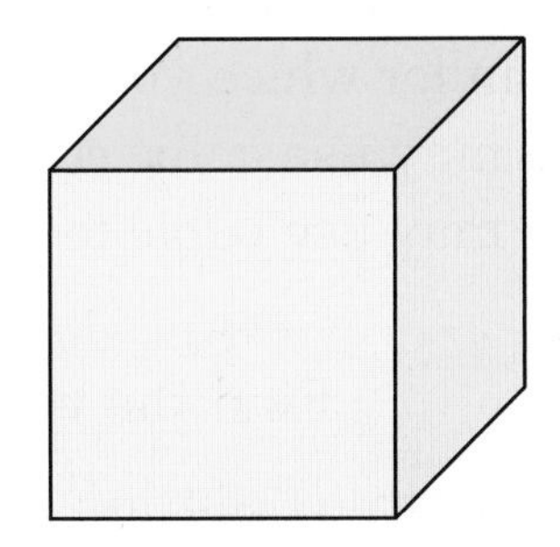

1 cubic inch
(actual size)

Volume

The **volume** of a solid object such as a brick or a ball is a measure of *how much space the object takes up.* The volume of a container such as a freezer is a measure of *how much the container will hold.*

Volume is measured in **cubic units.** A base-10 cube has sides that are 1 centimeter long; it is called a **cubic centimeter.** A cube with 1-inch sides is called a **cubic inch.**

Other cubic units are used to measure large volumes. A **cubic foot** has 1-foot sides. A **cubic yard** has 1-yard sides and can hold 27 cubic feet. A **cubic meter** has 1-meter sides and can hold more than 35 cubic feet.

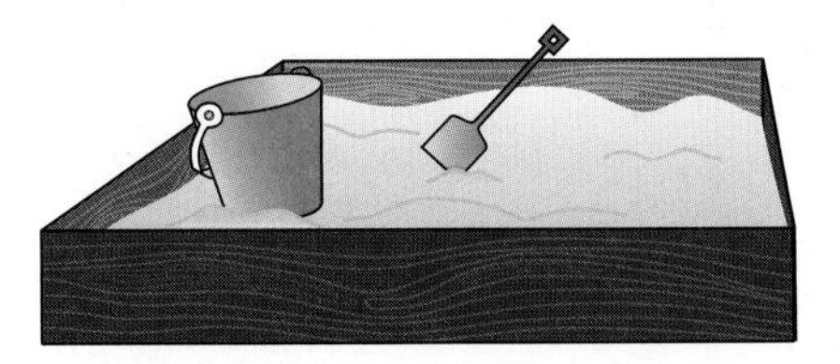

The volume of an object can be very useful to know. Suppose you wanted to buy sand to fill an empty sandbox. To estimate how much sand to buy, you would measure the length, width, and height of the sandbox. The length, width, and height are called the **dimensions** of the box. You would then use these dimensions to calculate how many cubic feet (or cubic yards) of sand to order. You could do similar calculations to determine how much concrete would be needed to build a patio, or how much gravel to buy for a path in the backyard.

Capacity

We often measure things that are poured into or out of containers such as liquids, grains, salt, and so on. The volume of a container that is filled with a liquid or a solid that can be poured is often called its **capacity.**

Capacity is usually measured in units such as **gallons, quarts, pints, cups, fluid ounces, liters,** and **milliliters.** These are standard units, but they are not cubic units.

The tables at the right compare different units of capacity.

Did You Know?

Jupiter is the largest planet in the solar system. The volume of Jupiter is 1,300 times the volume of Earth.

U.S. Customary Units
1 gallon (gal) = 4 quarts (qt)
1 gallon = 2 half-gallons
1 half-gallon = 2 quarts
1 quart = 2 pints (pt)
1 pint = 2 cups (c)
1 cup = 8 fluid ounces (fl oz)
1 pint = 16 fluid ounces
1 quart = 32 fluid ounces
1 half-gallon = 64 fluid ounces
1 gallon = 128 fluid ounces

Metric Units
1 liter (L) = 1,000 milliliters (mL)
1 milliliter = $\frac{1}{1,000}$ liter

Volume of a Rectangular Prism

The 3-dimensional shape shown at the right is a rectangular prism. It looks like a box. You can think of the volume of the prism as the total number of unit cubes needed to fill the interior of the prism. The size of the unit cube used will depend on the size of the prism for which you want to find the volume. For smaller prisms, you may use cubic centimeters or cubic inches; for larger prisms, you may use cubic feet, cubic yards, or cubic meters.

3 in.
2 in.
4 in.

8 cubes fill 1 layer

Example Find the volume of the prism at the right.

length (l) = 4 in.; width (w) = 2 in.; height (h) = 3 in.

Use cubic inches as the unit cubes. The cubes can be arranged in layers.

There are 8 cubes in the bottom layer.

There are 3 layers with 8 cubes in each layer for a total of 24 cubes.

The volume is 24 cubic inches (24 in.^{3}).

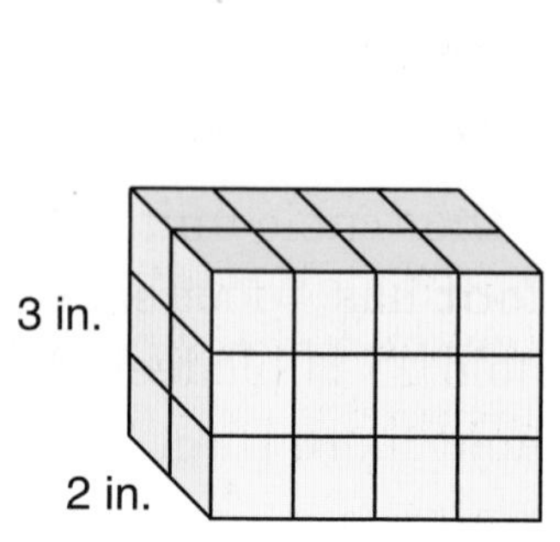

3 layers filled

You can find the volume of a prism without counting the number of unit cubes that fill it. If you know the dimensions of the prism, the formulas below can be used to calculate its volume.

Volume of a Rectangular Prism

$V = l * w * h$ or $V = B * h$

V is the volume of the prism.

l is the length of its base. w is the width of its base.

h is the height of the prism.
h is the shortest distance between the bases.

B is the area of the prism's base.

Example Use the formulas to find the volume of the prism.

Use the formula $V = l * w * h$

- length of base (l) = 4 in.
- width of base (w) = 2 in.
- height of prism (h) = 3 in.
- volume (V) =
  4 in. $*$ 2 in. $*$ 3 in. = 24 in.^{3}

Use the formula $V = B * h$

- area of base (B) = 4 in. $*$ 2 in. = 8 in.^{2}
- height of prism (h) = 3 in.
- volume (V) = 8 in.^{2} $*$ 3 in. = 24 in.^{3}

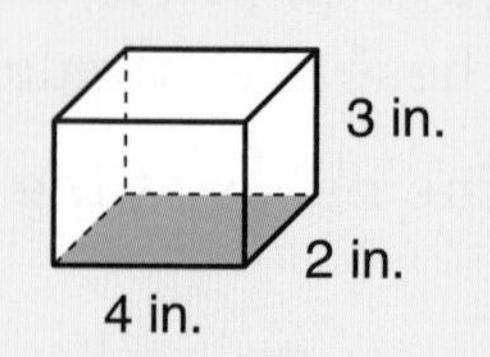

The volume is 24 in.^{3} (using either formula).

Temperature

Temperature is a measure of the hotness or coldness of something. To read a temperature in degrees, you need a reference frame that begins with a zero point and has a number-line scale. The two most commonly used temperature scales, Fahrenheit and Celsius, have different zero points.

Fahrenheit

This scale was invented in the early 1700s by the German physicist G.D. Fahrenheit. On the Fahrenheit scale, a saltwater solution freezes at 0°F (the zero point) at sea level. Pure water freezes at 32°F and boils at 212°F. The normal temperature for the human body is 98.6°F. The Fahrenheit scale is used primarily in the United States.

Celsius

This scale was developed in 1742 by the Swedish astronomer Anders Celsius. On the Celsius scale, the zero point (0 degrees Celsius or 0°C) is the freezing point of pure water. Pure water boils at 100°C. The Celsius scale divides the interval between these two points into 100 equal parts. For this reason, it is sometimes called the **centigrade** scale. The normal temperature for the human body is 37°C. The Celsius scale is the standard for most people outside of the United States and for scientists everywhere.

A **thermometer** measures temperature. The common thermometer is a glass tube that contains a liquid. When the temperature increases, the liquid expands and moves up the tube. When the temperature decreases, the liquid shrinks and moves down the tube.

Here are two formulas for converting from degrees Fahrenheit (°F) to degrees Celsius (°C) and vice versa:

$$F = \frac{9}{5} * C + 32 \text{ and } C = \frac{5}{9} * (F - 32)$$

Example Find the Celsius equivalent of 82°F.

Use the formula $C = \frac{5}{9} * (F - 32)$ and replace F with 82:

$C = \frac{5}{9} * (82 - 32) = \frac{5}{9} * (50) = 27.77$

The Celsius equivalent is about 28°C.

Did You Know?

The Fahrenheit temperature reading and the Celsius temperature reading are numerically the same at only one temperature:

−40°F = −40°C.

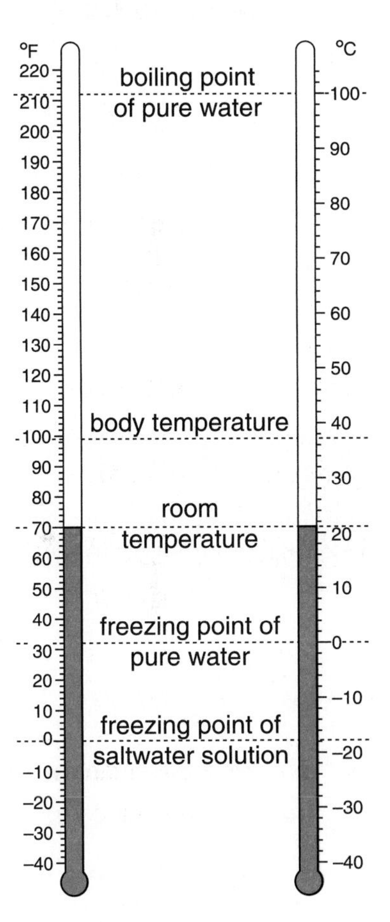

The thermometers show both the Fahrenheit and Celsius scales. Key reference temperatures, such as the boiling and freezing points of water, are indicated. A temperature of 70°F (or about 21°C) is normal room temperature.

Weight

The bicycle weighs 28 pounds 6 ounces

Today, in the United States, two different sets of standard units are used to measure weight.

- The standard unit for weight in the metric system is the **gram.** A small, plastic base-10 cube weighs about 1 gram. Heavier weights are measured in **kilograms.** One kilogram equals 1,000 grams.
- The standard units for weight in the U.S. customary system are the **ounce** and the **pound.** Heavier weights are measured in pounds. One pound equals 16 ounces. Some weights are reported in both pounds and ounces.

Metric Units	U.S. Customary Units
1 gram (g) = 1,000 milligrams (mg)	1 pound (lb) = 16 ounces (oz)
1 milligram = $\frac{1}{1,000}$ gram	1 ounce = $\frac{1}{16}$ pound
1 kilogram (kg) = 1,000 grams	1 ton (t) = 2,000 pounds
1 gram = $\frac{1}{1,000}$ kilogram	1 pound = $\frac{1}{2,000}$ ton
1 metric ton (t) = 1,000 kilograms	
1 kilogram = $\frac{1}{1,000}$ metric ton	

Rules of Thumb	Exact Equivalents
1 ounce equals about 30 grams	1 ounce = 28.35 grams
1 kilogram equals about 2 pounds	1 kilogram = 2.205 pounds

Example A bicycle weighs 14 kilograms. How many pounds is that?

Rough Solution: Use the Rule of Thumb. Since 1 kg equals about 2 lb, 14 kg equals about 14 * 2 lb = 28 lb.

Exact Solution: Use the exact equivalent. Since 1 kg = 2.205 lb, 14 kg = 14 * 2.205 lb = 30.87 lb.

Did You Know?

An adult male African elephant weighs about 6 tons (12,000 lb) and is the heaviest land animal.

Note

The "Rules of Thumb" table shows how units of weight in the metric system relate to units in the U.S. customary system. You can use this table to convert between ounces and grams, and between kilograms and pounds. For most everyday purposes, you need only remember the simple Rules of Thumb.

Check Your Understanding

Solve each problem.

1. A softball weighs 6 ounces. How many grams is that? Use both a Rule of Thumb and an exact equivalent.
2. Andy's brother weighs 25 pounds 12 ounces. How many ounces is that?

Check your answers on page 344.

Measuring and Drawing Angles

Angles are measured in **degrees.** When writing the measure of an angle, a small raised circle (°) is used as a symbol for the word *degree.*

Angles are measured with a tool called a **protractor.** You will find both a full-circle and a half-circle protractor on your Geometry Template. Since there are 360 degrees in a circle, a 1° angle marks off $\frac{1}{360}$ of a circle.

The **full-circle protractor** on the Geometry Template is marked off in 5° intervals from 0° to 360°. It can be used to measure angles, but it *cannot* be used to draw angles.

Sometimes you will use a full-circle protractor that is a paper cutout. This *can* be used to draw angles.

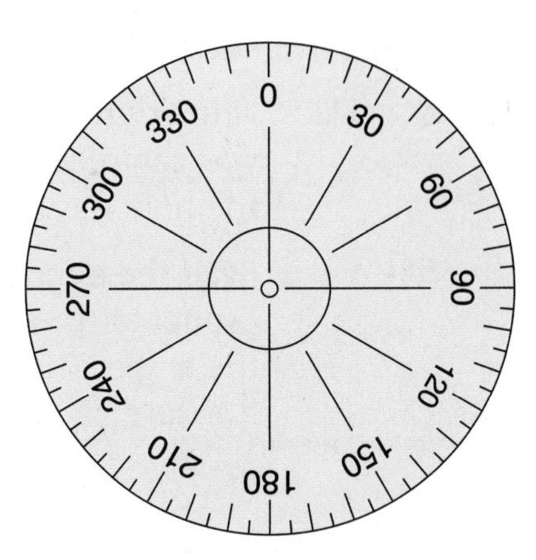

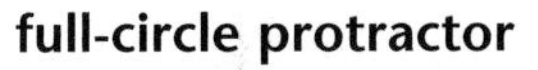
full-circle protractor

The **half-circle protractor** on the Geometry Template is marked off in 1° intervals from 0° to 180°.

It has two scales, each of which starts at 0°. One scale is read clockwise, and the other is read counterclockwise.

The half-circle protractor can be used both to measure *and* to draw angles.

Two rays starting from the same endpoint form two angles. The smaller angle measures between 0° and 180°. The larger angle is called a **reflex angle,** and it measures between 180° and 360°. The sum of the measures of the smaller angle and the reflex angle is 360°.

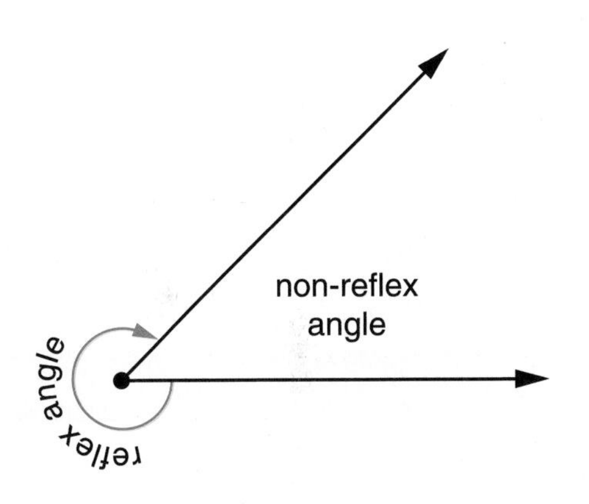

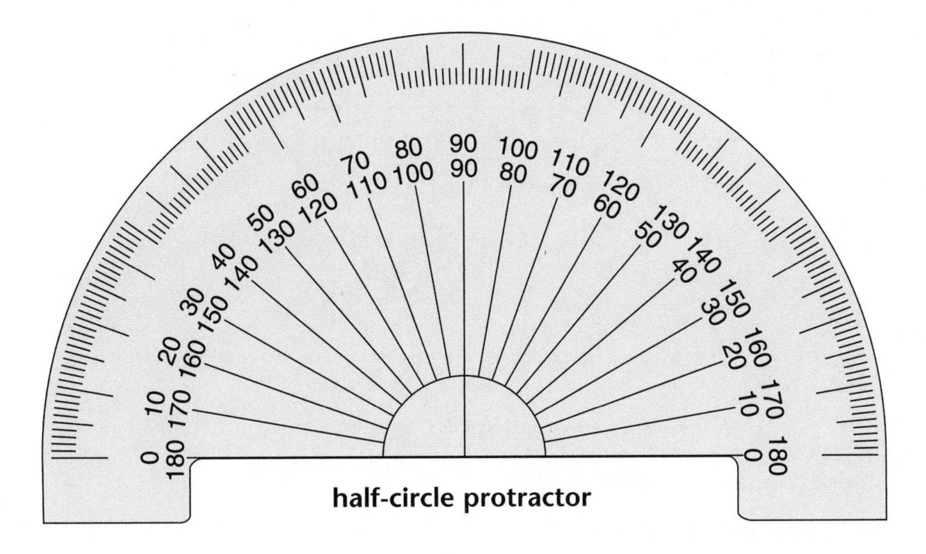

half-circle protractor

Measuring an Angle with a Full-Circle Protractor

Example Use the full-circle protractor to measure angle A.

Step 1: Place the hole in the center of the protractor over the vertex of the angle, point A.

Step 2: Line up the 0° mark with the side of the angle so that you can measure the angle clockwise. Make sure that the hole stays over the vertex.

Step 3: Read the degree measure at the mark on the protractor that lines up with the second side of the angle. This is the measure of the angle. The measure of ∠A is 45°.

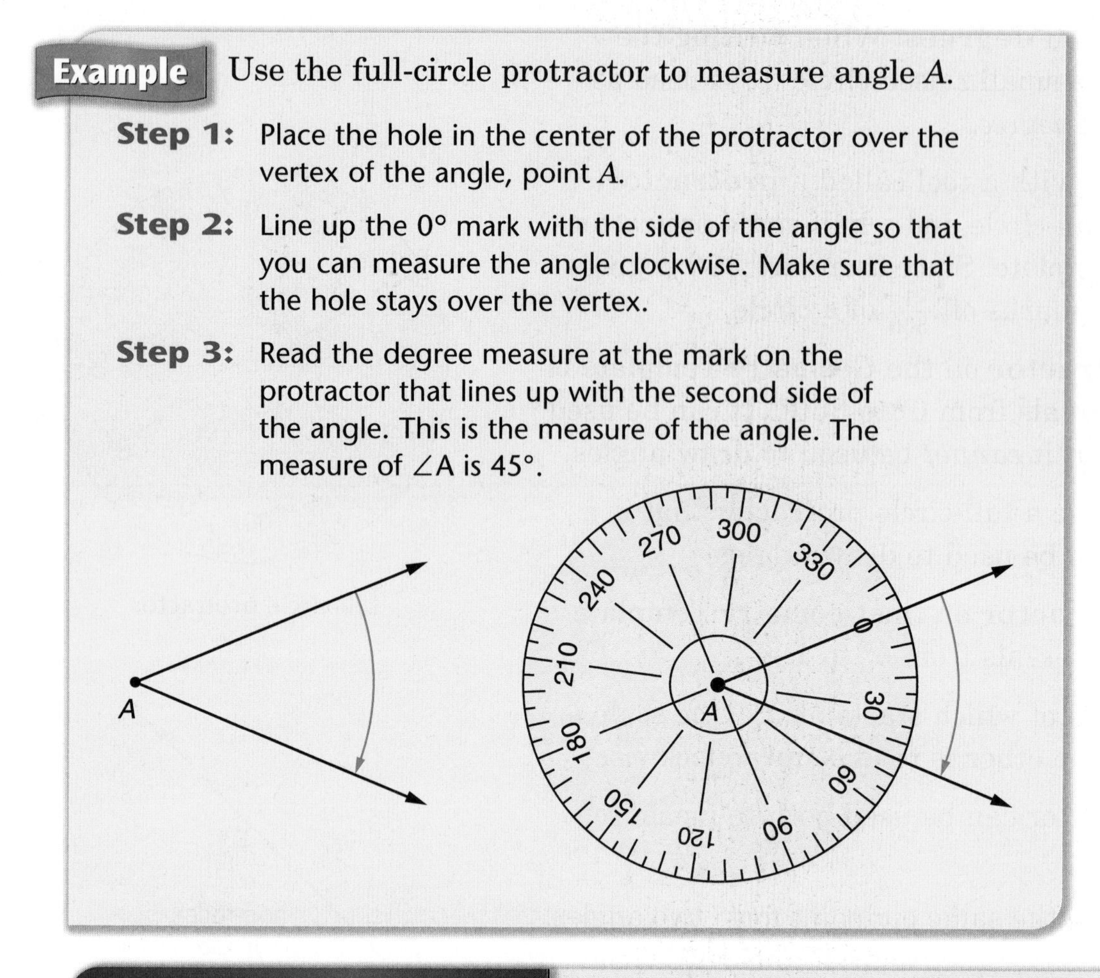

Check Your Understanding

Use your full-circle protractor to measure angles B and C to the nearest degree.

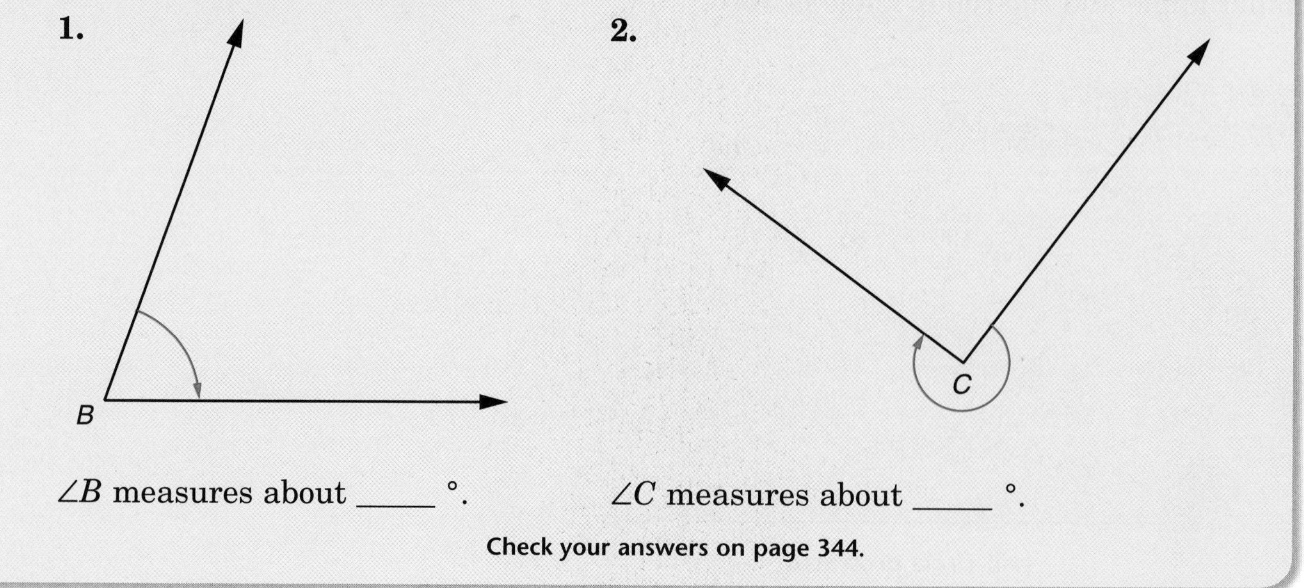

1. ∠B measures about _____ °.

2. ∠C measures about _____ °.

Check your answers on page 344.

Measuring an Angle with a Half-Circle Protractor

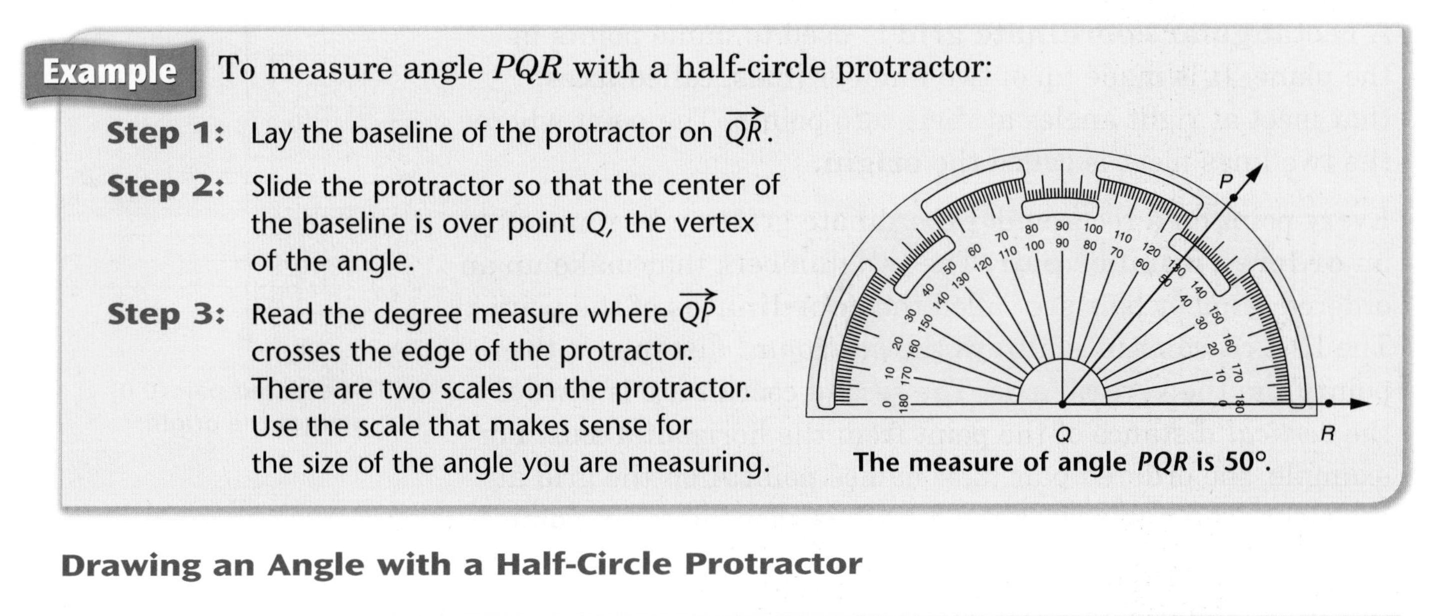

Example To measure angle PQR with a half-circle protractor:

Step 1: Lay the baseline of the protractor on $\overrightarrow{QR}$.

Step 2: Slide the protractor so that the center of the baseline is over point *Q*, the vertex of the angle.

Step 3: Read the degree measure where $\overrightarrow{QP}$ crosses the edge of the protractor. There are two scales on the protractor. Use the scale that makes sense for the size of the angle you are measuring.

The measure of angle *PQR* is 50°.

Drawing an Angle with a Half-Circle Protractor

Example To draw a 40° angle:

Step 1: Draw a ray from point *A*.

Step 2: Lay the baseline of the protractor on the ray.

Step 3: Slide the protractor so that the center of the baseline is over point *A*.

Step 4: Make a mark at 40° near the protractor. There are two scales on the protractor. Use the scale that makes sense for the size of the angle you are drawing.

Step 5: Draw a ray from point *A* through the mark.

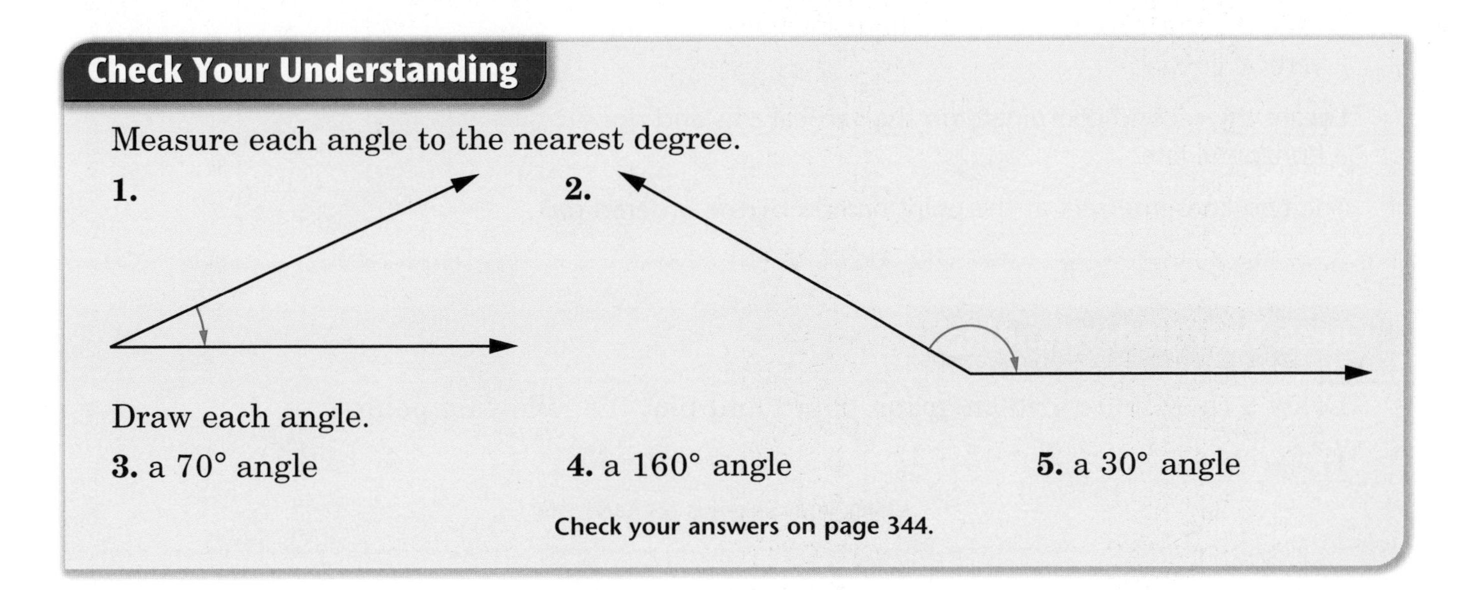

Check Your Understanding

Measure each angle to the nearest degree.

1.

2.

Draw each angle.

3. a 70° angle

4. a 160° angle

5. a 30° angle

Check your answers on page 344.

Plotting Ordered Number Pairs

A **rectangular coordinate grid** is used to name points in the plane. It is made up of two number lines, called **axes,** that meet at right angles at their zero points. The point where the two lines meet is called the **origin.**

Every point on a rectangular coordinate grid can be named by an **ordered number pair.** The two numbers that make up an ordered number pair are called the **coordinates** of the point. The first coordinate is always the *horizontal* distance of the point from the vertical axis. The second coordinate is always the *vertical* distance of the point from the horizontal axis. For example, the ordered pair (3,5) names point A on the grid at the right. The numbers 3 and 5 are the coordinates of point A.

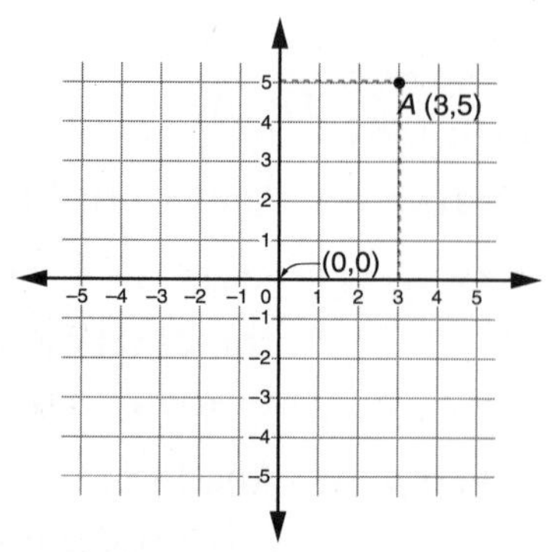

The ordered pair (0,0) names the origin.

Example Plot the ordered pair (5,3).

Step 1: Locate 5 on the horizontal axis. Draw a vertical line.

Step 2: Locate 3 on the vertical axis. Draw a horizontal line.

Step 3: The point (5,3) is located at the intersection of the two lines. The order of the numbers in an ordered pair is important. The ordered pair (5,3) does not name the same point as the ordered pair (3,5).

Example Locate (−2,3), (−4,−1), and $\left(3\frac{1}{2},0\right)$.

For each ordered pair:

Locate the first coordinate on the horizontal axis and draw a vertical line.

Locate the second coordinate on the vertical axis and draw a horizontal line.

The two lines intersect at the point named by the ordered pair.

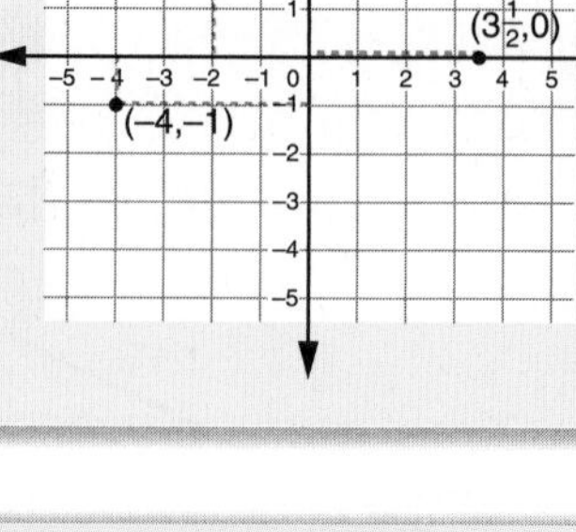

Check Your Understanding

Draw a coordinate grid on graph paper and plot the following points.

1. (2,4) **2.** (−3,−3) **3.** (0,−4) **4.** (−2,1)

Check your answers on page 344.

Map Scales and Distances

Map Scales

Mapmakers show large areas of land and water on small pieces of paper. Places that are actually thousands of miles apart may be only inches apart on a map. When you use a map, you can estimate real distances by using a **map scale.**

Different maps use different scales for distances. On one map, 1 inch may represent 10 miles in the real world, while on another map, 1 inch may represent 100 miles.

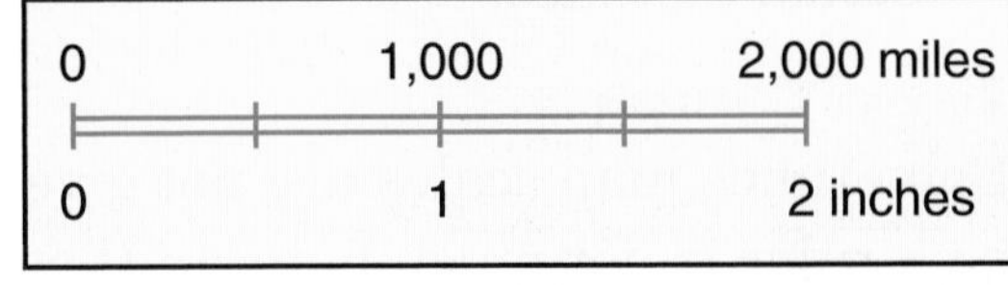

On the map scale shown here, the bar is 2 inches long. Two inches on the map represent 2,000 actual miles. One inch on the map represents 1,000 actual miles.

Sometimes you see a map scale written as "2 inches = 2,000 miles." This statement is not mathematically correct because 2 inches is not equal to 2,000 miles. What is meant is that a 2-inch distance on the map represents 2,000 miles in the real world.

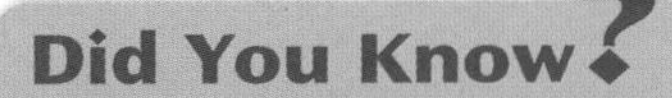

Interstate highway numbers are 1-, 2-, or 3-digit numbers. East-West Interstate highway numbers are even numbers. North-South Interstate highway numbers are odd numbers.

Measuring Distances on a Map

Here are some ways to measure distances on a map.

Use a Ruler

Sometimes the distance you want to measure is along a straight line. Measure the straight-line distance with a ruler, then use the map scale to change the map distance to the actual distance.

Example Use the map and scale below to find the air distance from Denver to Chicago. The air distance is the straight-line distance between the two cities.

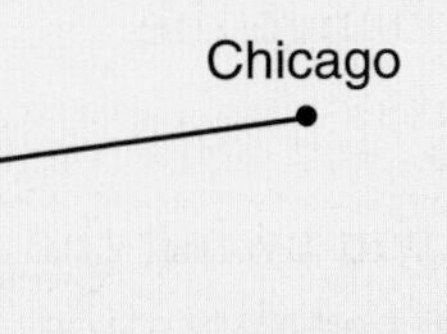

Denver

0 300 600 miles

0 1 2 inches

The line segment connecting Denver and Chicago is 3 inches long. The map scale shows that 1 inch represents 300 miles. So 3 inches must represent 3 * 300 miles, or 900 miles.

The air distance is 900 miles.

Use String and a Ruler

Sometimes you may need to find the length of a curved path such as a road or river. You can use a piece of string, a ruler, and the map scale to find the length.

- Lay a string along the path you want to measure. Mark the beginning and ending points on the string.
- Straighten out the string. Be careful not to stretch it. Use a ruler to measure between the beginning and ending points.
- Use the map scale to change the map distance into the actual distance.

Use a Compass

Sometimes map scales are not given in inches or centimeters, so a ruler is not much help. In these cases you can use a compass to find distances. Using a compass can also be easier than using a ruler, especially if you are measuring a curved path and you do not have string.

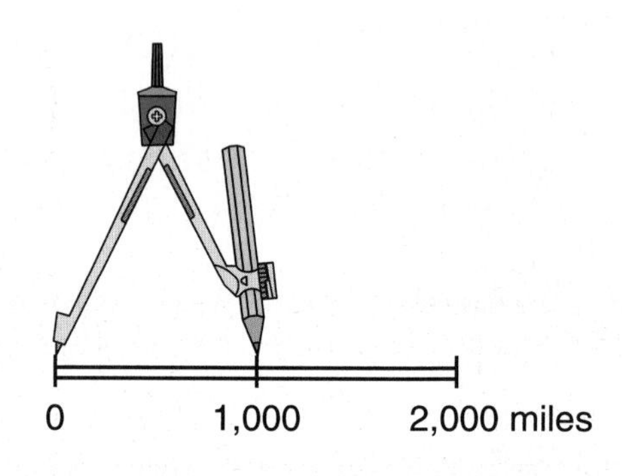

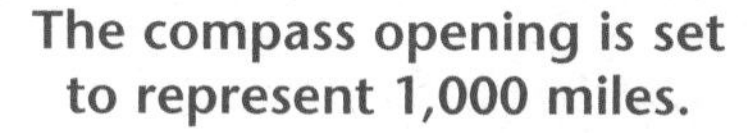

The compass opening is set to represent 1,000 miles.

Step 1: Adjust the compass so that the distance between the anchor point and the pencil point is the same as a distance on the map scale.

Step 2: Imagine a path connecting the starting point and ending point of the distance you want to measure. Put the anchor point of the compass at the starting point. Use the pencil point to make an arc on the path. Move the anchor point to the spot where the arc and the path meet. Continue moving the compass along the path and making arcs until you reach or pass the ending point. Be careful not to change the size of the opening of the compass.

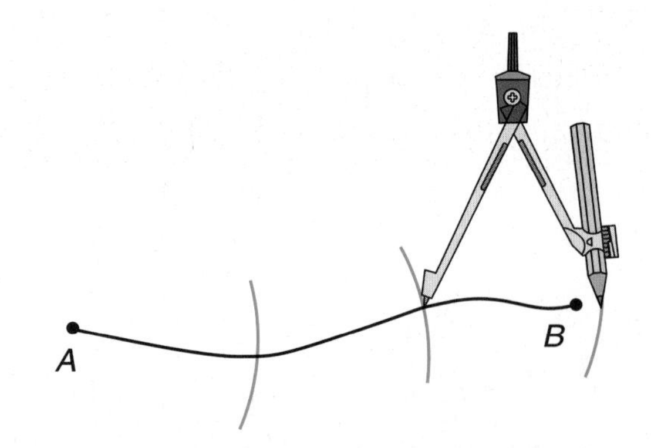

The real length of the curve is about 3,000 miles.

Step 3: Keep track of how many times you swing the compass. Each swing stands for the distance on the map scale. To estimate total distance, multiply the number of swings by the distance each swing stands for.

Algebra

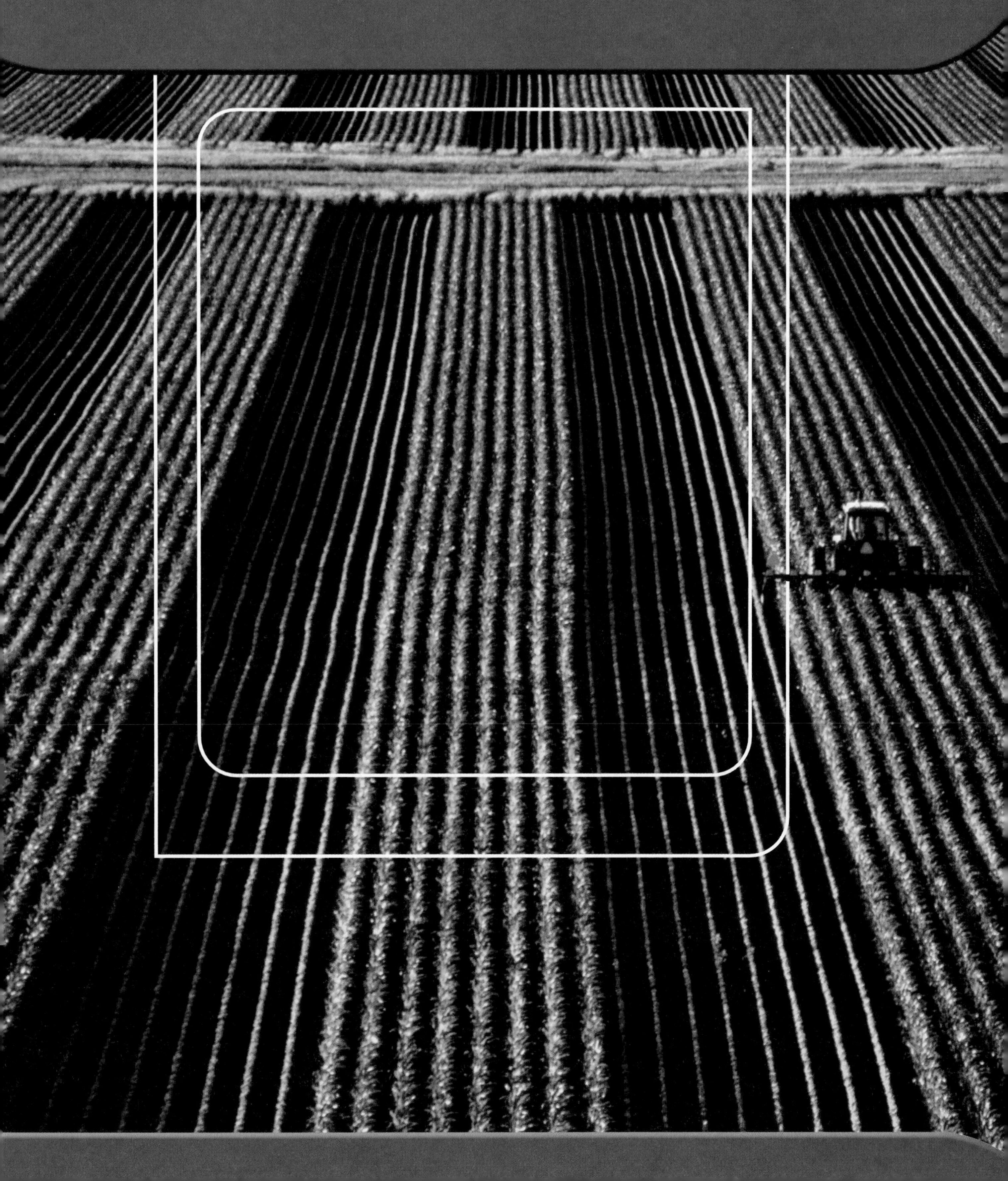

Algebra

Algebra is a type of arithmetic that uses letters (or other symbols such as blanks or question marks) as well as numbers.

Number Sentences

Number sentences are like English sentences, except that they use math symbols instead of words. Using symbols makes the sentences easier to write and work with.

So far, you have seen number sentences that contain **digits, operation symbols,** and **relation symbols.** A number sentence *must* contain numbers and a relation symbol. (It does not have to contain any operation symbols.) A number sentence may be **true** or **false.** For example, the number sentence $14 + 3 = 17$ is true; the number sentence $12 = 9$ is false.

Digits are 0, 1, 2, 3, 4, 5, 6, 7, 8, and 9.

Relation symbols are $=, \neq, <$, and $>$.

Operation symbols are +, −, * or ×, and ÷ or /.

Open Sentences

In some number sentences, one or more numbers may be missing. In place of each missing number there is a letter, a blank, a question mark, or some other symbol. These number sentences are called **open sentences.** A symbol that stands for a missing number is called a **variable.** For most open sentences, you can't tell whether the sentence is true or false until you know which number replaces the variable. For example, $9 + x = 15$ is an open sentence in which x stands for some number.

- If you replace x with 10 in the number sentence $9 + x = 15$, you get the number sentence $9 + 10 = 15$, which is false.
- If you replace x with 6 in the open sentence $9 + x = 15$, you get the number sentence $9 + 6 = 15$, which is true.

If a number used in place of a variable makes the number sentence true, the number is a **solution** of the open sentence. The number 6 is a solution of the open sentence $9 + x = 15$, because $9 + 6 = 15$ is a true number sentence.

Did You Know?

The first use in print of + and − symbols for the operations of addition and subtraction can be traced to a book by Johannes Widmann in 1489. At first, the symbols + and − were used only in algebra. It was not until 1630 that the symbols were used generally as they are today.

Note

Finding a solution for an open number sentence is called **solving the number sentence.**

Check Your Understanding

True or false?

1. $4 + 5 = 9$ **2.** $4 + 8 > 10$ **3.** $7 + 2 < 6 + 1$

Find a solution

4. $5 + y = 12$ **5.** $48 = z * 6$ **6.** $10 - w = 4$

Check your answers on page 344.

Relations

A **relation** tells how two things compare. The table at the right shows common relations that compare numbers and their symbols.

Symbol	Relation
$=$	is equal to
$\neq$	is not equal to
$<$	is less than
$>$	is greater than

Equations

A number has many different names. For example, $3 * 4$ is equal to 12, and $1.2 * 10$ is also equal to 12. So 12 and $3 * 4$ and $1.2 * 10$ are just different ways to express the same number. We say that 12 and $3 * 4$ and $1.2 * 10$ are **equivalent names** because they all name the same number.

Everyday Mathematics uses **name-collection boxes** as a simple way to show equality.

- The label at the top of the box shows a number.
- The names written inside the box are equivalent names for the number on the label. Each of these names is equal to 12.

12
$1.2 * 10$
$12 - 0$
$3 * 4$
$\frac{1}{2} * 24$
$6 \div \frac{1}{2}$

a name-collection box for 12

Another way to state that two things are equal is to write a number sentence using the $=$ symbol. Number sentences containing the $=$ symbol are called **equations.**

Examples Here are some equations:

$4 + 5 = 9$ $\quad$ $12 - 6 = 8$ $\quad$ $5 = 5$

Note

An equation that does not contain a variable is either true or false. Can you find the false equation in the examples at the left?

Inequalities

Number sentences that do not contain an $=$ symbol are called **inequalities.** Common relations in inequalities are $>$ (is greater than), $<$ (is less than), and $\neq$ (is not equal to).

Examples Here are some inequalities. The middle inequality is false.

$5 + 6 < 15$ $\quad$ $25 > 12 * 3$ $\quad$ $36 \neq 7 * 6$

Check Your Understanding

Compare. Use $=$, $<$, or $>$ to make each number sentence true.

1. $100 \square 55 + 45$
2. $\frac{1}{2} \square 0.5$
3. $\frac{3}{4} \square \frac{1}{4}$
4. $3 * 50 \square 200$
5. $\frac{1}{2} * 100 \square 50$
6. $4.3 \square 4.15$

Check your answers on page 344.

Parentheses

The meaning of a number sentence is not always clear. Here is an example: $15 - 3 * 2 = n$. Should you subtract first? Or should you multiply first? Parentheses tell you which operation to do first.

Example Solve. $(15 - 3) * 2 = n$

The parentheses tell you to subtract 15 − 3 first.	(15 − 3) ∗ 2
Then multiply by 2.	12 ∗ 2
The answer is 24. (15 − 3) ∗ 2 = 24	24

So $n = 24$.

Example Solve. $15 - (3 * 2) = n$

The parentheses tell you to multiply 3 ∗ 2 first.	15 − (3 ∗ 2)
Then subtract the result.	15 − 6
The answer is 9. 15 − (3 ∗ 2) = 9	9

So $n = 9$.

Example Make this number sentence true by inserting parentheses: $18 = 6 + 3 * 4$

There are two ways to insert parentheses, but only one will result in a true sentence.

18 = (6 + 3) ∗ 4	18 = 6 + (3 ∗ 4)
18 = 9 ∗ 4	18 = 6 + 12
18 = 36	18 = 18
18 is not equal to 36, so this statement is false.	This statement is true.

The correct way to insert parentheses is $18 = 6 + (3 * 4)$.

Check Your Understanding

Solve.

1. $(3 * 2) + 20 = x$
2. $n = (4 - 4) / 6$
3. ___ $= (5 + 5) * (50 + 50)$

Insert parentheses to make these number sentences true.

4. $5 - 1 + 4 = 0$
5. $1 = 4 * 5 / 4 * 5$
6. $10 = 3 + 2 * 2$

Check your answers on page 344.

Order of Operations

In many situations, the order in which things are done is important. For example, you always put on your socks before your shoes.

In mathematics, too, certain things must be done in a certain order. For example, to solve $8 + 4 * 3 =$ ___, you must know whether to add first or to multiply first. In arithmetic and algebra, there are rules that tell you what to do first and what to do next.

Rules for the Order of Operations

1. Do operations inside **parentheses** first.
   Follow rules 2–4 when you are computing inside parentheses.
2. Calculate all expressions with **exponents.**
3. **Multiply** and **divide** in order, from left to right.
4. **Add** and **subtract** in order, from left to right.

Did You Know?

The largest known human foot fit into a shoe that was about 18.5 inches long.

Some people find it is easier to remember the order of operations by memorizing this sentence:

Please **E**xcuse **M**y **D**ear **A**unt **S**ally.

Parentheses **E**xponents **M**ultiplication **D**ivision **A**ddition **S**ubtraction

Example Solve. $15 - 3 * 2 = ?$

Multiply first.	$15 - 3 * 2$
Then subtract.	$15 - 6$
The answer is 9.	9

$15 - 3 * 2 = 9$

Example Solve. $18 - (5 - 3 + 8) / 2 = ?$

Parentheses first.	$18 - (5 - 3 + 8) / 2$
Then divide.	$18 - 10 / 2$
Then subtract.	$18 - 5$
The answer is 13.	13

$18 - (5 - 3 + 8) / 2 = 13$

Check Your Understanding

Solve.

1. $14 - 4 / 2 + 2 = ?$

2. $5 * 7 / (5 + 2) = ?$

3. $6 + 3 * 4 - 20 / 5 = ?$

4. $9 + 9 * 5 = ?$

Check your answers on page 344.

Mathematical Models

A good way to learn about something is to work with a model of it. For example, a computer model of a city can help you understand how real cities grow and change. A scale model of the human body can help you understand how the different systems in your own body work together.

Models are important in mathematics too. A mathematical model can be as simple as acting out a problem with chips or blocks. Other mathematical models use drawings or symbols. Mathematical models can help you understand and solve problems.

Situation Diagrams

Examples Here are some examples of how you can use diagrams to model simple problems.

Problem	Diagram
Parts-and-Total Situation Samantha's class has 15 girls and 9 boys. How many students are there in all?	Total: ? Part: 15 | Part: 9
Change Situation Brian had $20 and spent $12.89 on a CD. How much money did he have left?	Start: $20 → Change: −$12.89 → End: ?
Comparison Situation The average summer high temperature in Cairo, Egypt, is 95°F. The average summer high temperature in Reykjavik, Iceland, is 56°F. How much warmer is it in Cairo?	Quantity: 95°F Quantity: 56°F | Difference: ?

These diagrams work well for many simple problems. But for harder problems you need to use more powerful tools such as graphs, tables, and number models.

Number Models

Number models provide another way to model situations. A **number model** is a number sentence (or part of a number sentence) that describes some situation. Often, two or more number models can fit a given situation.

Problem	Number Models
Samantha's class has 15 girls and 9 boys. How many students are there in all?	$15 + 9 = n$ (number sentence) or $15 + 9$ (part of a number sentence)
Brian had \$20 and spent \$12.89 on a CD. How much money did he have left?	$r = \$20 - \$12.89$ or $\$20 = \$12.89 + r$ (number sentences) or $\$20 - \$12.89$ (part of a number sentence)
The average summer high temperature in Cairo, Egypt, is 95°F. The average summer high temperature in Reykjavik, Iceland, is 56°F. How much warmer is it in Cairo?	$d = 95°F - 56°F$ or $95°F = 56°F + d$ (number sentences) or $95°F - 56°F$ (part of a number sentence)

Number models can help you solve problems. For example, the number sentence $\$20 = \$12.89 + r$ suggests counting up to find Brian's change from buying a \$12.89 CD with a \$20 bill.

Number models can also help you show the answer after you have solved the problem: $\$20 = \$12.89 + \$7.11$.

Did You Know?

In 1916, Browning, Montana, experienced the greatest recorded change in temperature in a single day. The temperature fell from 44°F to −56°F!

Check Your Understanding

Draw a diagram and write a number model for each problem. Then solve each problem.

1. Ella ran 4.5 miles on Monday, 7 miles on Tuesday, 5 miles on Wednesday, and 6 miles on Friday. How many miles did she run in all?
2. The Eagles scored 17 points in the first half. By the end of the game, they had scored 31 points. How many points did they score in the second half?
3. Maurice had \$37 and Monica had \$49. How much more money did Monica have?

Check your answers on page 344.

Variables and Unknowns

In mathematics, letters, blanks, question marks, or other symbols are often used in place of missing information. These symbols are called **variables.**

Variables are used in several different ways:

Variables Can Be Used to Stand for Unknown Numbers
For example, in the number sentence $5 + n = 8$, the variable n stands for an unknown number. To make the sentence true, the correct number has to be found for n. Finding the correct number is called "solving the number sentence."

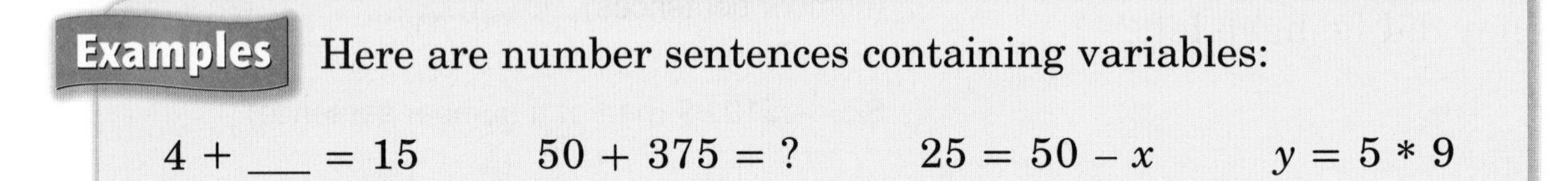

Examples Here are number sentences containing variables:

$4 + \_\_\_ = 15$ $50 + 375 = ?$ $25 = 50 - x$ $y = 5 * 9$

Variables Can Be Used to State Properties of the Number System
Properties of the number system are things that are true for all numbers. For example, any number multiplied by 1 is equal to itself. Variables are often used in statements that describe properties.

Examples

Property	Number sentence examples of the property
$a + b = b + a$	$50 + 80 = 80 + 50$ $6 + 3 = 3 + 6$
$a * b = b * a$	$6 * 2 = 2 * 6$ $30 * 4 = 4 * 30$
$1 * a = a$	$1 * 61 = 61$ $1 * 2.5 = 2.5$
$a = a$	$24 = 24$ $\frac{3}{4} = \frac{3}{4}$
$0 + a = a$	$0 + 8 = 8$ $0 + 1.3 = 1.3$

Variables Can Be Used in Formulas

Formulas are used in everyday life, in science, in business, and in many other situations as an easy way to describe relationships. The formula for the area of a rectangle, for example, is $A = b * h$, where A is the area, b is the length of the base, and h is the height.

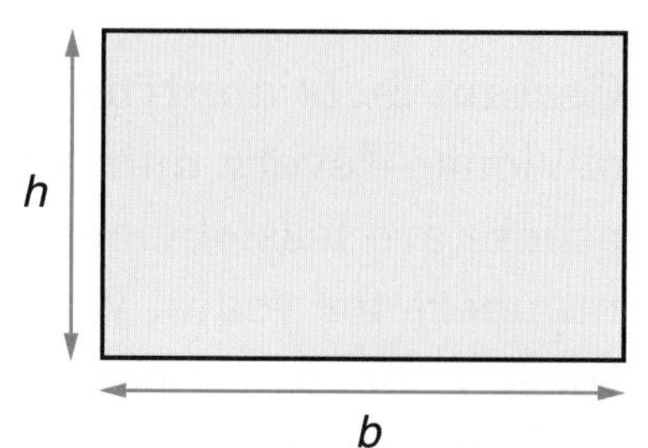

Variables Can Be Used to Express Rules or Functions

Function machines and "What's My Rule?" tables have rules that tell you how to get the "out" numbers from the "in" numbers. For example, a doubling machine might have the rule, "Double the 'in' number." This rule can be written as $y = 2 * x$ by using variables.

Rule

$y = 2 * x$

in	out
x	y
0	0
1	2
2	4
3	6
...	...

Note

The formula $A = b * h$ can also be written without a multiplication symbol: $A = bh$. Putting variables next to each other means they are to be multiplied.

In higher mathematics, in which there are many variables, multiplication symbols are normally left out. In *Everyday Mathematics,* however, the multiplication symbol is usually shown because this makes expressions easier to understand.

Check Your Understanding

For each problem, write a number sentence using a letter for the unknown number.

1. Some number plus 5 equals 8.

2. Some number times 3 equals 15.

Find the area of the parallelograms below.
Use the formula $A = b * h$.

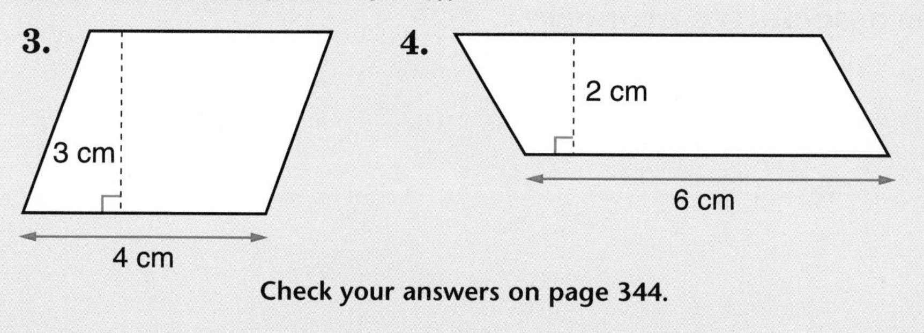

Check your answers on page 344.

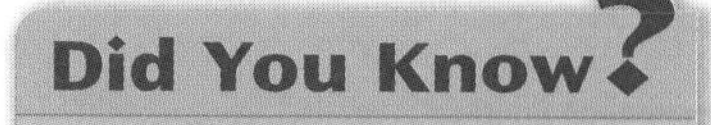

If you know your weight, you can use this formula to estimate the volume of blood in your body:

$B = 0.036 * W$.

W = weight (in pounds), and B = liters of blood.

For example, if you weigh 80 pounds, $B = 0.036 * 80 =$ 2.88 liters.

Some Properties of Arithmetic

Certain facts are true of all numbers. Some of them are obvious—"every number equals itself," for example—but others are less obvious. Since you have been working with numbers for years, you probably already know most of these facts, or properties. But you probably don't know their mathematical names.

The Identity Properties

The sum of any number and 0 is that number. For example, $15 + 0 = 15$. The **identity for addition** is 0. $a + 0 = a$ $0 + a = a$

The product of any number and 1 is that number. For example, $75 * 1 = 75$. The **identity for multiplication** is 1. $a * 1 = a$ $1 * a = a$

The Commutative Properties

When two numbers are added, the order of the numbers makes no difference. For example, $8 + 5 = 5 + 8$. This is known as the **commutative property of addition.** $a + b = b + a$

When two numbers are multiplied, the order of the numbers makes no difference. For example, $7 * 2 = 2 * 7$. This is known as the **commutative property of multiplication.** $a * b = b * a$

The Associative Properties

When three numbers are added, it makes no difference which two are added first. For example, $(3 + 4) + 5 = 3 + (4 + 5)$. This is known as the **associative property of addition.** $(a + b) + c = a + (b + c)$

When three numbers are multiplied, it makes no difference which two are multiplied first. For example, $(3 * 4) * 5 = 3 * (4 * 5)$. This is known as the **associative property of multiplication.** $(a * b) * c = a * (b * c)$

The Distributive Property

When you play *Multiplication Wrestling* or multiply with the partial-products method, you use the **distributive property:** $a * (b + c) = (a * b) + (a * c)$

For example, when you solve 5 ∗ 28 with partial products, you think of 28 as 20 + 8 and multiply each part by 5.

The distributive property says: $5 * (20 + 8) = (5 * 20) + (5 * 8)$.

		28
		∗ 5
5 ∗ 20	=	100
5 ∗ 8	=	40
5 ∗ 28	=	140

Example Show how the distributive property works by finding the area of Rectangle A in two different ways.

Method 1: To find the area of Rectangle A, you could find the total width (3 + 4) and multiply that by the height.

$$5 * (3 + 4) = 5 * 7$$
$$= 35$$

Method 2: Another way to find the area of Rectangle A would be to first find the area of each smaller rectangle, and then add these areas.

$$(5 * 3) + (5 * 4) = 15 + 20$$
$$= 35$$

A

5

3 4

Both methods show that the area of Rectangle A is 35 square units.

$5 * (3 + 4) = (5 * 3) + (5 * 4)$

The distributive property works with subtraction too. $a * (b - c) = (a * b) - (a * c)$

Example Find the area of the shaded rectangle in two different ways.

$5 * (7 - 4) = 5 * 3 = 15$, or

$(5 * 7) - (5 * 4) = 35 - 20 = 15$

Both methods show that the area of the shaded rectangle is 15 square units.

$5 * (7 - 4) = (5 * 7) - (5 * 4)$

5

4

7

Check Your Understanding

Use the distributive property to fill in the blanks.

1. 7 ∗ (10 + 8) = (7 ∗ ___) + (7 ∗ ___)

2. (6 ∗ 13) + (6 ∗ 16) = 6 ∗ (___ + ___)

3. 4 ∗ (___ – ___) = (4 ∗ 15) – (4 ∗ 8)

4. (___ ∗ 2) + (___ ∗ 4) = 8 ∗ (2 + 4)

Check your answers on page 345.

Number Patterns

You can use dot pictures to explore number patterns.

Even Numbers

Even numbers are counting numbers that have a remainder of 0 when they are divided by 2. An even number has a dot picture with 2 equal rows.

2 4 6 8 10 12

> **Note**
>
> All of the dot pictures shown here are for counting numbers. The *counting numbers* are 1, 2, 3, and so on.

Odd Numbers

Odd numbers are counting numbers that have a remainder of 1 when they are divided by 2. An odd number has a dot picture with 2 equal rows plus 1 extra dot.

1 3 5 7 9 11

> **Note**
>
> The number 0 and the opposites of even numbers (−2, −4, −6, and so on) are usually said to be even. The opposites of odd numbers (−1, −3, −5, and so on) are usually said to be odd.

Triangular Numbers

Each dot picture below has a triangular shape with the same number of dots on each side. Each row has 1 more dot than the row above it. Any number that has a dot picture like one of these is called a **triangular number.**

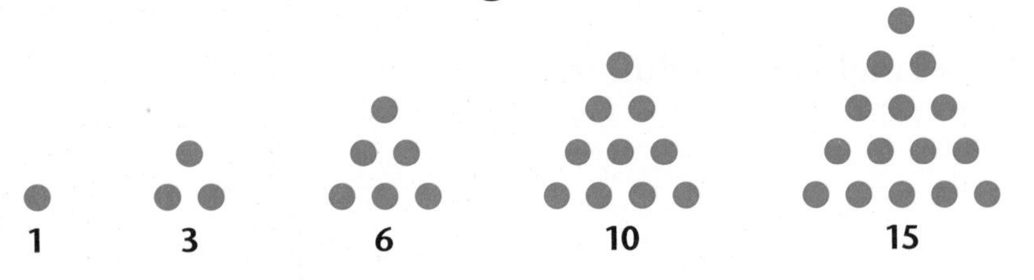

Square Numbers

A **square number** is the product of a counting number multiplied by itself. For example, 16 is a square number because 16 equals $4 * 4$, or 4^2. A square number has a dot picture with a square shape and the same number of dots in each row and column.

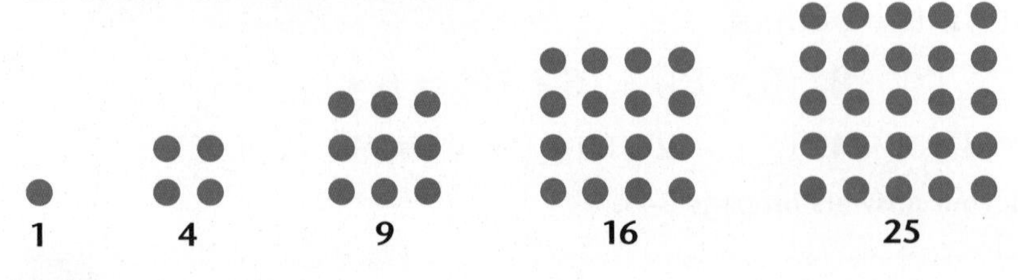

> **Did You Know?**
>
> The early Greeks used pebbles to represent square numbers. They represented triangular numbers by removing pebbles above the diagonals of square numbers.

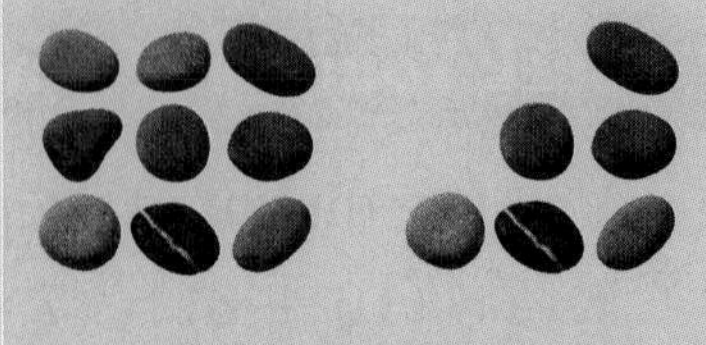

Rectangular Numbers

A **rectangular number** is a counting number that is the product of two smaller counting numbers. For example, 12 is a rectangular number because 12 = 3 * 4. A rectangular number has a dot picture with a rectangular shape, with at least 2 rows and at least 2 columns.

4 6 8 9 10 12

Did You Know?

If you look carefully, you can find 16 rectangular shapes with different dimensions (length and width) in the design of the United States flag.

Prime Numbers

A **prime number** is a counting number greater than 1 that is *not* equal to the product of two smaller counting numbers. So a prime number cannot be a rectangular number. This means that a prime number cannot be fit into a rectangular shape (with at least 2 rows and at least 2 columns).

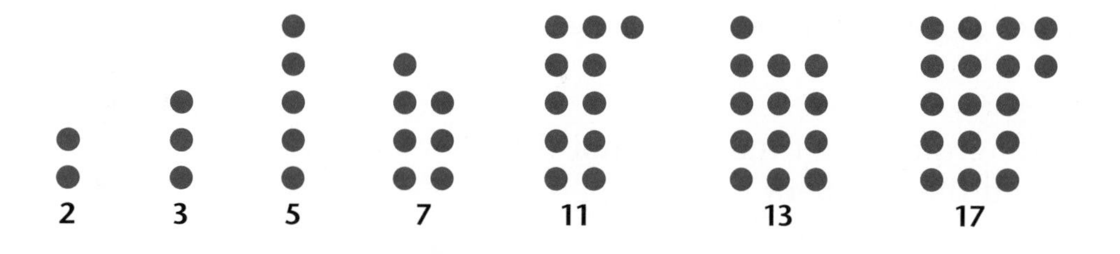

A Baker's Dozen is a group of 13. In the past, bakers would add an extra roll as a safeguard against the possibility of 12 rolls weighing too little.

Check Your Understanding

Draw a dot picture for each number and tell what kind of number it is. There may be more than one correct answer.

1. 12 **2.** 36 **3.** 11 **4.** 21

Is the number even or odd? Draw a dot picture to explain your reasoning.

5. 16 **6.** 19 **7.** 24 **8.** 23

9. List all the square numbers that are less than 100.

Check your answers on page 345.

Frames-and-Arrows Diagrams: Number Sequences

A **Frames-and-Arrows diagram** is one way to show a number pattern. There are three parts to a Frames-and-Arrows diagram:

- a set of **frames** that contains numbers
- **arrows** that show the path from one frame to the next
- a box with an arrow below it. The box has a **rule** inside. The rule tells how to get from one frame to the next.

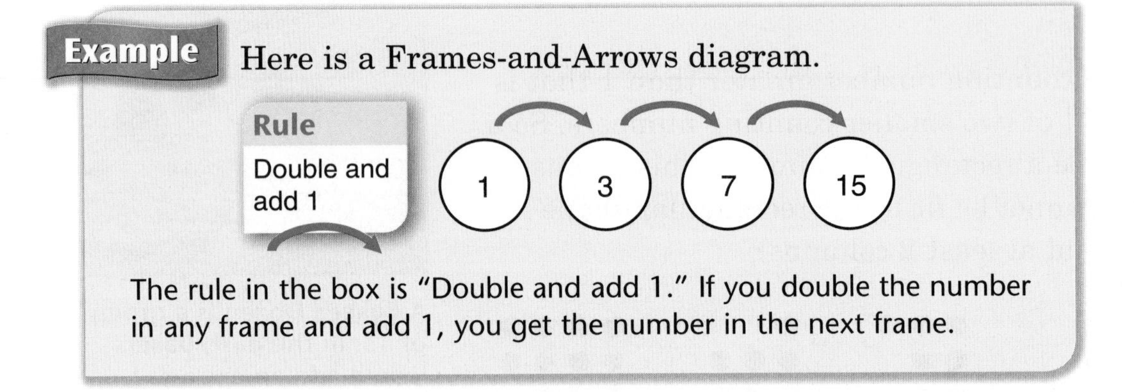

The rule in the box is "Double and add 1." If you double the number in any frame and add 1, you get the number in the next frame.

In some Frames-and-Arrows diagrams, the frames are not filled in. You must use the rule to find the numbers.

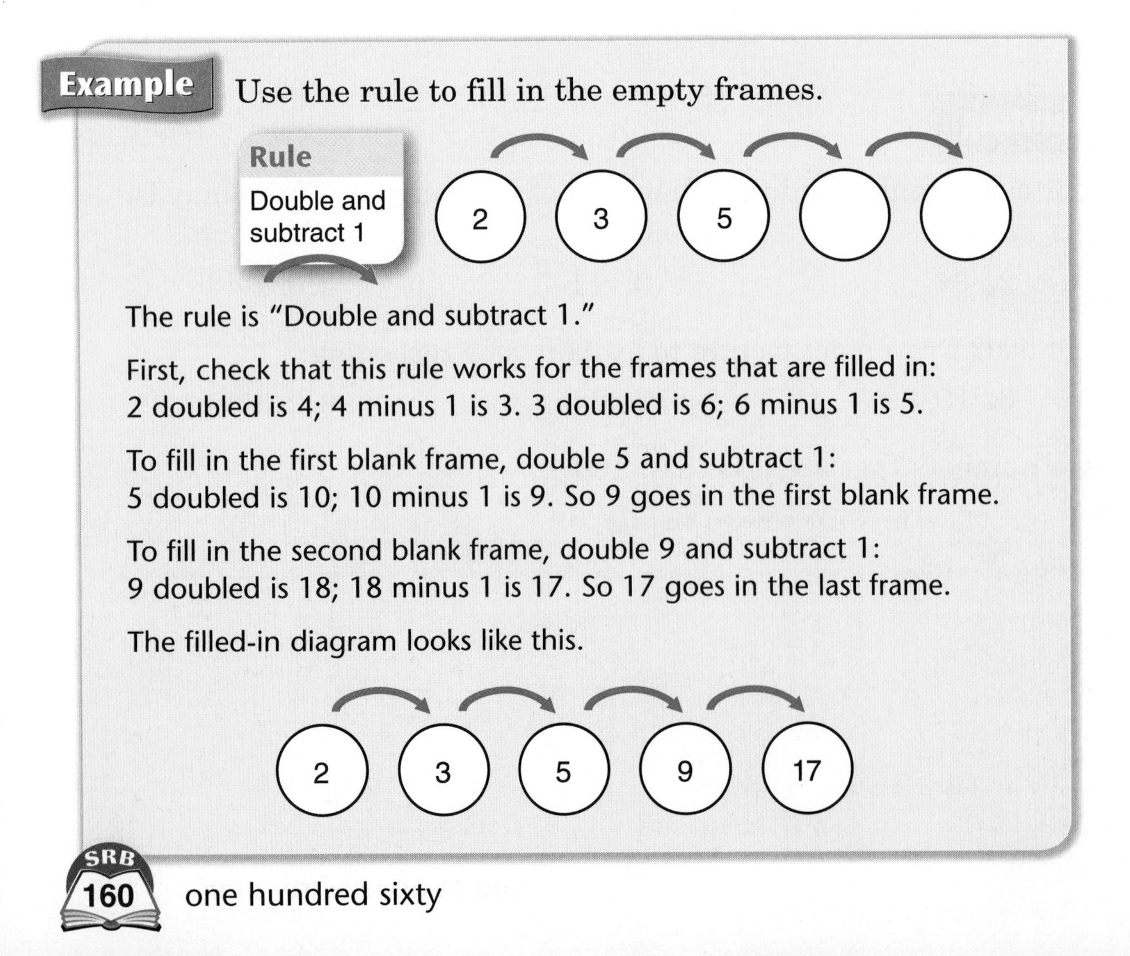

The rule is "Double and subtract 1."

First, check that this rule works for the frames that are filled in: 2 doubled is 4; 4 minus 1 is 3. 3 doubled is 6; 6 minus 1 is 5.

To fill in the first blank frame, double 5 and subtract 1: 5 doubled is 10; 10 minus 1 is 9. So 9 goes in the first blank frame.

To fill in the second blank frame, double 9 and subtract 1: 9 doubled is 18; 18 minus 1 is 17. So 17 goes in the last frame.

The filled-in diagram looks like this.

2 → 3 → 5 → 9 → 17

You can use the numbers in the frames to find the rule.

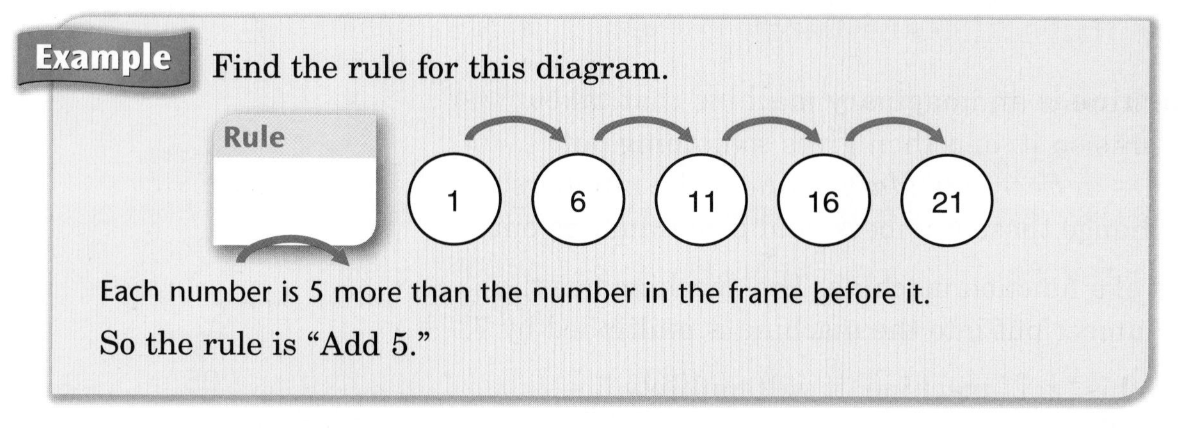

Find the rule for this diagram.

Each number is 5 more than the number in the frame before it.

So the rule is "Add 5."

Two-Rule Frames-and-Arrows Diagrams

A Frames-and-Arrows diagram can have two different rules. Each rule uses a different color arrow.

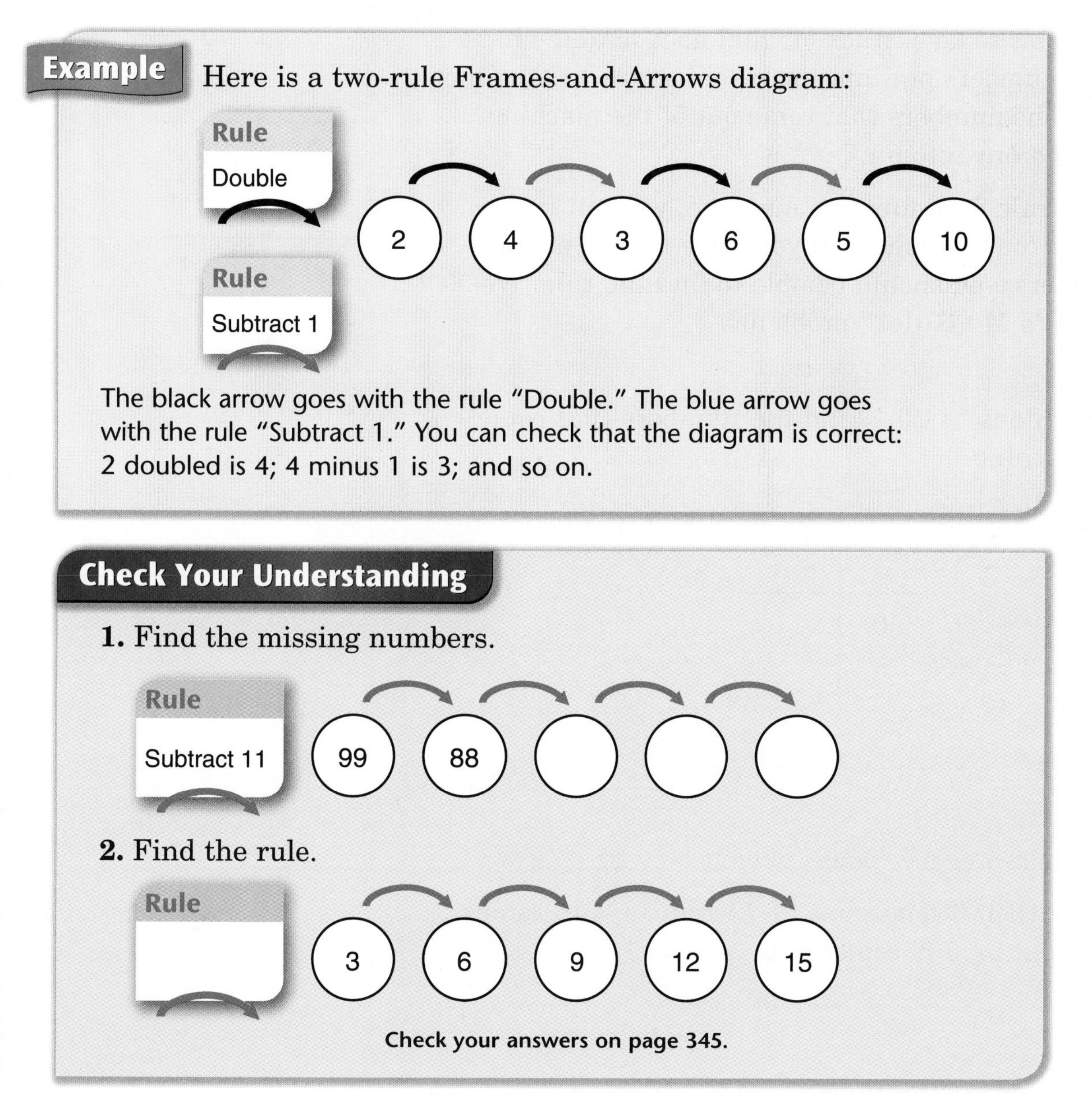

Here is a two-rule Frames-and-Arrows diagram:

The black arrow goes with the rule "Double." The blue arrow goes with the rule "Subtract 1." You can check that the diagram is correct: 2 doubled is 4; 4 minus 1 is 3; and so on.

Check Your Understanding

1. Find the missing numbers.

2. Find the rule.

Check your answers on page 345.

Function Machines and "What's My Rule?" Problems

A **function machine** is an imaginary machine that takes something in, works on it, and then gives something out. Function machines in *Everyday Mathematics* take numbers in, use rules to change those numbers, and give numbers out.

Here is a picture of a function machine. This machine has the rule "* 7." Any number put into the machine is multiplied by 7.

If you put 5 into this "* 7" machine, it will multiply 5 * 7. The number 35 will come out.

If you put 60 into the machine, it will multiply 60 * 7. The number 420 will come out.

in	out
0	0
1	7
2	14
3	21
n	*n* * 7

You can use a table to keep track of what goes in and what comes out. The numbers put into the machine are written in the **in** column. The numbers that come out of the machine are written in the **out** column.

If you know the rule for a function machine, you can make a table of "in" and "out" numbers. If you have a table of "in" and "out" numbers, you should be able to find the rule. We call these **"What's My Rule?"** problems.

Example The rule is "* 20." Find the numbers that come out of the machine.

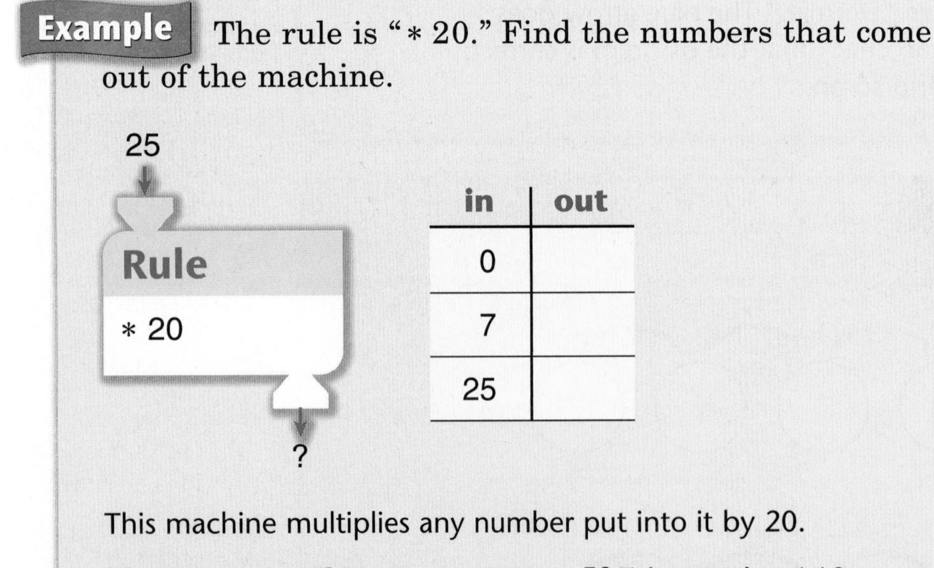

in	out
0	
7	
25	

This machine multiplies any number put into it by 20.

If 0 is put in, then 0 comes out. If 7 is put in, 140 comes out. If 25 is put in, 500 comes out.

Example The rule and the "in" numbers are given. Find the "out" numbers.

9 → **Rule**: Subtract 7 → ?

in	out	
9	2	[9 – 7 = 2]
27	20	[27 – 7 = 20]
0	–7	[0 – 7 = –7]
–5	–12	[–5 – 7 = –12]

The solutions (the missing information) appear in color.

If you know the rule and the "out" numbers, then you should be able to find the "in" numbers.

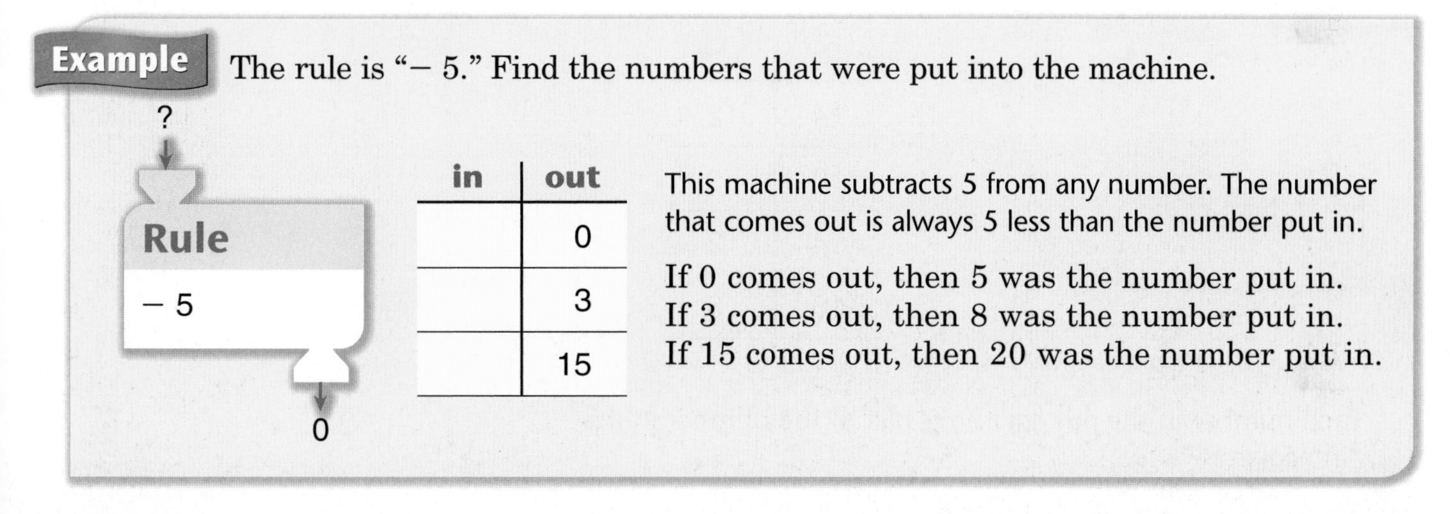

Example The rule is "– 5." Find the numbers that were put into the machine.

? → **Rule**: – 5 → 0

in	out
	0
	3
	15

This machine subtracts 5 from any number. The number that comes out is always 5 less than the number put in.

If 0 comes out, then 5 was the number put in.
If 3 comes out, then 8 was the number put in.
If 15 comes out, then 20 was the number put in.

Example The rule and the "out" numbers are given. Find the "in" numbers.

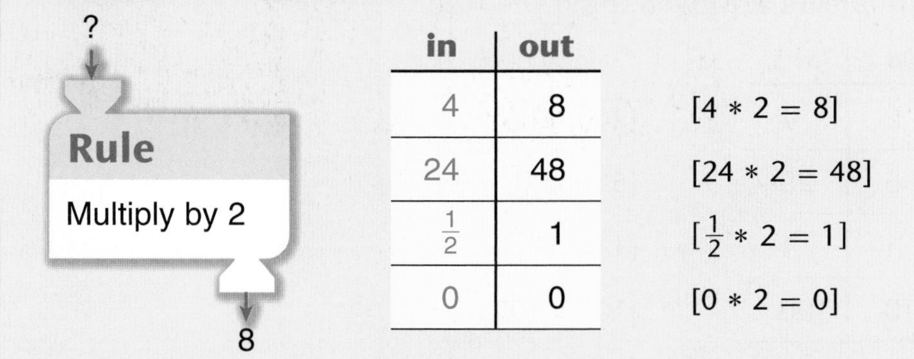

? → **Rule**: Multiply by 2 → 8

in	out	
4	8	[4 * 2 = 8]
24	48	[24 * 2 = 48]
$\frac{1}{2}$	1	[$\frac{1}{2}$ * 2 = 1]
0	0	[0 * 2 = 0]

The solutions (the missing information) appear in color.

Did You Know?

The ENIAC (Electronic Numerical Integrator and Computer) was the first electronic computer. It was built in 1946 to do calculations for the U.S. Army and weighed more than 30 tons.

Example The rule and some “in” and “out” numbers are given. Find the missing numbers.

4 → **Rule** Add 5 → ?

in	out	
4	9	[4 + 5 = 9]
7	12	[7 + 5 = 12]
53	58	[53 + 5 = 58]
−6	−1	[−6 + 5 = −1]

The solutions (the missing information) appear in color.

If you have a table of “in” and “out” numbers, then you should be able to find the rule.

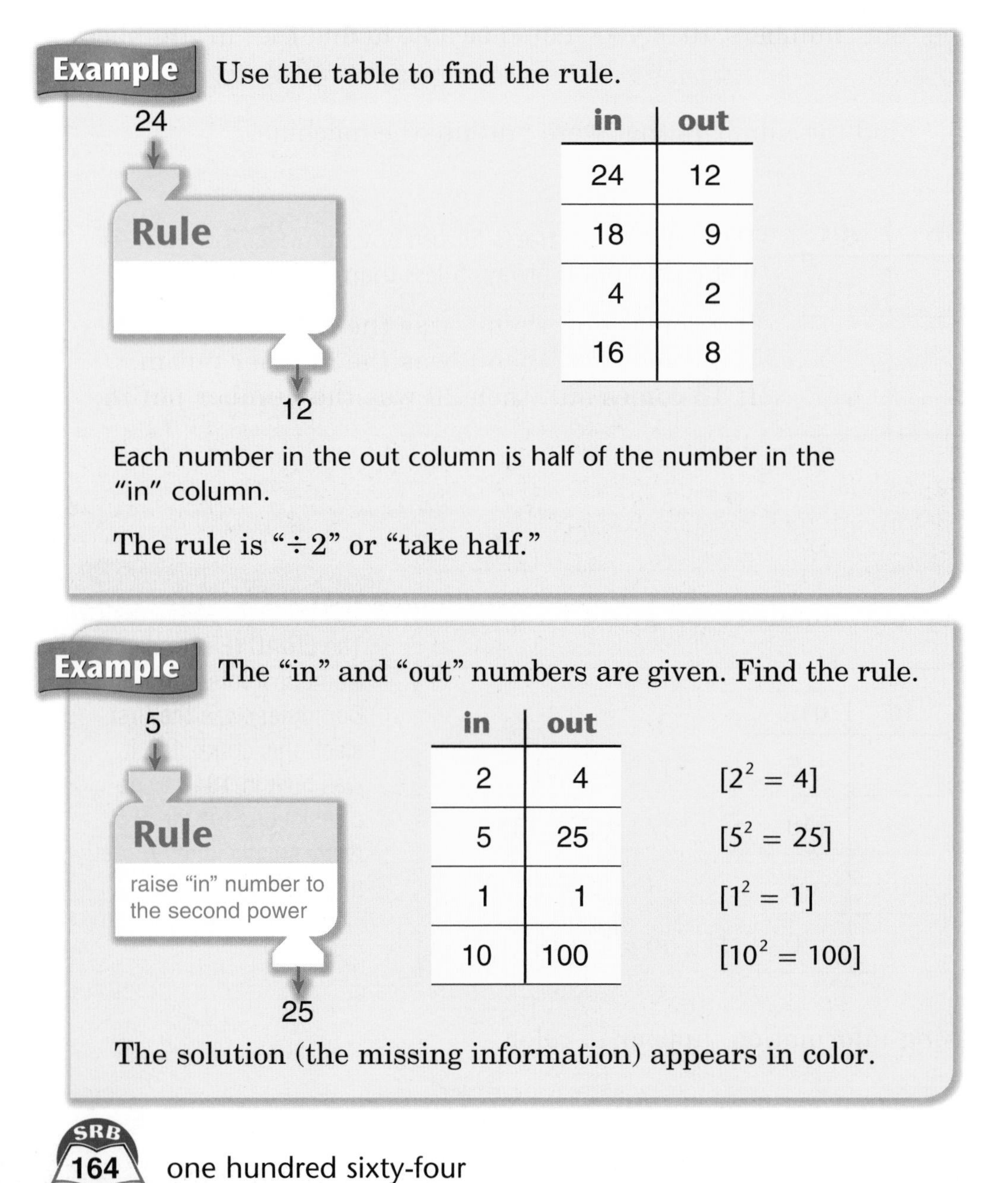

Example Use the table to find the rule.

24 → **Rule** → 12

in	out
24	12
18	9
4	2
16	8

Each number in the out column is half of the number in the “in” column.

The rule is “÷ 2” or “take half.”

Example The “in” and “out” numbers are given. Find the rule.

5 → **Rule** raise “in” number to the second power → 25

in	out	
2	4	[$2^2 = 4$]
5	25	[$5^2 = 25$]
1	1	[$1^2 = 1$]
10	100	[$10^2 = 100$]

The solution (the missing information) appears in color.

Another way to show the workings of a function machine is to draw a graph. The "in" and "out" numbers in the table form number pairs. To draw a graph, you must plot these number pairs on a coordinate grid.

Example Make a graph of this function machine.

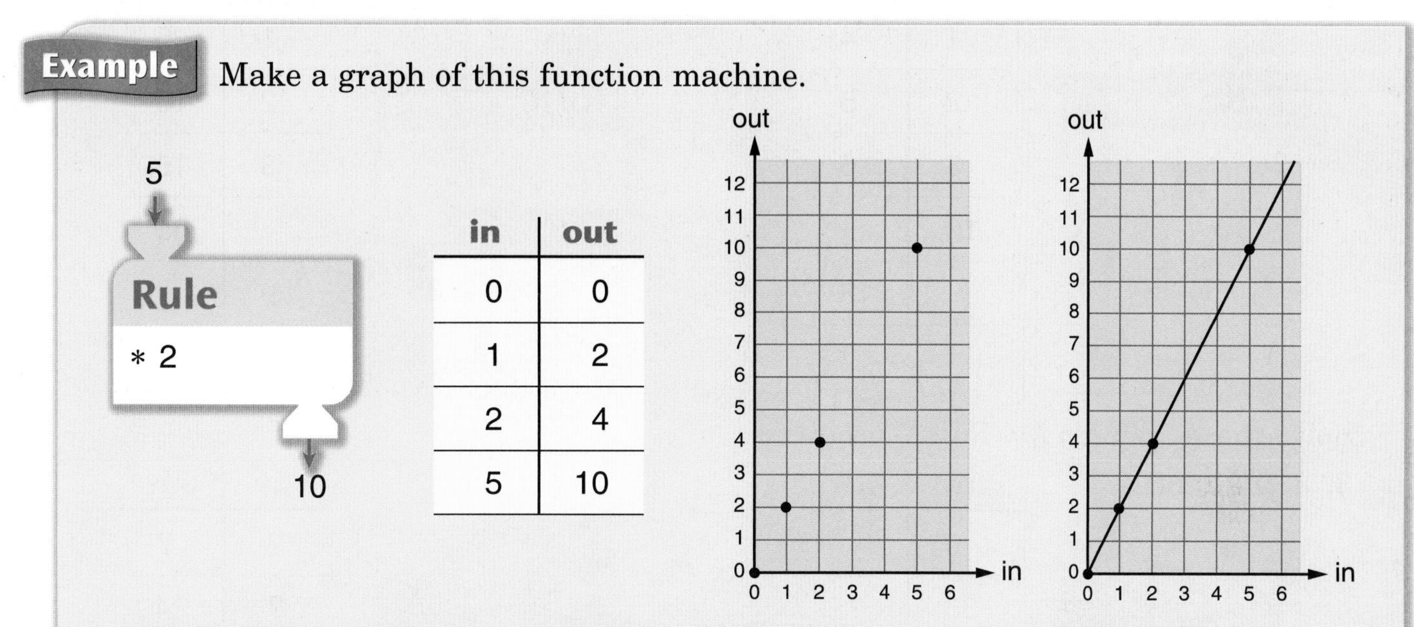

in	out
0	0
1	2
2	4
5	10

To draw a graph, plot the number pairs from the table: (0,0), (1,2), (2,4), (5,10). Plot each number pair as a point on the coordinate grid.

Connect the points on the grid with a straight line. Extend this line to complete the graph.

Check Your Understanding

1. Use the rule to find the missing "in" numbers.

2. Use the rule to find the missing "out" numbers. Then draw a graph of the function machine.

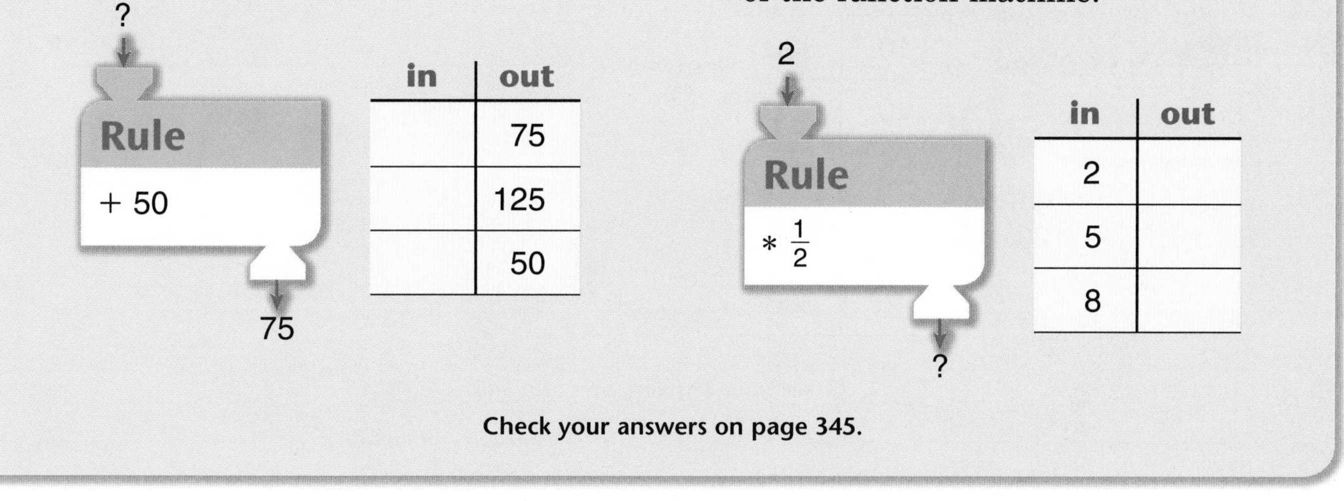

in	out
	75
	125
	50

in	out
2	
5	
8	

Check your answers on page 345.

Check Your Understanding

Find the missing rules. Then find the missing "in" and "out" numbers.

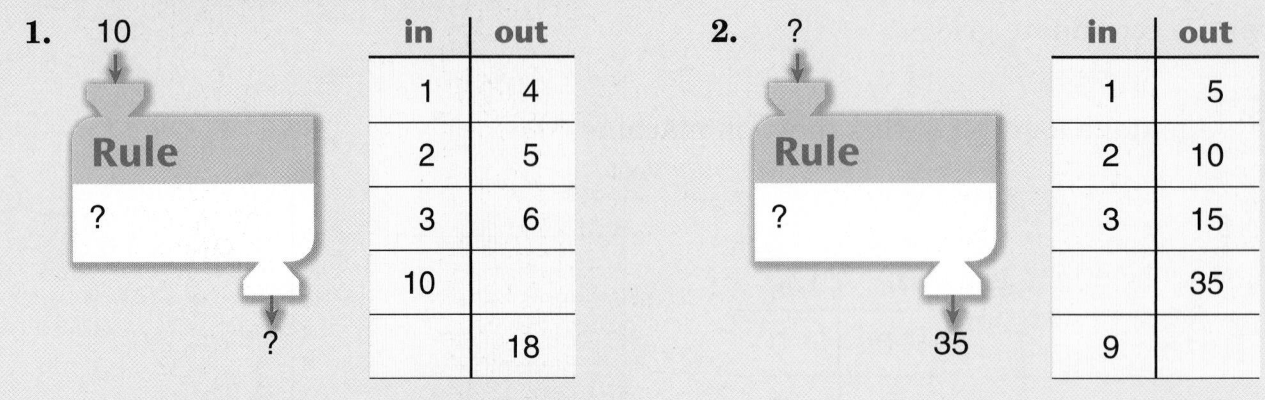

1.

in	out
1	4
2	5
3	6
10	
	18

2.

in	out
1	5
2	10
3	15
	35
9	

Solve these "What's My Rule?" problems.

3.

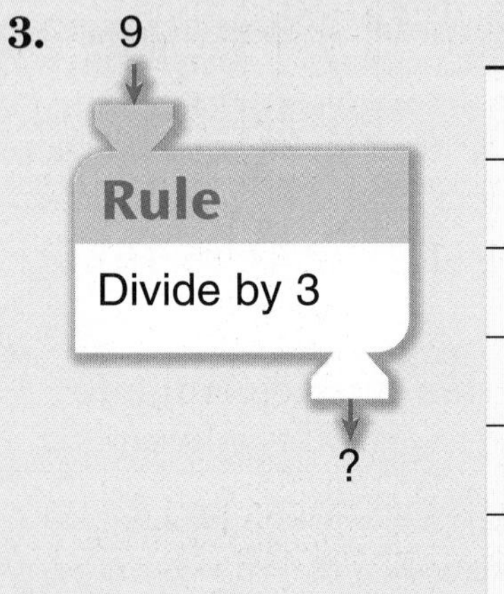

in	out
9	?
36	?
1	?
1.5	?
123	?
390	?

4. ?

Rule

Subtract 4

7

in	out
?	7
?	24
?	−4
?	0
?	50
?	118

5.

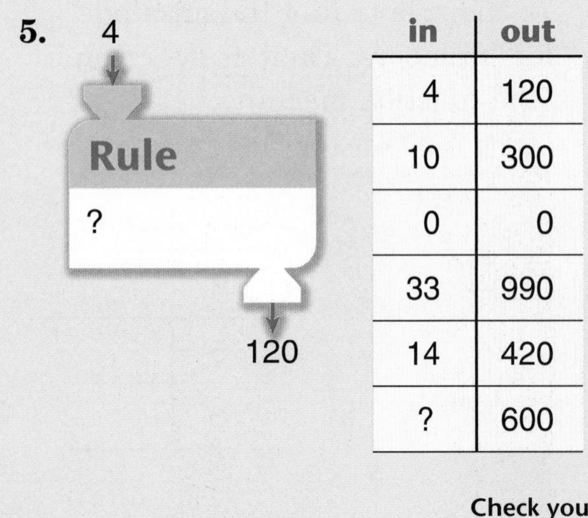

in	out
4	120
10	300
0	0
33	990
14	420
?	600

Check your answers on page 345.

Making and Using Maps

Mapmakers, or cartographers, attempt to show the world in a way they can hold in their hands. Mapping the earth's curved surface onto a flat plane is a very difficult task, but as technology advances, mapmakers are able to portray the world more and more accurately.

A Look at Early Maps

Mapmakers use their understanding of the world to create maps. Each type of map is unique and serves a specific purpose.

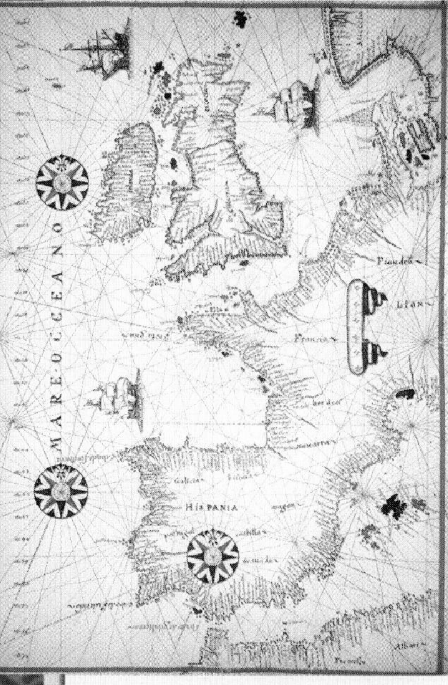

Medieval maps, like this one, often showed Jerusalem at the center of the world. This reflects the importance of Christianity at that time in Europe.

A "Portolan" chart is a type of navigation map used in Europe starting in the 14th century. Fine detail of the coastline and lines representing compass headings were useful to sailors.

Maps of constellations, like this one from *The Celestial Atlas* of 1660, were used by sailors to help with navigation.

Maps that show a "bird's eye view" of a city or region are common. They help people see places in three dimensions. This view of the city of Damascus was created in the late 1500s.

Map Scale

Before creating a map, a cartographer must determine the map's scale. The cartographer must figure out how much of the world to show, and how much detail is needed. As the scale of the map gets larger, the area it represents gets smaller.

This view of Earth was made from satellite photographs. The width of this view is roughly the circumference of the earth—about 25,000 miles. ➤

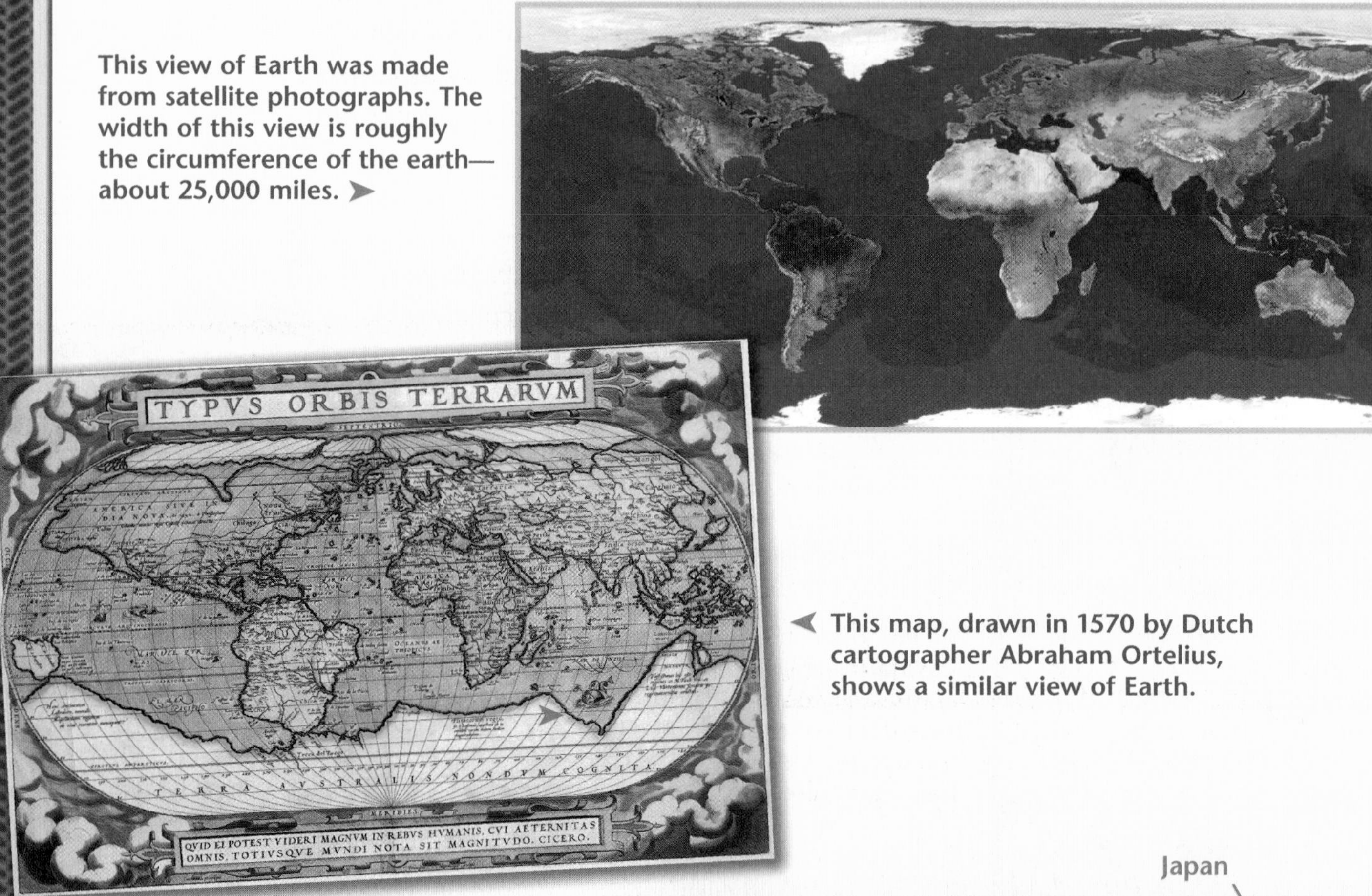

➤ This map, drawn in 1570 by Dutch cartographer Abraham Ortelius, shows a similar view of Earth.

This is a larger scale view of Earth that zooms in on South Asia. The scale in the bottom right corner shows that Japan is about 1,000 miles long. ➤

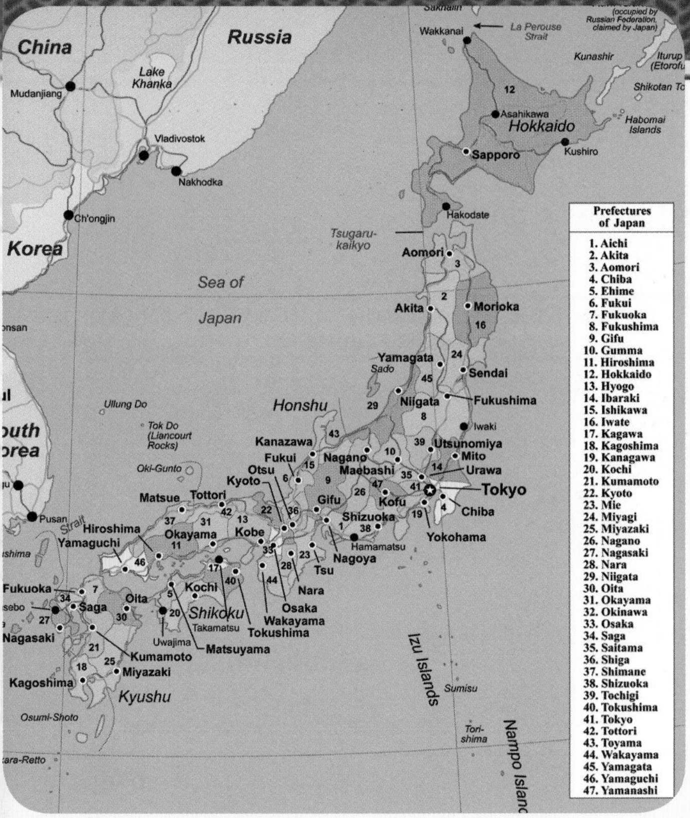

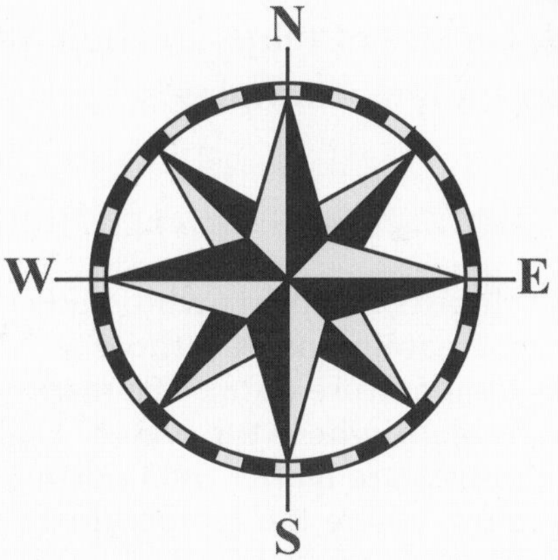

Mathematics ... Every Day

Japan is the focus of this larger scale map. The legend at the top left of the map explains the symbols and markings that represent cities, roads, borders and railroads of the country.

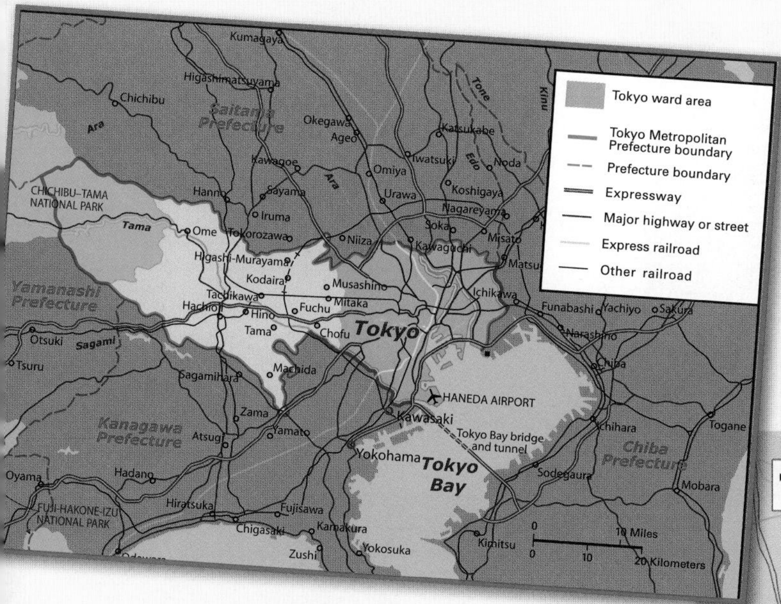

This even larger scale map shows just the city of Tokyo, the largest city in Japan. The map shows streets, highways, important landmarks, parks, bodies of water, and other features.

This very large scale map shows a small portion of Tokyo. It features the parks and landmarks near the Omiya Imperial Palace. According to the scale, the width of this map represents a distance of about 4 miles.

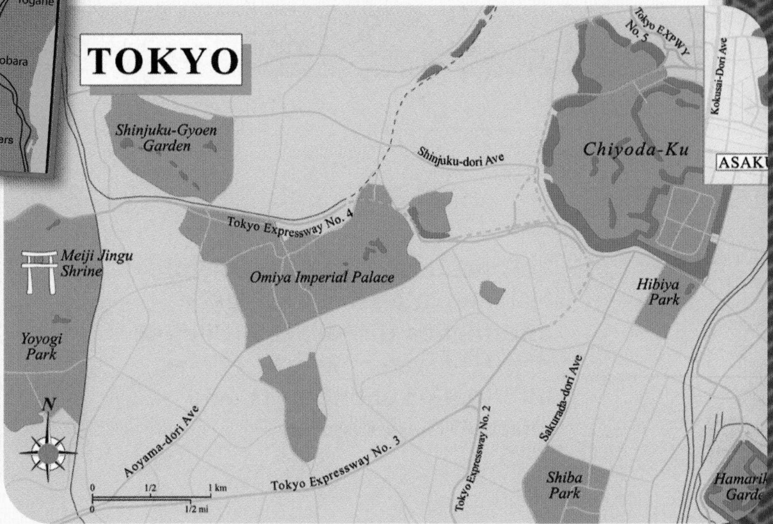

Map Uses

People use a variety of maps in their work and in their daily lives. Here are a few examples.

Climatologists, scientists who study climate, create maps that show ocean temperature data. This map shows regions where the water temperatures are higher than normal and regions where the temperatures are lower than normal. ➤

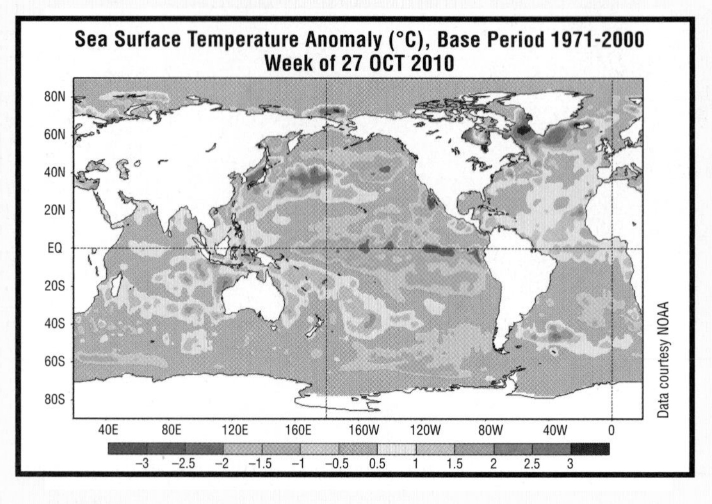

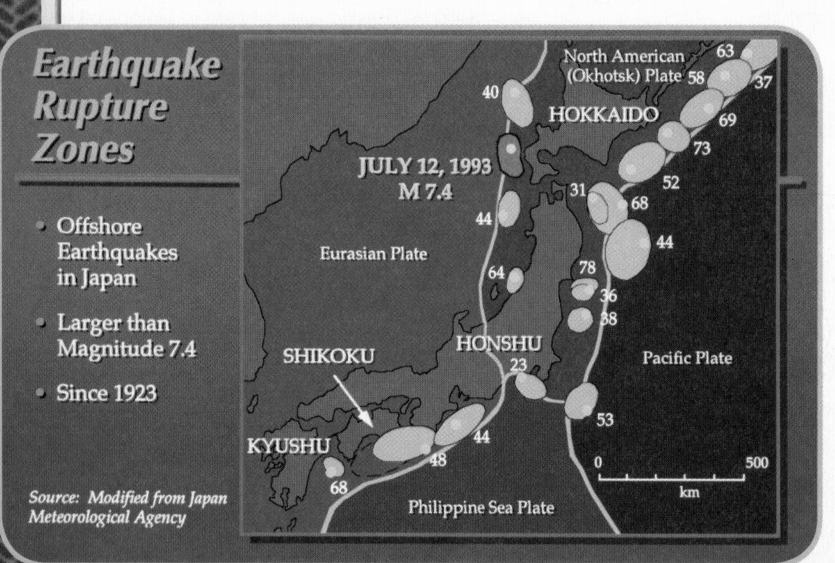

➤ This map shows the major earthquakes centered near Japan since 1923. Seismologists, scientists who study earthquakes, might use this kind of map.

This nautical chart of Wrangell Narrows in Alaska provides information for sailors, including: depth of water, buoy locations, information about ports, and channel markings. ➤

Mathematics ... Every Day

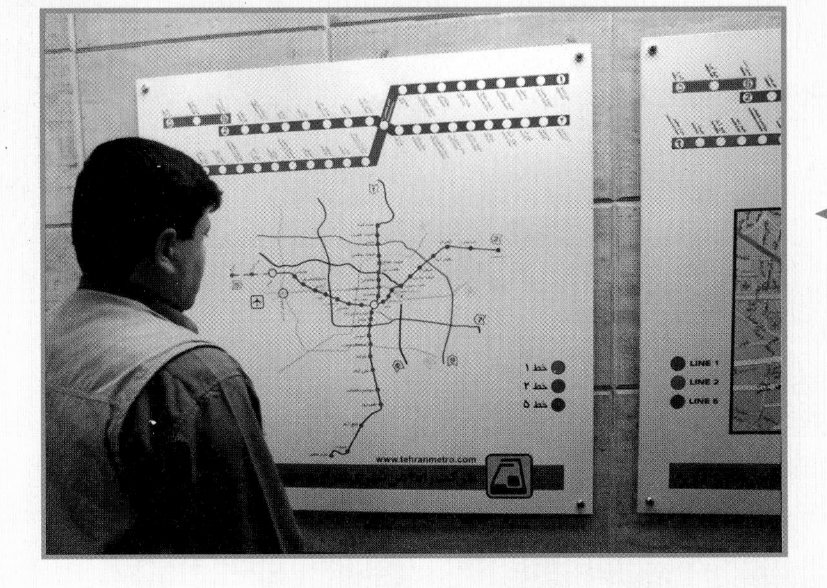

This is a map of the subway system for the city of Tehran, Iran. Subway maps are often not drawn to scale. This is because their purpose is to help people decide what trains to take, not to show exactly how far the train travels.

Many maps give information about historical events. This map provides details about the Historic Santa Fe Trail.

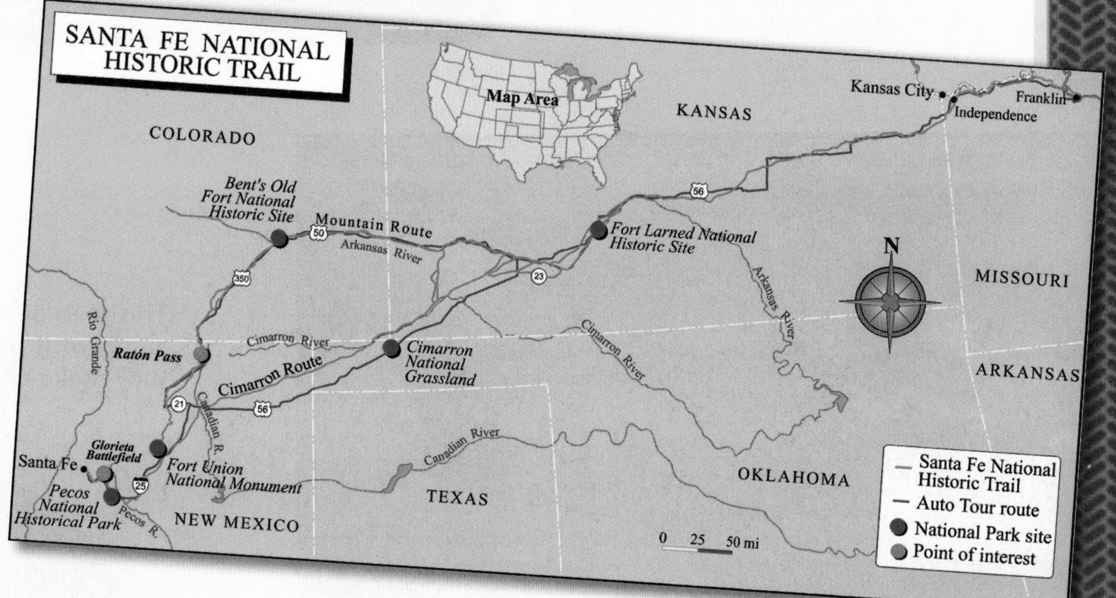

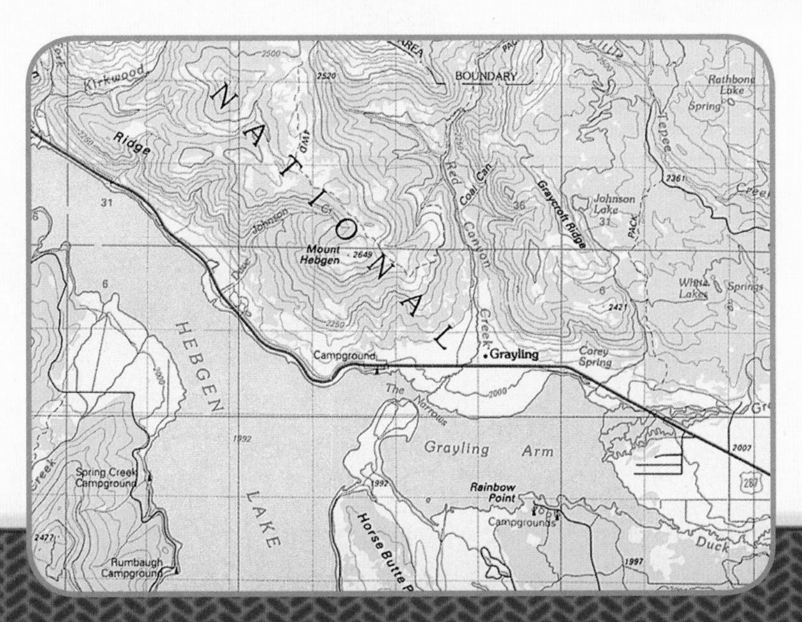

A hiker might use a topographic map to plan a trek in the mountains. The numbers on the curved lines on this type of map tell the elevation of the land as compared to sea level. Lines that are close together mean the land is steep. Lines that are far apart mean the land is flatter.

Maps of Earth and Beyond

People study maps of the earth and beyond. Here are some examples:

This world map shows daytime and nighttime on Earth at one particular moment. Since the North Pole is dark and the South Pole is light, it must be winter in the Northern Hemisphere and summer in the Southern Hemisphere. ➤

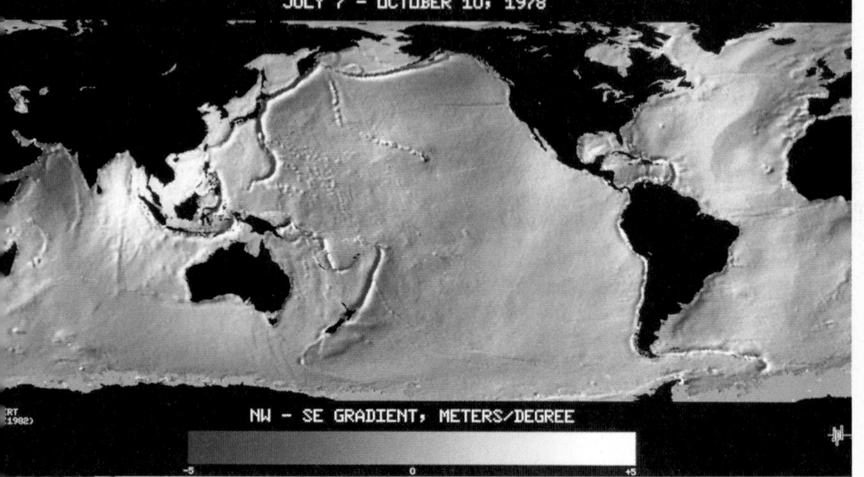

Oceanographers use maps to study underwater terrain. This map shows deep trenches found in the ocean floor.

This map of the sky shows stars along with the constellations they make up.

There are many different types of maps with many different purposes. If you were a cartographer, what kind of map would you create?

Problem Solving

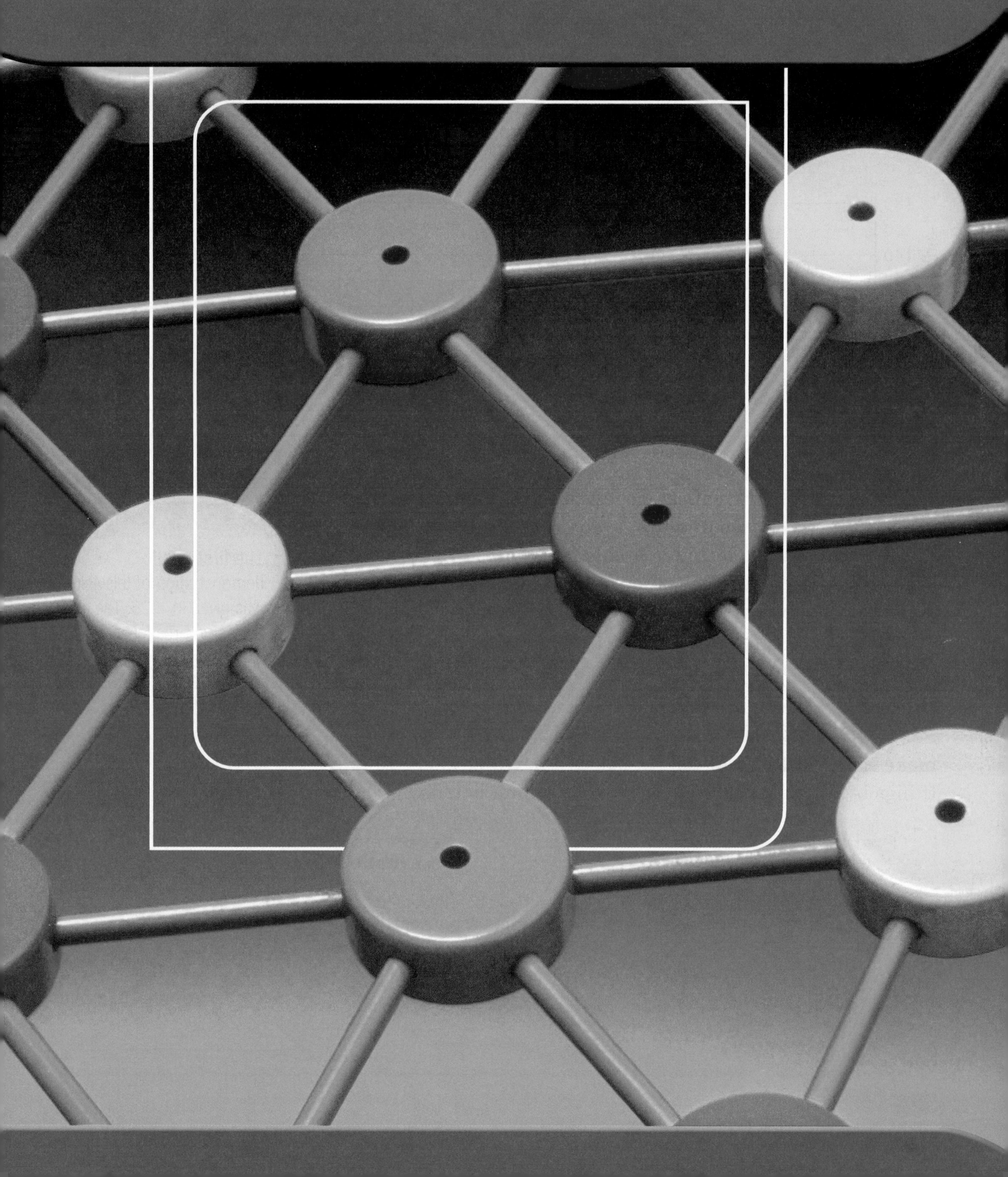

Mathematical Modeling

A **mathematical model** is something mathematical that describes something in the real world. A sphere, for example, is a model of a basketball. The number sentence $1.00 - 0.79 = 0.21$ is a model for buying a bottle of juice that costs 79¢ with a $1 bill and getting 21¢ change. The graph below is a model for the number of TV sets in several African countries.

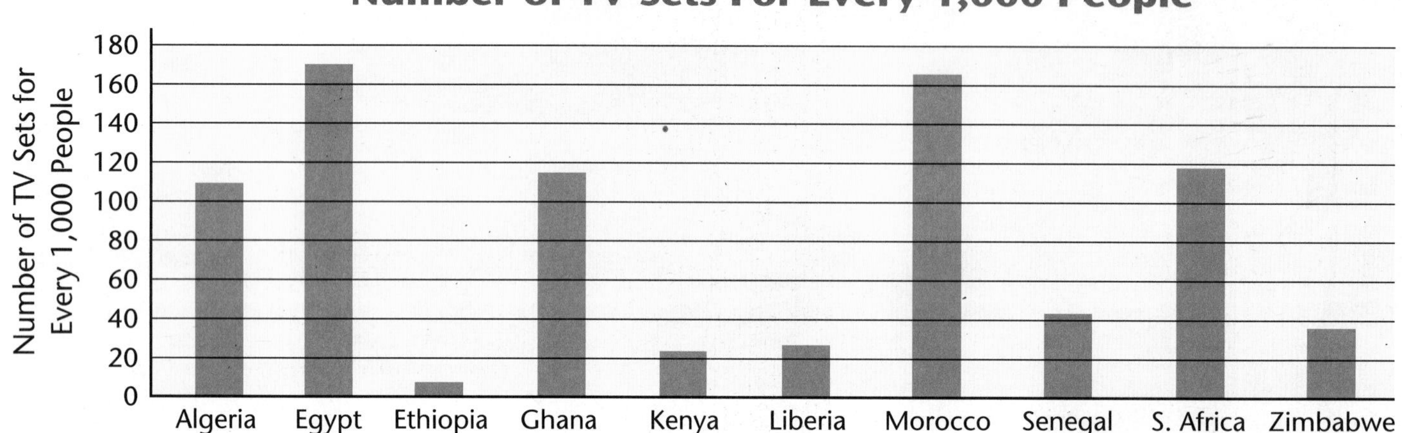

You have used mathematical models to solve problems for many years. In kindergarten and first grade, you used counters and drew pictures to solve simple problems. In second and third grades, you learned to use other models, such as situation diagrams and number models, to solve number stories. As you continue studying mathematics, you will learn to make and use more powerful mathematical models.

Everyday Mathematics has many different kinds of problems. Some problems ask you to find something. Others ask you to make something. When you get older, you will be asked to prove things by giving good reasons why something is true or correct.

Did You Know?

The first public demonstration of television was given in 1926. The first television broadcasting service was opened in Great Britain, in 1936.

Problems that ask you to find something	Problems that ask you to make something
What is the total cost of 5 pounds of apples at $1.49 a pound and a gallon of milk at $3.49?	Use straws and connectors to make a square-base pyramid.
What are the missing numbers? 1, 4, 9, 16, ______, 36, ______, 64	Use a compass and a straightedge to copy this triangle:

A Guide for Solving Number Stories

Learning to solve problems is the main reason for studying mathematics. One way you learn to solve problems is by solving number stories. A **number story** is a story with a problem that can be solved with arithmetic.

1. Understand the problem.
- Read the problem. Can you retell it in your own words?
- What do you want to find out?
- What do you know?
- Do you have all the information needed to solve the problem?

2. Plan what to do.
- Is the problem like one that you solved before?
- Is there a pattern you can use?
- Can you draw a picture or a diagram?
- Can you write a number model or make a table?
- Can you use counters, base-10 blocks, or some other tool?
- Can you estimate the answer and check if you are right?

3. Carry out the plan.
- After you decide what to do, do it. Be careful.
- Make a written record of what you do.
- Answer the question.

4. Look back.
- Does your answer make sense?
- Does your answer agree with your estimate?
- Can you write a number model for the problem?
- Can you solve the problem in another way?

A Guide for Number Stories

1. Understand the problem.
2. Plan what to do.
3. Carry out the plan.
4. Look back.

Note
Understanding the problem is a very important step. Good problem solvers take time to make sure they really understand the problem.

Note
Sometimes it's easy to know what to do. Other times you need to be creative.

Check Your Understanding

Use the *Guide for Solving Number Stories* to help you solve the following problems. Explain your thinking at each step. Also explain your answer(s).

1. Mr. Cline walked 2 miles per day for 2 days in a row, 3 miles per day for the next 3 days, and so on, until he walked 6 miles per day. On which day did he first walk 6 miles?

2. Lisa cut a rectangle with a perimeter of 24 inches into two squares. What were the length and width of the original rectangle?

Check your answers on page 346.

A Problem-Solving Diagram

Problems from everyday life, science, and business are often more complicated than the number stories you solve in school. Sometimes the steps in the *Guide for Solving Number Stories* may not be helpful.

The diagram below shows another way to think about problem solving. This diagram is more complicated than a list, but it is more like what people do when they solve problems in science and business. The arrows connecting the boxes are meant to show that you don't always do things in the same order.

Did You Know?

A company will often form a committee to solve a business problem. Each member of the group has a skill in a particular area. The members combine their skills and ideas to understand and solve the problem. Group members often sit together and share ideas and possible solutions to the problem. This problem-solving method is called *brainstorming.*

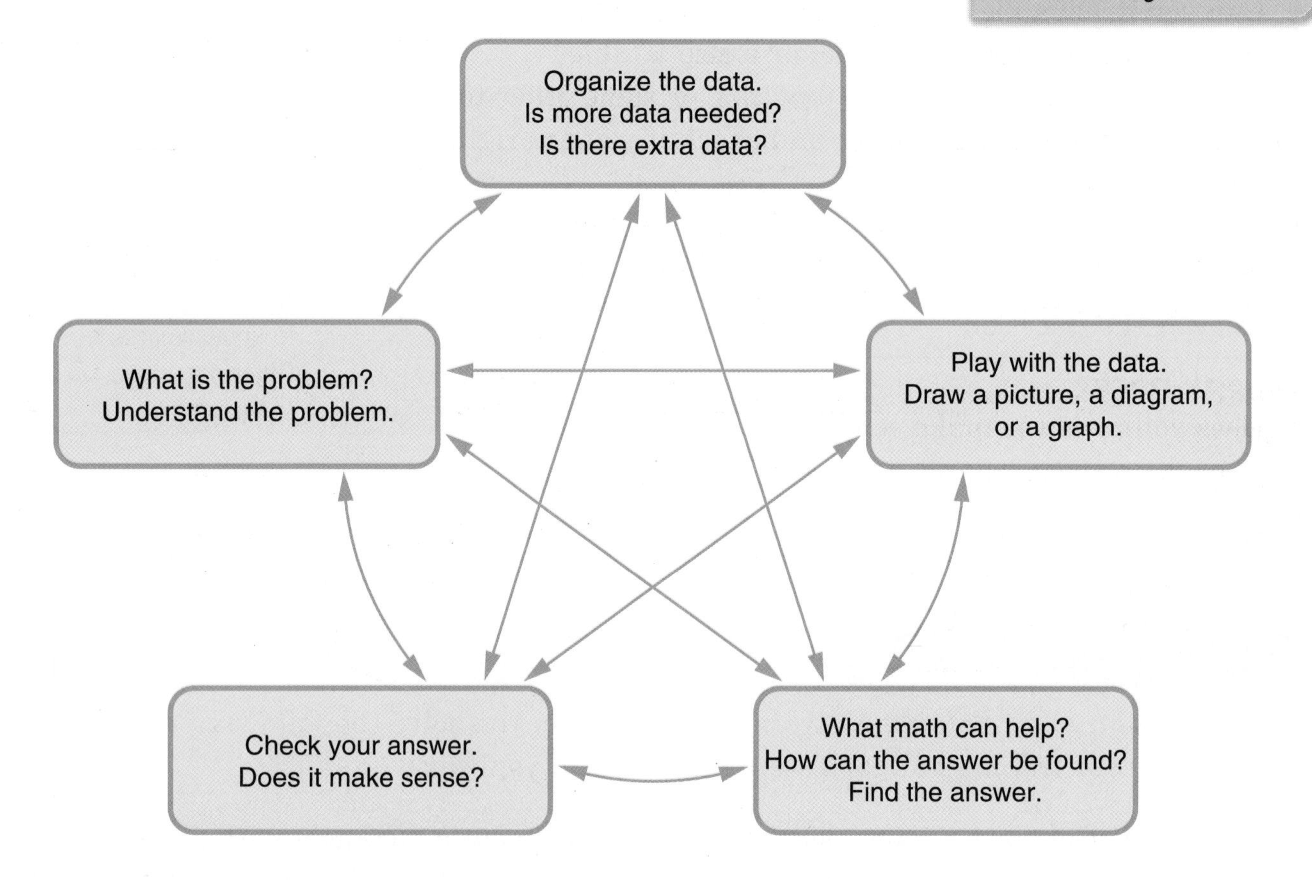

Using the diagram on the previous page as you solve problems may help you be a better problem solver. Here are some things to try for each of the boxes in the diagram. Remember, these are not rules. They are only suggestions that might help.

- **What is the problem?** Try to understand the problem. Can you retell it in your own words? What do you want to find out? What do you know? Try to imagine what an answer might look like.
- **Organize the data.** Study the data you have and organize it in a list or in some other way. Look for more data if you need it. Get rid of any data that you don't need.
- **Play with the data.** Try drawing a picture, a diagram, or a graph. Can you write a number model? Can you model the problem with counters or blocks?
- **What math can help?** Can you use arithmetic, geometry, or other mathematics to find the answer? Do the math. Label the answer with units.
- **Check your answer.** Does it make sense? Compare your answer to a friend's answer. Try the answer in the problem. Can you solve the problem another way?

Check Your Understanding

1. About how long would it take to count to one million?
2. On a piece of paper, draw the diagram below. Put the digits 1–6 in the circles so that the sum along each line is the same.

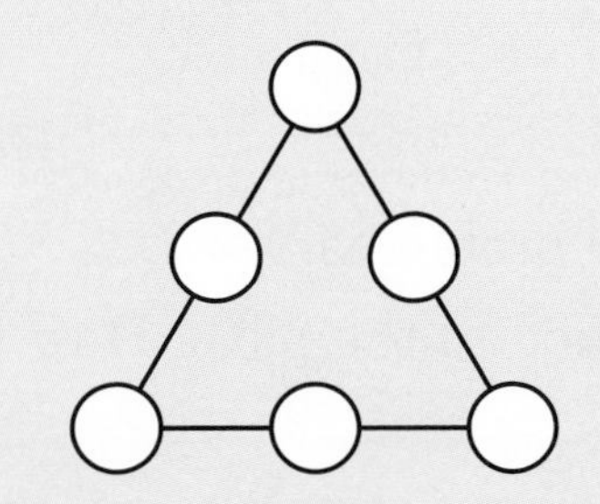

Can you find another way?

3. What is the area of the shaded figure below?

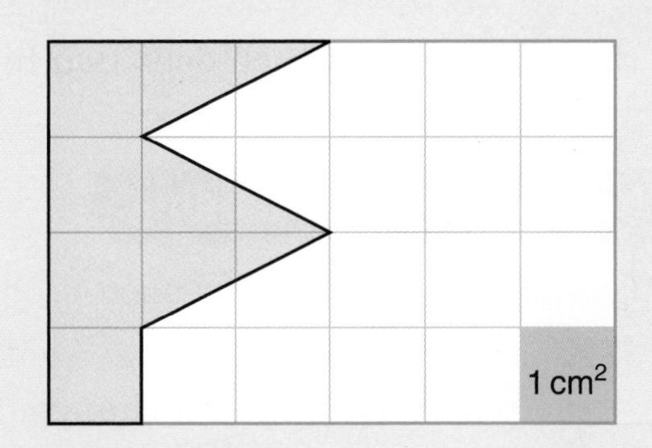

Check your answers on page 346.

Situation Diagrams

There are three basic kinds of addition and subtraction situations: change situations, parts-and-total situations, and comparison situations.

Everyday Mathematics uses diagrams for each of these types of situations. The diagrams help you organize the information in a problem. The diagrams can help you decide what to do. You can also use the diagrams to show the answer after you solve a problem.

Change Situations

In a change situation, there is a starting quantity, then a change, and finally an ending quantity. The change can be either change-to-more or change-to-less.

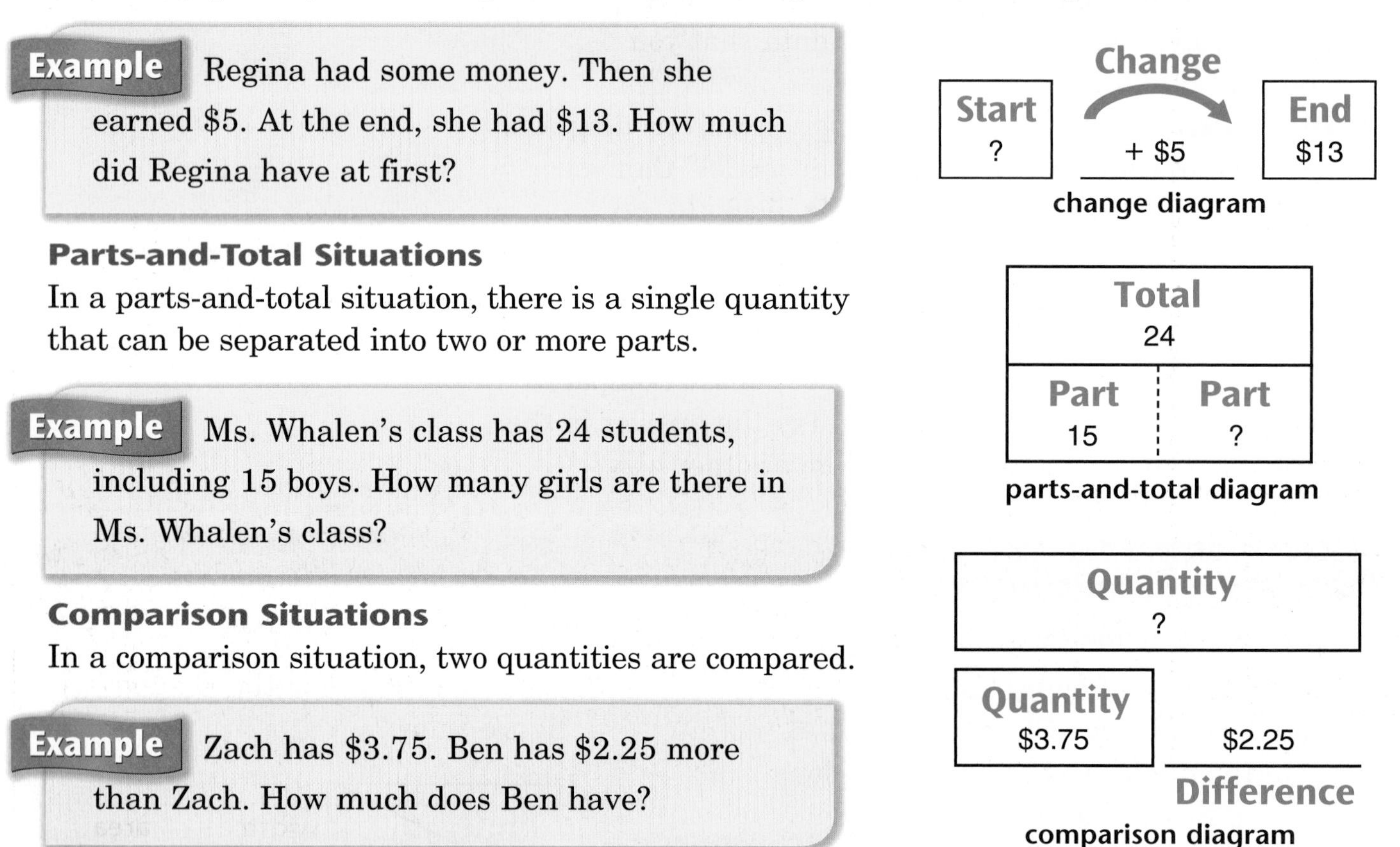

Example Regina had some money. Then she earned $5. At the end, she had $13. How much did Regina have at first?

change diagram

Parts-and-Total Situations

In a parts-and-total situation, there is a single quantity that can be separated into two or more parts.

Example Ms. Whalen's class has 24 students, including 15 boys. How many girls are there in Ms. Whalen's class?

parts-and-total diagram

Comparison Situations

In a comparison situation, two quantities are compared.

Example Zach has $3.75. Ben has $2.25 more than Zach. How much does Ben have?

comparison diagram

Check Your Understanding

Complete a diagram for each problem. Then solve the problems.

1. A path was 12 miles long. Another 9 miles was added to the path. How long is the path? (*Hint:* Use a change diagram.)
2. Gary is 137 cm tall. His sister Pat is 78 cm tall. How much taller is Gary than Pat? (*Hint:* Use a comparison diagram.)
3. Rita and Sam collect coins. Rita has 218 coins and Sam has 125 coins. How many coins do they have in all? (*Hint:* Use a parts-and-total diagram.)

Check your answers on page 346.

Multiplication and division situations include equal-grouping situations, array and area situations, and two comparison situations: rate and ratio, and scaling. You can also make diagrams to decide what to do in these kinds of problems.

Equal-Grouping Situations

In equal-grouping situations, there are several groups of things with the same number of things in each group. You can find the total number of things by multiplying.

Example A wedding party orders 3 dozen roses. There are 12 roses in a dozen. How many roses are there in all?

dozens of roses	roses per dozen	total roses
3	12	r

Array and Area Situations

If equal groups are arranged in rows and columns, then a rectangular *array* is formed. You can multiply to find the total number of items in the array.

Example For seating at a wedding, there are 6 rows of chairs with 10 chairs in each row. How many chairs are there in all?

rows	chairs per row	total chairs
6	10	c

Arrays are closely related to *area*. You can find the area of a 6 cm by 8 cm rectangle by tiling it with square-centimeter tiles.

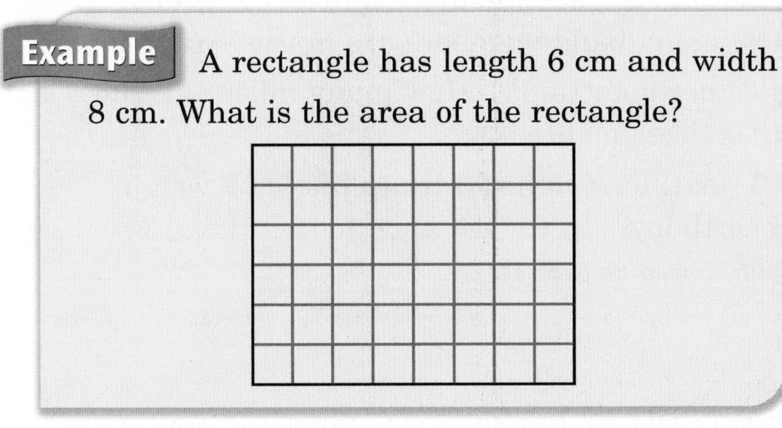

Example A rectangle has length 6 cm and width 8 cm. What is the area of the rectangle?

length (cm)	width (cm)	area (cm^2)
6	8	A

Unlike equal-grouping and array situations, you can use an area model to multiply in situations with fractions, decimals, or mixed numbers.

Multiplication Comparison: Rate and Ratio Situations

Rates and *ratios* are comparisons of quantities by division. But in situations where the rate or ratio is known, you can solve problems using multiplication.

Example A car gets an average of 34 miles per gallon (mpg) of gas. It uses 9 gallons on a trip. How many miles did it go on the trip?

gallons	miles per gallon	total miles
9	34	m

Multiplication Comparison: Scaling Situations

In scaling comparison situations, something becomes larger or smaller by a *scale comparison factor*. For example, if you double a recipe, you are scaling by a factor of 2. If the scale factor is less than 1, then scaling makes something smaller. For example, if you halve a recipe, you are scaling by a factor of $\frac{1}{2}$.

Example A granola recipe uses 3 cups of rolled oats. How many cups of oats should you use to make 2 times as much granola as in the recipe?

oats per recipe (cups)	scale comparison factor	total oats (cups)
3	2	t

Check Your Understanding

Complete a diagram for each problem. Then solve each problem.

1. A box of granola bars contains 10 bars. How many bars total are there in 7 boxes?
2. A marching band in a parade is in a rectangular array. There are 4 band members per row and 7 rows. How many band members are in the array?
3. A garden snail averages 0.03 miles per hour (mph). How many miles will the snail travel in 3 hours (assuming it takes no breaks)?
4. Hector weighed 6 lb at birth. At 1 year, he weighed 3 times his birth weight. What was his weight on his first birthday?

Check your answers on page 347.

Interpreting a Remainder in Division

Some number stories are solved by dividing whole numbers. You may need to decide what to do when there is a non-zero remainder.

There are three possible choices:

- Ignore the remainder. Use the quotient as the answer.
- Round the quotient up to the next whole number.
- Rewrite the remainder as a fraction or decimal.
  Use this fraction or decimal as part of the answer.

To rewrite a remainder as a fraction:

1. Make the remainder the *numerator* of the fraction.
2. Make the divisor the *denominator* of the fraction.

Problem	Answer	Remainder rewritten as a fraction	Answer written as a mixed number	Answer written as a decimal
371 / 4	92 R3	$\frac{3}{4}$	$92\frac{3}{4}$	92.75

Example

- Suppose 3 people share 14 counters equally.
  How many counters will each person get?

14 / 3 → 4 R2

Ignore the remainder. Use the quotient as the answer.

Each person will have 4 counters, with 2 counters left over.

- Suppose 14 photos are placed in a photo album. How many pages are needed if 3 photos can fit on a page?

14 / 3 → 4 R2

You must round the quotient up to the next whole number. The album will have 4 pages filled and another page only partially filled.

So, 5 pages are needed.

- Suppose three friends share a 14-inch string of licorice. How long is each piece if the friends receive equal shares?

14 / 3 → 4 R2

The answer, 4 R2, shows that if each friend receives 4 inches of licorice, 2 inches remain to be divided. Imagine that this 2-inch remainder is divided into pieces that are $\frac{1}{3}$ inch long. Each friend receives 2 of these $\frac{1}{3}$ inch pieces, or $\frac{2}{3}$ inch.

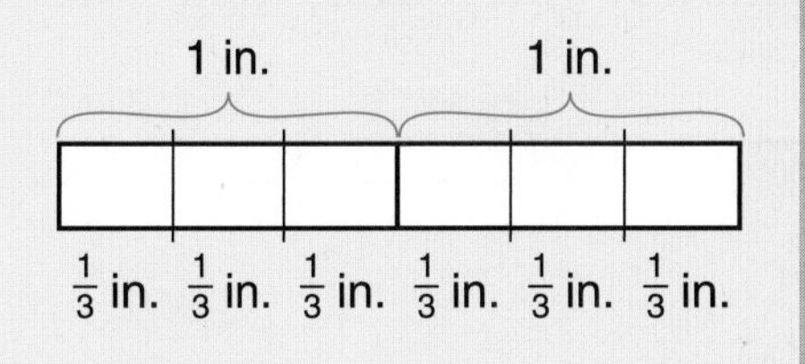

The remainder (2) has been rewritten as a fraction $\left(\frac{2}{3}\right)$. Use this fraction as part of the answer.

Each friend will get a $4\frac{2}{3}$-inch piece of licorice.

Estimation

An **estimate** is an answer that should be close to an exact answer. You make estimates every day.

- You estimate how long it will take to walk to school.
- You estimate how much money you can save by the end of the year.
- You estimate how long it will take you to do your homework.

Sometimes you must estimate because it is impossible to know the exact answer. When you predict an event in the future, for example, you have to estimate since it is impossible to know exactly what will happen. A weather forecaster's prediction is an estimate of what will happen in the future.

Sometimes you estimate because finding an exact answer is not practical. For example, you might estimate the number of books in your school library. You could count the books and find an exact answer. But that would not be practical.

Sometimes you estimate because finding an exact answer is not worth the trouble. For example, you might estimate the cost of several items at the store to be sure you have enough money. There is no need to find an exact answer until you pay for the items.

Estimation in Problem Solving

Estimation is useful even when you need to find an exact answer. Making an estimate when you first start working on a problem may help you understand the problem better. Estimating before you solve a problem is like making a rough draft of a writing assignment.

Estimation is also useful after you have found an answer for a problem. You can use the estimate to check whether your answer makes sense. If your estimate is not close to the exact answer you found, then you need to check your work.

Did You Know?

In 1993, Australia began the first Walking Bus to cut down on automobile traffic. Volunteer adults walk a specified route to school, and school children join them along the way.

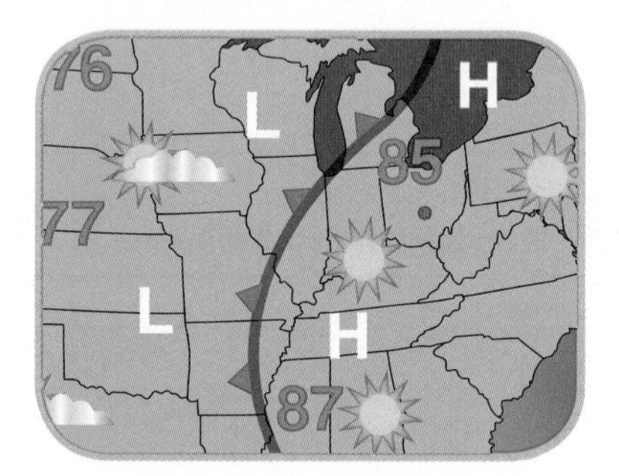

"Columbus, Ohio may *expect* sunny weather tomorrow. A high temperature of *about* 85 degrees is *predicted*."

You may want to estimate the total cost of your items before paying for them.

Leading-Digit Estimation

The best estimators are often people who are professionals. Someone who lays carpets for a living, for example, would be good at estimating the size of rooms. A waiter would be good at estimating the proper amount for a tip.

One way to estimate is to adjust each number in a problem *before* you estimate. The adjusted numbers should be simple and easy to work with.

For example, you might adjust each number as follows:

1. Keep the first non-zero digit of the number.

2. Replace the other digits of the number with zeros.

Then make your estimate using these adjusted numbers. This way of estimating is called **leading-digit estimation.**

Exact number from a problem	Adjusted number to make an estimate
0.67	0.60
6	6
78	70
349	300
8,765	8,000
128,871	100,000

Example Estimate the area of the rectangle.

Use leading-digit estimation.
The width is about 10 cm, and the length is about 60 cm.
The adjusted numbers are 10 and 60.

67.5 cm

13.2 cm

So, the area of the rectangle is about 10 * 60 or 600 sq cm.

Leading-digit estimates are usually not accurate. But they can be useful for checking calculations. If the estimate and the exact answer are not close, you should look for a mistake in your work.

Example Erica added 526 + 348 and got 984. Was she correct?

The adjusted numbers are 500 and 300. The estimate using these adjusted numbers is 800. Since 800 is not close to 984, Erica is probably not correct. She should check her work.

Exact numbers
526 + 348

Adjusted numbers and leading-digit estimate
500 + 300 = 800

Check Your Understanding

1. David multiplied 621 * 33 and got 2,043. Use leading-digit estimation to decide if this seems correct.

2. Caroline said there are 168 hours in a week. Adjust the numbers and then estimate to decide if this seems correct.

Check your answers on page 346.

Rounding

Rounding is another way of adjusting numbers to make them simpler and easier to work with. Estimation with rounded numbers is usually more accurate than leading-digit estimation.

This example shows the steps to follow in rounding a number.

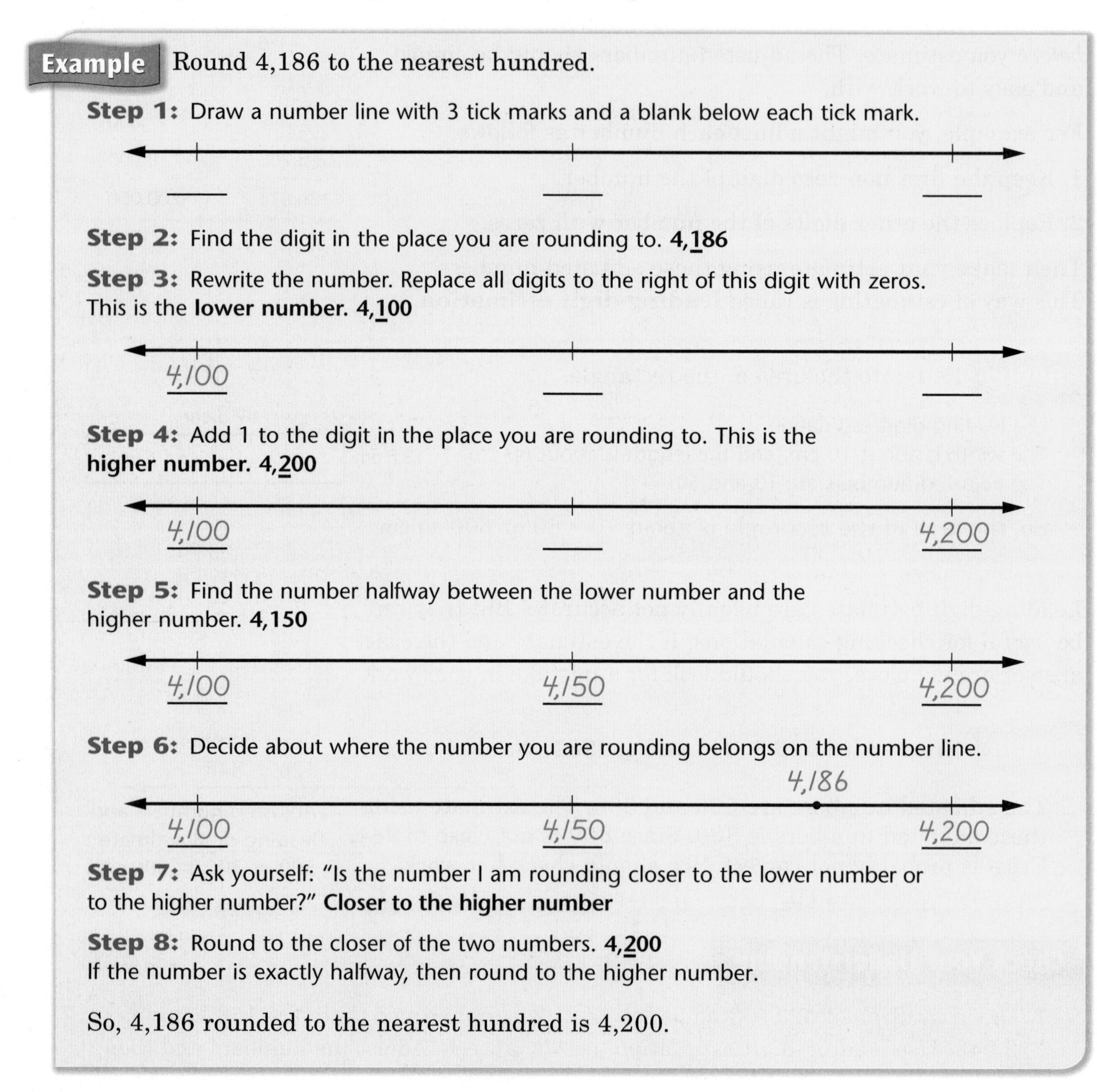

Example Round 4,186 to the nearest hundred.

Step 1: Draw a number line with 3 tick marks and a blank below each tick mark.

Step 2: Find the digit in the place you are rounding to. **4,186**

Step 3: Rewrite the number. Replace all digits to the right of this digit with zeros. This is the **lower number. 4,100**

Step 4: Add 1 to the digit in the place you are rounding to. This is the **higher number. 4,200**

Step 5: Find the number halfway between the lower number and the higher number. **4,150**

Step 6: Decide about where the number you are rounding belongs on the number line.

Step 7: Ask yourself: "Is the number I am rounding closer to the lower number or to the higher number?" **Closer to the higher number**

Step 8: Round to the closer of the two numbers. **4,200**
If the number is exactly halfway, then round to the higher number.

So, 4,186 rounded to the nearest hundred is 4,200.

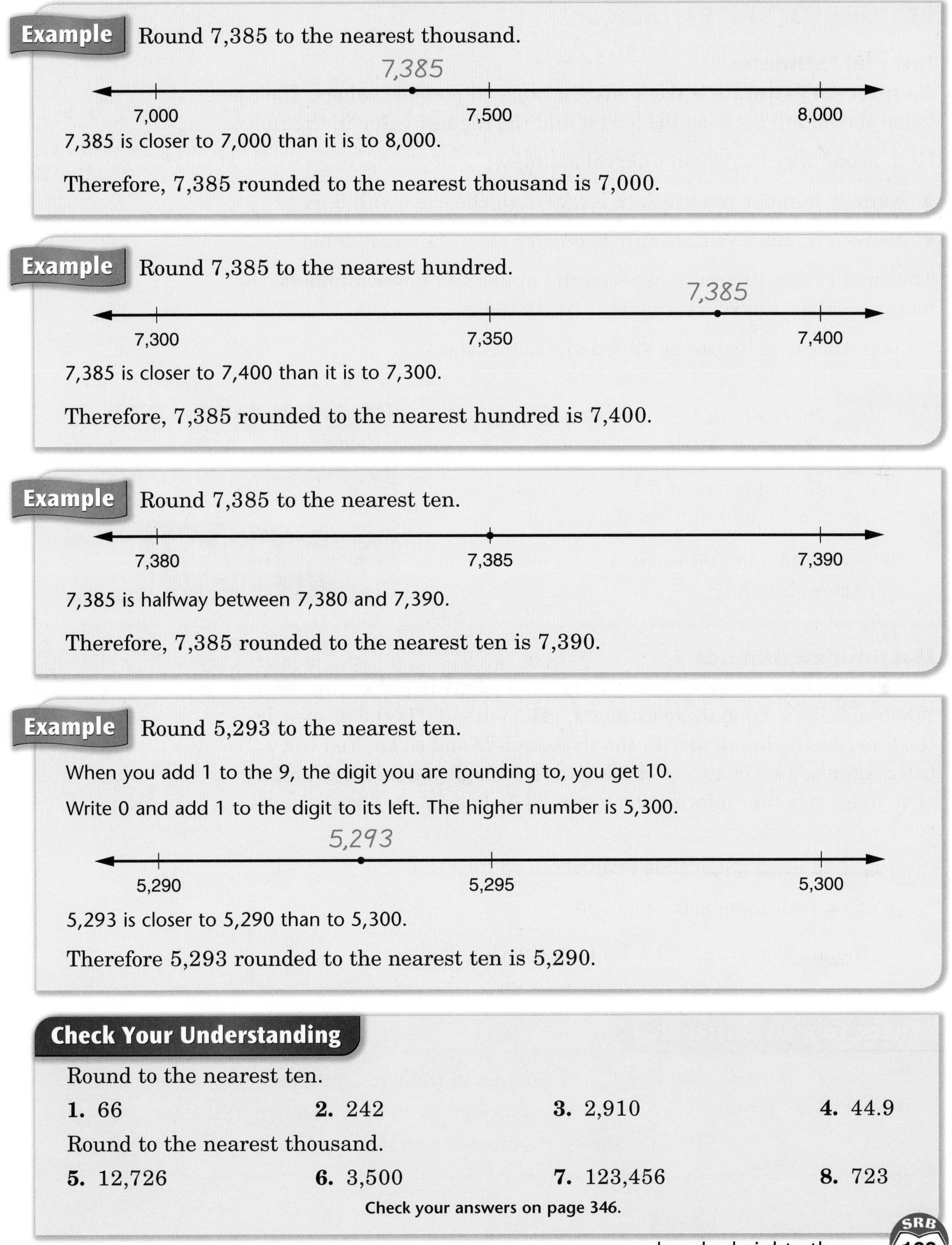

Example Round 7,385 to the nearest thousand.

7,385 is closer to 7,000 than it is to 8,000.

Therefore, 7,385 rounded to the nearest thousand is 7,000.

Example Round 7,385 to the nearest hundred.

7,385 is closer to 7,400 than it is to 7,300.

Therefore, 7,385 rounded to the nearest hundred is 7,400.

Example Round 7,385 to the nearest ten.

7,385 is halfway between 7,380 and 7,390.

Therefore, 7,385 rounded to the nearest ten is 7,390.

Example Round 5,293 to the nearest ten.

When you add 1 to the 9, the digit you are rounding to, you get 10.

Write 0 and add 1 to the digit to its left. The higher number is 5,300.

5,293 is closer to 5,290 than to 5,300.

Therefore 5,293 rounded to the nearest ten is 5,290.

Check Your Understanding

Round to the nearest ten.

1. 66 **2.** 242 **3.** 2,910 **4.** 44.9

Round to the nearest thousand.

5. 12,726 **6.** 3,500 **7.** 123,456 **8.** 723

Check your answers on page 346.

Making Other Estimates

Interval Estimates

An **interval estimate** is made up of a range of possible values. The exact value should fall between the lowest and the highest value in the range.

Here is one way to give an interval estimate:

- Name a number you are sure is *less than* the exact value.
- Name a number you are sure is *greater than* the exact value.

The smaller the difference between the upper and lower numbers, the more useful an interval estimate is likely to be.

An interval estimate can be stated in various ways.

Examples There are *at least* 30 [12s] in 400, but *not more than* 40 [12s].

The number of books in the school library is *greater than* 6,000 but *less than* 6,500.

Between 125 and 200 people live in my apartment building.

Magnitude Estimates

One kind of very rough estimate is called a **magnitude estimate.** When making a magnitude estimate, ask yourself: "Is the answer in the tens? In the hundreds? In the thousands?" and so on. You can use magnitude estimates to check answers displayed on a calculator or to judge whether information you read or hear makes sense.

Example Make a magnitude estimate. 783.29 * 3.9

783.29 * 3.9 is about 800 * 4 or 3,200.

So, the answer to 783.29 * 3.9 is in the thousands.

Check Your Understanding

Make a magnitude estimate. Is the answer in the tens, hundreds, or thousands?

1. 7.41 * 42.3 **2.** 7,914 / 4 **3.** 4,316 − 3,729

Check your answers on page 347.

Machines that Calculate

Mathematics ... Every Day

For thousands of years, people have used tools to calculate. Over time, advances in technology have greatly influenced these tools.

The abacus, which can be used to add, subtract, multiply, and divide, probably started out as pebbles on lines in the dirt. The word calculate, in fact, comes from the Latin word for pebble. The wire and bead abacus, invented in China in about 1200 A.D., is still used in many parts of the world.

Napier's Bones **are portable tools for multiplying and dividing. They were invented by John Napier, a Scottish mathematician, in the late 1500s. The "bones" were often made of wood, heavy paper, ivory, or bone.**

Multiplication with Napier's Bones is similar to the lattice method used in ***Everyday Mathematics.***

Mechanical Calculators

Beginning in the 1600s, calculating machines that used levers, wheels, and gears were invented.

In 1623, William Shickard, a German professor, built the first known mechanical calculator for his friend, the astronomer Johannes Kepler. The top part of the machine consisted of a cylindrical set of Napier's Bones. The dials in the lower part added all the partial products and displayed the final product. ➤

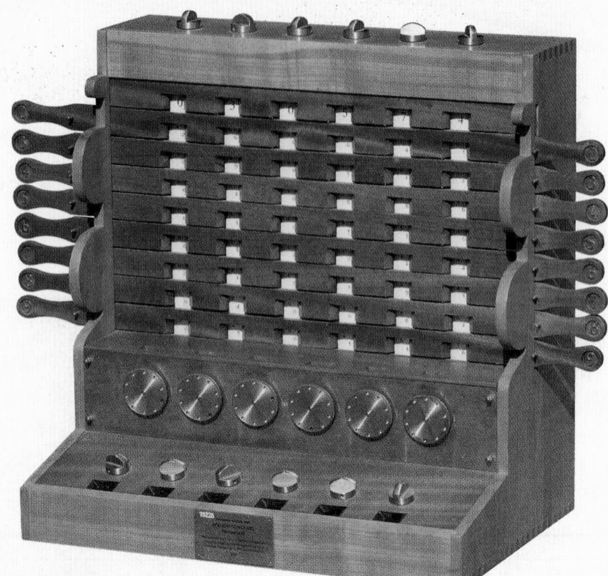

In 1674 another German, Gottfried Leibniz, invented a mechanical calculator that could multiply numbers without using Napier's Bones.

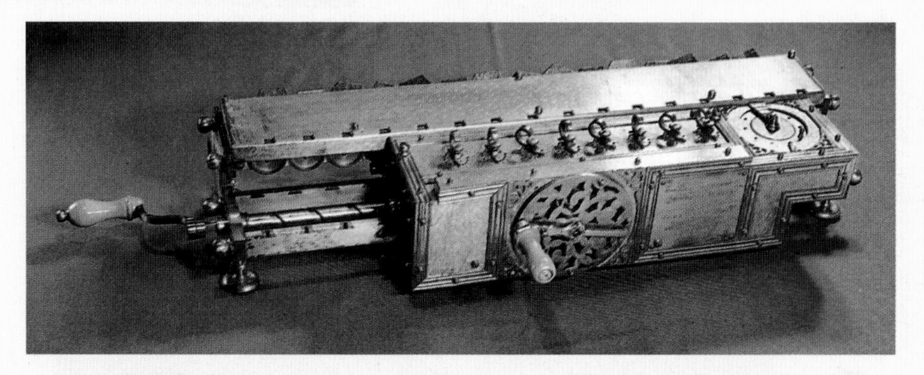

In 1886, Dorr E. Felt invented the *Comptometer,* which used keys to input numbers much more quickly than dials or levers. The speed with which long lists of figures could be added (other operations were trickier) made the Comptometer useful for business. Comptometers were made in large numbers until the 1970s.

Mechanical Computers

In 1833, Charles Babbage, a brilliant scientist, began designing the first computer, which he called the Analytic Engine.

The Analytic Engine would have been powered by steam and programmed by punched cards. It would have had separate memory and processing parts, just as computers do today. Unfortunately, Babbage was only able to build a portion of the processing part and the printing mechanism of this trial model before he died.

In 1936, Konrad Zuse built the first working automatic computer, the Z1, in his parents' living room. The Z1 was completely mechanical and contained about 20,000 parts. It used a small electric motor to move the mechanical parts. Instructions for the Z1 were punched into used movie film.

Electronic Computers

The first all-purpose electronic computer, called ENIAC, was built in 1946. ENIAC, or Electrical Numerical Integrator and Calculator, was huge. It weighed 30 tons, used about 18,000 vacuum tubes, and took up an entire room.

ENIAC could compute about 1,000 times faster than any previous device.

A computer called UNIVAC, which stands for Universal Automatic Computer, was first sold in 1951. "Universal" meant it could solve problems for scientists, engineers, and businesses. "Automatic" meant the instructions in the computer could be stored in the computer's memory.

In 1964, IBM introduced the System/360. This was a series of computers that took advantage of new electronic inventions, such as transistors. The System/360 design was a leader among large computers for the next 25 years.

Personal Computers

By the 1970s, personal computers were invented. The personal computer was made for a single person to use. It fit on a desktop and was much less expensive than previous computers.

The Apple II was introduced in 1977. The computer could use a standard television set for its monitor.

Apple co-founder, Steve Jobs, points to a modern iMAC computer built by his company.

The IBM Personal Computer was first sold in 1981. It was fully assembled, and included hardware and software made by other companies. Within ten years, there were 50 million computers modeled after the PC.

The "laptop" changed people's lives by making it possible to take computer work almost anywhere.

Modern Electronic Products

Many of the electronic products designed for convenience and entertainment were made possible as a result of advances in computer technology.

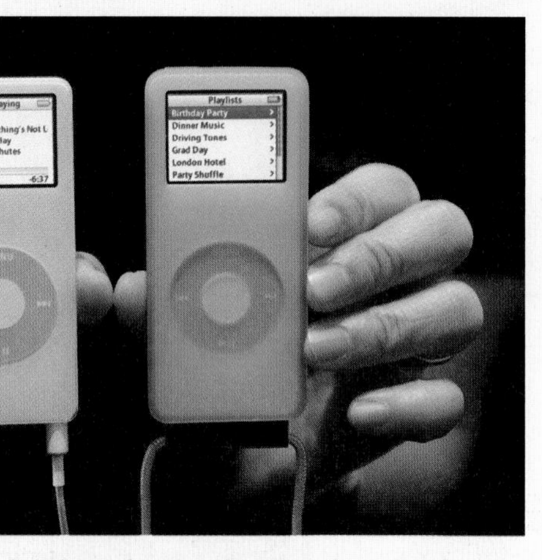

Digital sound recorders and players can store huge amounts of data. With products like these, we can listen for hours to music or audio books.

Thanks to great computing power and sophisticated graphics, we can play fast-moving, life-like video games on machines.

We can use portable DVD players to watch our favorite movies almost anywhere.

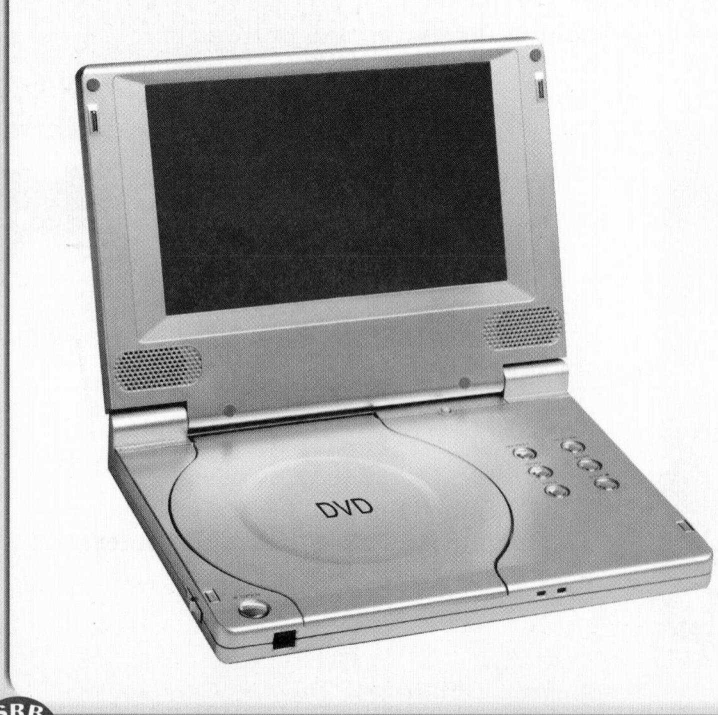

Cell phones are more than portable telephones. Many cell phones have watches, calculators, sound recorders and players, web browsers, and cameras built into them.

What machines do you use for calculating, convenience, and entertainment?

Calculators

About Calculators

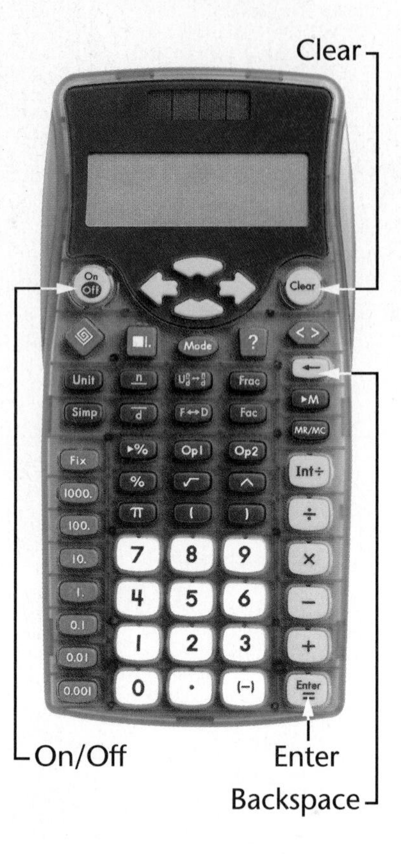

You have used calculators to help you learn to count. Now you can use them for working with whole numbers, fractions, decimals, and percents.

As with any mathematical tool or strategy, you need to think about when and how to use a calculator. It can help you compute quickly and accurately when you have many problems to do in a short time. Calculators can help you compute with very large and very small numbers that may be hard to do in your head or with pencil and paper. Whenever you use a calculator, estimation should be part of your work. Always ask yourself if the number in the display makes sense.

There are many different kinds of calculators. **Four-function calculators** do little more than add, subtract, multiply, and divide whole numbers and decimals. More advanced **scientific calculators** let you find powers and reciprocals, and some perform operations with fractions. After elementary school, you may use **graphic calculators** that draw graphs, find data landmarks, and do even more complicated mathematics.

There are many calculators that work well with *Everyday Mathematics.* If the instructions in this book don't work for your calculator, or the keys on your calculator are not explained, you should refer to the directions that came with the calculator, or ask your teacher for help.

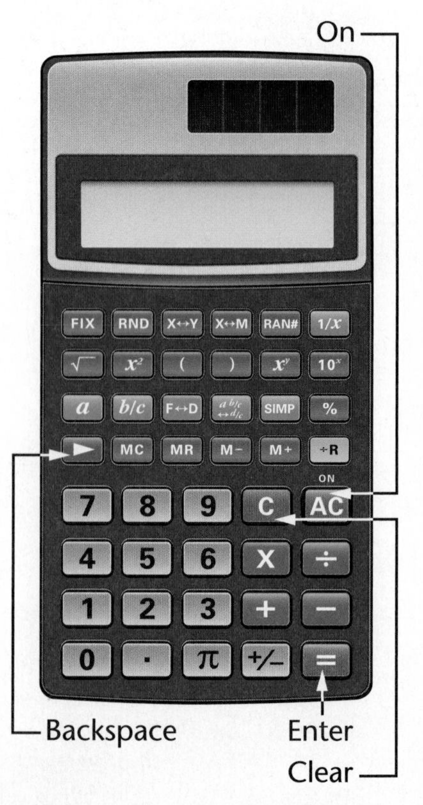

Basic Operations

You must take care of your calculator. Dropping it, leaving it in the sun, or other kinds of carelessness may break it or make it less reliable.

Many four-function and scientific calculators use light cells for power. If you press the ON key and see nothing on the display, hold the front of the calculator toward a light or a sunny window for a moment and then press ON again.

> **Note**
>
> Calculators have two kinds of memory. **Short-term memory** is for the last number entered. The keys with an "M" are for **long-term memory** and are explained on pages 218–220.

Entering and Clearing

Pressing a key on a calculator is called **keying in** or **entering.** In this book, calculator keys, except numbers and decimal points, are shown in rectangular boxes: [+], [=], [×], and so on. A set of instructions for performing a calculation is called a **key sequence.**

The simplest key sequences turn the calculator on and enter or clear numbers or other characters. These keys are labeled on the photos and summarized below.

Calculator A	Key Sequence	Purpose
	[On/Off]	Turn the display on.
	[Clear] and [On/Off]	Clear the display and the short-term memory.
	[Clear]	Clear only the display.
	[←]	Clear the last digit.

Calculator B	Key Sequence	Purpose
	[ON/AC]	Turn the display on.
	[AC]	Clear the display and the short-term memory.
	[C]	Clear the number if you make a mistake while entering a calculation.
	[►]	Clear the last digit.

Always clear both the display and the memory each time you turn on the calculator.

Many calculators have a backspace key that will clear the last digit or digits you entered without re-entering the whole number.

Example Enter 123.444. Change it to 123.456.

Calculator A	Key Sequence	Display
	123.444	123.444
	⟵ ⟵	123.4
	56	123.456

Calculator B	Key Sequence	Display
	123.444	123.444
	► ►	123.4
	56	123.456

Try using the backspace key on your calculator.

Order of Operations and Parentheses

When you use a calculator for basic arithmetic operations, you enter the numbers and operations and press [=] or [Enter] to see the answer.

Try your calculator to see if it follows the rules for the order of operations. Key in 5 [+] 6 [×] 2 [=].

- If your calculator follows the order of operations, it will display 17.
- If it does not follow the order of operations, it will probably do the operations in the order they were entered: adding and then multiplying, displaying 22.

If you want the calculator to do operations in an order different from the order of operations, use the parentheses keys [(] and [)].

Note

Examples for arithmetic calculations are given in earlier sections of this book.

Example Evaluate. 7 − (2 + 1)

Calculator A	Key Sequence	Display
	7 [−] [(] 2 [+] 1 [)] [Enter]	7−(2+1)= 4

Calculator B	Key Sequence	Display
	7 [−]	7.−
	[(]	( ) (01 0.
	2 [+]	( ) 2.+
	1	( ) 1.+
	[)]	3.−
	[=]	= 4.

7 − (2 + 1) = 4

Note

If you see a tiny up or down arrow on the calculator display, you can use the up or down arrows to scroll the screen.

Sometimes expressions are shown without all of the multiplication signs. *Remember to press the multiplication key even when it is not shown in an expression.*

Example Evaluate. $9 - 2(1 + 2)$

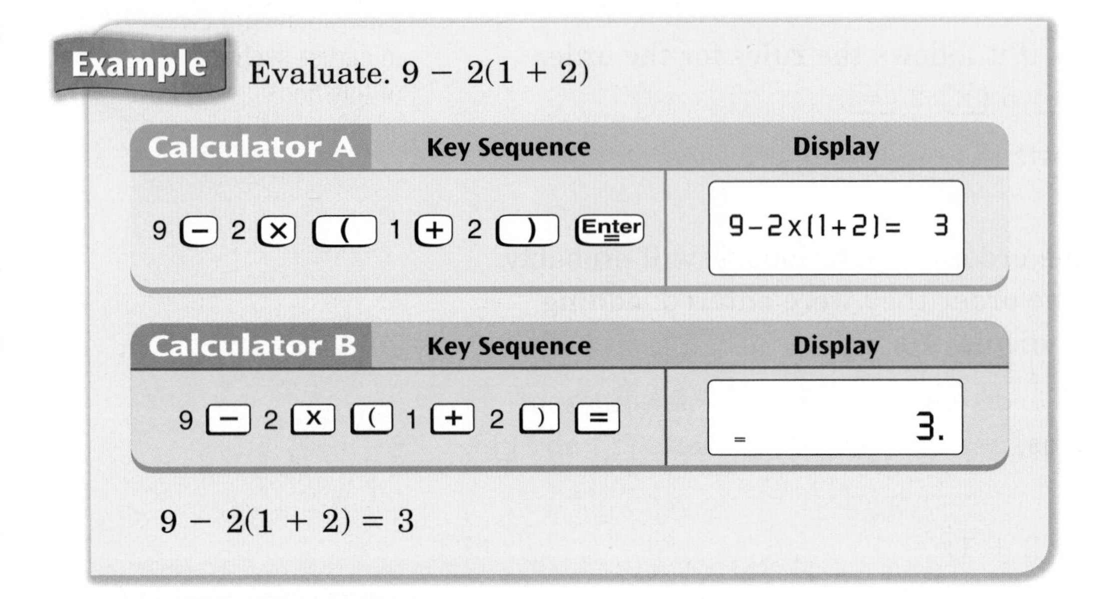

$9 - 2(1 + 2) = 3$

Check Your Understanding

Use your calculator to evaluate each expression.

1. $48 - (8 + 5)$ **2.** $81 - 7(4 + 2)$ **3.** $8(4 + 6.2) - 24$ **4.** $(64 - 16)/4 + 31$

Check your answers on page 347.

Negative Numbers

How you enter a negative number depends on your calculator. You will use the change sign key, either [+/-] or [(-)] depending on your calculator. Both keys change the sign of the number.

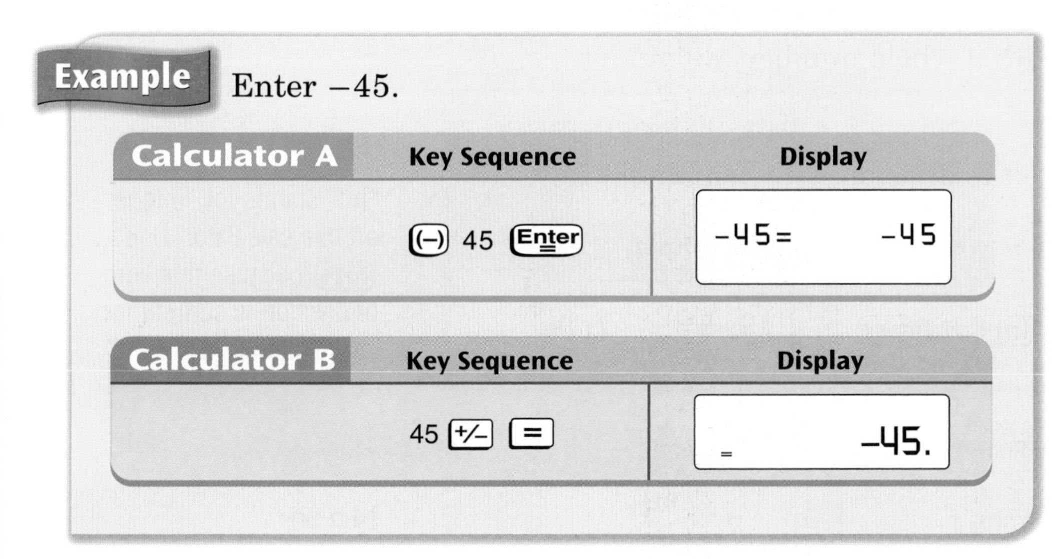

Example Enter -45.

Calculator A	Key Sequence	Display
	[(-)] 45 [Enter]	-45= -45

Calculator B	Key Sequence	Display
	45 [+/-] [=]	-45.

Note

If the number on the display is positive, it becomes negative after you press [+/-]. If the number on the display is negative, it becomes positive after pressing [+/-]. Keys like this are called **toggles.**

Example What happens if you try to subtract with [(-)] or [+/-]? Try it for $38 - 9 = ?$

Calculator A	Key Sequence	Display
	38 [(-)] 9 [Enter]	SYN ERROR

Calculator B	Key Sequence	Display
	38 [+/-] 9 [=]	-389.

Note

"SYN" is short for "syntax," which means the ordering and meaning of keys in a sequence.

Note

If you try to subtract using [+/-] on this calculator, it just changes the sign of the first number and adds the digits of the second number to it.

Division with Remainders

The answer to a division problem with whole numbers does not always result in whole number answers. When this happens, most calculators display the answer as a decimal. Some calculators also have a second division key that displays the whole number quotient with a whole number remainder.

Example $39 \div 5 = ?$ Use the division with remainder key.

Calculator A	Key Sequence	Display
	39 (Int÷) 5 (Enter)	39÷5= 7r 4

Calculator B	Key Sequence	Display
	39 [÷R] 5 [=]	7 R4

$39 \div 5 \rightarrow 7\ \text{R}4$

Note

"Int" stands for "integer" on this calculator. Use (Int÷) because this kind of division is sometimes called "integer division."

Note

[÷R] means "divide with remainder." You can also divide positive fractions and decimals with [÷R].

Try the division with remainder in the previous example to see how your calculator works.

Check Your Understanding

Divide with remainder.

1. $422 \div 7$ **2.** $231 \div 13$ **3.** $11{,}111 \div 43$

Check your answers on page 347.

Fractions and Percent

Some calculators let you enter, rewrite, and do operations with fractions. Once you know how to enter a fraction, you can add, subtract, multiply, or divide them just like whole numbers and decimals.

Entering Fractions and Mixed Numbers

Most calculators that let you enter fractions use similar key sequences. For proper fractions, always start by entering the numerator. Then press a key to tell the calculator to begin writing a fraction.

Example Enter $\frac{5}{8}$ as a fraction in your calculator.

Calculator A	Key Sequence	Display
	5 [n] 8 [d] [Enter]	$\frac{5}{8} = \frac{5}{8}$

Calculator B	Key Sequence	Display
	5 [b/c] 8 [=]	$\frac{5}{8}$

Note

Pressing [d] after you enter the denominator is optional.

To enter a mixed number, enter the whole number part and then press a key to tell the calculator what you did.

Example Enter $73\frac{2}{5}$ as a fraction in your calculator.

Calculator A	Key Sequence	Display
	73 [Unit] 2 [n] 5 [d] [Enter]	$73\frac{2}{5} = 73\frac{2}{5}$

Calculator B	Key Sequence	Display
	73 [a] 2 [b/c] 5 [=]	$73\frac{2}{5}$

Try entering a mixed number on your calculator.

The keys to convert between mixed numbers and improper fractions are similar on all fraction calculators.

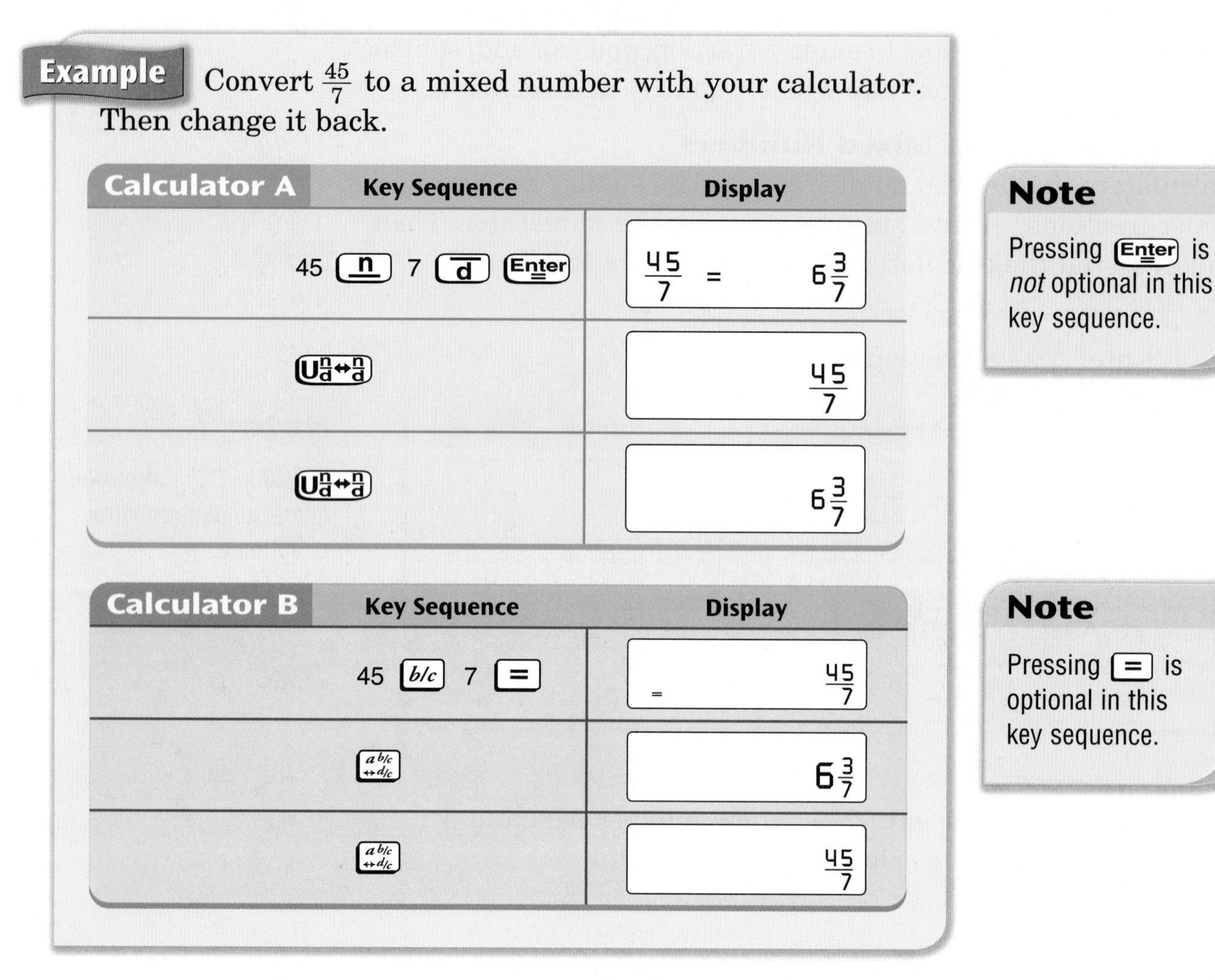

Example Convert $\frac{45}{7}$ to a mixed number with your calculator. Then change it back.

Calculator A Key Sequence	Display
45 [n] 7 [d] [Enter]	$\frac{45}{7} = 6\frac{3}{7}$
[U n/d ↔ n/d]	$\frac{45}{7}$
[U n/d ↔ n/d]	$6\frac{3}{7}$

Calculator B Key Sequence	Display
45 [b/c] 7 [=]	= $\frac{45}{7}$
[a b/c ↔ d/c]	$6\frac{3}{7}$
[a b/c ↔ d/c]	$\frac{45}{7}$

Note

Pressing [Enter] is *not* optional in this key sequence.

Note

Pressing [=] is optional in this key sequence.

Both [U n/d ↔ n/d] and [a b/c ↔ d/c] toggle between mixed number and improper fraction notation.

Simplifying Fractions

Ordinarily, calculators do not simplify fractions on their own. The steps for simplifying fractions are similar for many calculators, but the order of the steps varies. Approaches for two calculators are shown on the next three pages depending on the keys you have on your calculator. Read the approaches for the calculator having keys most like yours.

Simplifying Fractions on Calculator A

This calculator lets you simplify a fraction in two ways. Each way divides the numerator and the denominator by a common factor. The first approach uses [Simp] to automatically divide by the smallest common factor, and [Fac] to display the factor.

Example Convert $\frac{18}{24}$ to simplest form using smallest common factors.

Calculator A Key Sequence	Display
18 [n] 24 [d] [Simp] [Enter]	$\frac{18}{24}$ ▸ S $\frac{9}{12}$ (N→n / D→d)
[Fac]	2
[Fac] [Simp] [Enter]	$\frac{9}{12}$ ▸ S $\frac{3}{4}$
[Fac]	3

$\frac{18}{24} = \frac{3}{4}$

Calculator A

[Fac] displays the common factor used to simplify a fraction.

[Simp] simplifies fractions.

Note

Pressing [Fac] toggles between the display of the factor and the display of the fraction.

In the second approach, you can simplify the fraction in one step by telling the calculator to divide by the greatest common factor of the numerator and the denominator.

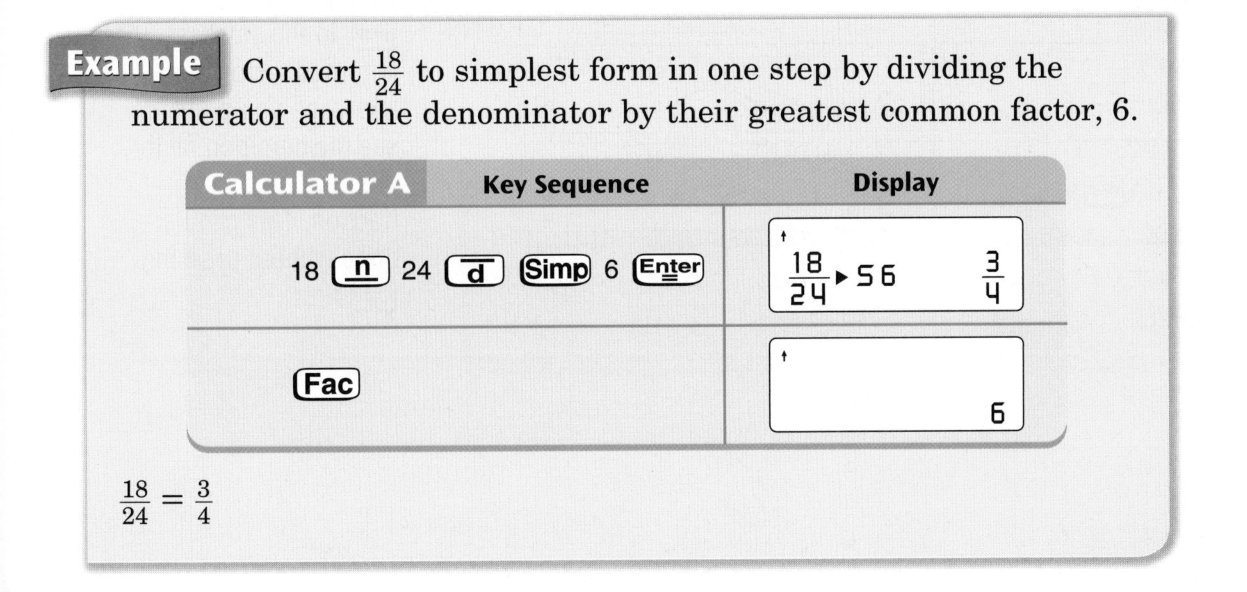

Example Convert $\frac{18}{24}$ to simplest form in one step by dividing the numerator and the denominator by their greatest common factor, 6.

Calculator A Key Sequence	Display
18 [n] 24 [d] [Simp] 6 [Enter]	$\frac{18}{24}$ ▸ S6 $\frac{3}{4}$
[Fac]	6

$\frac{18}{24} = \frac{3}{4}$

Simplifying Fractions on Calculator B

Calculators like this one let you simplify fractions three different ways. Each way divides the numerator and the denominator by a common factor. The first approach uses [=] to give the simplest form in one step. The word *Simp* in the display means that the fraction shown is not in simplest form.

Calculator B

[SIMP] simplifies a fraction by a common factor.

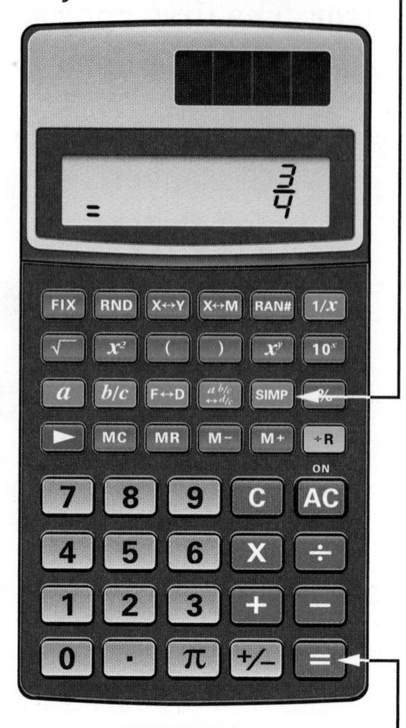

Press [=] [=] to write in simplest form.

Example Convert $\frac{18}{24}$ to simplest form in one step.

Calculator B	Key Sequence	Display
	18 [b/c] 24 [=]	Simp = $\frac{18}{24}$
	[=]	= $\frac{3}{4}$

$\frac{18}{24} = \frac{3}{4}$

If you enter a fraction that is already in simplest form, you will not see Simp on the display. The one-step approach does not tell you the common factor as the next two approaches do using [SIMP].

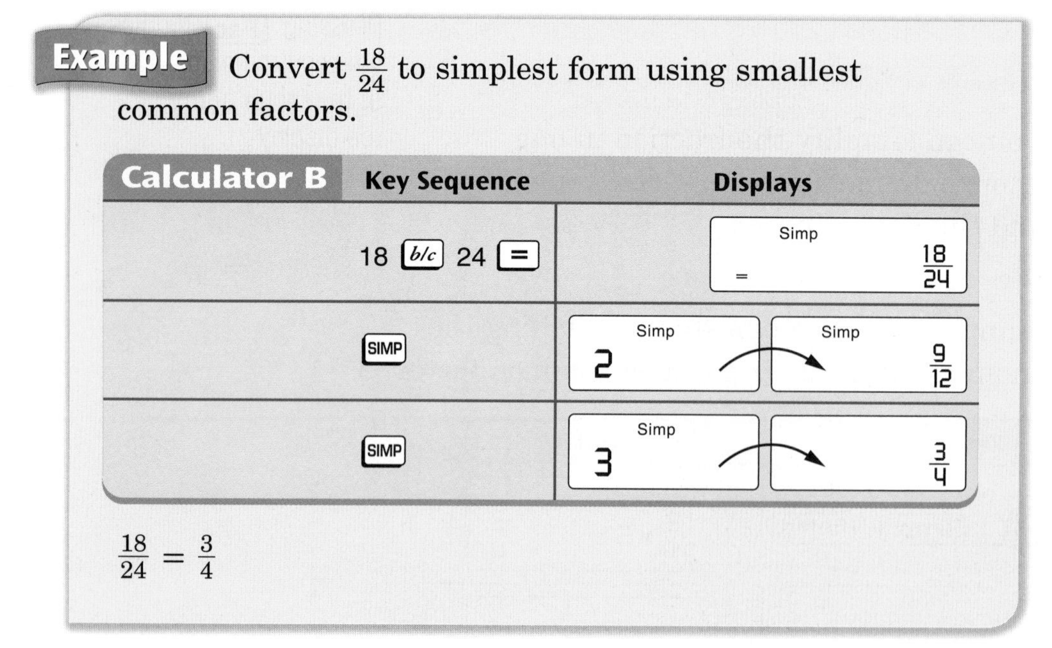

Example Convert $\frac{18}{24}$ to simplest form using smallest common factors.

Calculator B	Key Sequence	Displays
	18 [b/c] 24 [=]	Simp = $\frac{18}{24}$
	[SIMP]	Simp 2 → Simp $\frac{9}{12}$
	[SIMP]	Simp 3 → $\frac{3}{4}$

$\frac{18}{24} = \frac{3}{4}$

Note

Each time you press [SIMP] in the smallest common factor approach you briefly see the common factor, then the simplified fraction. This can be done without pressing [=] first.

In the last approach to simplifying fractions with this type of calculator, you tell it what common factor to divide by. If you use the greatest common factor of the numerator and the denominator, you can simplify the fraction in one step.

Example Convert $\frac{18}{24}$ to simplest form by dividing the numerator and the denominator by their greatest common factor, 6.

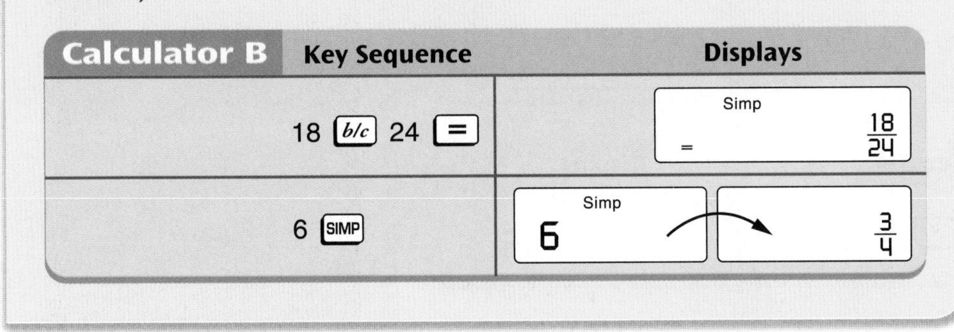

Calculator B Key Sequence	Displays
18 b/c 24 =	Simp = 18/24
6 SIMP	Simp 6 → 3/4

Note

If you enter a number that is not a common factor of the numerator and the denominator, you will get an error symbol "E" in the display with the unchanged fraction.

Try simplifying the fractions in the previous examples to see how your calculator works.

Percent

Many calculators have a [%] key, but it is likely that they work differently. The best way to learn what your calculator does with percents is to read its manual.

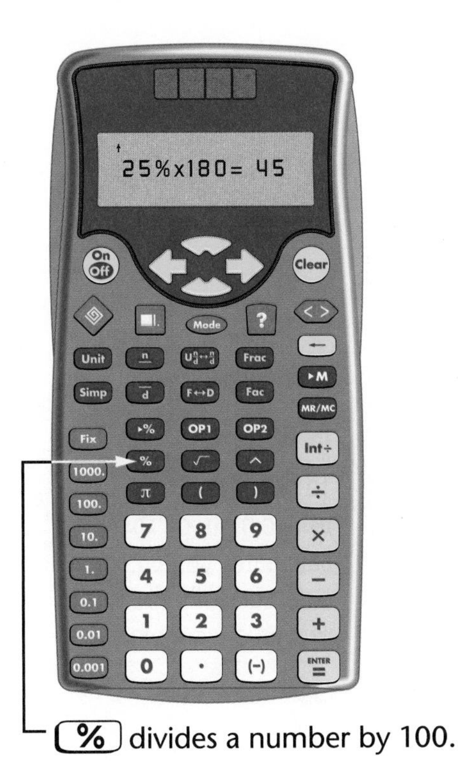

[%] divides a number by 100.

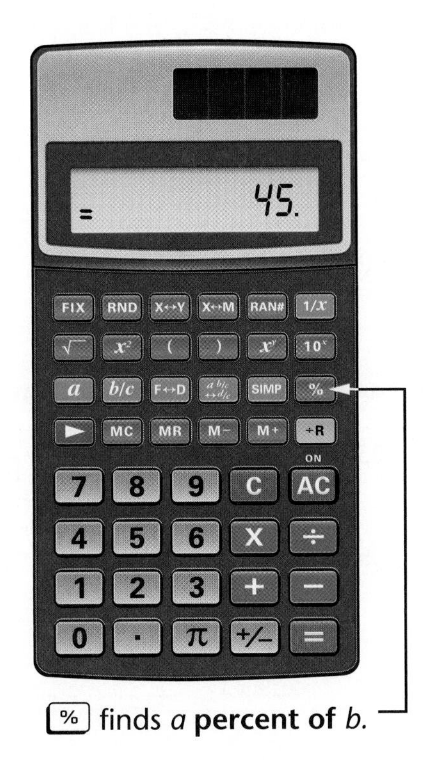

[%] finds *a* **percent of** *b*.

Most calculators include [%] to solve "percent of" problems.

Example Calculate 25% of 180.

Calculator A	Key Sequence	Display
	180 [×] 25 [%] [Enter]	180x25%= 45
	25 [%] [×] 180 [Enter]	25%x180= 45

Calculator B	Key Sequence	Display
	180 [X] 25 [%]	45.

25% of 180 is 45.

You can change percents to decimals with [%].

Examples Display 85%, 250%, and 1% as decimals.

Calculator A	Key Sequence	Display
	85 [%] [Enter]	85%= 0.85
	250 [%] [Enter]	250%= 2.5
	1 [%] [Enter]	1%= 0.01

Calculator B	Key Sequence	Display
	1 [X] 85 [%]	0.85
	1 [X] 250 [%]	2.5
	1 [X] 1 [%]	0.01

Note

To convert percents to decimals on this calculator you calculate a percent of 1, as in the previous example.

You can also use [%] to convert percents to fractions.

On many calculators, first change the percent to a decimal as in the previous examples, then use [F↔D] to change to a fraction.

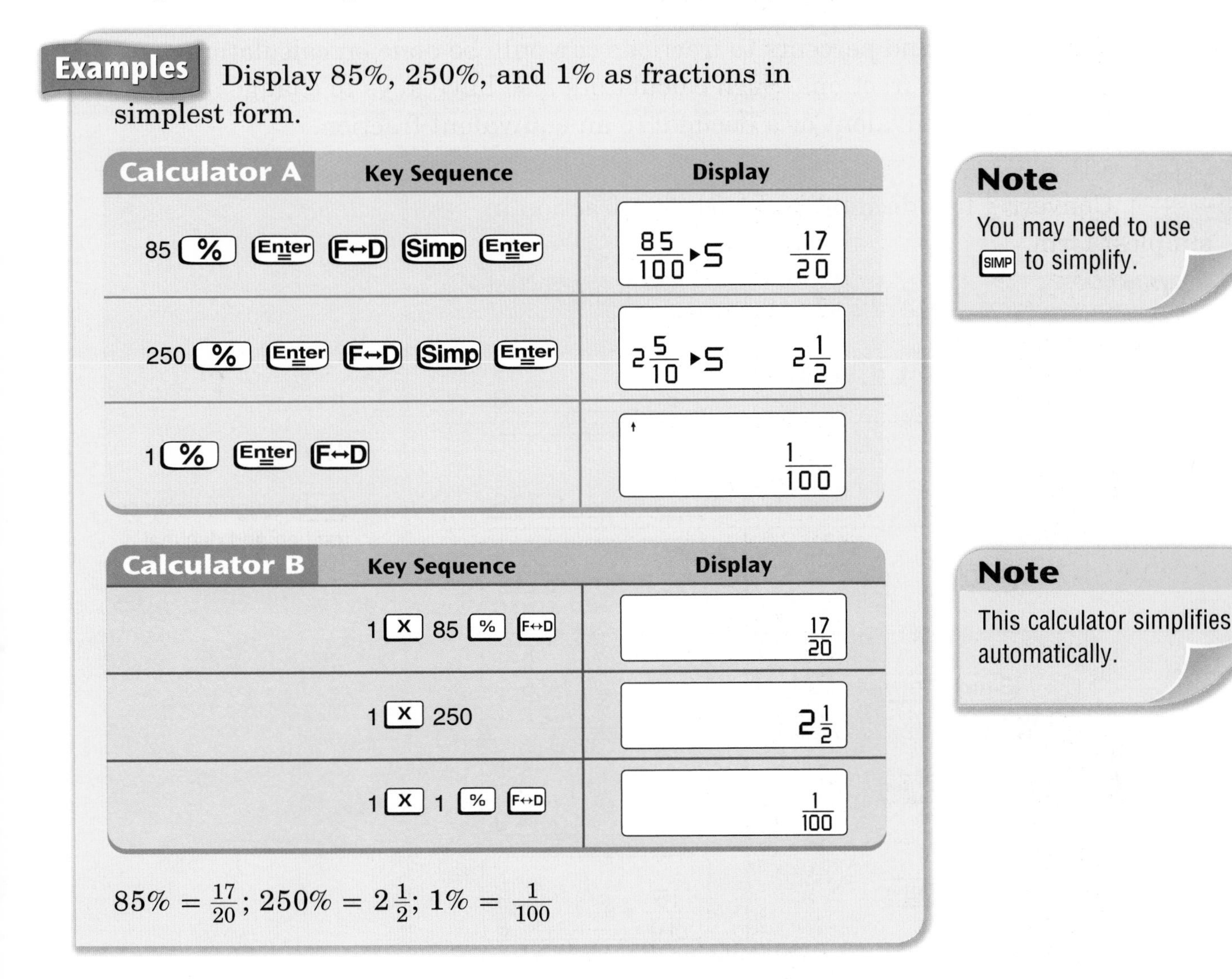

Examples Display 85%, 250%, and 1% as fractions in simplest form.

Calculator A Key Sequence	Display
85 [%] [Enter] [F↔D] [Simp] [Enter]	$\frac{85}{100}$ ▸S $\frac{17}{20}$
250 [%] [Enter] [F↔D] [Simp] [Enter]	$2\frac{5}{10}$ ▸S $2\frac{1}{2}$
1 [%] [Enter] [F↔D]	$\frac{1}{100}$

Calculator B Key Sequence	Display
1 [X] 85 [%] [F↔D]	$\frac{17}{20}$
1 [X] 250	$2\frac{1}{2}$
1 [X] 1 [%] [F↔D]	$\frac{1}{100}$

$85\% = \frac{17}{20}$; $250\% = 2\frac{1}{2}$; $1\% = \frac{1}{100}$

Note

You may need to use [SIMP] to simplify.

Note

This calculator simplifies automatically.

Try displaying some percents as fractions and decimals on your calculator.

Fraction/Decimal/Percent Conversions

Conversions of fractions to decimals and percents can be done on any calculator. For example, to rename $\frac{3}{5}$ as a decimal, simply enter 3 [÷] 5 [=]. The display will show 0.6. To rename a decimal as a percent, just multiply by 100.

Conversions of decimals and percents to fractions can only be done on calculators that have special keys for fractions. Such calculators also have keys to change a fraction to its decimal equivalent or a decimal to an equivalent fraction.

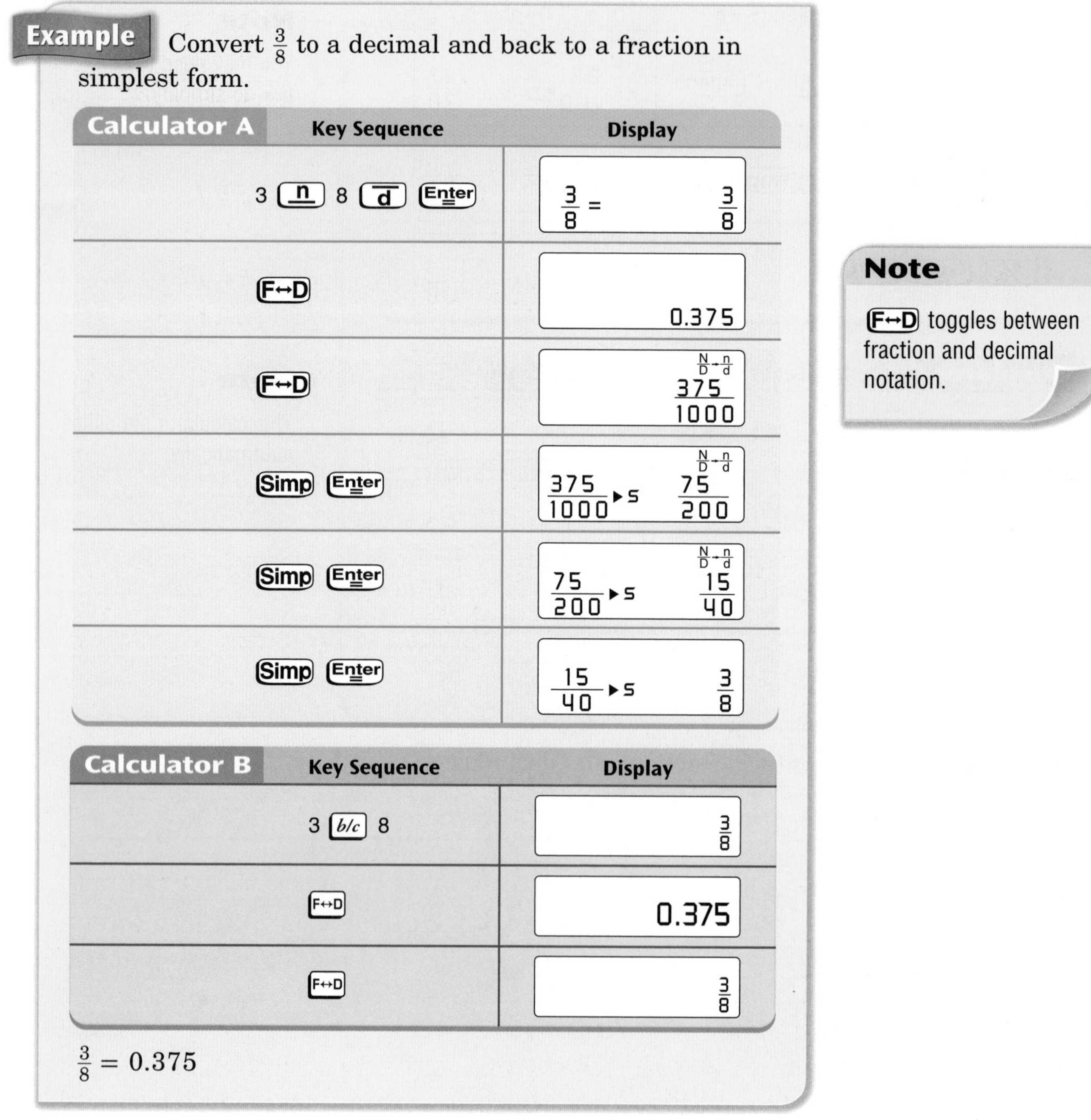

Example Convert $\frac{3}{8}$ to a decimal and back to a fraction in simplest form.

Calculator A Key Sequence	Display
3 [n] 8 [d] [Enter]	$\frac{3}{8}$ = $\frac{3}{8}$
[F↔D]	0.375
[F↔D]	N/D→n/d $\frac{375}{1000}$
[Simp] [Enter]	$\frac{375}{1000}$ ▶S N/D→n/d $\frac{75}{200}$
[Simp] [Enter]	$\frac{75}{200}$ ▶S N/D→n/d $\frac{15}{40}$
[Simp] [Enter]	$\frac{15}{40}$ ▶S $\frac{3}{8}$

Calculator B Key Sequence	Display
3 [b/c] 8	$\frac{3}{8}$
[F↔D]	0.375
[F↔D]	$\frac{3}{8}$

$\frac{3}{8} = 0.375$

Note

[F↔D] toggles between fraction and decimal notation.

See how your calculator changes fractions to decimals.

The tables below show examples of various conversions. Although only one key sequence is shown for each conversion, there are often other key sequences that work as well.

Conversion	Starting Number	Calculator A Key Sequence	Display
Fraction to decimal	$\frac{3}{5}$	3 [n] 5 [d] [Enter] [F↔D]	0.6
Decimal to fraction	0.125	.125 [Enter] [F↔D]	N/D→n/d $\frac{125}{1000}$
Decimal to percent	0.75	.75 [▶%] [Enter]	0.75▶% 75%
Percent to decimal	125%	125 [%] [Enter]	125%= 1.25
Fraction to percent	$\frac{5}{8}$	5 [n] 8 [d] [▶%] [Enter]	$\frac{5}{8}$▶% 62.5%
Percent to fraction	35%	35 [%] [Enter] [F↔D]	N/D→n/d $\frac{35}{100}$

Conversion	Starting Number	Calculator B Key Sequence	Display
Fraction to decimal	$\frac{3}{5}$	3 [b/c] 5 [F↔D]	0.6
Decimal to fraction	0.125	.125 [F↔D]	$\frac{1}{8}$
Decimal to percent	0.75	.75 [X] 100 [=]	= 75.
Percent to decimal	125%	1 [X] 125 [%]	1.25
Fraction to percent	$\frac{5}{8}$	5 [b/c] 8 [F↔D] [X] 100 [=]	= 62.5
Percent to fraction	35%	1 [X] 35 [%] [F↔D]	$\frac{7}{20}$

Check Your Understanding

Use your calculator to convert between fractions, decimals, and percents.

1. $\frac{15}{16}$ to a decimal
2. 0.565 to a fraction
3. 0.007 to a percent
4. 291% to a decimal
5. $\frac{11}{32}$ to a percent
6. 72% to a fraction

Check your answers on page 347.

Other Operations

Your calculator can do more than simple arithmetic with whole numbers, fractions, and decimals. Each kind of calculator does some things that other calculators cannot, or does them in different ways. See your owner's manual, or ask your teacher, to help you explore these things. The following pages explain some other things that many calculators can do.

Rounding

All calculators can round decimals. Decimals must be rounded to fit on the display. For example, if you key in 2 [÷] 3 [=]

- Calculator A shows 11 digits and rounds to the nearest value: 0.6666666667.
- Calculator B shows 8 digits and rounds down to 0.6666666.

Try 2 [÷] 3 [=] on your calculator to see how big the display is and how it rounds.

All scientific calculators have a [FIX] key to set, or **fix,** the place value of decimals on the display. Fixing always rounds to the nearest value.

Note

To turn off fixed rounding on a calculator, press [FIX] [•].

Examples Clear your calculator and fix it to round to tenths. Round each number, 1.34; 812.79; and 0.06, to the nearest tenth.

Calculator A Key Sequence	Display
[Clear] [Fix] [0.1]	Fix
1.34 [Enter]	Fix 1.34= 1.3
812.79 [Enter]	Fix 812.79= 812.8
.06 [Enter]	Fix .06= 0.1

1.34 rounds to 1.3; 812.79 rounds to 812.8; 0.06 rounds to 0.1.

Note

On this calculator you can fix the decimal places to the left of the decimal point but are limited to the right of the decimal point to thousandths (0.001).

Examples Clear your calculator and fix it to round to tenths. Round each number, 1.34; 812.79; and 0.06, to the nearest tenth.

Calculator B Key Sequence	Display
AC FIX	0 – 7
1	FIX 0.0
1.34 =	FIX = 1.3
812.79 =	FIX = 812.8
.06 =	FIX = 0.1

1.34 rounds to 1.3; 812.79 rounds to 812.8; 0.06 rounds to 0.1.

Note

This calculator only lets you fix places to the right of the decimal point.

Note

You can fix either calculator to round without clearing the display first. It will round the number on the display.

Check Your Understanding

Use your calculator to round to the indicated place.

1. 0.37 to tenths **2.** 521.73 to ones **3.** 6174.6138 to thousandths

4. 0.8989 to hundredths

Check your answers on page 347.

Fixing the display to round to hundredths is helpful for solving problems about dollars and cents.

Example One CD costs $11.23 and another costs $14.67. Set your calculator to round to the nearest cent and calculate the total cost of the CDs.

Calculator A Key Sequence	Display
Fix 0.01	Fix
11.23 + 14.67 Enter	Fix 11.23 + 14.67 = 25.90

Calculator B Key Sequence	Display
FIX	0 - 7
2	FIX 0.00
11.23 + 14.67 =	FIX = 25.90

Together, the CDs cost $25.90.

On most calculators, if you find the total in the Example above with the "fix" turned off, the display reads 25.9. To show the answer in dollars and cents, fix the display to round to hundredths and you will see 25.90.

Powers, Reciprocals, and Square Roots

Powers of numbers can be calculated on all scientific calculators.

Look at your calculator to see which key it has for finding powers of numbers.

- The key may look like [x^y] and is read "x to the y."
- The key may look like [∧], and is called a **caret.**

To compute a number to a negative power, be sure to use the change-sign key [(–)] or [+/–], not the subtraction key [–].

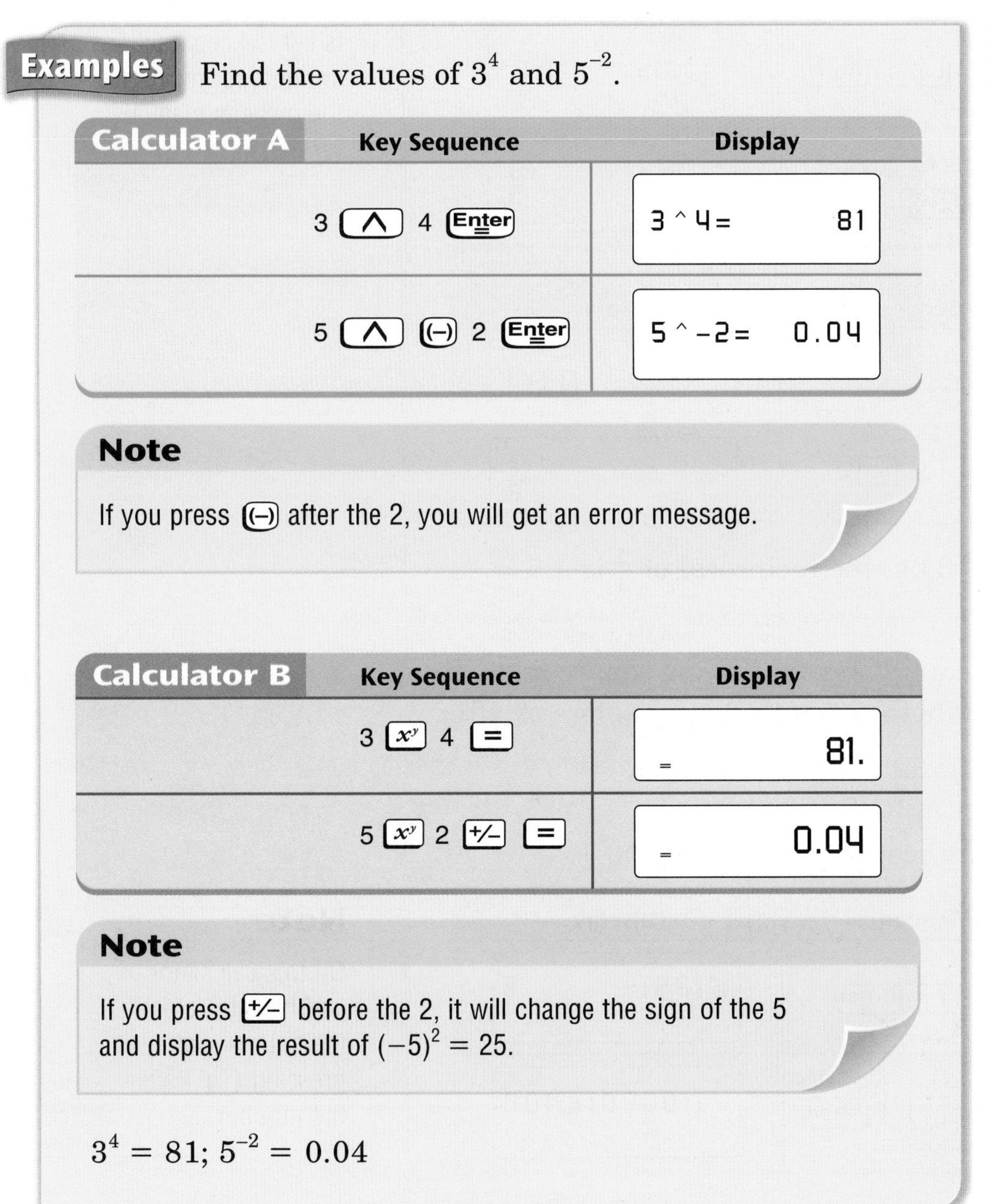

Examples Find the values of 3^4 and 5^{-2}.

Calculator A	Key Sequence	Display
	3 [∧] 4 [Enter]	3 ^ 4 = 81
	5 [∧] [(–)] 2 [Enter]	5 ^ –2 = 0.04

Note

If you press [(–)] after the 2, you will get an error message.

Calculator B	Key Sequence	Display
	3 [x^y] 4 [=]	81.
	5 [x^y] 2 [+/–] [=]	0.04

Note

If you press [+/–] before the 2, it will change the sign of the 5 and display the result of $(-5)^2 = 25$.

$3^4 = 81$; $5^{-2} = 0.04$

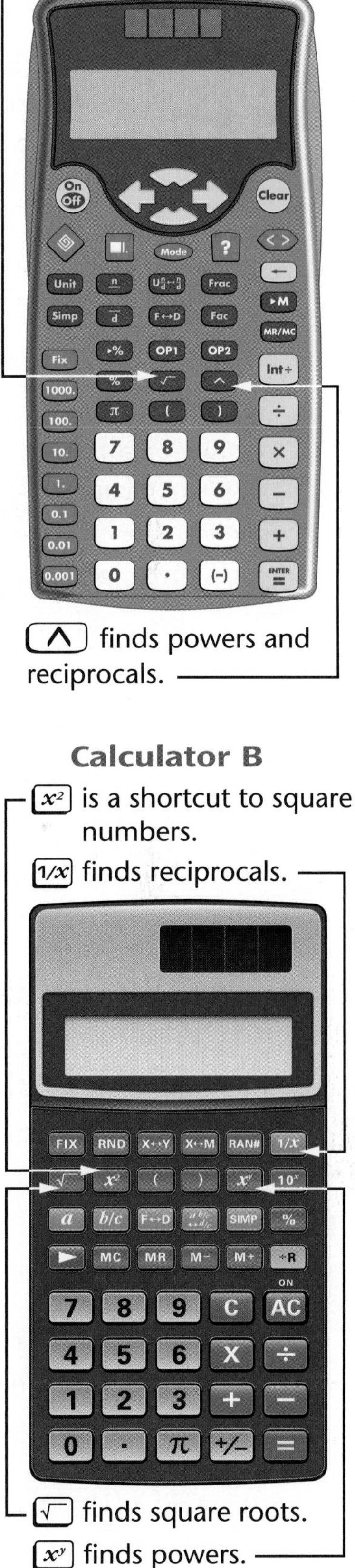

Most scientific calculators have a reciprocal key [1/x]. On all scientific calculators you can find a reciprocal of a number by raising it to the −1 power.

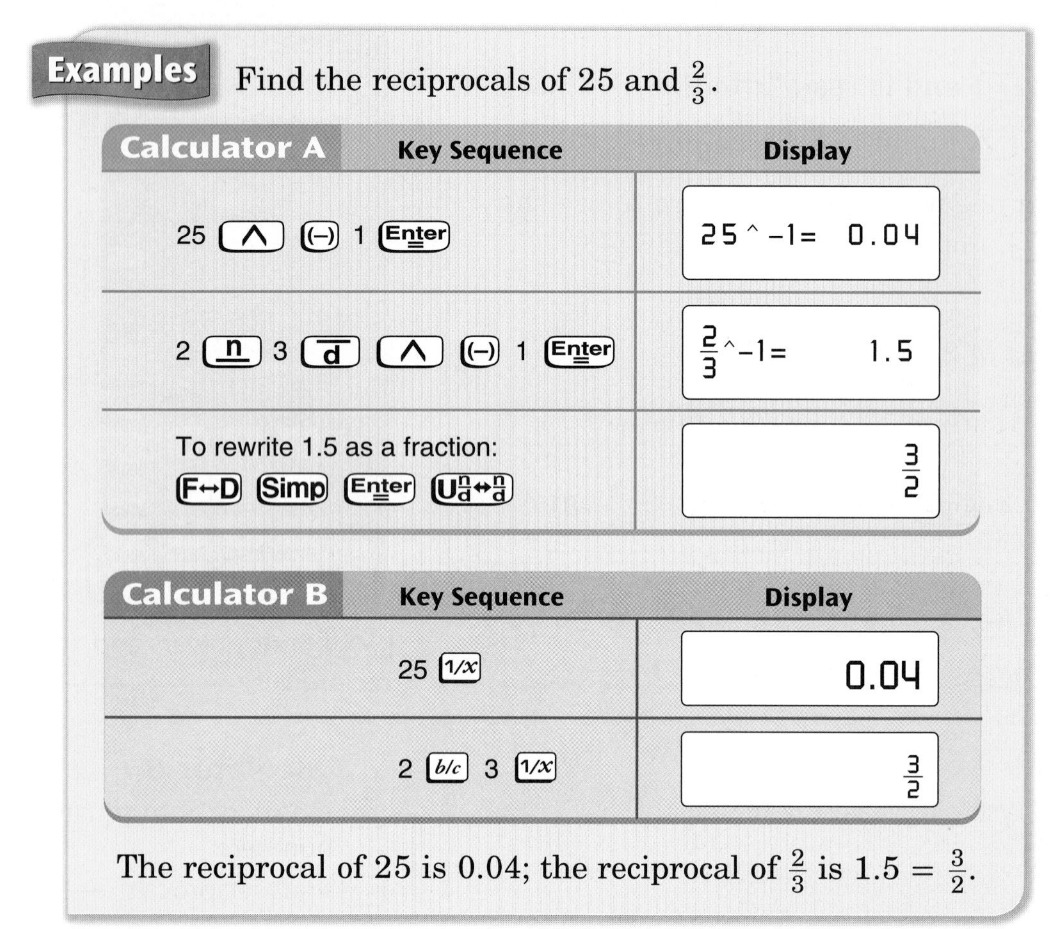

Examples Find the reciprocals of 25 and $\frac{2}{3}$.

Calculator A Key Sequence	Display
25 [^] [(−)] 1 [Enter]	25 ^ −1= 0.04
2 [n] 3 [d] [^] [(−)] 1 [Enter]	$\frac{2}{3}$ ^ −1= 1.5
To rewrite 1.5 as a fraction: [F↔D] [Simp] [Enter] [U$\frac{n}{d}$↔$\frac{n}{d}$]	$\frac{3}{2}$

Calculator B Key Sequence	Display
25 [1/x]	0.04
2 [b/c] 3 [1/x]	$\frac{3}{2}$

The reciprocal of 25 is 0.04; the reciprocal of $\frac{2}{3}$ is $1.5 = \frac{3}{2}$.

Note

You don't need to press the last key in the final step if your calculator is set to keep fractions in improper form. See the owner's manual for details.

Almost all calculators have a [√] key for finding square roots. It depends on the calculator whether you press [√] before or after entering a number.

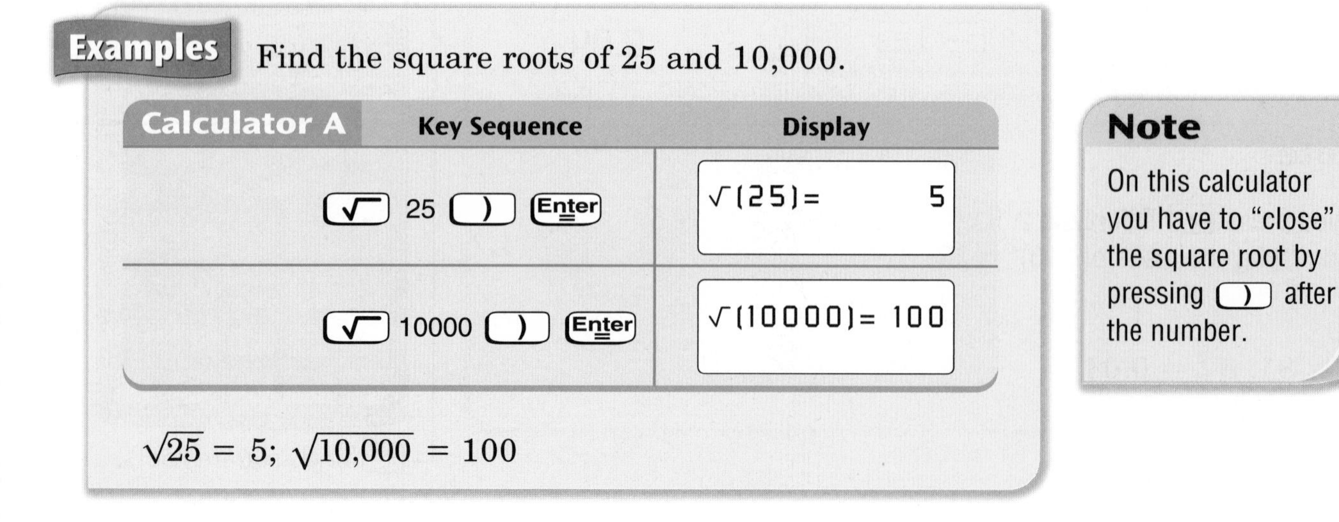

Examples Find the square roots of 25 and 10,000.

Calculator A Key Sequence	Display
[√] 25 [)] [Enter]	√(25)= 5
[√] 10000 [)] [Enter]	√(10000)= 100

$\sqrt{25} = 5$; $\sqrt{10{,}000} = 100$

Note

On this calculator you have to "close" the square root by pressing [)] after the number.

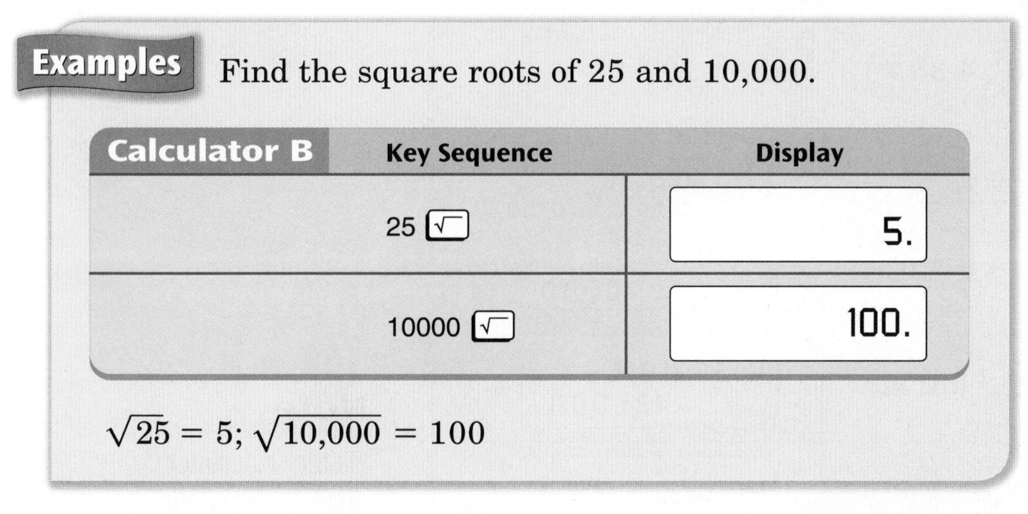

Examples Find the square roots of 25 and 10,000.

Calculator B	Key Sequence	Display
	25 [√]	5.
	10000 [√]	100.

$\sqrt{25} = 5$; $\sqrt{10{,}000} = 100$

Try finding the square roots of a few numbers.

Scientific Notation

Scientific notation is a way of writing very large or very small numbers. A number in scientific notation is shown as a product of a number between 1 and 10 and a power of 10. In scientific notation, the 9,000,000,000 bytes of memory on a 9-gigabyte hard drive is written $9 * 10^9$. On scientific calculators, numbers with too many digits to fit on the display are automatically shown in scientific notation like the calculator below on the right.

Different calculators use different symbols for scientific notation. Your calculator may display raised exponents of 10, although most do not. Since the base of the power is always 10, most calculators leave out the 10 and simply put a space between the number and the exponent.

This calculator uses the caret ^ to display scientific notation.

This calculator shows $9 * 10^9$

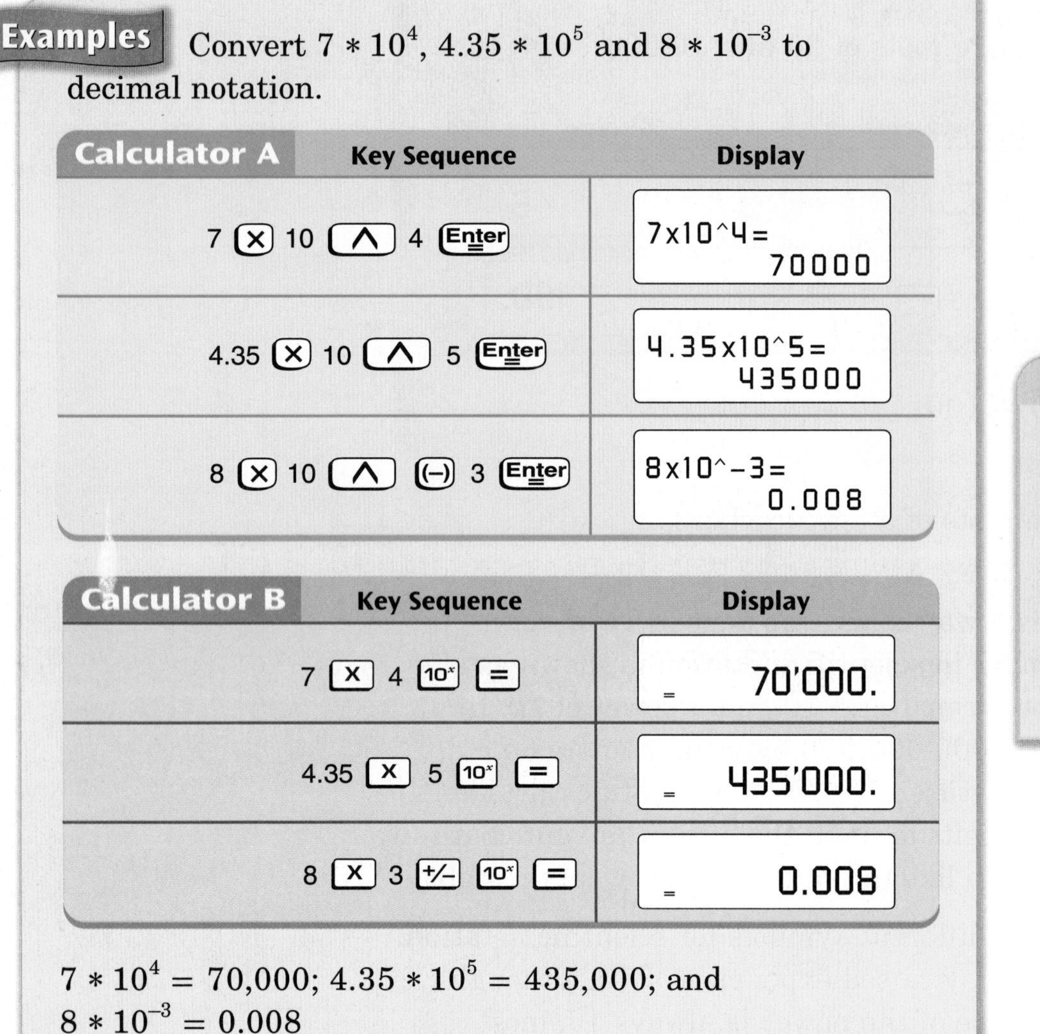

Examples Convert $7 * 10^4$, $4.35 * 10^5$ and $8 * 10^{-3}$ to decimal notation.

Calculator A Key Sequence	Display
7 [×] 10 [^] 4 [Enter]	7x10^4= 70000
4.35 [×] 10 [^] 5 [Enter]	4.35x10^5= 435000
8 [×] 10 [^] [(–)] 3 [Enter]	8x10^-3= 0.008

Calculator B Key Sequence	Display
7 [X] 4 [10ˣ] [=]	70'000.
4.35 [X] 5 [10ˣ] [=]	435'000.
8 [X] 3 [+/–] [10ˣ] [=]	0.008

$7 * 10^4 = 70{,}000$; $4.35 * 10^5 = 435{,}000$; and $8 * 10^{-3} = 0.008$

Note

Neither calculator displays large numbers in decimal notation with a comma like you do with pencil and paper. One uses an apostrophe; the other uses no symbol at all.

Check Your Understanding

Use your calculator to convert the following to standard notation:

1. $4.9 * 10^{-4}$ **2.** $3.3 * 10^7$ **3.** $6.147 * 10^{-6}$ **4.** $-7.8 * 10^{-5}$

Check your answers on page 347.

Calculators have different limits to the numbers they can display without scientific notation.

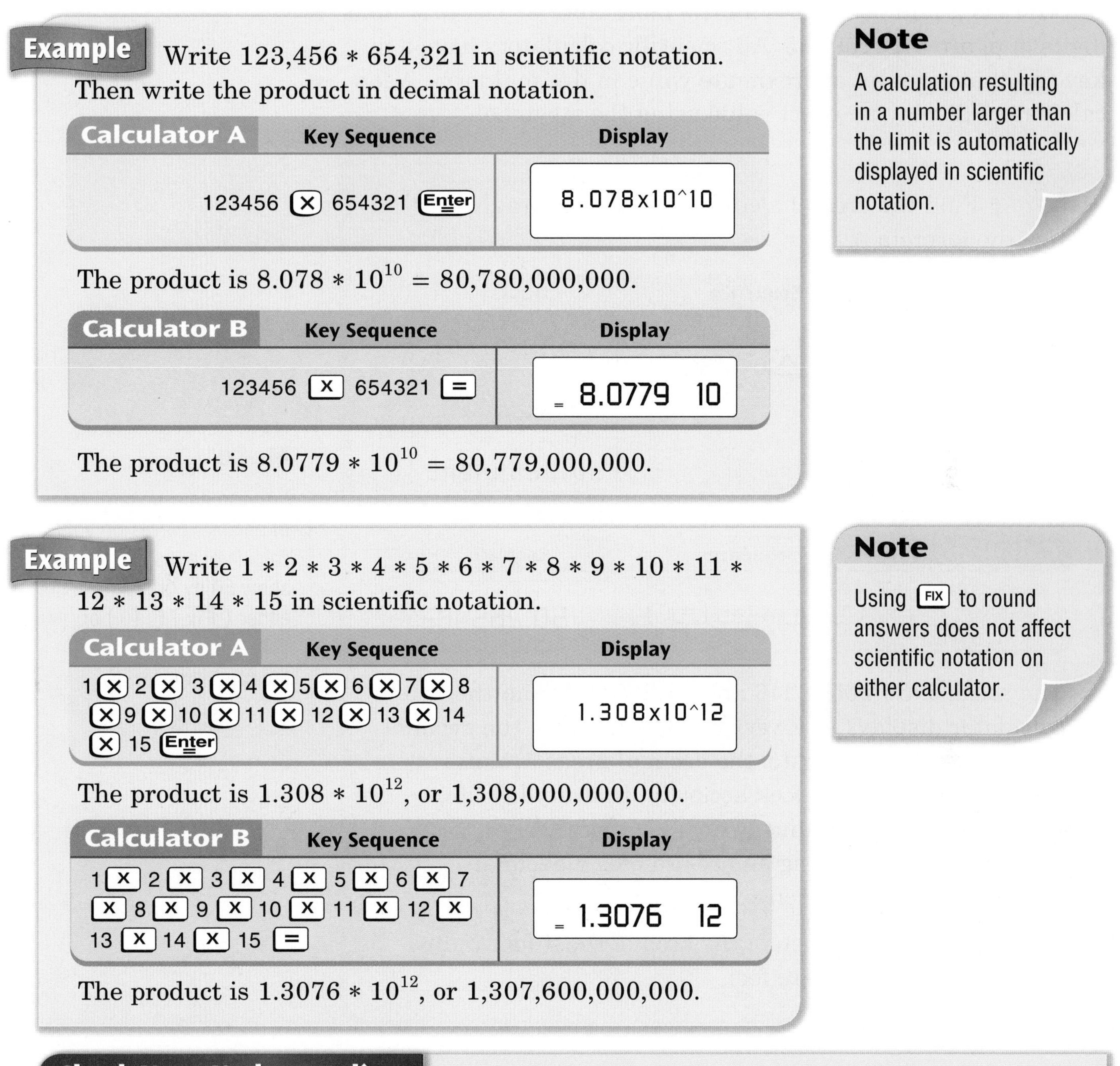

Example Write 123,456 * 654,321 in scientific notation. Then write the product in decimal notation.

Calculator A	Key Sequence	Display
	123456 [×] 654321 [Enter]	8.078x10^10

The product is $8.078 * 10^{10} = 80{,}780{,}000{,}000$.

Calculator B	Key Sequence	Display
	123456 [×] 654321 [=]	8.0779 10

The product is $8.0779 * 10^{10} = 80{,}779{,}000{,}000$.

Note

A calculation resulting in a number larger than the limit is automatically displayed in scientific notation.

Example Write 1 * 2 * 3 * 4 * 5 * 6 * 7 * 8 * 9 * 10 * 11 * 12 * 13 * 14 * 15 in scientific notation.

Calculator A	Key Sequence	Display
	1 [×] 2 [×] 3 [×] 4 [×] 5 [×] 6 [×] 7 [×] 8 [×] 9 [×] 10 [×] 11 [×] 12 [×] 13 [×] 14 [×] 15 [Enter]	1.308x10^12

The product is $1.308 * 10^{12}$, or 1,308,000,000,000.

Calculator B	Key Sequence	Display
	1 [×] 2 [×] 3 [×] 4 [×] 5 [×] 6 [×] 7 [×] 8 [×] 9 [×] 10 [×] 11 [×] 12 [×] 13 [×] 14 [×] 15 [=]	1.3076 12

The product is $1.3076 * 10^{12}$, or 1,307,600,000,000.

Note

Using [FIX] to round answers does not affect scientific notation on either calculator.

Check Your Understanding

Write in scientific notation.

1. 995 * 7 * 54 * 65 * 659 * 807 * 468
2. 956 * 859 * 760 * 862
3. 527 * 32 * 987 * 424 * 77 * 145 * 195
4. 15^9 * 13 * 996 * 558
5. The number of different 5-card hands that can be drawn from a standard deck of 52 cards is: 52 * 51 * 50 * 49 * 48. How many hands is this in scientific notation? In decimal notation?

Check your answers on page 347.

Pi (π)

The formulas for the circumference and area of circles involve **pi (π).** Pi is a number that is a little more than 3. The first nine digits of pi are 3.14159265. All scientific calculators have a pi key [π] that gives an approximate value in decimal form. A few calculators display an exact value using the π symbol.

Example Find the area of a circle with a 4-foot radius. Use the formula $A = \pi r^2$.

Calculator A	Key Sequence	Display
	[π] [×] 4 [∧] 2 [Enter]	πx4^2= 16π
	[F↔D]	50.26548246

Calculator B	Key Sequence	Display
	[π] [×] 4 [x²] [=]	50.265482

The areas of 50.26548246 and 50.265482 from the two calculator displays look very precise. Because the decimal value of π is approximate, the decimal areas are also approximate, but still look accurate. In everyday life, the measure of the radius of a circle is probably approximate, and giving an area to 6 or 8 decimal places does not make sense.

So a good approximation of the area of the 4-foot radius circle is about 50 square feet.

Note

You can set the number of decimal places on your calculator's display to show 50 by pressing either [Fix] [1.] or [FIX] 0 depending on the calculator.

Example Find the circumference of a circle with a 15-centimeter diameter to the nearest tenth of a centimeter. Use the formula $C = \pi d$.

Calculator A	Key Sequence	Display
	Fix 0.1	Fix
	π × 15 Enter	Fix π x15= 15π
	F↔D	Fix 47.1

Calculator B	Key Sequence	Display
	FIX 1	FIX 0.0
	π × 15 =	FIX = 47.1

The circumference is about 47.1 centimeters.

Note

When you are finished, remember to turn off the fixed rounding by pressing FIX •.

Check Your Understanding

1. Find the area of a circle with an 18 foot radius. Display your answer to the nearest square foot.

2. To the nearest tenth of a centimeter, find the circumference of a circle with a 26.7 centimeter diameter.

Check your answers on page 347.

Using Calculator Memory

Many calculators let you save a number in **long-term memory** using keys with "M" on them. Later on, when you need the number, you can recall it from memory. Most calculators display an "M" or similar symbol when there is a number other than 0 in the memory.

Memory Basics

There are two main ways to enter numbers into long-term memory. Some calculators, including most 4-function calculators, have the keys in the table on page 220. If your calculator does not have at least the M+ and M- keys, see the examples on this page.

Memory on Calculator A

One way that calculators can put numbers in memory is using a key to **store** a value. On the first calculator, the store key is ▶M and only works on numbers that have been entered into the display with Enter.

Calculator A

▶M stores the displayed number in memory.

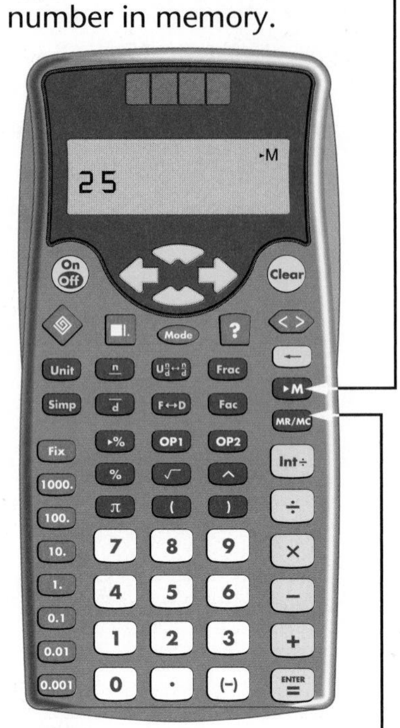

MR/MC recalls and displays the number in memory. Press it twice to clear memory.

Calculator A	Key Sequence	Purpose
	MR/MC MR/MC	Clear the long-term memory. This should always be the first step to any key sequence using the memory. Afterward, there will be no "M" in the display. This tells you there is no number in memory.
	▶M Enter	Store the number entered in the display in memory.
	MR/MC	Recall the number stored in memory and show it in the display.

Note

If you press MR/MC more than twice, you will recall and display the 0 that is now in memory. Press Clear to clear the display.

The following example first shows what happens if you don't enter a number before trying to store it.

Example Store 25 in memory and recall it to show that it was saved.

Calculator A	Key Sequence	Display
	MR/MC MR/MC Clear	
	25 ▶M Enter	MEM ERROR

Oops. Start again. First, press Clear twice.

Calculator A	Key Sequence	Display
	MR/MC MR/MC Clear	
	25 Enter ▶M Enter	M 25= 25
	Clear	M
	MR/MC	M 25

If your calculator is like this one, try the Check Your Understanding problem on page 220.

Memory on Calculator B

Calculators put a 0 into memory when MC is pressed. To store a single number in a cleared memory, simply enter the number and press M+.

Calculator B Key Sequence	Display
MC	Clear the long-term memory. This should always be the first step to any key sequence using the memory. Afterward, there will be no "M" in the display. This tells you there is no number in memory.
MR	Recall the number stored in memory and show it in the display.
M+	Add the number on the display to the number in the memory.
M−	Subtract the number on the display from the number in memory.

Calculator B

MC clears the memory.

M+ adds the number.

MR recalls the number in memory and displays it.

M− subtracts the number on the display from the number in memory.

Example Store 25 in memory and recall it to show that it was saved.

Calculator B Key Sequence	Display
AC MC	0.
25 M+	M 25.
AC	M 0.
MR	M 25.

Note

When this calculator turns off, the display clears, but a value in memory is **not** erased.

Check Your Understanding

Store π in the long-term memory. Clear the display. Then compute the area A of a circle whose radius r is 12 feet without pressing the π key. ($A = \pi r^2$)

Check your answer on page 347.

Using Memory in Problem Solving

A common use of memory in calculators is to solve problems that have two or more steps in the solution.

Example Compute a 15% tip on a $25 bill. Store the tip in the memory, then find the total bill.

Calculator A Key Sequence	Display
MR/MC MR/MC Clear	
15 % × 25 Enter	15%x25= 3.75
▶M Enter	M 15%x25= 3.75
25 + MR/MC Enter	M 25+3.75= 28.75

Calculator B Key Sequence	Display
AC MC	0.
25 X 15 % =	3.75
M+	M 3.75
25 + MR =	M = 28.75

Calculator B Key Sequence	Display
AC MC	0.
25 X 15 % +	28.75

The total bill is $28.75.

Note

Always be sure to clear the memory after solving one problem and before beginning another.

Note

The second solution shows how this calculator solves the problem by using memory automatically.

Check Your Understanding

Compute an 18% tip on an $85 bill. Then find the total bill.

Check your answer on page 347.

Example Marguerite ordered the following food at the food court: 2 hamburgers at $1.49 each and 3 hot dogs at $0.89 each. How much change will she receive from a $10 bill?

Calculator A	Key Sequence	Display
	(MR/MC) (MR/MC) (Clear)	
	2 (×) 1.49 (Enter) (▶M) (Enter)	M 2x1.49= 2.98
	3 (×) .89 (Enter) (▶M) (+)	M 3x.89= 2.67
	10 (−) (MR/MC) (Enter)	M 10-5.65= 4.35

Calculator B	Key Sequence	Display
	(AC) (MC)	0.
	2 (×) 1.49 (=) (M+)	M 2.98
	3 (×) .89 (=) (M+)	M 2.67
	10 (−) (MR) (=)	M = 4.35

Marguerite will receive $4.35 in change.

Note

The key sequence (▶M) (+) is a shortcut to add the displayed number to memory. Similarly, (▶M) (−) subtracts a number from memory.

Example Mr. Beckman bought 2 adult tickets at $8.25 each and 3 child tickets at $4.75 each. He redeemed a $5 gift certificate. How much did he pay for the tickets?

Calculator A Key Sequence	Display
MR/MC MR/MC Clear	
2 × 8.25 Enter ▶M Enter	M 2x8.25= 16.5
3 × 4.75 Enter ▶M +	M 3x4.75= 14.25
MR/MC − 5 Enter	M 30.75−5= 25.75

Calculator B Key Sequence	Display
AC MC	0.
2 × 8.25 = M+	M 16.5
3 × 4.75 = M+	M 14.25
MR − 5 =	M = 25.75

Mr. Beckman paid $25.75 for the tickets.

Note

If you fix the rounding to hundredths, all the values will be displayed as dollars and cents.

Example Juan bought the following tickets to a baseball game for himself and 6 friends: 2 bleacher seats at $15.25 each and 5 mezzanine seats at $27.50 each. If everyone intends to split the costs evenly 7 ways so that they can swap seats now and then, how much does each person owe Juan?

Calculator A Key Sequence	Display
MR/MC MR/MC Clear	
2 × 15.25 Enter ▶M Enter	M 2x15.25= 30.5
5 × 27.50 Enter ▶M +	M 5x27.50= 137.5
MR/MC ÷ 7 Enter	M 168÷7= 24

Calculator B Key Sequence	Display
AC MC	0.
2 × 15.25 = M+	M 30.5
5 × 27.50 = M+	M 137.5
MR ÷ 7 =	M = 24.

Each friend owes Juan $24.00 for the tickets.

Note

If you fix the rounding to hundredths, all the values will be displayed as dollars and cents.

Check Your Understanding

1. How much would 3 movie tickets and 2 popcorns cost if movie tickets cost $13.75 each and popcorns cost $4.25 each?
2. How much would it cost to take a group of 2 adults and 6 children to a bowling alley if games cost $3.75 for adults and $2.50 for children?

Check your answers on page 347.

Skip Counting on a Calculator

In earlier grades, you may have used a 4-function calculator to skip-count.

Recall that the program needs to tell the calculator:

1. What number to count by;
2. Whether to count up or down;
3. What number to start at; and
4. When to count.

Here's how to program each calculator.

Calculator A

Op1 and Op2 allow you to program and repeat operations.

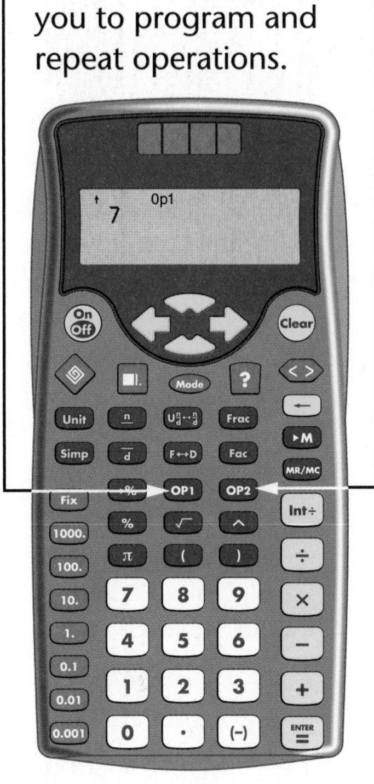

Example Starting at 3, count by 7s on this calculator.

Calculator A		
Purpose	**Key Sequence**	**Display**
Tell the calculator to count up by 7. Op1 is programmed to do any operation with any number that you enter between presses of Op1.	Op1 + 7 Op1	Op1 +7
Tell the calculator to start at 3 and do the first count.	3 Op1	Op1 3+7 1 10
Tell the calculator to count again.	Op1	Op1 10+7 2 17
Keep counting by pressing Op1.	Op1	Op1 17+7 3 24

To count back by 7, begin with Op1 − 7 Op1.

Note

You can use Op1 to define a second constant operation. Op2 works in exactly the same way as Op1.

Note

The number in the lower left corner of the display shows how many counts you have made.

Example Starting at 3, count by 7s on this calculator.

Purpose	Key Sequence	Display
Tell the calculator to count up by 7. The "K" on the display means you have successfully programmed the "constant," as the count-by number is sometimes called.	7 [+] [+]	K 7.+
Tell the calculator to start at 3 and do the first count.	3 [=]	K 10.+
Tell the calculator to count again.	[=]	K 17.+
Keep counting by pressing [=].	[=]	K 24.+

To count back by 7, begin with 7 [−] [−].

Check Your Understanding

Use your calculator to do the following counts. Write 5 counts each.

1. Starting at 22, count on by 8s.
2. Starting at 146, count back by 16s.

Check your answers on page 347.

Games

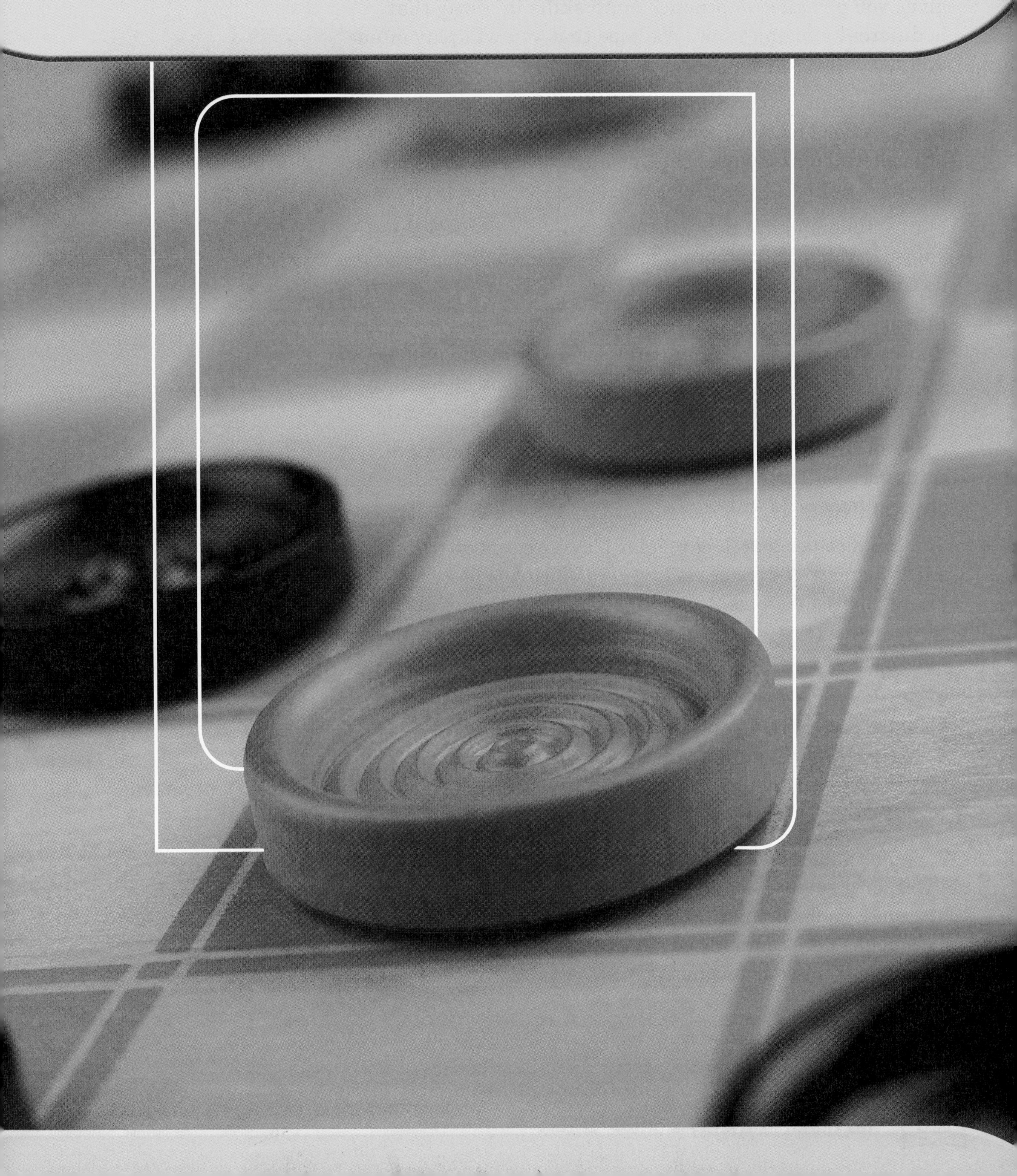

Games

Throughout the year, you will play games that help you practice important math skills. Playing mathematics games gives you a chance to practice math skills in a way that is different and enjoyable. We hope that you will play often and have fun!

In this section of your *Student Reference Book,* you will find the directions for many games. The numbers in most games are generated randomly. This means that the games can be played over and over without repeating the same problems.

Many students have created their own variations of these games to make them more interesting. We encourage you to do this too.

Materials

You need a deck of number cards for many of the games. You can use an Everything Math Deck, a deck of regular playing cards, or make your own deck out of index cards.

An Everything Math Deck includes 54 cards. There are 4 cards each for the numbers 0–10. And there is 1 card for each of the numbers 11–20.

You can also use a deck of regular playing cards after making a few changes. A deck of playing cards includes 54 cards (52 regular cards, plus 2 jokers). To create a deck of number cards, use a permanent marker to mark the cards in the following way:

- Mark each of the 4 aces with the number 1.
- Mark each of the 4 queens with the number 0.
- Mark the 4 jacks and 4 kings with the numbers 11 through 18.
- Mark the 2 jokers with the numbers 19 and 20.

For some games you will have to make a gameboard, a score sheet, or a set of cards that are not number cards. The instructions for doing this are included with the game directions. More complicated gameboards and card decks are available from your teacher.

SRB
229

Angle Tangle

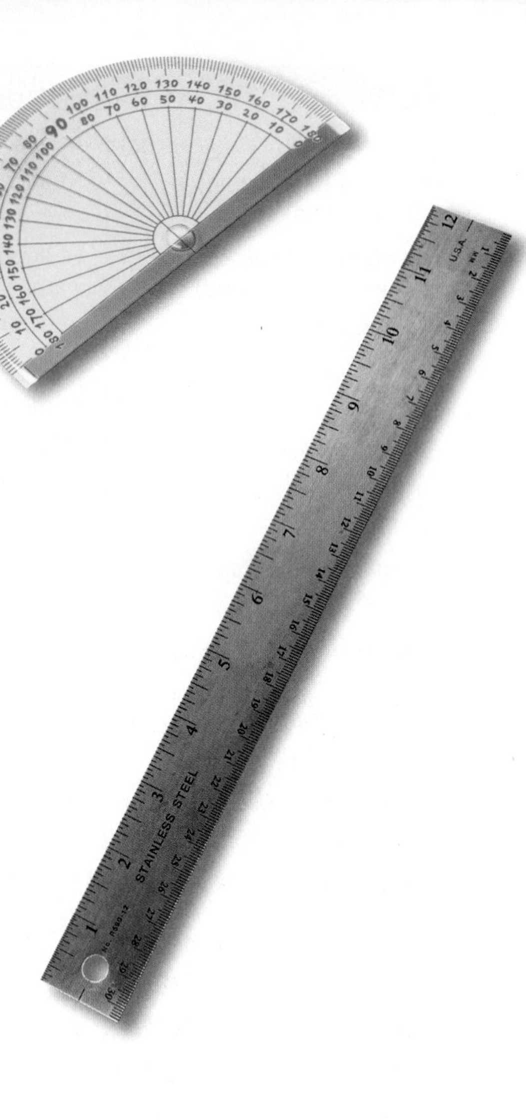

Materials ☐ 1 protractor

☐ 1 straightedge

☐ several blank sheets of paper

Players 2

Skill Estimating and measuring angle size

Object of the game To estimate angle sizes accurately and have the lower total score.

Directions

In each round:

1. Player 1 uses a straightedge to draw an angle on a sheet of paper.
2. Player 2 estimates the degree measure of the angle.
3. Player 1 measures the angle with a protractor. Players agree on the measure.
4. Player 2's score is the difference between the estimate and the actual measure of the angle. (The difference will be 0 or a positive number.)
5. Players trade roles and repeat Steps 1–4.

Players add their scores at the end of five rounds. The player with the lower total score wins the game.

Example

	Player 1			Player 2		
	Estimate	Actual	Score	Estimate	Actual	Score
Round 1	120°	108°	12	50°	37°	13
Round 2	75°	86°	11	85°	87°	2
Round 3	40°	44°	4	15°	19°	4
Round 4	60°	69°	9	40°	56°	16
Round 5	135°	123°	12	150°	141°	9
Total Score			48			44

Player 2 has the lower total score. Player 2 wins the game.

Baseball Multiplication (1 to 6 Facts)

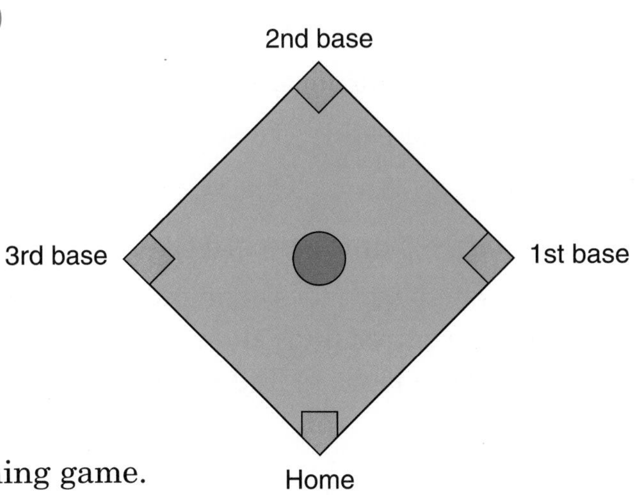

Materials
- ☐ 1 *Baseball Multiplication* Game Mat (*Math Masters,* p. 460)
- ☐ 2 six-sided dice
- ☐ 4 counters
- ☐ 1 calculator or a multiplication/division table

Players 2 teams of one or more players each.

Skill Multiplication facts 1 to 6

Object of the game To score more runs in a 3-inning game.

Directions

1. Draw a diamond and label *Home, 1st base, 2nd base,* and *3rd base.* Make a Scoreboard sheet like the one shown at the right.
2. Teams take turns being the *pitcher* and the *batter.* The rules are similar to the rules of baseball, but this game lasts only 3 innings.
3. The batter puts a counter on home plate. The pitcher rolls the dice. The batter multiplies the numbers rolled and gives the answer. The pitcher checks the answer and may use a calculator to do so.
4. If the answer is correct, the batter looks up the product in the Hitting Table at the right. If it is a hit, the batter moves all counters on the field the number of bases shown in the table. The pitcher tallies each out on the scoreboard.
5. An incorrect answer is a strike and another pitch (dice roll) is thrown. Three strikes make an out.
6. A run is scored each time a counter crosses home plate. The batter tallies each run scored on the Scoreboard.
7. After each hit or out, the batter puts a counter on home plate. The batting and pitching teams switch roles after the batting team has made 3 outs. The inning is over when both teams have made 3 outs.

Scoreboard

Inning		1	2	3	Total
Team 1	outs				
	runs				
Team 2	outs				
	runs				

Hitting Table
1 to 6 Facts

1 to 9	Out
10 to 19	Single (1 base)
20 to 29	Double (2 bases)
30 to 35	Triple (3 bases)
36	Home Run (4 bases)

The team with more runs at the end of 3 innings wins the game. If the game is tied at the end of 3 innings, play continues into extra innings until one team wins.

Baseball Multiplication (Advanced Versions)

Skill Multiplication facts through 12s, extended facts

Object of the game To score more runs in a 3-inning game.

1 to 10 Facts

Materials □ number cards 1–10 (4 of each)

Follow the basic rules. The pitcher draws 2 cards from the deck. The batter finds their product and uses the Hitting Table at the right to find out how to move the counters.

Hitting Table 1 to 10 Facts	
1 to 21	Out
24 to 45	Single (1 base)
48 to 70	Double (2 bases)
72 to 81	Triple (3 bases)
90 to 100	Home Run (4 bases)

2 to 12 Facts

Materials □ 4 six-sided dice

Follow the basic rules. The pitcher rolls 4 dice. The batter separates them into 2 pairs, adds the numbers in each pair, and multiplies the sums. Use the Hitting Table at the right.

Hitting Table 2 to 12 Facts	
4 to 24	Out
25 to 49	Single (1 base)
50 to 64	Double (2 bases)
66 to 77	Triple (3 bases)
80 to 144	Home Run (4 bases)

How you pair the numbers can determine the kind of hit you get or whether you get an out. For example, suppose you roll a 1, 2, 3, and 5. You could add pairs in different ways and multiply as follows:

one way	a second way	a third way
1 + 2 = 3 3 + 5 = 8 3 * 8 = 24 Out	1 + 3 = 4 2 + 5 = 7 4 * 7 = 28 Single	1 + 5 = 6 2 + 3 = 5 6 * 5 = 30 Single

Three-Factors Game

Materials □ 3 six-sided dice

The pitcher rolls 3 dice. The batter multiplies the 3 numbers (factors) and uses the Hitting Table at the right.

Hitting Table Three-Factors Game	
1 to 54	Out
60 to 90	Single (1 base)
96 to 120	Double (2 bases)
125 to 150	Triple (3 bases)
180 to 216	Home Run (4 bases)

*10s * 10s Game*

Materials □ 4 six-sided dice

The rules for this game are the same as for the *2 to 12 Facts* game with two exceptions:

1. A sum of 2 through 9 represents 20 through 90.
   A sum of 10 through 12 represents itself. For example:
   Roll 1, 2, 3, and 5. Get sums 6 and 5. Multiply 60 * 50.
   Roll 3, 4, 6, and 6. Get sums 12 and 7. Multiply 12 * 70.
2. Use the Hitting Table at the right.

Hitting Table 10 * 10s Game	
100 to 2,000	Out
2,100 to 4,000	Single (1 base)
4,200 to 5,400	Double (2 bases)
5,600 to 6,400	Triple (3 bases)
7,200 to 8,100	Home Run (4 bases)

Beat the Calculator

Multiplication Facts

Materials □ number cards 1–10 (4 of each)

□ 1 calculator

□ *Beat the Calculator* Gameboard (optional) (*Math Masters,* p. 461)

Players 3

Skill Mental multiplication skills

Object of the game To multiply numbers without a calculator faster than a player using one.

Directions

1. One player is the "Caller," one is the "Calculator," and one is the "Brain."
2. Shuffle the deck and place it number-side down on the table.
3. The Caller draws 2 cards from the number deck and asks for their product.
4. The Calculator solves the problem using a calculator. The Brain solves it without a calculator. The Caller decides who got the answer first.
5. The Caller continues to draw 2 cards at a time from the number deck and ask for their product.
6. Players trade roles every 10 turns or so.

Example The Caller draws a 10 and a 5 and calls out "10 times 5." The Brain and the Calculator each solve the problem. The Caller decides who got the answer first.

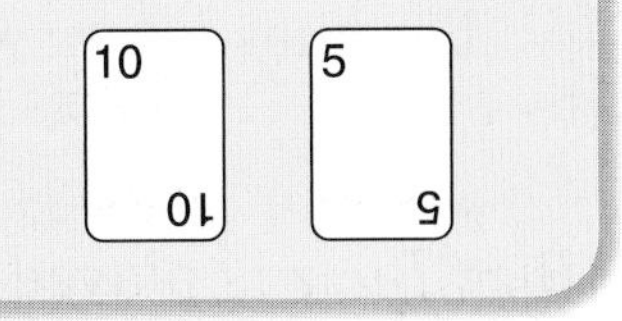

Extended Multiplication Facts

In this version of the game, the Caller:

- Draws 2 cards from the number deck.
- Attaches a 0 to either one of the factors, or to both factors, before asking for the product.

Example If the Caller turns over a 4 and an 8, he or she may make up any one of the following problems:

4 * 80 40 * 8 40 * 80

Buzz Games

Buzz

Materials none

Players 5–10

Skill Finding multiples of a number and common multiples of two numbers

Object of the game To correctly say either "BUZZ" or the next number when it is your turn.

Directions

1. Players sit in a circle and choose a leader. The leader names any whole number from 3 to 9. This number is the BUZZ number. The leader also chooses the STOP number. The STOP number should be at least 30.
2. The player to the left of the leader begins the game by saying "one." Play continues clockwise with each player saying either the next whole number or "BUZZ."
3. A player must say "BUZZ" instead of the next number if:
   - ♦ The number is the BUZZ number or a multiple of the BUZZ number; or
   - ♦ The number contains the BUZZ number as one of its digits.
4. If a player makes an error, the next player starts with 1.
5. Play continues until the STOP number is reached.
6. For the next round, the player to the right of the leader becomes the new leader.

Example The BUZZ number is 4. Play should proceed as follows: 1, 2, 3, BUZZ, 5, 6, 7, BUZZ, 9, 10, 11, BUZZ, 13, BUZZ, 15, and so on.

Bizz-Buzz

Bizz-Buzz is played like *Buzz*, except the leader names 2 numbers: a BUZZ number and a BIZZ number.

Players say:

1. "BUZZ" if the number is a multiple of the BUZZ number.
2. "BIZZ" if the number is a multiple of the BIZZ number.
3. "BIZZ–BUZZ" if the number is a multiple of both the BUZZ number and the BIZZ number.

Example The BUZZ number is 6, and the BIZZ number is 3. Play should proceed as follows: 1, 2, BIZZ, 4, 5, BIZZ-BUZZ, 7, 8, BIZZ, 10, 11, BIZZ-BUZZ, 13, 14, BIZZ, 16, and so on. The numbers 6 and 12 are replaced by "BIZZ-BUZZ" since 6 and 12 are multiples of both 6 and 3.

Calculator 10,000

Materials ☐ 1 calculator

Skill Practice with all four arithmetic operations, estimation

Object of the game To get from a starting number to 10,000, or as close as possible. Each operation—addition, subtraction, multiplication, and division—must be used exactly one time.

One-Player Game

1. Create a starting number. Pick any number from 1 to 12 and cube it. For example, pick 5 and cube it.
   5 [×] 5 [×] 5 [=] 125. 125 is the starting number.
2. Pick a number. Add, subtract, multiply, or divide your starting number with this number. For example, pick 100. Multiply your starting number with 100. 125 [×] 100 [=] 12,500.
3. Pick a different number. Add, subtract, multiply, or divide your result in Step 2 with this number. Use a **different operation from** the one in Step 2. For example, pick 2. Divide the result in Step 2 by 2. 12,500 [÷] 2 [=] 6,250.
4. Continue to pick numbers and use operations until you have used each of the 4 operations once. You can use the operations in any order, but may *use each operation only one time.* You must pick a **different number** for each operation.
5. You can pick the numbers you use to add, subtract, multiply, and divide with from either Level 1 or Level 2. Decide which level to use before playing. (Level 2 is harder.)
   - Level 1: any number except 0
   - Level 2: only numbers from 2 to 100
6. Record what you did for each game. A sample game record (using Level 1) is shown at the right. The final result is 10,150.

Pick a Number	Key In	Result
5	5 [×] 5 [×] 5 [=]	125 (starting number)
100	[×] 100 [=]	12,500
2	[÷] 2 [=]	6,250
3,000	[−] 3000 [=]	3,250
6,900	[+] 6900 [=]	10,150

Two-Player Game

Each player plays 3 games using the rules above. Find the difference between your final result and 10,000 for each game. Add the 3 differences to find your total score. The player with the lower total score wins the game.

Chances Are

Materials ☐ 1 deck of *Chances Are* Event Cards (*Math Masters,* pp. 462–463)

☐ 1 deck of *Chances Are* Probability Cards (*Math Masters,* pp. 465–466)

☐ 1 *Chances Are* Gameboard for each player (*Math Masters,* p. 464)

Players 2

Skill Calculating probabilities for events

Object of the game To match all of the cards on your gameboard.

Directions

1. Shuffle the Event Cards and Probability Cards together. Deal 5 cards faceup in front of each player. Then place the deck facedown on the table. Players arrange the Event Cards and Probability Cards on their gameboards.

Example Gameboard for a player at the beginning of a game will include 5 cards

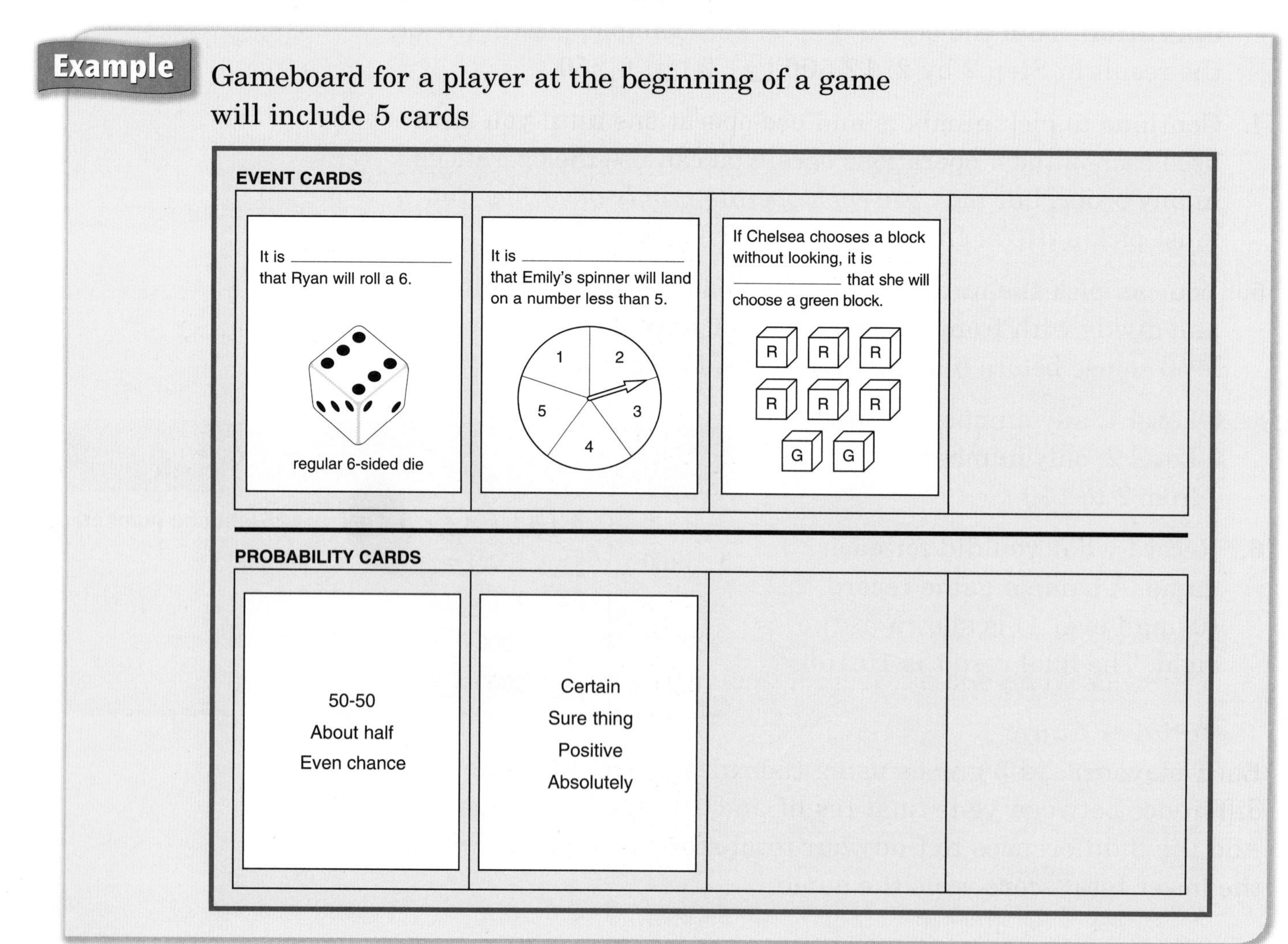

2. Players take turns. At each turn, a player draws one card from the deck and places it faceup in the correct section of his or her gameboard. The player tries to match a Probability Card with an Event Card. A match is made when a player finds a Probability Card with the best description of the chance that the event on the Event Card will happen.

3. If a match is made, the player removes the matching cards from the gameboard and the turn is over. **Only one match may be made per turn.**
If no match is made, the turn is over.

Example Suppose that a player's Gameboard looks like the one on page 236, and the player draws a Probability Card that says "Unlikely".

The player may match this pair of cards and remove them from the Gameboard.

It is ______ that Ryan will roll a 6.

regular 6-sided die

Unlikely
Less than half
Slight chance
Not much

Or the player may match this pair of cards and remove them from the Gameboard.

If Chelsea chooses a block without looking, it is ______ that she will choose a green block.

R R R
R R R
G G

Unlikely
Less than half
Slight chance
Not much

4. Play ends when one player is out of cards or when the deck is gone.

5. The player with more matches wins.

Credits/Debits Game

Materials ☐ 1 complete deck of number cards

☐ 1 *Credits/Debits Game* Record Sheet for each player (*Math Masters,* p. 468)

Players 2

Skill Addition of positive and negative numbers

Object of the game To have more money after adjusting for credits and debits.

Directions

Record Sheet			
	Start	Change	End, and next start
1	+$10		
2			
3			
4			
5			
6			
7			
8			
9			
10			

Each player uses one Record Sheet.

You are an accountant for a business. Your job is to keep track of the company's current balance. The current balance is also called the "bottom line." As credits and debits are reported, you will record them and then adjust the bottom line.

1. Shuffle the deck and lay it number-side down between the players.
2. The black-numbered cards are the "credits," and the blue- or red-numbered cards are the "debits."
3. Each player begins with a bottom line of +$10.
4. Players take turns. On your turn, do the following:
   - Draw a card. The card tells you the dollar amount and whether it is a credit or debit to the bottom line. Record the credit or debit in your "Change" column.
   - Add the credit or debit to adjust your bottom line.
   - Record the result in your table.
5. At the end of 10 draws each, the player with more money is the winner of the round.

Note

If both players have negative dollar amounts at the end of the round, the player whose amount is closer to 0 wins.

Examples Beth has a "Start" balance of +$20. She draws a black 4. This is a credit of $4, so she records +$4 in the "Change" column. She adds $4 to the bottom line: $20 + $4 = $24. She records +$24 in the "End" column, and +$24 in the "Start" column on the next line.

Alex has a "Start" balance of +$10. He draws a red 12. This is a debit of $12, so he records −$12 in the "Change" column. He adds −$12 to the bottom line: $10+(−$12) = −$2. Alex records −$2 in the "End" column. He also records −$2 in the "Start" column on the next line.

Credits/Debits Game (Advanced Version)

Materials ☐ 1 complete deck of number cards

☐ 1 penny

☐ 1 *Credits/Debits Game* (Advanced Version) Record Sheet for each player (*Math Masters,* p. 469)

Players 2

Skill Addition and subtraction of positive and negative numbers

Object of the game To have more money after adding and subtracting credits and debits.

Record Sheet				
		Change		
	Start	Addition or Subtraction	Credit or Debit	End, and next start
1	+$10			
2				
3				
4				
5				
6				
7				
8				
9				
10				

Each player uses one Record Sheet.

Directions

1. Shuffle the deck and lay it number-side down between the players.
2. The black-numbered cards are the "credits," and the blue- or red-numbered cards are the "debits."
3. The heads side of the coin tells you to **add** a credit or debit to the bottom line. The tails side of the coin tells you to **subtract** a credit or debit from the bottom line.
4. Each player begins with a bottom line of +$10.
5. Players take turns. On your turn, do the following:
   - ♦ Flip the coin. This tells you whether to add or subtract.
   - ♦ Draw a card. The card tells you what amount in dollars (positive or negative) to add or subtract from your bottom line. Red or blue numbers are negative numbers. Black numbers are positive numbers.
   - ♦ Record the results in your table.
6. Scoring is the same as in the *Credits/Debits Game*.

Examples Max has a "Start" balance of $5. His coin lands heads up and he records + in the "Addition or Subtraction" column. He draws a red 9 and records −$9 in the "Credit or Debit" column. Max adds: $5 + (−$9) = −$4. He records −$4 in the "End" balance column and also in the "Start" column on the next line.

Beth has a "Start" balance of −$20. Her coin lands tails up, which means subtract. She draws a black 2 (+$2). She subtracts: −$20 − (+$2) = −$22. Her "End" balance is −$22.

Division Arrays

Materials ☐ number cards 6–18 (1 of each)

☐ 1 six-sided die

☐ 18 counters

Players 2 to 4

Skill Division and equal shares

Object of the game To have the highest total score.

Directions

1. Shuffle the cards. Place the deck number-side down on the table.

2. Players take turns. When it is your turn, draw a card and take the number of counters shown on the card. You will use the counters to make an array.

   - Roll the die. The number on the die is the number of equal rows you must have in your array.
   - Make an array with the counters.
   - Your score is the number of counters in 1 row. If there are no leftover counters, your score is double the number of counters in 1 row.

3. Keep track of your scores. The player with the highest total score at the end of 5 rounds wins.

Example Dave draws a 14-card and takes 14 counters. He rolls a 3 and makes an array with 3 rows by putting 4 counters in each row. Two counters are left over.

Dave scores 4 because there are 4 counters in each row.

Example Marsha draws a 15-card and takes 15 counters. She rolls a 3 and makes an array with 3 rows by putting 5 counters in each row.

Her score is $5 * 2 = 10$ because there are 5 counters in each row, with none left over.

Division Dash

Player 1		Player 2	
Quotient	Score	Quotient	Score

Materials □ number cards 1–9 (4 of each)

□ 1 score sheet

Players 1 or 2

Skill Division of 2-digit by 1-digit numbers

Object of the game To reach 100 in the fewest divisions possible.

Directions

1. Prepare a score sheet like the one shown at the right.
2. Shuffle the cards and place the deck number-side down on the table.
3. Each player follows the instructions below:
   - ♦ Turn over 3 cards and lay them down in a row, from left to right. Use the 3 cards to generate a division problem. The 2 cards on the left form a 2-digit number. This is the *dividend*. The number on the card at the right is the *divisor*.
   - ♦ Divide the 2-digit number by the 1-digit number and record the result. This result is your quotient. Remainders are ignored. Calculate mentally or on paper.
   - ♦ Add your quotient to your previous score and record your new score. (If this is your first turn, your previous score was 0.)
4. Players repeat Step 3 until one player's score is 100 or more. The first player to reach at least 100 wins. If there is only one player, the object of the game is to reach 100 in as few turns as possible.

Example

Turn 1: Bob draws 6, 4, and 5.
He divides 64 by 5. Quotient = 12.
Remainder is ignored. The score is 12 + 0 = 12.

64 is the dividend. 5 is the divisor.

Turn 2: Bob then draws 8, 2, and 1.
He divides 82 by 1. Quotient = 82.
The score is 82 + 12 = 94.

Turn 3: Bob then draws 5, 7, and 8.
He divides 57 by 8. Quotient = 7.
Remainder is ignored. The score is 7 + 94 = 101.
Bob has reached 100 in 3 turns and the game ends.

Quotient	Score
12	12
82	94
7	101

Fishing for Digits

Materials ☐ 1 *Fishing for Digits* Record Sheet for each player (*Math Masters,* p. 472; optional)

☐ 1 calculator for each player

Players 2

Skill Place value; developing a winning game strategy

Object of the game To have the larger number after 5 rounds.

Name Date Time

Fishing for Digits Record Sheet

millions	hundred thousands	ten thousands	thousands	hundreds	tens	ones
1,000,000	100,000	10,000	1,000	100	10	1

	Beginning Number	X						
1	New Number							
	New Number							
2	New Number							
	New Number							
3	New Number							
	New Number							
4	New Number							
	New Number							
5	New Number							
	Final Number							

Directions

1. Each player secretly enters a 6-digit number into their calculator. Zeros may not be used.

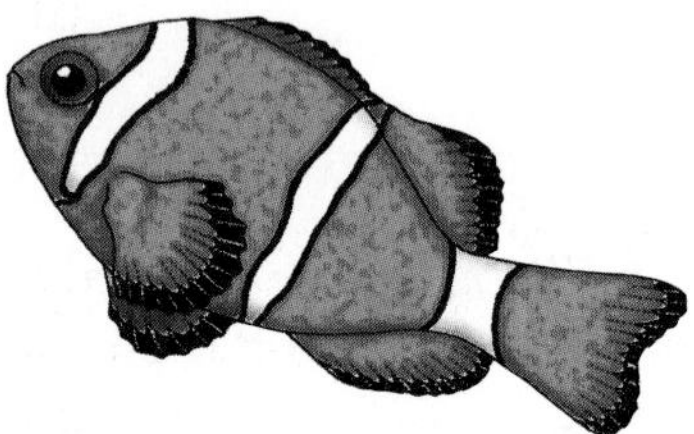

2. Player 1 goes "fishing" for a digit in Player 2's number by naming a digit.
3. If the digit named is one of the digits in Player 2's number:
   - ♦ Player 2 reports the value of the digit. If the digit appears more than once in Player 2's number, Player 2 reports the largest value of that digit in the number. For example, if Player 1 names the digit 7, and Player 2's number is 987,675, then Player 2 would report the value 7,000, rather than the value 70.
   - ♦ Player 1 *adds* the value of that digit to his or her number.
   - ♦ Player 2 *subtracts* the value of that digit from his or her number.
4. If the digit named is not one of the digits in Player 2's number, Player 1 adds 0 and Player 2 subtracts 0 for that turn.
5. It is now Player 2's turn to "fish." Reverse the roles of Player 1 and 2 and repeat Steps 2, 3, and 4. When each player has "fished" once, the round is over.
6. The player whose calculator displays the larger number at the end of 5 rounds is the winner.

Example Player 1's calculator shows 813,296. Player 2's calculator shows 328,479.
Player 1 asks: "Do you have the digit 4?"
Player 2 replies: "Yes, the digit 4 is in the hundreds place."

Player 1 adds 400: 813296 [+] 400 [=] Calculator shows 813696.
Player 2 subtracts 400: 328479 [−] 400 [=] Calculator shows 328079.

Variation Begin with a number having fewer than 6 digits.

Fraction Match

Materials ☐ 1 deck of *Fraction Match* Cards (*Math Masters,* pp. 473–476)

Players 2 to 4

Skill Recognizing equivalent fractions

Object of the game To match all of your cards and have none left.

Directions

1. Shuffle the deck and deal 7 cards to each player. Place the remaining cards facedown on the table. Turn over the top card and place it beside the deck. This is the *target card.* If a WILD card is drawn, return it to the deck and continue drawing until the first target card is a fraction.

2. Players take turns trying to match the target card with a card from their hand in one of 3 possible ways:

   ♦ a card with an equivalent fraction
   ♦ a card with a like denominator
   ♦ a WILD card.

Example $\frac{2}{3}$ is the target card. It can be matched with:

♦ an equivalent fraction card such as $\frac{4}{6}$, $\frac{6}{9}$, or $\frac{8}{12}$, or

♦ a like denominator card such as $\frac{0}{3}$, $\frac{1}{3}$, or $\frac{3}{3}$, or

♦ a WILD card. The player names any fraction (with a denominator of 2, 3, 4, 5, 6, 8, 9, 10, or 12) that is equivalent to the target card. The player can match $\frac{2}{3}$ by saying $\frac{4}{6}$, $\frac{6}{9}$, or $\frac{8}{12}$. The player may not match $\frac{2}{3}$ by saying $\frac{2}{3}$.

3. If a match is made, the player's matching card is placed on top of the pile and becomes the new target card. It is now the next player's turn. When a WILD card is played, the next player uses the fraction just stated for the new target card.

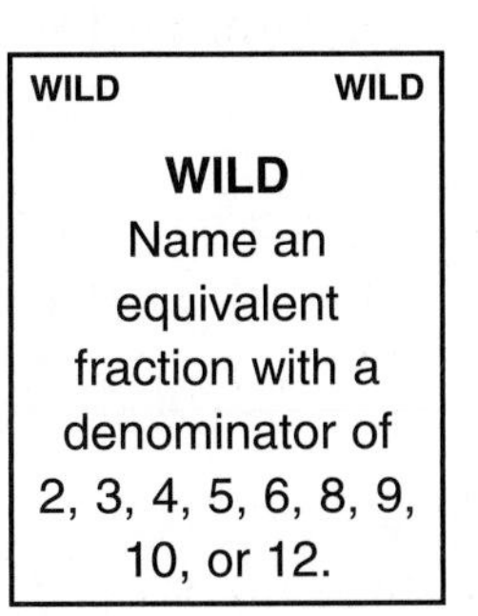

4. If no match can be made, the player takes 1 card from the deck. If the card drawn matches the target card, it may be played. If not, the player keeps the card and the turn ends.

5. The game is over when one of the players runs out of cards, when there are no cards left in the *Fraction Match* deck, or time runs out. The player with the fewest cards wins.

Fraction Of

Materials □ 1 deck of *Fraction Of* Fraction Cards (*Math Masters,* pp. 477 and 478)

□ 1 deck of *Fraction Of* Set Cards (*Math Masters,* pp. 480)

□ 1 *Fraction Of* Gameboard and Record Sheet for each player (*Math Masters,* p. 479)

Players 2

Skill Multiplication of fractions and whole numbers

Object of the game To score more points by solving "fraction of" problems.

Directions

$\frac{5}{10}$ $\frac{10}{10}$

$\frac{2}{3}$ $\frac{3}{3}$

Examples of Fraction Cards

1. Shuffle each deck separately. Place the decks facedown on the table.
2. Players take turns. At each turn, a player draws 1 card from each deck. The player uses the cards to create a "fraction of" problem on their gameboard.
   - The Fraction Card indicates what fraction of the set the player must find.
   - The Set Card offers 3 possible choices. The player must choose a set that will result in a "fraction of" problem with a whole number solution.

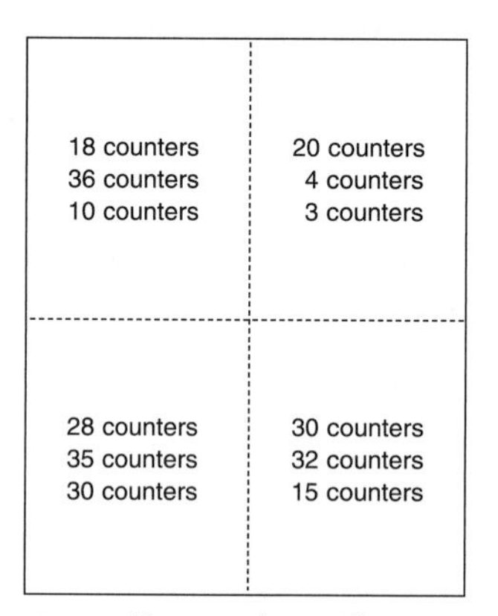

Examples of Set Cards

Example Player 1 draws the Fraction Card and Set Card shown here.

$\frac{1}{10}$ of 28 will *not* result in a whole-number solution.

$\frac{1}{10}$ of 28 counters is 2.8 counters.

$\frac{1}{10}$ of 35 will *not* result in a whole-number solution.

$\frac{1}{10}$ of 35 counters is 3.5 counters.

$\frac{1}{10}$ of 30 *will* result in a whole-number solution.

$\frac{1}{10}$ of 30 counters is 3 counters.

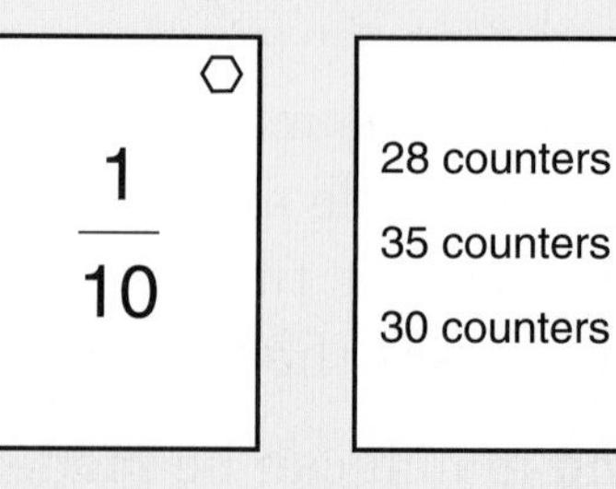

Player 1 chooses 30 counters as the set for the "fraction of" problem.

3. The player solves the "fraction of" problem and sets the 2 cards aside. The solution is the player's point score for the turn.

Example Player 2 draws the Fraction Card and Set Card shown here.

Player 2 could choose 12 or 30 counters as the set.

Player 2 chooses 30 counters (more points than 12), finds $\frac{1}{2}$ of 30, and earns 15 points.

$\frac{1}{2}$	12 counters 30 counters 25 counters

4. Play continues until all of the cards in the fraction pile or set pile have been used. The player with more points wins.

Variation For a basic version of the game, use only the fraction cards with a hexagon in the corner.

Name Date Time

***Fraction Of* Gameboard and Record Sheet**

Fraction card **of** WHOLE (Choose 1 of these sets.) Set card

Round	"Fraction-of" Problem	Points
Sample	$\frac{1}{5}$ of 25	5
1		
2		
3		
4		
5		
6		
7		
8		
	Total score	

Fraction Of Gameboard and Record Sheet

Fraction/Percent Concentration

Materials ☐ 1 set of Fraction/Percent Tiles (*Math Masters,* pp. 481 and 482)

☐ 1 calculator

Players 2 or 3

Skill Recognizing fractions and percents that are equivalent

Object of the game To collect the most tiles by matching equivalent fraction and percent tiles.

Directions

Advance Preparation Before beginning the game, write the letter "F" on the back of each fraction tile. And write the letter "P" on the back of each percent tile.

1. Spread the tiles out number-side down on the table. Create 2 separate piles—a fraction pile and a percent pile. Mix up the tiles in each pile. The 12 fraction tiles should have the letter "F" showing. The 12 percent tiles should have the letter "P" showing.
2. Players take turns. At each turn, a player turns over both a fraction tile and a percent tile. If the fraction and the percent are equivalent, the player keeps the tiles. If the fraction and the percent are not equivalent, the player turns the tiles number-side down.
3. Players may use a calculator to check each other's matches.
4. The game ends when all tiles have been taken. The player with the most tiles wins.

Variations Write the letter "D" on the back of each decimal tile. Play the game using only the "F" and "D" tiles. Or, play the game using only the "P" and "D" tiles.

10%	20%	25%	30%
40%	50%	60%	70%
75%	80%	90%	100%
$\frac{1}{2}$	$\frac{1}{4}$	$\frac{3}{4}$	$\frac{1}{5}$
$\frac{2}{5}$	$\frac{3}{5}$	$\frac{4}{5}$	$\frac{1}{10}$
$\frac{3}{10}$	$\frac{7}{10}$	$\frac{9}{10}$	$\frac{2}{2}$

Fraction/Percent Tiles
(number-side up)

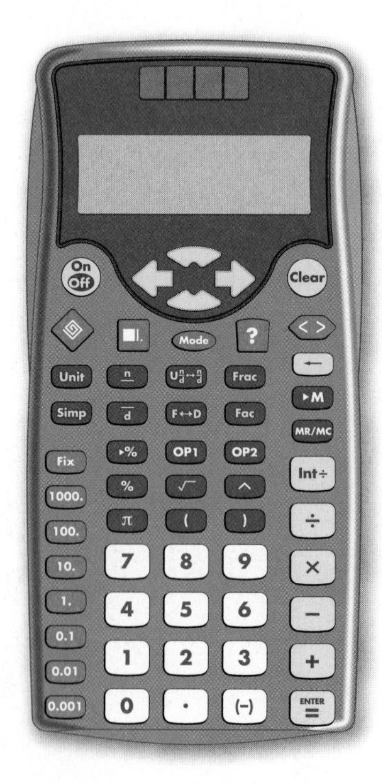

Fraction Top-It

Materials ☐ 1 set of Fraction Cards 1 and 2 (*Math Journal* 2, Activity Sheets 5 and 6)

Players 2 to 4

Skill Comparing fractions

Object of the game To collect the most cards.

Directions

Advance Preparation Before beginning the game, write the fraction for the shaded part on the back of each card.

1. Deal the same number of cards, fraction-side up, to each player:
   - If there are 2 players, 16 cards each.
   - If there are 3 players, 10 cards each.
   - If there are 4 players, 8 cards each.
2. Players spread their cards out, fraction-side up, so that all of the cards may be seen.
3. Starting with the dealer and going in a clockwise direction, each player plays one card. Place the cards fraction-side up on the table.
4. The player with the largest fraction wins the round and takes the cards. Players may check who has the largest fraction by turning over the cards and comparing the amounts shaded.
5. If there is a tie for the largest fraction, each player plays another card. The player with the largest fraction takes all the cards from both plays.
6. The player who takes the cards starts the next round.
7. The game is over when all cards have been played. The player who takes the most cards wins.

Fraction Cards 1

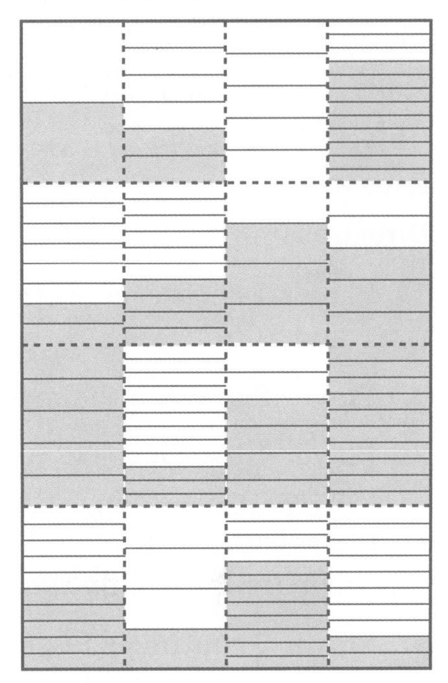

Fraction Cards 2

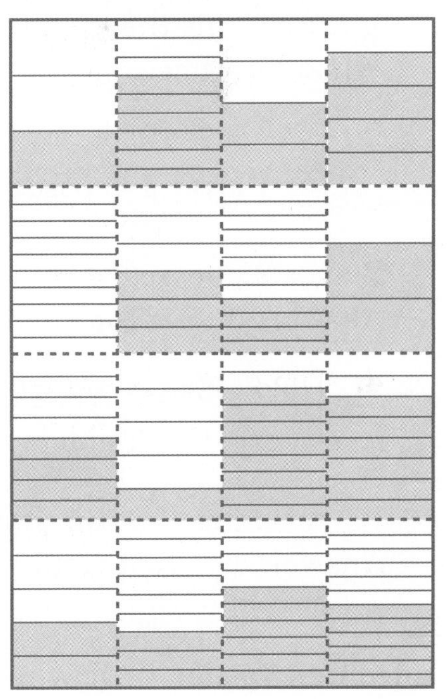

Getting to One

Materials □ 1 calculator

Players 2

Skill Estimation

Object of the game To guess a mystery number in as few tries as possible.

Directions

1. Player 1 chooses a mystery number that is between 1 and 100.
2. Player 2 guesses the mystery number.
3. Player 1 uses a calculator to divide Player 2's guess by the mystery number. Player 1 then reads the answer in the calculator display. If the answer has more than 2 decimal places, only the first 2 decimal places are read.
4. Player 2 continues to guess until the calculator result is 1. Player 2 keeps track of the number of guesses.
5. When Player 2 has guessed the mystery number, players trade roles and follow Steps 1–4 again. The player who guesses his or her mystery number in the fewest number of guesses wins the round. The first player to win 3 rounds wins the game.

Example Player 1 chooses the mystery number 65.

Player 2 guesses: 45. Player 1 keys in: 45 [÷] 65 [=].
Answer: 0.69 Too small.

Player 2 guesses: 73. Player 1 keys in: 73 [÷] 65 [=].
Answer: 1.12 Too big.

Player 2 guesses: 65. Player 1 keys in: 65 [÷] 65 [=].
Answer: 1. Just right!

Advanced Version Allow mystery numbers up to 1,000.

Grab Bag

Materials ☐ 1 deck of *Grab Bag* Cards (*Math Masters,* pp. 483–484)

☐ 1 *Grab Bag* Record Sheet for each player or team (*Math Masters,* p. 485)

☐ 3 six-sided dice

Players 2, or two teams of 2

Skill Variable substitution; calculating probabilities of events

Object of the game To score more points by calculating the probabilities of events.

Directions

1. Shuffle the deck of *Grab Bag* cards and place it problem-side down on the table.
2. Players (or teams) take turns. When it is your turn:
   - Draw a card and place it problem-side up on the table. Two quantities are missing from each card. They are shown with the variables x and y.
   - Roll the 3 dice and substitute the numbers rolled for the variables x and y in the following way:
     Replace x with the number shown on 1 of the dice.
     Replace y with the sum of the numbers on the other 2 dice.
   - Solve the problem and give an answer. The opposing player (or team) checks the answer. Your score for the round is calculated as follows:

     10 points: if the event is unlikely (probability is less than $\frac{1}{2}$).
     30 points: if the event is likely (probability is greater than $\frac{1}{2}$).
     50 points: if the event has a 50–50 chance (probability exactly $\frac{1}{2}$).
3. The player (or team) with the higher score after 5 rounds wins.

Note

Use a strategy when replacing x and y by the dice numbers to earn the most points possible for that turn.

Example Paul draws the card shown to the right.

He rolls 6, 1, and 4, and substitutes 1 for x and $6 + 4 = 10$ for y.

Lina's grab bag has 2 red, 2 blue, 1 pink, and 10 green ribbons. The probability of Lina picking a green ribbon is 10 out of 15 or $\frac{10}{15}$ or $\frac{2}{3}$.

Picking a green ribbon is likely (probability is greater than $\frac{1}{2}$).

Paul scores 30 points.

> Lina has a bag of ribbons. She has 2 red, 2 blue, x pink, and y green ribbons.
>
> What are the chances she will pick a green ribbon without looking?

Grid Search

Materials ☐ 1 sheet of *Grid Search* Grids for each player (*Math Masters,* p. 486)

Players 2

Skill Deduction; developing a search strategy

Object of the game To locate the opponent's queen on a coordinate grid in the fewest turns possible.

Directions

Players sit so that they cannot see what the other player is doing. Each player uses 2 grids like those shown at the right.

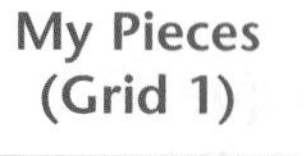

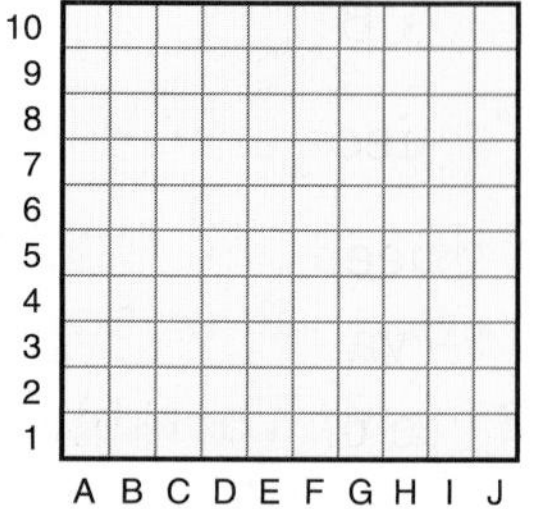

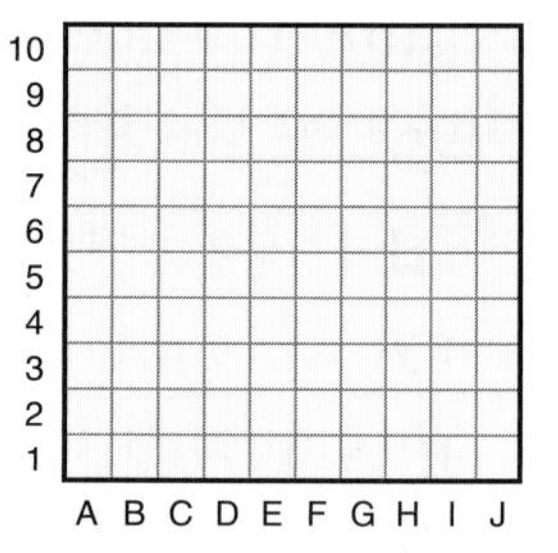

Advance Preparation Before the start of the game, each player secretly decides where to place a queen and 6 knights on their Grid 1. They write the letter Q to record the location of the queen and the letter K to record the location of each knight.

- ♦ The queen may be placed on any square.
- ♦ The knights may also be placed on any squares, as long as the queen and the knights can all be connected without skipping squares.

These are acceptable arrangements of the pieces:

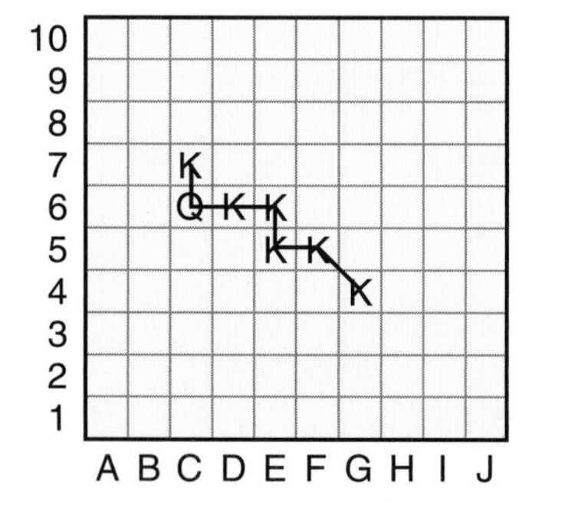

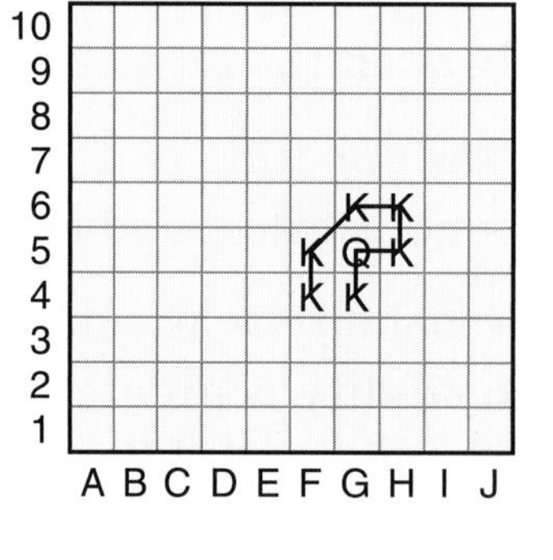

These are *not* acceptable arrangements because the pieces cannot be connected without skipping squares.

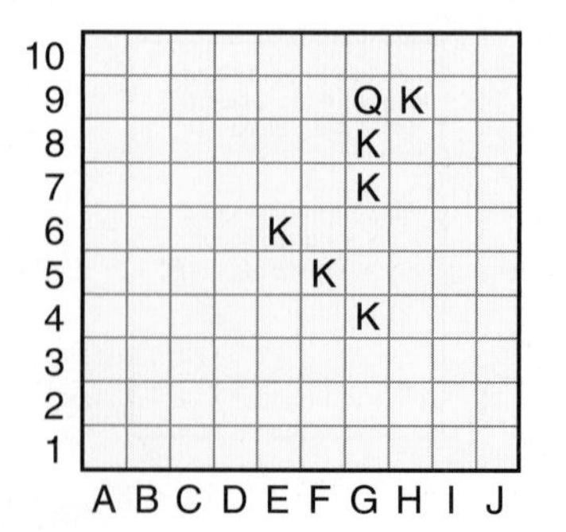

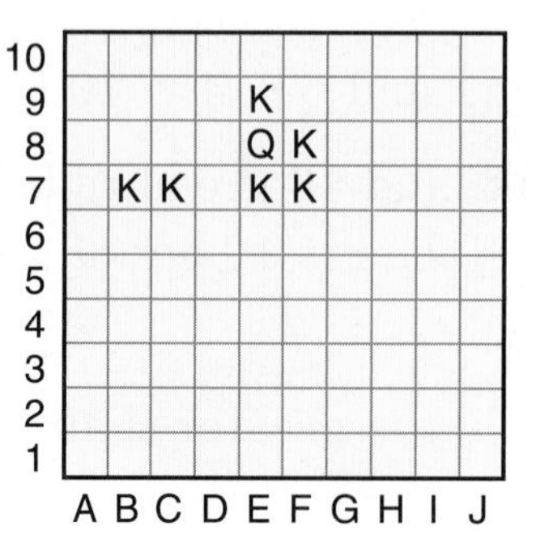

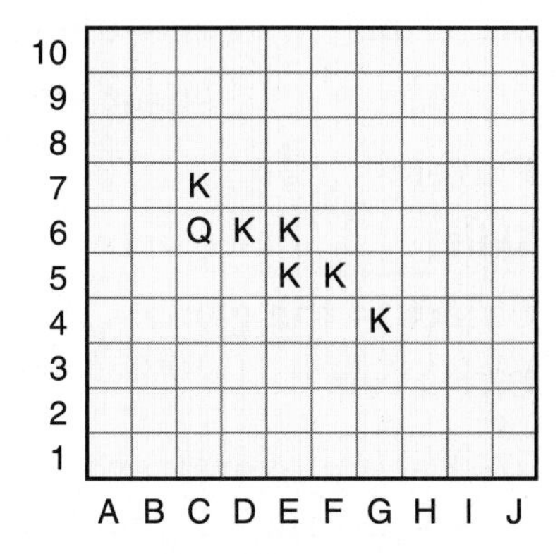

Each square is named by a letter-number coordinate pair such as B-3.

Each piece and each square has a value.

- ♦ A queen is worth 5 points; each knight is worth 1 point.
- ♦ The value of a square is equal to the sum of the values of the piece on the square itself and the pieces on the 8 surrounding squares, including squares on a diagonal.

Examples of square values for the grid above:

G-4 is worth 2:
$1 + 1 = 2$

E-6 is worth 4:
$1 + 1 + 1 + 1 = 4$

C-7 is worth 7:
$1 + 5 + 1 = 7$

B-6 is worth 6:
$5 + 1 = 6$

A-6 is worth 0

Searching for the opponent's queen Once players have recorded the location of their pieces on their Grid 1 sheets, they take turns searching for the other player's queen. They use Grid 2 to record the results of their search. The object is to develop a search strategy that will locate the queen in as few turns as possible.

1. At the start of the game: Player A names a square on Player B's Grid 1.
   - ♦ If Player B's queen is on that square, Player A wins the game.
   - ♦ If Player B's queen is not on that square, Player B tells the value of the square. Player A then records the value of the square on his or her Grid 2.
2. Players A and B reverse roles. Play proceeds as above.
3. Play continues until one player figures out where the other player's queen is located.

Example Suppose Player B has arranged the pieces as shown.

Player A: "I guess E-6."

Player B: "You didn't find the queen, but square E-6 is worth 4 points."

Player A writes 4 in square E-6 on his or her Grid 2.

Player B: Grid 1

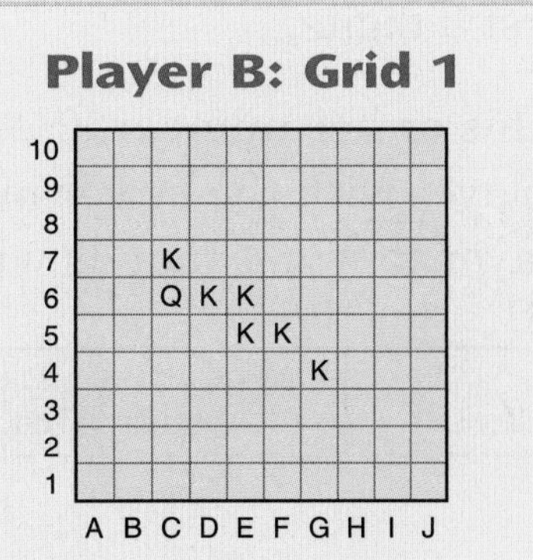

Player A: Grid 2

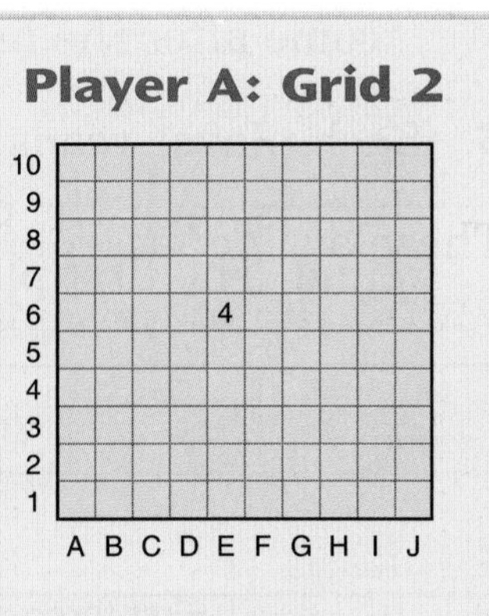

High-Number Toss

Materials ☐ 1 six-sided die

☐ 1 sheet of paper for each player

Players 2

Skill Place value; exponential notation

Object of the game To make the largest numbers possible.

Directions

1. Each player draws 4 blank lines on a sheet of paper to record the numbers that come up on the rolls of the die.

   Player 1: ______ ______ ______ | ______

   Player 2: ______ ______ ______ | ______

2. Player 1 rolls the die and writes the number on any of his or her 4 blank lines. It does not have to be the first blank—it can be any of them. *Keep in mind that the larger number wins!*
3. Player 2 rolls the die and writes the number on one of his or her blank lines.
4. Players take turns rolling the die and writing the number 3 more times each.
5. Each player then uses the 4 numbers on his or her blanks to build a number.
   - ♦ The numbers on the first 3 blanks are the first 3 digits of the number the player builds.
   - ♦ The number on the last blank tells the number of zeros that come after the first 3 digits.
6. Each player reads his or her number. (See the place-value chart below.) The player with the larger number wins the round. The first player to win 4 rounds wins the game.

Note

If you don't have a die, you can use a deck of number cards. Use all cards with the numbers 1 through 6. Instead of rolling the die, draw the top card from the facedown deck.

Hundred Millions	Ten Millions	Millions	,	Hundred Thousands	Ten Thousands	Thousands	,	Hundreds	Tens	Ones

Example

First three digits Number of zeros

Player 1: <u>1</u> <u>3</u> <u>2</u> | <u>6</u> = 132,000,000 (132 million)

Player 2: <u>3</u> <u>5</u> <u>6</u> | <u>4</u> = 3,560,000 (3 million, 560 thousand)

Player 1 wins.

Multiplication Wrestling

Materials ☐ 1 *Multiplication Wrestling* Worksheet for each player (*Math Masters,* p. 488)

☐ number cards 0–9 (4 of each) or 1 ten-sided die

Players 2

Skill Partial-products algorithm

Object of the game To get the larger product of two 2-digit numbers.

Directions

1. Shuffle the deck of cards and place it number-side down on the table.
2. Each player draws 4 cards and forms two 2-digit numbers. Players should form their 2 numbers so that their product is as large as possible.
3. Players create 2 "wrestling teams" by writing each of their numbers as a sum of 10s and 1s.
4. Each player's 2 teams wrestle. Each member of the first team (for example, 70 and 5) is multiplied by each member of the second team (for example, 80 and 4). Then the 4 products are added.
5. **Scoring:** The player with the larger product wins the round and receives 1 point.
6. To begin a new round, each player draws 4 new cards to form 2 new numbers. A game consists of 3 rounds.

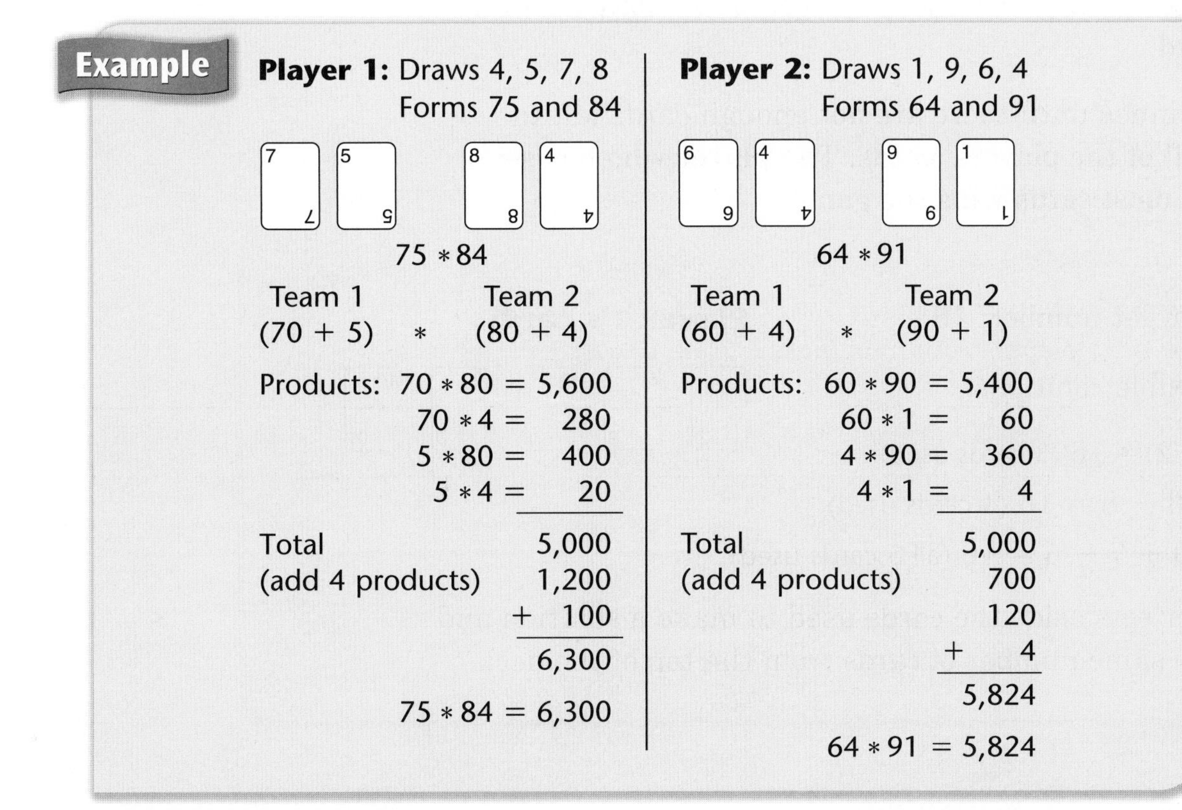

Name That Number

Materials □ 1 complete deck of number cards

Players 2 or 3

Skill Naming numbers with expressions

Object of the game To collect the most cards.

Directions

1. Shuffle the cards and deal 5 cards to each player. Place the remaining cards number-side down on the table between the players. Turn over the top card and place it beside the deck. This is the **target number** for the round.
2. Players try to match the target number by adding, subtracting, multiplying, or dividing the numbers on as many of their cards as possible. A card may only be used once.
3. Players write their solutions on a sheet of paper. When players have written their best solutions:
   - ♦ Each player sets aside the cards they used to match the target number.
   - ♦ Each player replaces the cards they set aside by drawing new cards from the top of the deck.
   - ♦ The old target number is placed on the bottom of the deck.
   - ♦ A new target number is turned over, and another round is played.
4. Play continues until there are not enough cards left to replace all of the players' cards. The player who has set aside the most cards wins the game.

Example Target number: 16

Player 1's cards:

7 5 8 2 10

Some possible solutions:

10 + 8 − 2 = 16 (3 cards used)

7 * 2 + 10 − 8 = 16 (4 cards used)

8 / 2 + 10 + 7 − 5 = 16 (all 5 cards used)

The player sets aside the cards used to make a solution and draws the same number of cards from the top of the deck.

Number Top-It (7-Digit Numbers)

Materials ☐ number cards 0–9 (4 of each)

☐ 1 Number Top-It Mat (7-Digit) (*Math Masters,* pp. 492 and 493)

Players 2 to 5

Skill Place value for whole numbers

Object of the game To make the largest 7-digit numbers.

Directions

1. Shuffle the cards and place the deck number-side down on the table.
2. Each player uses one row of boxes on the place-value game mat.
3. In each round, players take turns turning over the top card from the deck and placing it on any one of their empty boxes. Each player takes a total of 7 turns, and places 7 cards on his or her row of the game mat.
4. At the end of each round, players read their numbers aloud and compare them to the other players' numbers. The player with the largest number for the round scores 1 point. The player with the next-largest number scores 2 points, and so on.
5. Players play 5 rounds for a game. Shuffle the deck between each round. The player with the *smallest* total number of points at the end of five rounds wins the game.

Example Andy and Barb played 7-digit *Number Top-It.* Here is the result for one complete round of play.

Number Top-It Mat (7-Digit)

	Millions	Hundred Thousands	Ten Thousands	Thousands	Hundreds	Tens	Ones
Andy	7	6	4	5	2	0	1
Barb	4	9	7	3	5	2	4

Andy's number is larger than Barb's number. So Andy scores 1 point for this round. Barb scores 2 points.

Number Top-It (Decimals)

Materials □ number cards 0–9 (4 of each)

□ 1 Number Top-It Mat (Decimals) (*Math Masters,* pp. 490 or 491)

Players 2 to 5

Skill Place value for decimals

Object of the game To make the largest 2-digit (or 3-digit) decimal numbers.

Directions

1. This game is played using the same directions as those for *Number Top-It* (7-Digit Numbers). The only difference is that players use a place-value mat for decimals. Steps 2 and 3 give directions for a game played on a place-value mat for 2-place decimals.
2. In each round, players take turns turning over the top card from the deck and placing it on any one of their empty boxes. Each player takes 2 turns, and places 2 cards on his or her row of the game mat.
3. Players play 5 rounds for a game. Shuffle the deck between each round. The player with the *smallest* total number of points at the end of the 5 rounds wins the game.

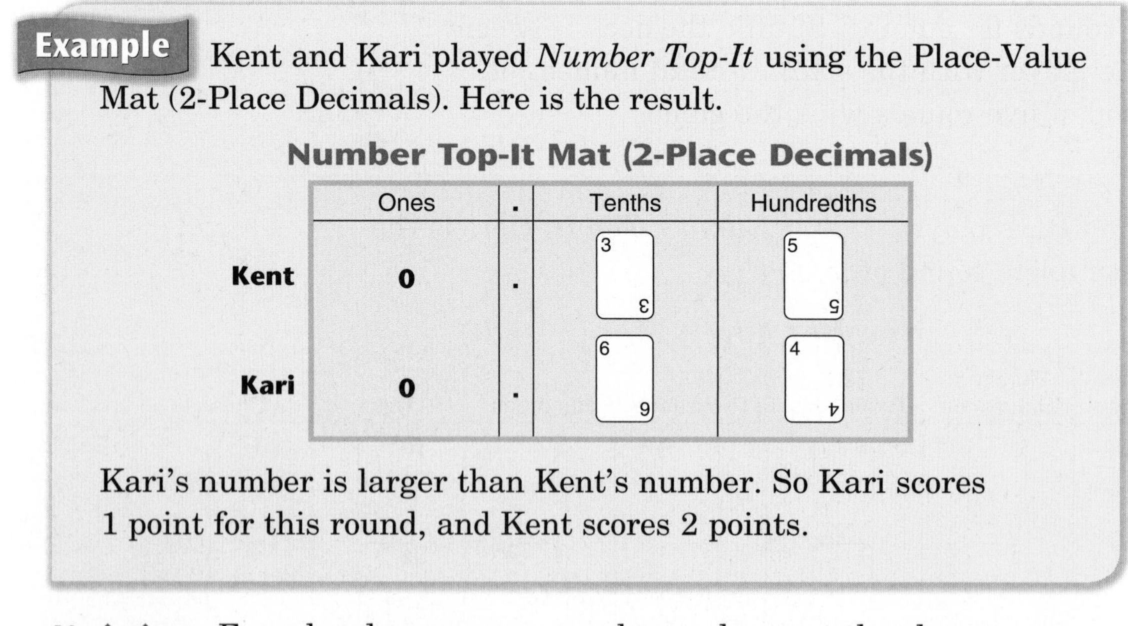

Example Kent and Kari played *Number Top-It* using the Place-Value Mat (2-Place Decimals). Here is the result.

Number Top-It Mat (2-Place Decimals)

	Ones	.	Tenths	Hundredths
Kent	**0**	.	3	5
Kari	**0**	.	6	4

Kari's number is larger than Kent's number. So Kari scores 1 point for this round, and Kent scores 2 points.

Variation For a harder game, use a place-value mat that has empty boxes in the tenths, hundredths, and thousandths places. Each player takes 3 turns, and places 3 cards on his or her row of the game mat.

Over and Up Squares

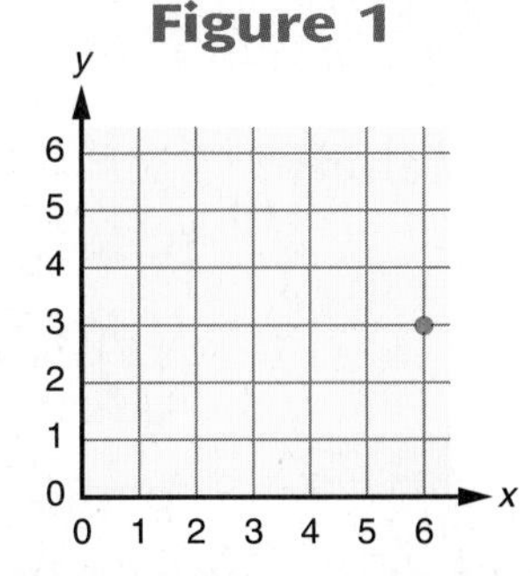

Player 1 rolls a 3 and a 6. The point (6,3) is marked on the grid.

Materials ☐ 1 *Over and Up Squares* Gameboard and Record Sheet (*Math Masters,* p. 494)

☐ 1 colored pencil per player (different colors)

☐ 2 six-sided dice

Players 2

Skill Plotting ordered pairs; developing a winning game strategy

Object of the game To score more points by connecting ordered pairs on a coordinate grid.

Directions

1. Player 1 rolls 2 dice and uses the numbers to make an ordered pair. Either number can be used to name the x-coordinate (over) of the ordered pair. The other number is used to name the y-coordinate (up) of the ordered pair. After deciding which ordered pair to use, the player marks it on the grid with his or her colored pencil (See Figure 1.)
2. Player 1 records the ordered pair and the score in the first table. A player earns 10 points each time an ordered pair is marked correctly.
3. Player 2 rolls the dice and decides how to make an ordered pair. If both possible ordered pairs are already marked on the grid, the player rolls the dice again. (Variation: If both possible ordered pairs are already marked, the player can change one or both of the numbers to 0.)
4. Player 2 uses the other colored pencil to mark the ordered pair and records the ordered pair and score in the second table.
5. Players take turns rolling the dice, marking ordered pairs on the grid, and recording the results. On a player's turn, if 2 marked grid points are next to each other on the same side of one of the grid squares, the player connects them with a line segment he or she makes. Sometimes more than 1 line segment may be drawn in a single turn. (See Figure 2.) A player scores 10 points for each line segment drawn.
6. If a player draws a line segment that completes a grid square, (so that all 4 sides of the square are now drawn), that player colors in the square and earns 50 points. (See Figure 3.)
7. The player with more points after 10 rounds wins.

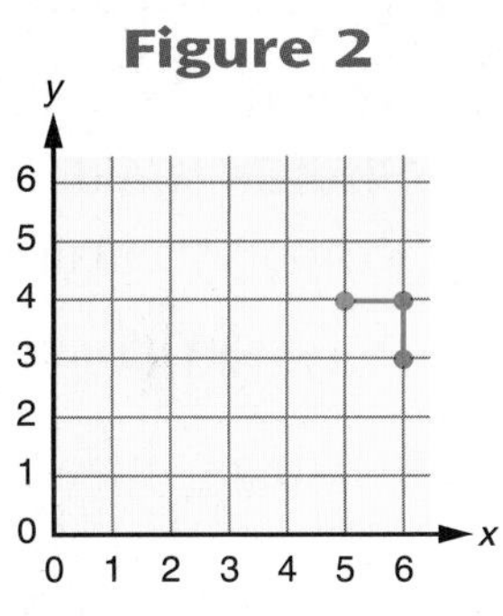

Player 1 marks (6,4) and scores 10 points. Player 1 draws 2 line segments and scores 20 points. The score for the round is 30 points.

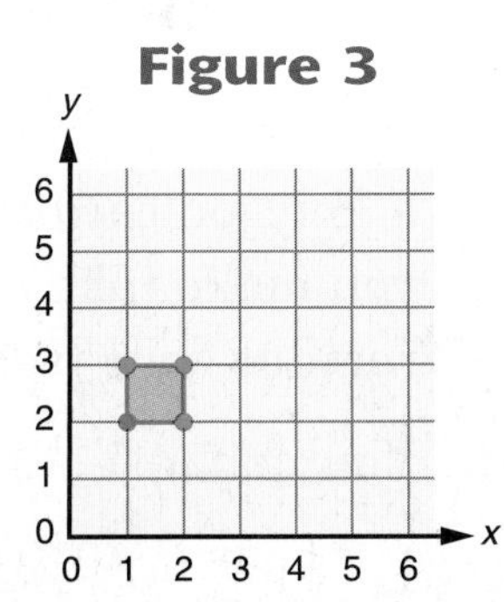

Player 1 marks (1,2) and scores 10 points. Player 1 draws 2 line segments and scores 20 points. The line segments complete a square. Player 1 colors in the square and scores 50 points. The score for the round is 80 points.

Polygon Pair-Up

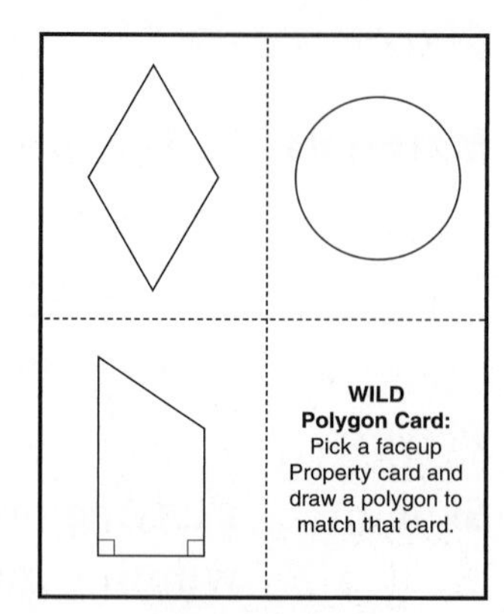

Examples of Polygon Cards

Materials ☐ 1 *Polygon Pair-Up* Polygon Deck (*Math Masters,* p. 496)

☐ 1 *Polygon Pair-Up* Property Deck (*Math Masters,* p. 497)

☐ paper and pencils for sketching

Players 2, or two teams of 2

Skill Properties of polygons

Object of the game To collect more cards by matching polygons with their properties.

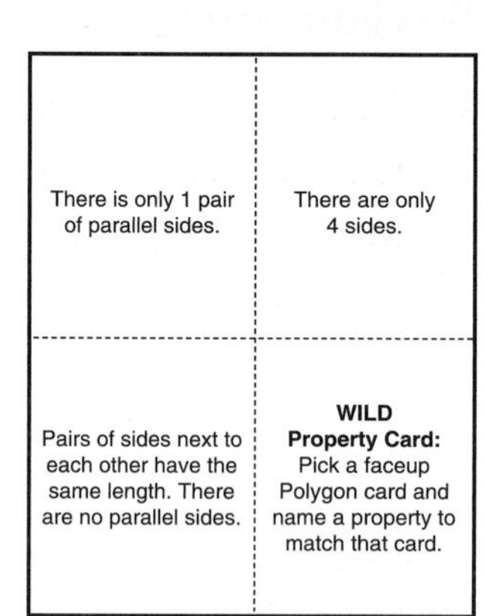

Examples of Property Cards

Directions

1. Shuffle the deck of Polygon cards. Then shuffle the deck of Property cards. Place the decks side by side and facedown.
2. Players take turns. When it is your turn:
   - ♦ Turn over one Polygon card and one Property card. Place these cards faceup and below the card decks.
   - ♦ If you are able to match a Polygon card with a Property card, say "Match!" and take those two cards. Your turn is over. (You may make only one match and take two cards per turn.)
   - ♦ If you are not able to match a Polygon card with a Property card, say "Done!" Your turn is over. All the cards that were faceup remain faceup for the next player's turn.
3. When you are ready to begin your turn, you may notice a Polygon card and Property card that match. If you say "Steal!" you make take those matching cards. Then continue with your regular turn (see Step 2).
4. You may use a WILD card to make a match during any turn.
   - ♦ To use a WILD PROPERTY card, pick any faceup Polygon card. If you name a property to match that Polygon card, you take both cards.
   - ♦ To use a WILD POLYGON card, pick any faceup Property card. If you sketch a polygon that matches that Property card, you take both cards.
5. The game is over when all the cards have been turned over and no more matches can be made. The player with the most cards wins.

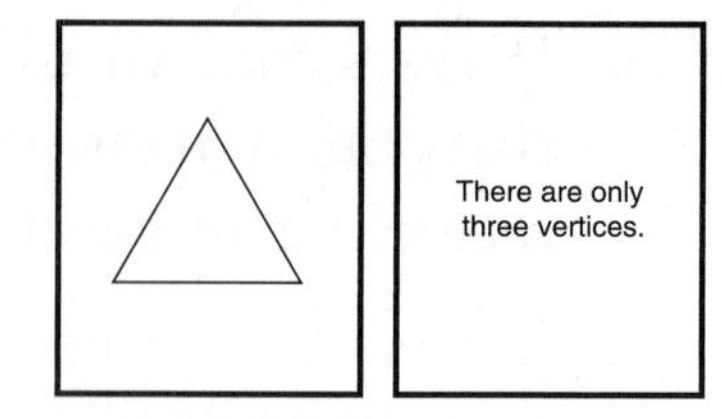

a Polygon card and a Property card that match

Product Pile-Up

Materials □ number cards 1–10 (8 of each)

Players 3 to 5

Skill Multiplication facts 1 to 10

Object of the game To play all of your cards and have none left.

Directions

1. Shuffle the cards and deal 12 cards to each player. Place the rest of the deck number-side down on the table.

2. The player to the left of the dealer begins. This player selects 2 of their cards, places them number-side up on the table, multiplies the numbers, and gives the product.

3. Play continues with each player selecting and playing 2 cards with a product that is *greater than* the product of the last 2 cards played.

Example Joe plays 3 and 6 and says, "3 times 6 equals 18."

The next player, Rachel, looks at her hand to find 2 cards with a product higher than 18. She plays 5 and 4 and says, "5 times 4 equals 20."

4. If a player is not able to play 2 cards with a greater product, the player must draw 2 cards from the deck. These 2 cards are added to the player's hand. If the player is now able to make a greater product, the 2 cards are played, and play continues.

5. If after drawing the 2 cards a player still cannot make a play, the player says "Pass." If all the other players say "Pass," the last player who was able to lay down 2 cards starts play again. That player may select any 2 cards to make *any* product and play continues.

6. If a player states an incorrect product, he or she must take back the 2 cards, draw 2 cards from the deck, and say "Pass." Play moves to the next person.

7. The winner is the first player to run out of cards, or the player with the fewest cards when there are no more cards to draw.

Rugs and Fences

Materials ☐ 1 *Rugs and Fences* Polygon Deck A, B, or C (*Math Masters,* page 499, 500, or 501)

☐ 1 *Rugs and Fences* Area and Perimeter Deck (*Math Masters,* page 498)

☐ 1 *Rugs and Fences* Record Sheet for each player (*Math Masters,* page 502)

Players 2

Skill Calculating area and perimeter

Object of the game To score more points by finding the perimeters and areas of polygons.

Directions

1. Select one of the Polygon Decks—A, B, or C. Shuffle the deck and place it picture-side down on the table. (Variation: Combine 2 or 3 Polygon decks.)
2. Shuffle the deck of Area and Perimeter cards and place it word-side down next to the Polygon Deck.
3. Players take turns. At each turn, a player draws 1 card from each deck and places them faceup on the table. The player finds the area (A) *or* the perimeter (P) of the polygon, as directed by the Area and Perimeter card.
   - ♦ If a "Player's Choice" card is drawn, the *player* may choose to find either the area or the perimeter of the polygon.
   - ♦ If an "Opponent's Choice" card is drawn, the *opposing player* chooses whether the area or the perimeter of the polygon will be found.
4. A player records a turn on his or her Record Sheet. The player records the polygon card number, circles A (area) or P (perimeter), and writes a number model used to calculate the area or perimeter. The solution is the player's score for the round.
5. The player with the higher total score at the end of 8 rounds is the winner.

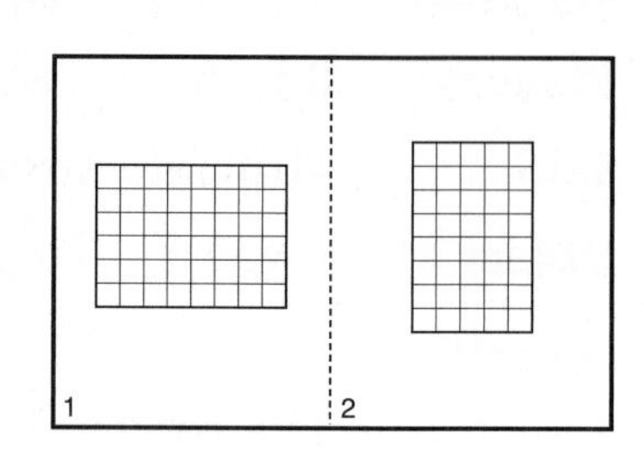

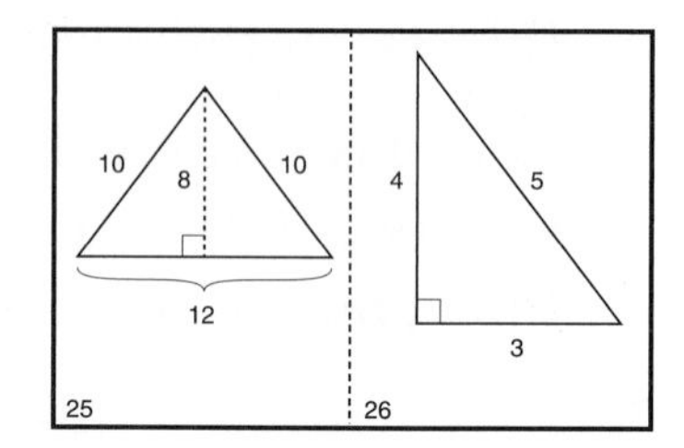

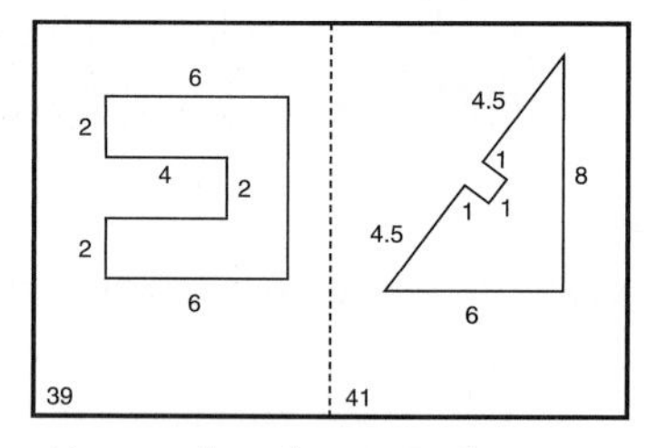

Examples from Polygon Decks A, B, and C

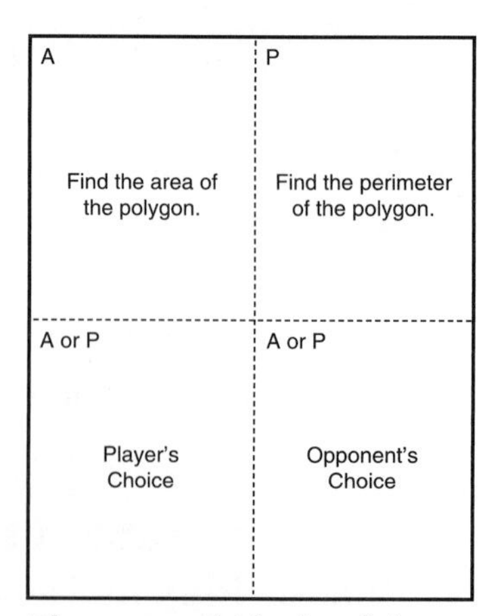

There are 4 kinds of Area and Perimeter cards.

Examples Ama draws the two cards shown here. Ama may choose to calculate the area or the perimeter.

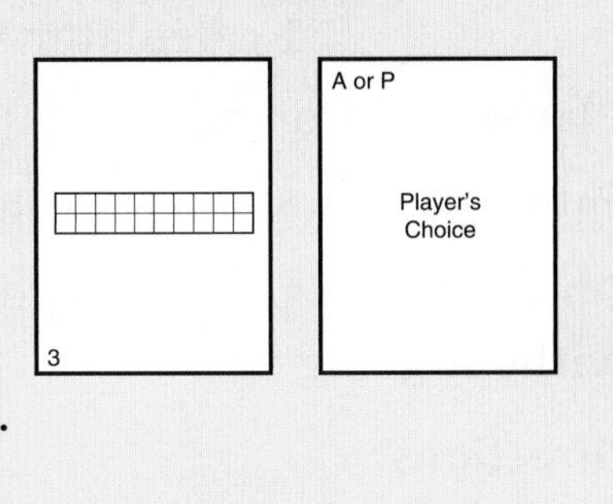

- ♦ Ama counts unit squares to find the area.
  Area = 20 square units.
- ♦ Ama counts unit lengths around the polygon to find the perimeter.
  Perimeter = 24 units.

Ama records card number 3 and circles P on her record sheet. She writes the number model 10 + 10 + 2 + 2 = 24, and earns 24 points.

Parker draws the 2 cards shown here.

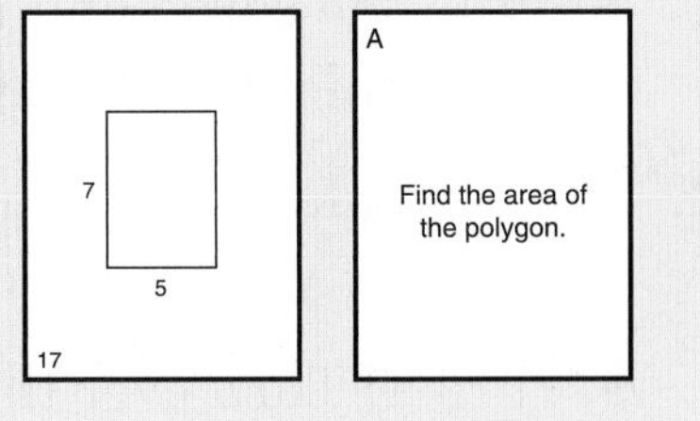

He finds the area of the polygon by using the formula $A = b * h$. He records card number 17 and circles A on his record sheet. He writes the number model $5 * 7 = 35$, and earns 35 points.

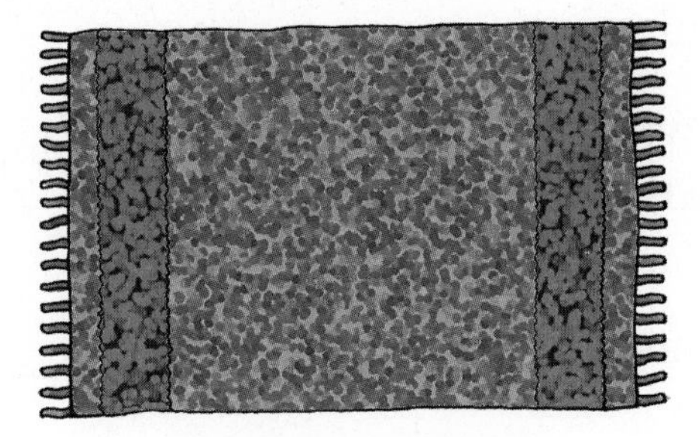
The rug covers an area.

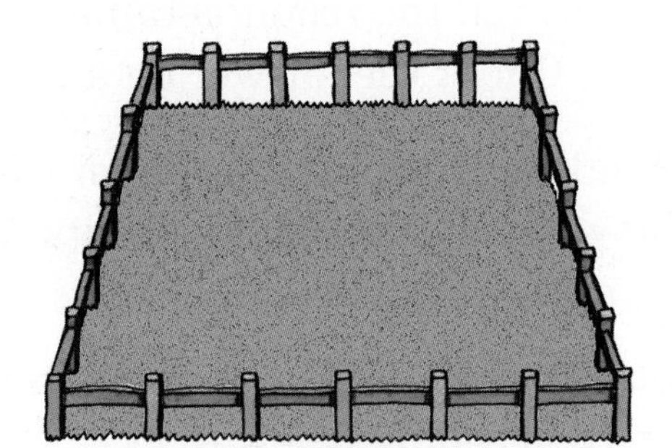
The fence goes around the perimeter.

Name Date Time

***Rugs and Fences* Record Sheet**

Round	Record the card number	Circle A (area) or P (perimeter)	Write a number model	Record your score
Sample	3	A or (P)	10 + 10 + 2 + 2 = 24	24
1		A or P		
2		A or P		
3		A or P		
4		A or P		
5		A or P		
6		A or P		
7		A or P		
8		A or P		
Total Score				

Subtraction Target Practice

Materials ☐ number cards 0–9 (4 of each)

☐ 1 calculator for each player

Players 1 or more

Skill 2- and 3-digit subtraction

Object of the game To get as close as possible to 0, without going below it.

Directions

1. Shuffle the cards and place the deck number-side down on the table. Each player starts at 250.
2. Players take turns. Each player has 5 turns in the game. When it is your turn, do the following:
   - ♦ Turn 1: Turn over the top 2 cards and make a 2-digit number. (You may place the cards in either order.) Subtract this number from 250 on scratch paper. Check the answer on a calculator.
   - ♦ Turns 2–5: Take 2 cards and make a 2-digit number. Subtract this number from the result obtained in your previous subtraction. Check the answer on a calculator.
3. The player with a final result that is is closest to 0, without going below 0, is the winner. If the final results for all players are below 0, no one wins.

If there is only 1 player, the object of the game is to get as close to 0 as possible, without going below 0.

Example

Turn 1: Draw 4 and 5. Subtract 45 or 54. 250 – 45 = 205

Turn 2: Draw 0 and 6. Subtract 6 or 60. 205 – 60 = 145

Turn 3: Draw 4 and 1. Subtract 41 or 14. 145 – 41 = 104

Turn 4: Draw 3 and 2. Subtract 32 or 23. 104 – 23 = 81

Turn 5: Draw 6 and 8. Subtract 68 or 86. 81 – 68 = 13

Variation Each player starts at 100 instead of 250.

Top-It Games

The materials, number of players, and object of the game are the same for all *Top-It Games*.

Materials ☐ number cards 1–10 (4 of each)

☐ 1 calculator (optional)

Players 2 to 4

Skill Addition, subtraction, multiplication, and division facts

Object of the game To collect the most cards.

Addition Top-It

Directions

1. Shuffle the cards and place the deck number-side down on the table.
2. Each player turns over 2 cards and calls out the sum of the numbers. The player with the largest sum takes all the cards. In case of a tie for the largest sum, each tied player turns over 2 more cards and calls out the sum of the numbers. The player with the largest sum takes all the cards from both plays.

3. Check answers using an Addition Table or a calculator.
4. The game ends when there are not enough cards left for each player to have another turn.
5. The player with the most cards wins.

Variation Each player turns over 3 cards and finds their sum.

Advanced Version Use only the number cards 1–9. Each player turns over 4 cards, forms two 2-digit numbers, and finds the sum. Players should carefully consider how they form their numbers since different arrangements have different sums. For example, 74 + 52 has a greater sum than 47 + 25.

Subtraction Top-It

Directions

1. Each player turns over 3 cards, finds the sum of any 2 of the numbers, then finds the difference between the sum and the third number.
2. The player with the largest difference takes all the cards.

Example A 4, an 8, and a 3 are turned over. There are three ways to form the numbers. Always subtract the smaller number from the larger one.

$4 + 8 = 12$	or	$3 + 8 = 11$	or	$3 + 4 = 7$
$12 - 3 = 9$		$11 - 4 = 7$		$8 - 7 = 1$

Advanced Version Use only the number cards 1–9. Each player turns over 4 cards, forms two 2-digit numbers, and finds their difference. Players should carefully consider how they form their numbers. For example, $75 - 24$ has a greater difference than $57 - 42$ or $74 - 25$.

Multiplication Top-It

Directions

1. The rules are the same as for *Addition Top-It,* except that players find the product of the numbers instead of the sum.
2. The player with the largest product takes all the cards. Answers can be checked with a Multiplication Table or a calculator.

Variation Use only the number cards 1–9. Each player turns over 3 cards, forms a 2-digit number, then multiplies the 2-digit number by the remaining number.

Division Top-It

Directions

1. Use only the number cards 1–9. Each player turns over 3 cards and uses them to generate a division problem as follows:
   - Choose 2 cards to form the dividend.
   - Use the remaining card as the divisor.
   - Divide and drop any remainder.

The player with the largest quotient takes all the cards.

Advanced Version Use only the number cards 1–9. Each player turns over 4 cards, chooses 3 of them to form a 3-digit number, then divides the 3-digit number by the remaining number. Players should carefully consider how they form their 3-digit numbers. For example, 462 / 5 is greater than 256 / 4.

World Tour

Bd. Edgar Quinet
Bd. Pasteur
Avenue du
Rue Raymond Losserand
R. de Gergovie
Rue d'Alésia
Rue de Vouillé
R. Yvart
R. Brancion
R. Labrouste
Place Falguière
Pernety
Plaisance
R. Vercingétorix
R. Dutot
Rue Lecourbe
R. de Javel
Entrepreneurs
R. P. Larousse

Introduction

About the World Tour

For the rest of the school year, you and your classmates will go on an imaginary tour of the world. As you visit various countries, you will learn about the customs of people in other parts of the world. You will practice globe and map skills to help you locate the places you visit. As you collect and examine numerical information for the countries you visit, you will have many opportunities to apply your knowledge of mathematics.

A Chinese farmer waters crops by hand.

How the World Tour Is Organized

You will first fly from your hometown to Washington, D.C. The class will then visit five regions of the world: Africa, Europe, South America, Asia and Australia, and, finally, North America.

The class will first visit the continent of Africa, landing in Cairo, Egypt. From there, you will fly to the second region, Europe, landing in Budapest, Hungary. This pattern will be repeated for each of the other regions. The class will complete the tour by flying back to Washington, D.C.

Women in Guatemala sell fruits and vegetables at a market.

As you visit each country, you will collect information about that country and record it in a set of Country Notes pages in your journal. The World Tour section of your *Student Reference Book* will serve as a major source for that information.

As you make your world tour, there will be opportunities to learn about countries you have not visited. This World Tour section includes detailed information for 10 selected countries within each of the five regions of the world. It includes a Fascinating Facts insert that lists interesting facts about world geography, population, and climate. And it includes a collection of games played in different parts of the world that require logical or mathematical thinking. You can also look up additional information in a world almanac, *National Geographic* magazines, travel brochures, guidebooks, and newspapers.

Travel brochures include both country facts and travel information.

Washington, D.C. Facts

Washington, D.C., is the **capital** of the United States of America. The capital is where our country's laws are made. It is also where our president lives.

Washington, D.C., has been the capital of the United States since 1800. Before that, the capital was in Philadelphia, Pennsylvania. The capital city was named "Washington, the District of Columbia" in honor of George Washington and Christopher Columbus. "District of Columbia" is usually abbreviated as D.C.

Washington, D.C., has an area of 68 square miles. With a population of about 600,000 people, more people live in our capital city than in the state of Wyoming!

There are many interesting things to do in Washington, D.C. The facts below will help you plan your visit.

Average High/Low Temperatures (°F)			Average Precipitation (in.)	
Month	High	Low	Month	Precipitation
Jan	42	25	Jan	3.1
Feb	46	28	Feb	2.7
Mar	55	35	Mar	3.6
Apr	66	43	Apr	3.0
May	75	53	May	4.0
Jun	83	62	Jun	3.6
Jul	88	67	Jul	3.6
Aug	86	66	Aug	3.6
Sep	79	59	Sep	3.8
Oct	68	46	Oct	3.3
Nov	57	37	Nov	3.2
Dec	46	29	Dec	3.1

The White House

Every U.S. president except George Washington has lived here. The White House sits on 18 acres of land. It has 132 rooms, five of which can be seen on a 20-minute public tour. Every year more than 1,500,000 people tour the White House. Web site: http://www.whitehouse.gov

Washington, D.C. Facts

Washington Metrorail

This system of underground electric trains opened in 1976. There are 83 stations in the Washington area; some are decorated with beautiful artwork. They are connected by more than 103 miles of train lines. On an average weekday, about 500,000 people ride the Metro trains.
Web site: http://www.wmata.com

Washington Monument

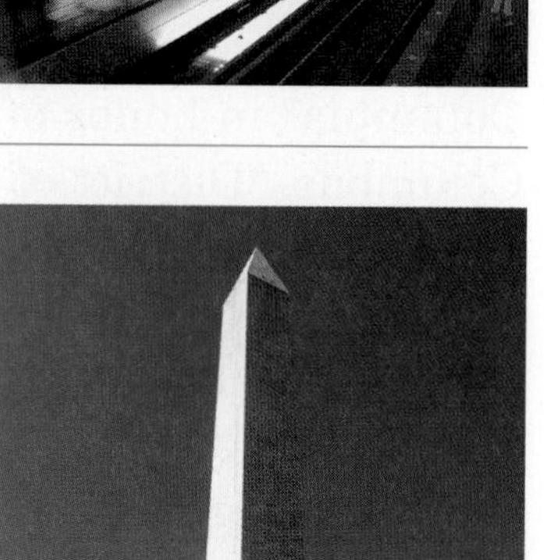

This 555-foot-tall monument was built to honor George Washington, the first president of the United States. It is one of the tallest masonry structures in the world. You can take an elevator to a viewing area at the 500-foot level. The monument's cornerstone was laid in 1848, but building was interrupted by the Civil War (1861–1865). Construction workers started building again in 1880 and completed it in 1884. The Washington Monument receives more than 800,000 visitors each year.
Web site: http://www.nps.gov/wamo

Jefferson Memorial

This memorial was built in honor of Thomas Jefferson. Jefferson was the third president of the United States and the author of the *Declaration of Independence*. The memorial was dedicated in 1943, exactly 200 years after Jefferson was born. Inside the memorial, a statue of Jefferson stands 19 feet tall.
Web site: http://www.nps.gov/thje/home.htm

Lincoln Memorial

This memorial was built in 1922 in honor of Abraham Lincoln. Lincoln was the 16th U.S. president. Inside the memorial is a 19-foot-high statue of Lincoln seated in a large armchair. The statue is made of 28 blocks of white marble from the state of Georgia.
Web site: http://www.nps.gov/linc/home.htm

The United States Capitol

The Capitol Building is where the senators and representatives in Congress meet to make laws. George Washington laid its cornerstone in 1793. The building's cast iron dome weighs 9 million pounds. Brass doors that weigh 10 tons lead to the rotunda, which is 180 feet high and 96 feet in diameter. Web site: http://www.aoc.gov

---

Library of Congress

The Library of Congress is the world's largest library. It contains more than 128 million items in 450 languages. It has about 535 miles of bookshelves. The Library is composed of three buildings. One of these, the James Madison Building, encloses an area greater than 35 football fields. The Library has more than 18 million books, 12 million photographs, 3 million sound recordings, and 5 million maps. It also has about 125,000 telephone books and 100,000 comic books. The Library of Congress collection grows by more than 10,000 items every day. Web site: http://www.loc.gov

---

National Museum of Natural History

Exhibits in the National Museum of Natural History include an 8-ton African bull elephant, the 45.5-carat Hope Diamond, a 360 million-year-old fossilized fish, a 90-foot-long skeleton of a diplodocus, a 30-foot-long giant squid, moon rocks, a mural of a 3.5 billion-year-old shoreline, and a life-size 92-foot model of a blue whale. The Insect Zoo has live insects as well as scorpions and tarantulas. In the Discovery Room, you can try on costumes from around the world. The museum is part of the Smithsonian Institution. Web site: http://www.mnh.si.edu

---

National Air and Space Museum

The National Air and Space Museum is the most popular museum in the world. Every year more than 10 million people visit it. Its collection includes the *Wright Flyer,* which is the original plane flown by the Wright brothers at Kitty Hawk, North Carolina, in 1903. It also includes the *Spirit of St. Louis,* the plane in which Charles Lindbergh made the first nonstop flight across the Atlantic in 1927. *Columbia*, the Apollo 11 command module that brought back the first men to walk on the moon in 1969, is also on display.
Web site: http://www.nasm.si.edu

Map of the National Mall in Washington, D.C.

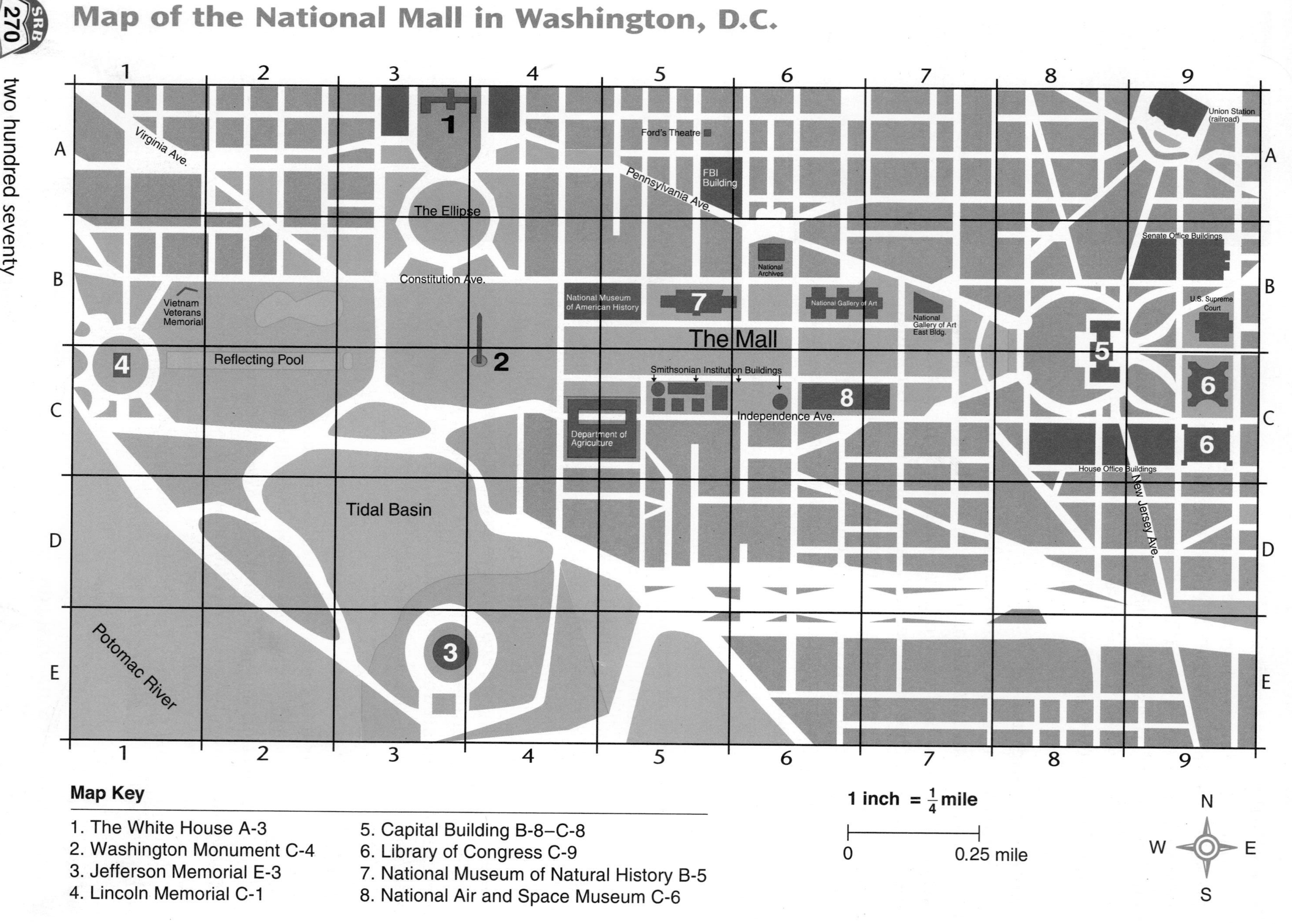

Map Key

1. The White House A-3
2. Washington Monument C-4
3. Jefferson Memorial E-3
4. Lincoln Memorial C-1
5. Capital Building B-8–C-8
6. Library of Congress C-9
7. National Museum of Natural History B-5
8. National Air and Space Museum C-6

Facts About the World

Continents are large land masses. There are seven continents on the Earth, although Europe and Asia are sometimes thought of as one continent. Most continents contain many countries, but there are no countries at all in Antarctica.

A **country** is a territory and the people who live there under one government. The number of countries in the world often changes as countries split apart or join with other countries. At this time, there are about 200 countries in the world.

Population is the number of people who live in a certain region. Population growth is the change in the population every year after all births and deaths are accounted for. The **population growth rate** is the increase (or decrease) in population per year, written as a percent.

Dimensions of the Earth

Equatorial circumference*:
about 24,900 miles
(40,000 kilometers)

Equatorial diameter:**
about 7,930 miles
(12,760 kilometers)

Volume: 2.6×10^{11} cubic miles
(1.1×10^{12} cubic kilometers)

Weight (mass): 6.6×10^{21} tons
(6.0×10^{21} metric tons)

Total world water area:
about 139,433,000 square miles
(361,129,000 square kilometers)

*Circumference is the distance around a circle or sphere.

**Diameter is the distance measured by a straight line passing from one side of a circle or sphere, through the center, to the other side.

The world's population is now increasing by about 200,000 people per day, or about 75 million people per year. Over the last 40 years, the world's population has about doubled. It reached the 6 billion mark in 1999. World population is expected to reach about 9 billion people by the year 2050.

The Continents

Continent	Population*	Percent of World Population	Area (sq miles)	Percent of Land Area
North America	509,000,000	8.0%	8,300,000	14.8%
South America	367,000,000	5.8	6,800,000	12.1
Europe	799,000,000	12.5	4,100,000	7.3
Asia	3,797,000,000	59.5	16,700,000	29.8
Africa	874,000,000	13.7	11,500,000	20.5
Australia	32,000,000	0.5	3,300,000	5.9
Antarctica	0	0.0	5,400,000	9.6
World Totals	**6,378,000,000 (about 6.4 billion)**	**100.0%**	**56,100,000**	**100.0%**

*Data are for the year 2004. World population growth rate for the year 2004: about 1.2% per year

Latitude and Longitude

You sometimes use a world globe or a flat map to locate countries, cities, rivers, and so forth. Reference lines are drawn on globes and maps to make places easier to find.

Latitude

Lines that go east and west around the Earth are called **lines of latitude.** The **equator** is a special line of latitude. Every point on the equator is the same distance from the North Pole and the South Pole. Lines of latitude are called **parallels** because each one is a circle that is parallel to the equator.

Latitude is measured in **degrees.** The symbol for degrees is (°). Lines north of the equator are labeled °N. Lines south of the equator are labeled °S. The number of degrees tells how far north or south of the equator a place is. The area north of the equator is called the **Northern Hemisphere.** The area south of the equator is called the **Southern Hemisphere.**

Examples The latitude of Cairo, Egypt, is 30°N. We say that Cairo is 30 degrees north of the equator.

The latitude of Durban, South Africa, is 30°S. Durban is in the Southern Hemisphere.

The latitude of the North Pole is 90°N. The latitude of the South Pole is 90°S. The poles are the points farthest north and farthest south on Earth.

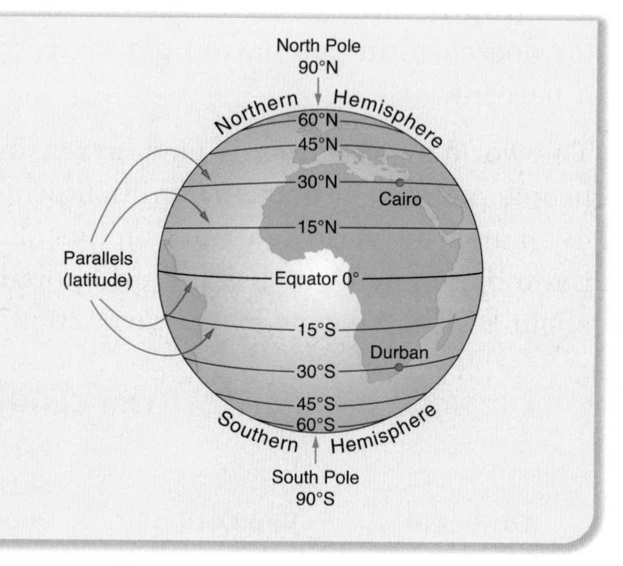

Longitude

A second set of lines runs from north to south. These are semicircles (half-circles) that connect the poles. They are called **lines of longitude** or **meridians.** The meridians are not parallel since they meet at the poles.

The **prime meridian** is the special meridian labeled 0°. The prime meridian passes through Greenwich, near London, England. Another special meridian falls on, or close to, the **International Date Line.** This meridian is labeled 180° and is exactly opposite the prime meridian, on the other side of the world.

Longitude is measured in degrees. Lines west of the prime meridian are labeled °W. Lines east of the prime meridian are labeled °E. The number of degrees tells how far west or east of the prime meridian a place is located. The area west of the prime meridian is called the **Western Hemisphere.** The area east of the prime meridian is called the **Eastern Hemisphere.**

Examples The longitude of London is 0° because London lies close to the prime meridian.

The longitude of Durban, South Africa, is 30°E. Durban is in the Eastern Hemisphere.

The longitude of Gambia (a small country in Africa) is about 15°W. We say that Gambia is 15 degrees west of the prime meridian.

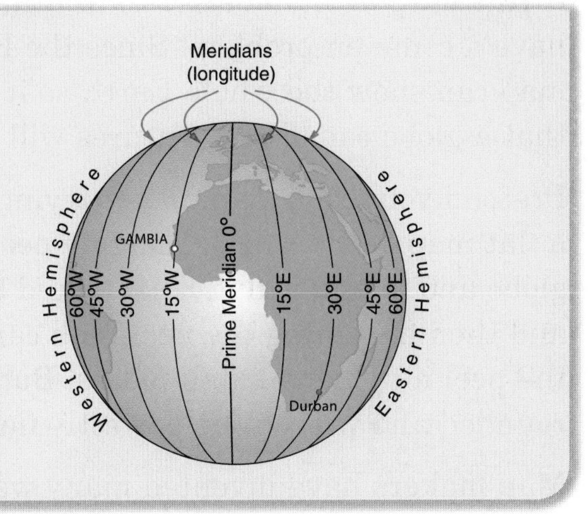

When both latitude and longitude lines are shown on a globe or map, they form a pattern of crossing lines called a **grid.** The grid can help you locate places on the globe or map. Any place on the map can be located by naming its latitude and longitude.

Check Your Understanding

Use the grid below to find the approximate latitude and longitude for the cities shown on the map. For example, Denver, Colorado, is about 40° North and 105° West.

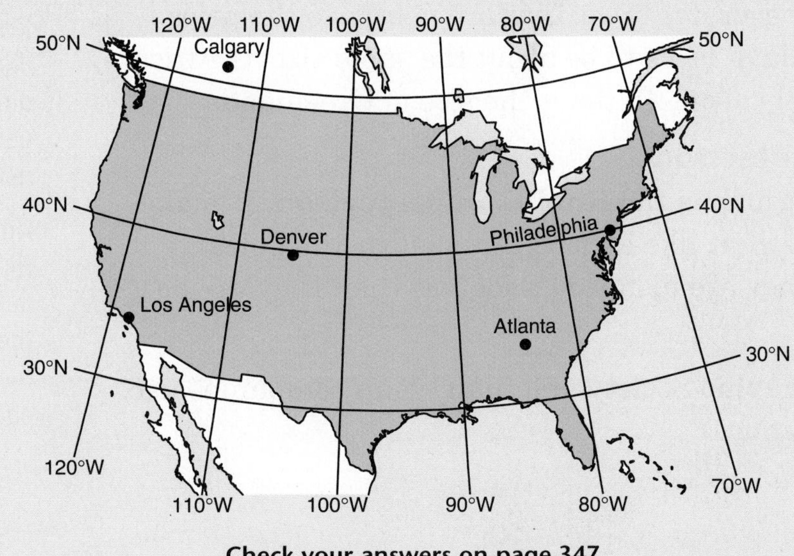

Check your answers on page 347.

Map Projections

A **world globe** shows the Earth accurately. All of the distances and directions are shown correctly. The sizes and shapes of countries, bodies of water, and other features are all as correct as possible.

A **flat map** is often easier to use than a globe. But all flat maps have a common problem. Since the Earth is a sphere, no flat map can show the whole Earth as it really is in the same way that a globe can. Some features will be distorted on a flat map.

Pretend you are a map maker trying to turn a world globe into a flat map of the world. Imagine peeling the Earth map off the globe and laying it flat. That would be like peeling an orange and then flattening the peel. You can almost do this if you tear the peel into many small pieces. But even those pieces are rounded and will not lie perfectly flat.

Map makers have invented many ways to show the spherical world as a flat map. These flat views of the Earth are called **map projections.** Every map projection has some distortions because the map maker must cut and stretch the shape of the globe to make it flat.

The Mercator Projection

One of the most common types of flat maps is the **Mercator projection.** It was invented by Gerardus Mercator in 1569. Mercator's map projection was useful to sailors, but as a picture of our world, it creates many false impressions.

A Mercator map exaggerates areas that are nearer to the poles. For example, Greenland looks to be about the same size as Africa. But, Africa is actually about 15 times the size of Greenland.

The Robinson Projection

Another kind of flat map is the **Robinson projection.** It makes the world look somewhat like a globe. Areas near the poles are distorted, but they are distorted a lot less than they are on a Mercator map.

The National Geographic Society and many map companies use the Robinson projection.

Note

Look at the Mercator map on the opposite page.

- Notice that the meridians (lines of longitude) are an equal distance apart. On a globe, the meridians get closer as they get near the poles.
- Notice that the parallels (lines of latitude) are farther apart toward the poles. On a globe, the parallels are an equal distance apart.

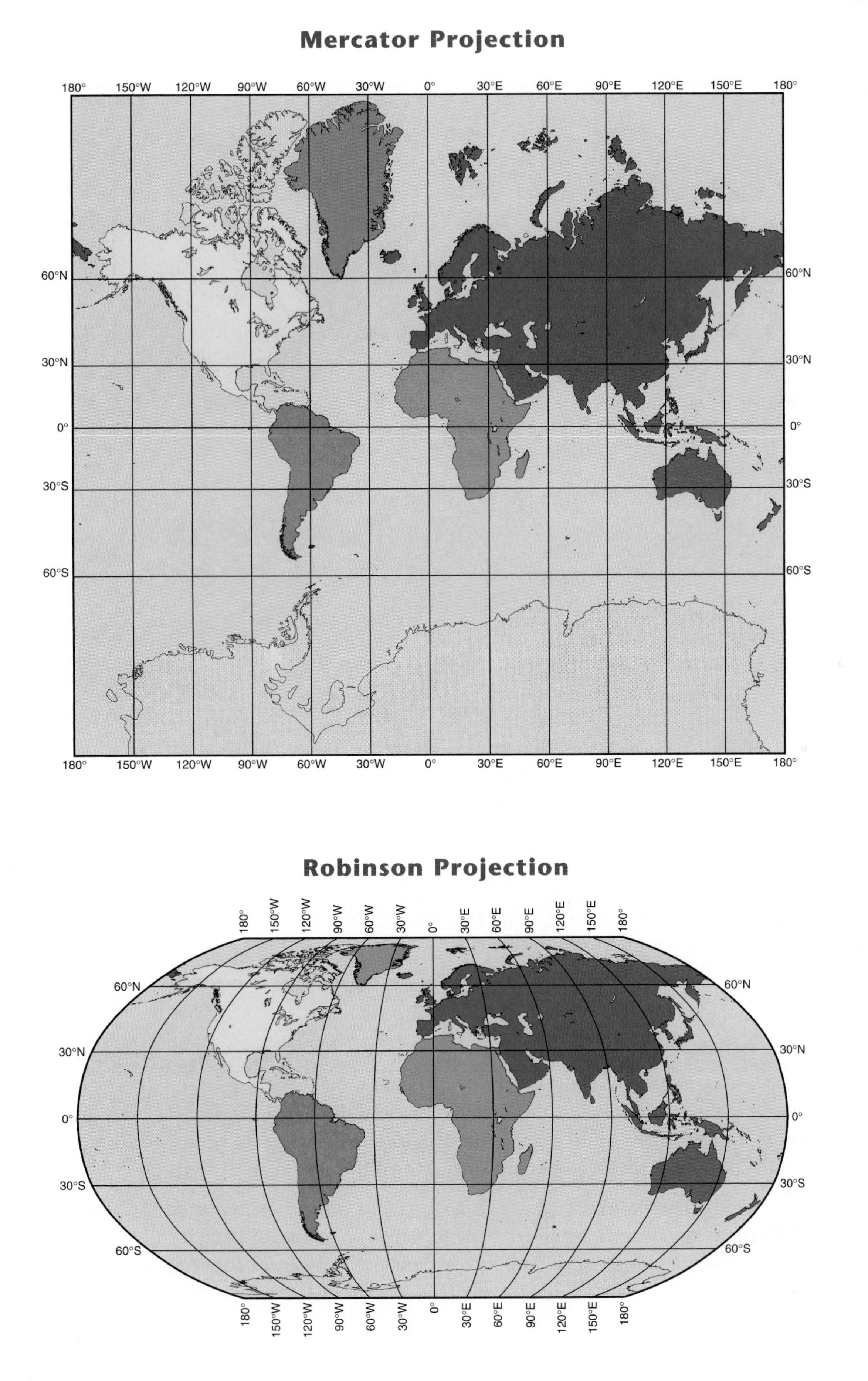
Mercator Projection
180°
150°W
120°W
90°W
60°W
30°W
0°
30°E
60°E
90°E
120°E
150°E
180°
60°N
30°N
0°
30°S
60°S
Robinson Projection
60°N
30°N
0°
30°S
60°S

Time Zones of the World

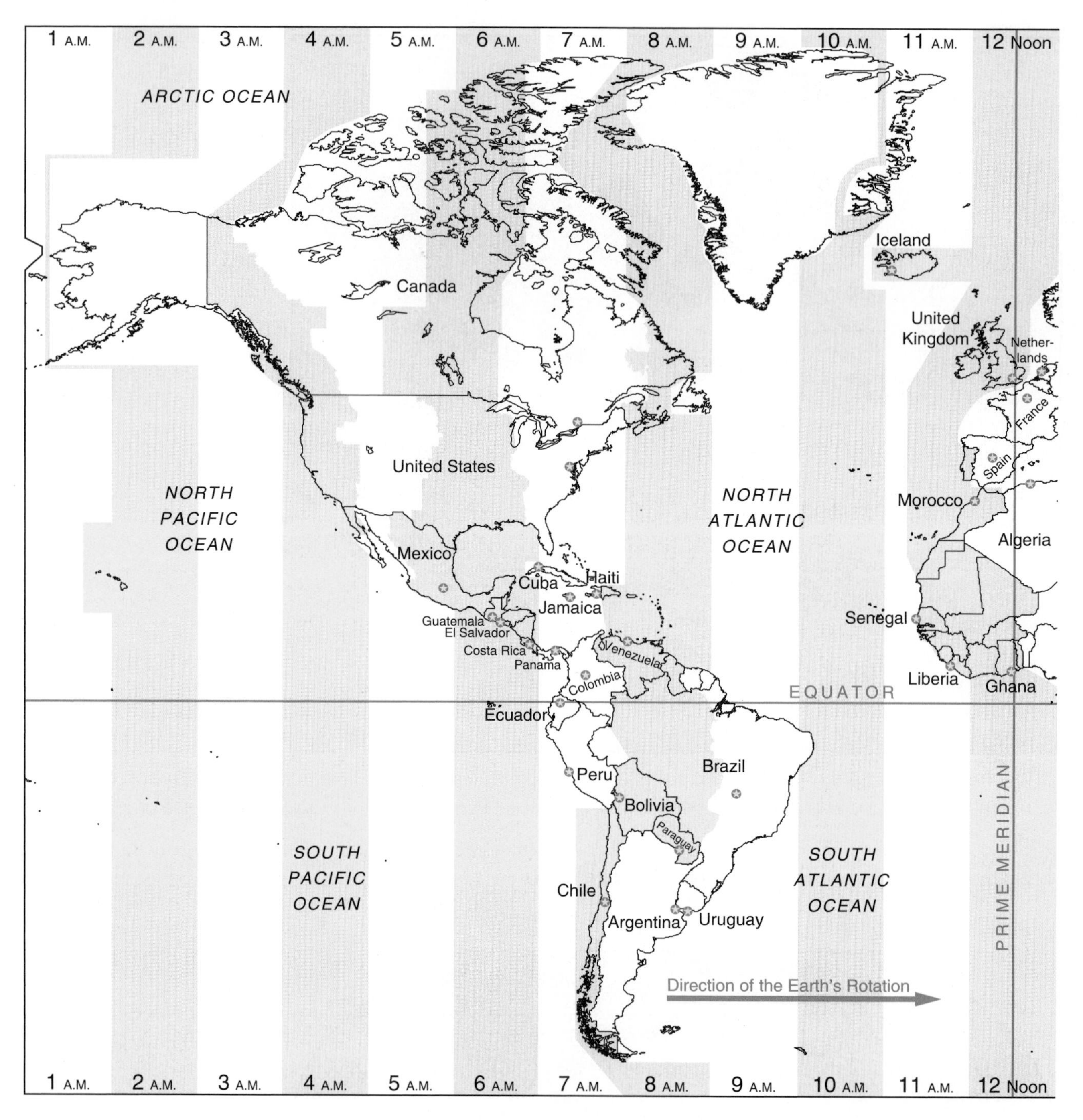

The Earth is divided into 24 time zones. We need time zones because the Earth is spinning, and the sun does not shine on all of the Earth at the same time. It takes one day, or 24 hours, for the Earth to make one complete spin. Each time zone represents one of the 24 hours of that day.

This map shows all 24 time zones. The times are given at the top and bottom of the map. As you read from left to right, the time is one hour later in each zone. This is because the Earth rotates toward the east, which is left to right on the map. The arrow near the bottom of the map shows the direction of the Earth's rotation.

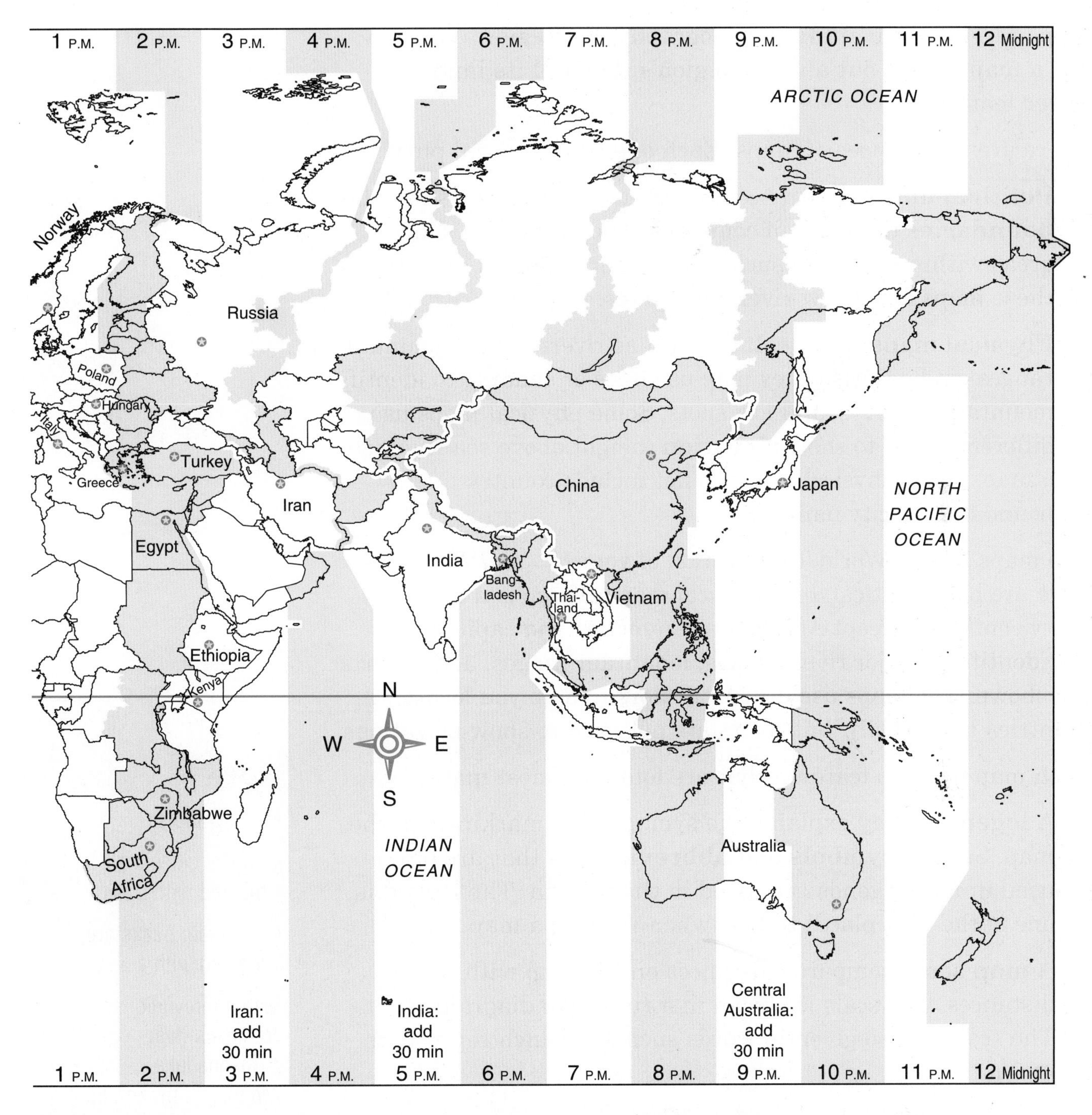

The map tells you what time it is in a location when it is 12 noon in London, England.

- Put your finger on the place you want to find the time for. See if it is in a gold strip or a white strip.
- Keep your finger in the strip you start in. Follow it up to the top of the map or down to the bottom. Read the time.

If you know the time in one location, you can find the time in another location.

- Put your finger on the place that is farther west.
- Slide your finger to the right until you reach the place that is farther east.
- Count the number of gold and white time zones you enter as you slide your finger.
- The number of time zones you enter is the number of hours that the place to the east is ahead of the place to the west.

Political and Physical Maps

Maps can help you study the geography of a region. You can use a map to find out about a region's size and its land and water features.

There are many types of maps. Each one has a special purpose.

- **Political maps** identify countries and cities. They show the **boundaries** (borders) of countries. They may also show areas within a country such as states or counties. Sometimes these maps also show rivers and lakes.
- **Physical maps** show features such as rivers, lakes, mountain ranges, and deserts. They may use lines or shading to identify mountains, valleys, and low spots. Some physical maps use different colors to show **elevation** (height above sea level) on a map. Many physical maps do not include country boundaries or city names.

The maps in this World Tour section of your *Student Reference Book* are both political *and* physical maps. There is a map for every continent except Antarctica. Countries and capital cities are identified. Major rivers, lakes, mountain ranges, and deserts are shown. The maps use different colors to help you locate countries more easily, but colors are not used to show elevations.

Each map has two features that are found on most maps:

- A **legend** or **key** explains the symbols and markings on the map. Several **symbols** and **abbreviations** that are often used in map legends are listed in the margin. The legend is one of the first places to look when reading a map.
- A **map scale** compares distances on the map with actual distances. The scale is shown in a ruler-like diagram. The scale is also given in words such as "1 inch represents 400 miles."

Note

Map symbols:

- Capital cities are marked with a ✪.
- Mountain peaks are marked with a △.

Map abbreviations:
R. means river.
L. means lake.
Mt. means mountain.

Example of a map scale:

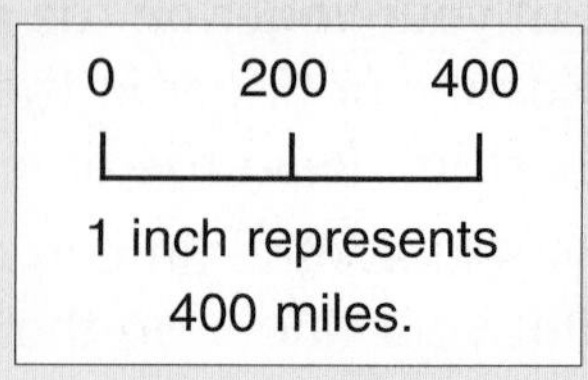

Country Profiles

The countries you can visit on your world tour are listed below by geographical region. Languages in bold type are the official languages of a country. Some countries have no official language, while others have more than one. All measurements in the country profiles are approximate.

REGION 1 Africa

Algeria
Area: 919,600 sq mi
Population: 32,129,000
Capital: Algiers (**Pop.** 3,060,000)
Languages: Arabic, French, Berber
Monetary unit: Dinar

Egypt
Area: 386,700 sq mi
Population: 76,117,000
Capital: Cairo (**Pop.** 10,834,000)
Languages: Arabic, English, French
Monetary unit: Pound

Ethiopia
Area: 435,200 sq mi
Population: 67,851,000
Capital: Addis Ababa (**Pop.** 2,723,000)
Languages: Amharic, Tigrinya, Orominga
Monetary unit: Birr

Ghana
Area: 92,500 sq mi
Population: 20,757,000
Capital: Accra (**Pop.** 1,847,000)
Languages: English, Akan, Ewe, Ga, Moshi-Dagomba
Monetary unit: Cedi

Kenya
Area: 225,000 sq mi
Population: 32,022,000
Capital: Nairobi (**Pop.** 2,575,000)
Languages: Swahili, English, numerous indigenous languages
Monetary unit: Shilling

Liberia
Area: 43,000 sq mi
Population: 3,391,000
Capital: Monrovia (**Pop.** 572,000)
Languages: English, tribal languages
Monetary unit: Liberian Dollar

Morocco
Area: 172,400 sq mi
Population: 32,209,000
Capital: Rabat (**Pop.** 1,759,000)
Languages: Arabic, Berber
Monetary unit: Dirham

Senegal
Area: 75,700 sq mi
Population: 10,852,000
Capital: Dakar (**Pop.** 2,167,000)
Languages: French, Pulaar, Wolof, Diola, Mandingo
Monetary unit: CFA franc

South Africa
Area: 471,000 sq mi
Population: 42,719,000
Capitals: Pretoria (**Pop.** 1,209,000), administrative; Cape Town (**Pop.** 2,967,000), legislative; Bloemfontein, judicial
Languages: 11 official languages including **Afrikaans, English, Ndebele, Sotho, Pedi**
Monetary unit: Rand

Zimbabwe
Area: 150,800 sq mi
Population: 12,672,000
Capital: Harare (**Pop.** 1,469,000)
Languages: English, Sindebele, Shona
Monetary unit: Zimbabwe Dollar

REGION 2 Europe

France
Area: 211,200 sq mi
Population: 60,424,000
Capital: Paris (**Pop.** 9,794,000)
Language: French
Monetary unit: Euro

Greece
Area: 50,900 sq mi
Population: 10,648,000
Capital: Athens (**Pop.** 3,215,000)
Languages: Greek, English, French
Monetary unit: Euro

Hungary
Area: 35,900 sq mi
Population: 10,032,000
Capital: Budapest (**Pop.** 1,708,000)
Language: Hungarian (Magyar)
Monetary unit: Forint

Iceland
Area: 40,000 sq mi
Population: 294,000
Capital: Reykjavik (**Pop.** 184,000)
Language: Icelandic
Monetary unit: Krona

Italy
Area: 116,300 sq mi
Population: 58,057,000
Capital: Rome (**Pop.** 2,665,000)
Languages: Italian, German, French, Slovene
Monetary unit: Euro

Netherlands
Area: 16,000 sq mi
Population: 16,318,000
Capital: Amsterdam (**Pop.** 1,145,000)
Language: Dutch
Monetary unit: Euro

Norway
Area: 125,200 sq mi
Population: 4,575,000
Capital: Oslo (**Pop.** 795,000)
Language: Norwegian
Monetary unit: Kroner

Poland
Area: 120,700 sq mi
Population: 38,626,000
Capital: Warsaw (**Pop.** 2,200,000)
Language: Polish
Monetary unit: Zloty

Spain
Area: 194,900 sq mi
Population: 40,281,000
Capital: Madrid (**Pop.** 5,103,000)
Languages: Castilian Spanish, Basque, Catalan, Galician
Monetary unit: Euro

United Kingdom
Area: 94,500 sq mi
Population: 60,271,000
Capital: London (**Pop.** 7,619,000)
Languages: English, Welsh, Scottish, Gaelic
Monetary unit: Pound

REGION 3 South America

Argentina
Area: 1,068,300 sq mi
Population: 39,145,000
Capital: Buenos Aires (**Pop.** 13,047,000)
Languages: Spanish, English, Italian
Monetary unit: Peso

Bolivia
Area: 424,200 sq mi
Population: 8,724,000
Capital: La Paz (**Pop.** 1,477,000)
Languages: Spanish, Quechua, Aymara
Monetary unit: Boliviano

Brazil
Area: 3,286,500 sq mi
Population: 184,101,000
Capital: Brasília (**Pop.** 3,099,000)
Languages: Portuguese, English, Spanish, French
Monetary unit: Real

Chile
Area: 292,300 sq mi
Population: 15,824,000
Capital: Santiago (**Pop.** 5,478,000)
Language: Spanish
Monetary unit: Peso

Colombia
Area: 439,700 sq mi
Population: 42,311,000
Capital: Bogota (**Pop.** 7,290,000)
Language: Spanish
Monetary unit: Peso

Ecuador
Area: 109,500 sq mi
Population: 13,213,000
Capital: Quito (**Pop.** 1,451,000)
Languages: Spanish, Quechua, other Amerindian
Monetary unit: U.S. Dollar

Paraguay
Area: 157,000 sq mi
Population: 6,191,000
Capital: Asunción (**Pop.** 1,639,000)
Languages: Spanish, Guarani
Monetary unit: Guarani

Peru
Area: 496,200 sq mi
Population: 27,544,000
Capital: Lima (**Pop.** 7,899,000)
Languages: Spanish, Quechua, Aymara
Monetary unit: Nuevo Sol

Uruguay
Area: 68,000 sq mi
Population: 3,399,000
Capital: Montevideo (**Pop.** 1,341,000)
Language: Spanish
Monetary unit: Peso

Venezuela
Area: 352,100 sq mi
Population: 25,017,000
Capital: Caracas (**Pop.** 3,226,000)
Language: Spanish
Monetary unit: Bolivar

REGION 4 Asia and Australia

Australia
Area: 2,967,900 sq mi
Population: 19,913,000
Capital: Canberra (**Pop.** 373,000)
Languages: English, aboriginal languages
Monetary unit: Australian Dollar

Bangladesh
Area: 55,600 sq mi
Population: 141,340,000
Capital: Dhaka (**Pop.** 11,560,000)
Languages: Bangla, English
Monetary unit: Taka

China
Area: 3,705,400 sq mi
Population: 1,298,848,000
Capital: Beijing (**Pop.** 10,848,000)
Languages: Mandarin, Gan, Wu, Haka, Yue, Minbei, Xiang, Minnan
Monetary unit: Renminbi (Yuan)

India
Area: 1,269,300 sq mi
Population: 1,065,071,000
Capital: New Delhi (**Pop.** 12,441,000)
Languages: Hindi, English, 14 regional languages
Monetary unit: Rupee

Iran
Area: 636,000 sq mi
Population: 67,503,000
Capital: Tehran (**Pop.** 7,190,000)
Languages: Farsi, Kurdish, Turkic, Luri
Monetary unit: Rial

Japan
Area: 145,900 sq mi
Population: 127,333,000
Capital: Tokyo (**Pop.** 34,997,000)
Language: Japanese
Monetary unit: Yen

Russia
Area: 6,592,800 sq mi
Population: 143,782,000
Capital: Moscow (**Pop.** 6,468,000)
Languages: Russian, many others
Monetary unit: Ruble

Thailand
Area: 198,500 sq mi
Population: 64,866,000
Capital: Bangkok (**Pop.** 6,486,000)
Languages: Thai, English
Monetary unit: Baht

Turkey
Area: 301,400 sq mi
Population: 68,894,000
Capital: Ankara (**Pop.** 3,428,000)
Languages: Turkish, Arabic, Kurdish
Monetary unit: Lira

Vietnam
Area: 127,200 sq mi
Population: 82,690,000
Capital: Hanoi (**Pop.** 3,977,000)
Languages: Vietnamese, Chinese, French, English
Monetary unit: Dong

REGION 5 North America

Canada
Area: 3,851,800 sq mi
Population: 32,508,000
Capital: Ottawa (**Pop.** 1,093,000)
Languages: English, French
Monetary unit: Dollar

Costa Rica
Area: 19,700 sq mi
Population: 3,957,000
Capital: San José (**Pop.** 1,085,000)
Language: Spanish
Monetary unit: Colon

Cuba
Area: 42,800 sq mi
Population: 11,309,000
Capital: Havana (**Pop.** 2,189,000)
Language: Spanish
Monetary unit: Peso

El Salvador
Area: 8,100 sq mi
Population: 6,588,000
Capital: San Salvador (**Pop.** 1,424,000)
Language: Spanish
Monetary unit: Colon

Guatemala
Area: 42,000 sq mi
Population: 14,281,000
Capital: Guatemala City (**Pop.** 951,000)
Languages: Spanish, Mayan languages
Monetary unit: Quetzal

Haiti
Area: 10,700 sq mi
Population: 7,656,000
Capital: Port-au-Prince (**Pop.** 1,961,000)
Languages: French, Haitian Creole
Monetary unit: Gourde

Jamaica
Area: 4,200 sq mi
Population: 2,713,000
Capital: Kingston (**Pop.** 575,000)
Languages: English, Jamaican Creole
Monetary unit: Jamaican Dollar

Mexico
Area: 761,600 sq mi
Population: 104,960,000
Capital: Mexico City (**Pop.** 18,660,000)
Languages: Spanish, Mayan dialects
Monetary unit: New Peso

Panama
Area: 30,200 sq mi
Population: 3,000,000
Capital: Panama City (**Pop.** 930,000)
Languages: Spanish, English
Monetary unit: Balboa

United States of America
Area: 3,717,800 sq mi
Population: 293,028,000
Capital: Washington, D.C. (**Pop.** 563,000)
Languages: English, Spanish
Monetary unit: Dollar

Region 1: Africa

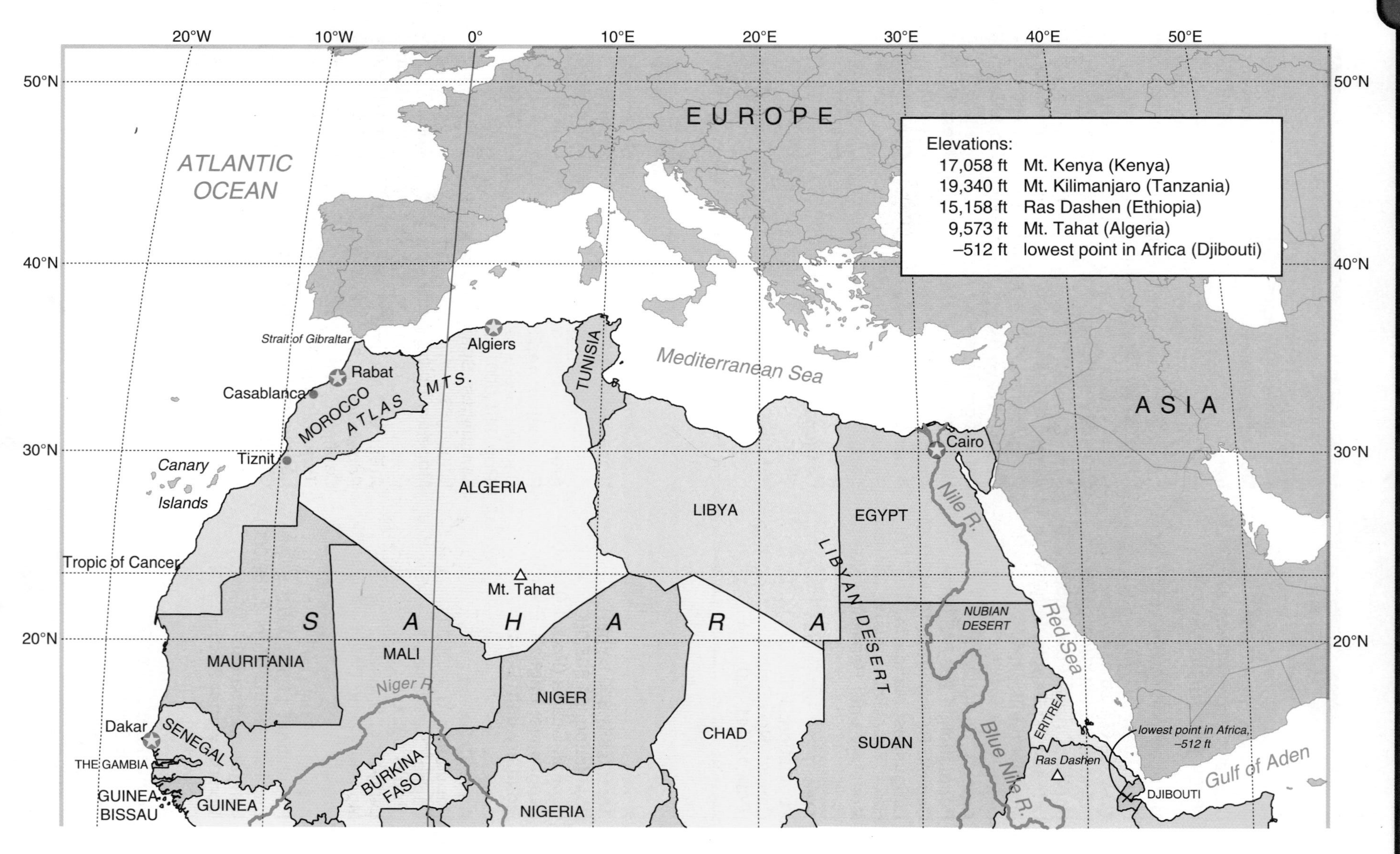

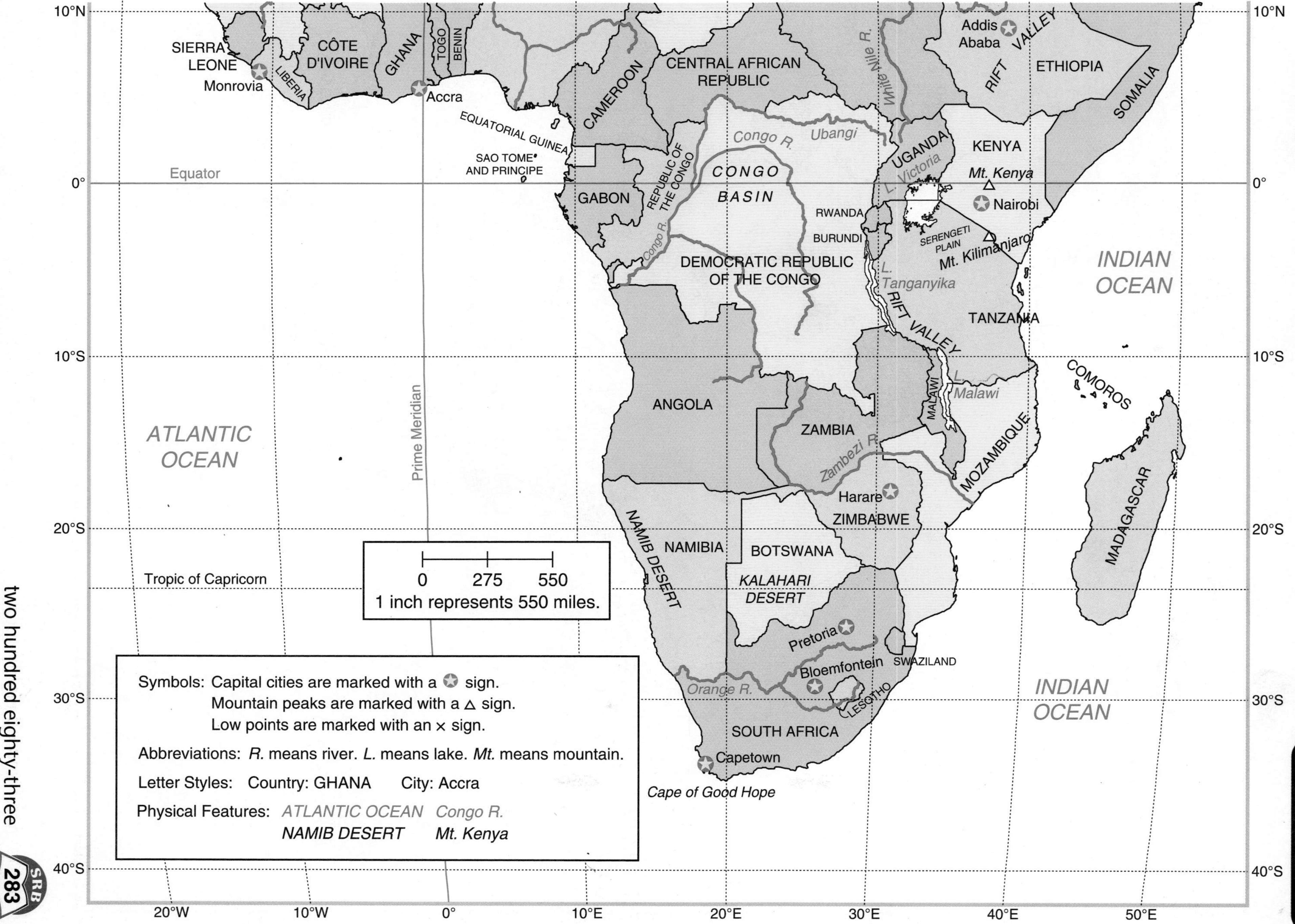

10°N
0°
10°S
20°S
30°S
40°S
20°W
10°W
0°
10°E
20°E
30°E
40°E
50°E
SIERRA LEONE
Monrovia
LIBERIA
CÔTE D'IVOIRE
GHANA
TOGO
BENIN
Accra
CAMEROON
CENTRAL AFRICAN REPUBLIC
White Nile R.
Addis Ababa
RIFT VALLEY
ETHIOPIA
SOMALIA
EQUATORIAL GUINEA
SAO TOME AND PRINCIPE
Equator
Congo R.
Ubangi
UGANDA
L. Victoria
KENYA
Mt. Kenya
Nairobi
GABON
REPUBLIC OF THE CONGO
CONGO BASIN
RWANDA
BURUNDI
SERENGETI PLAIN
Mt. Kilimanjaro
Congo R.
DEMOCRATIC REPUBLIC OF THE CONGO
L. Tanganyika
INDIAN OCEAN
RIFT VALLEY
TANZANIA
COMOROS
L. Malawi
MALAWI
ANGOLA
ATLANTIC OCEAN
Prime Meridian
ZAMBIA
Zambezi R.
MOZAMBIQUE
MADAGASCAR
Harare
ZIMBABWE
NAMIB DESERT
NAMIBIA
BOTSWANA
KALAHARI DESERT
Tropic of Capricorn
0 275 550
1 inch represents 550 miles.
Pretoria
Bloemfontein
SWAZILAND
Orange R.
LESOTHO
INDIAN OCEAN
SOUTH AFRICA
Capetown
Cape of Good Hope
Symbols: Capital cities are marked with a ✪ sign.
Mountain peaks are marked with a Δ sign.
Low points are marked with an × sign.
Abbreviations: R. means river. L. means lake. Mt. means mountain.
Letter Styles: Country: GHANA City: Accra
Physical Features: ATLANTIC OCEAN Congo R.
NAMIB DESERT Mt. Kenya

30°W
20°W
10°W
0°
10°E
20°E
30°E
40°E
Barents
70°N
60°N
50°N
40°N
Reykjavik
ICELAND
Hvannadalshnúkur
Arctic Circle
Norwegian Sea
Glittertind
Bergen
NORWAY
SWEDEN
FINLAND
Oslo
Helsinki
Stockholm
ESTONIA
LATVIA
LITHUANIA
RUSSIA
Daugava R.
BELARUS
Scotland
N. Ireland
UNITED KINGDOM
IRELAND
Dublin
Wales
England
London
North Sea
DENMARK
NETHERLANDS
Amsterdam
Baltic Sea
ATLANTIC OCEAN
GERMANY
Berlin
POLAND
Warsaw
Vistula R.
Oder R.
Elbe R.
Rhine R.
BELGIUM
LUXEMBOURG
Kiev
UKRAINE
Paris
Seine R.
Loire R.
FRANCE
CZECH REPUBLIC
CARPATHIAN MOUNTAINS
Driester R.
Danube R.
LIECHTENSTEIN
SLOVAKIA
MOLDOVA
SWITZERLAND
Vienna
AUSTRIA
Budapest
HUNGARY
ROMANIA
ALPS
Mont Blanc
Matterhorn
Po R.
SLOVENIA
CROATIA
Pico de Aneto
ANDORRA
Rhône R.
PORTUGAL
Ebro R.
PYRENEES
MONACO
SAN MARINO
DINARIC ALPS
BOSNIA AND HERZEGOVINA
Belgrade
SERBIA & MONTENEGRO
Danube R.
Madrid
Lisbon
Tagus R.
SPAIN
Corsica
APENNINES
Adriatic Sea
Rome
ITALY
BALKAN MTS.
BULGARIA
MACEDONIA
ALBANIA
VATICAN CITY
Vesuvius
Olympus
Majorca
Strait of Gibraltar
Mediterranean Sea
Sardinia
GREECE
Aegean Sea
Prime Meridian
Sicily
Etna
Athens
AFRICA
MALTA
Rhodes
Crete
0°
10°E
20°E
30°E

Region 2: Europe

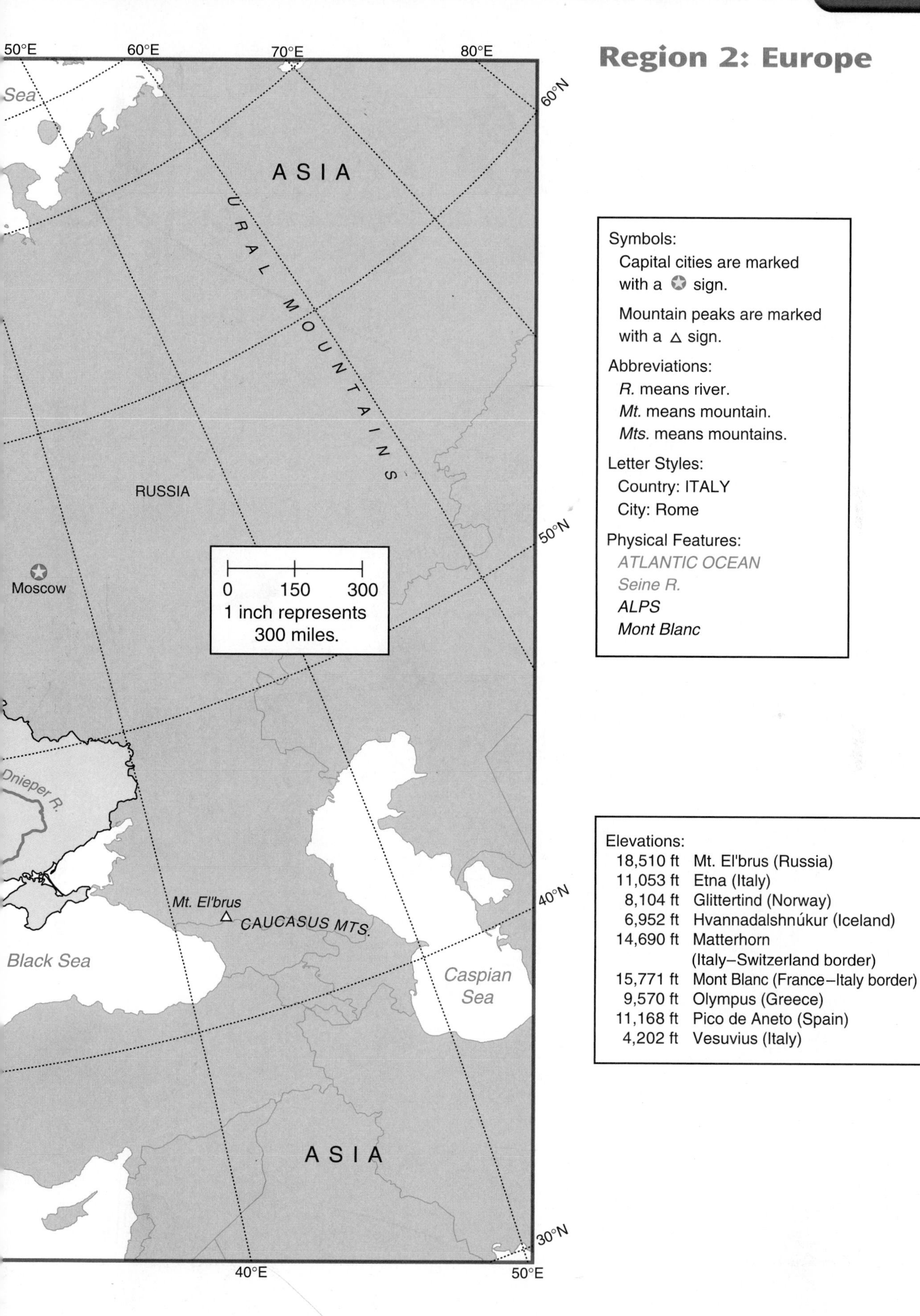

Symbols:

Capital cities are marked with a ✪ sign.

Mountain peaks are marked with a △ sign.

Abbreviations:

R. means river.
Mt. means mountain.
Mts. means mountains.

Letter Styles:

Country: ITALY
City: Rome

Physical Features:

ATLANTIC OCEAN
Seine R.
ALPS
Mont Blanc

Elevations:

18,510 ft	Mt. El'brus (Russia)
11,053 ft	Etna (Italy)
8,104 ft	Glittertind (Norway)
6,952 ft	Hvannadalshnúkur (Iceland)
14,690 ft	Matterhorn (Italy–Switzerland border)
15,771 ft	Mont Blanc (France–Italy border)
9,570 ft	Olympus (Greece)
11,168 ft	Pico de Aneto (Spain)
4,202 ft	Vesuvius (Italy)

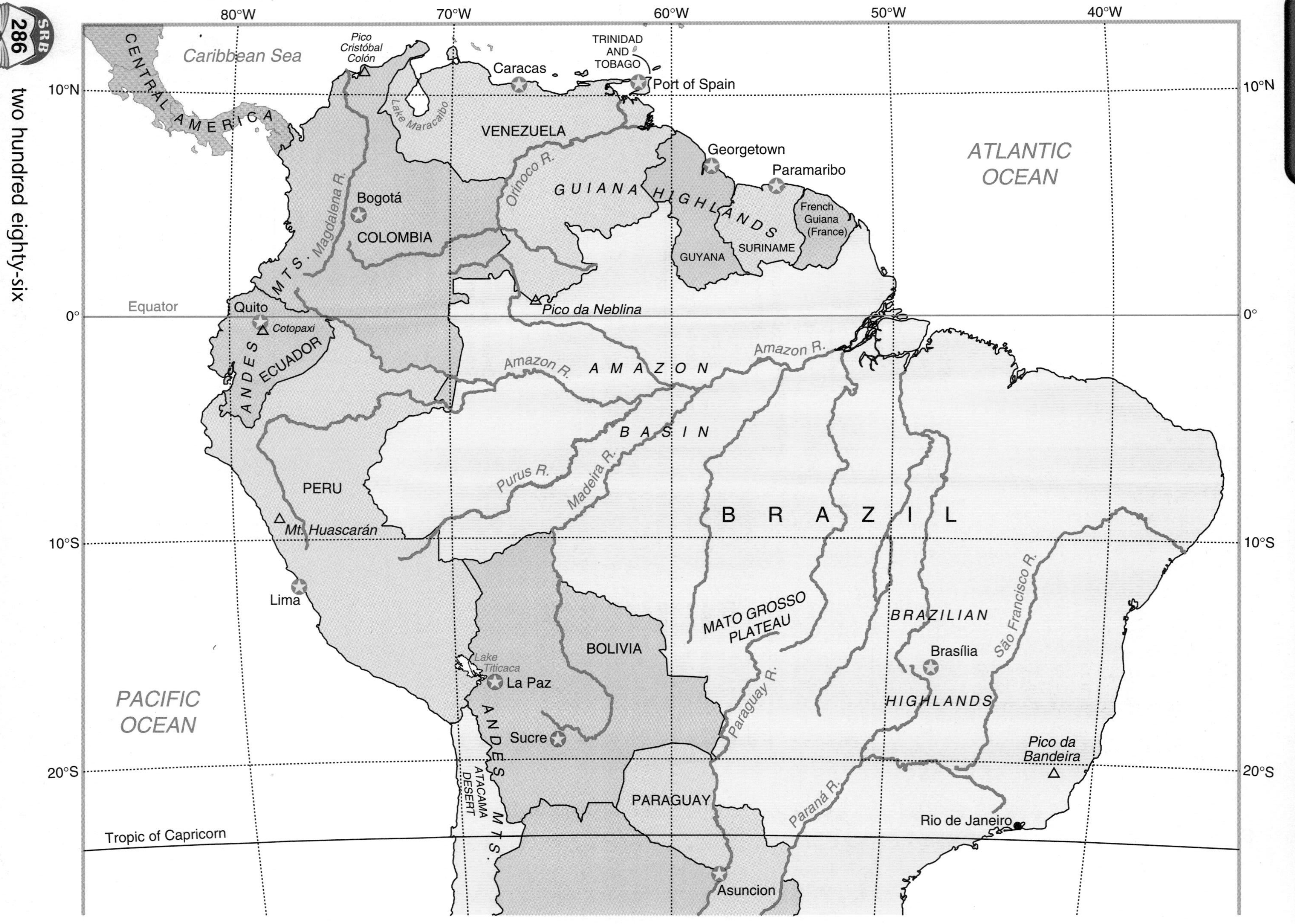

Caribbean Sea
CENTRAL AMERICA
Pico Cristóbal Colón
TRINIDAD AND TOBAGO
Caracas
Port of Spain
Lake Maracaibo
VENEZUELA
Georgetown
Paramaribo
ATLANTIC OCEAN
Magdalena R.
Bogotá
COLOMBIA
Orinoco R.
GUIANA HIGHLANDS
French Guiana (France)
SURINAME
GUYANA
MTS.
Equator
Quito
Cotopaxi
Pico da Neblina
ANDES
ECUADOR
Amazon R.
AMAZON BASIN
Purus R.
Madeira R.
PERU
Mt. Huascarán
BRAZIL
Lima
MATO GROSSO PLATEAU
BRAZILIAN HIGHLANDS
Brasília
São Francisco R.
BOLIVIA
Lake Titicaca
La Paz
PACIFIC OCEAN
Sucre
ANDES MTS.
Paraguay R.
Pico da Bandeira
ATACAMA DESERT
PARAGUAY
Paraná R.
Rio de Janeiro
Tropic of Capricorn
Asuncion
80°W
70°W
60°W
50°W
40°W
10°N
0°
10°S
20°S

Region 3: South America

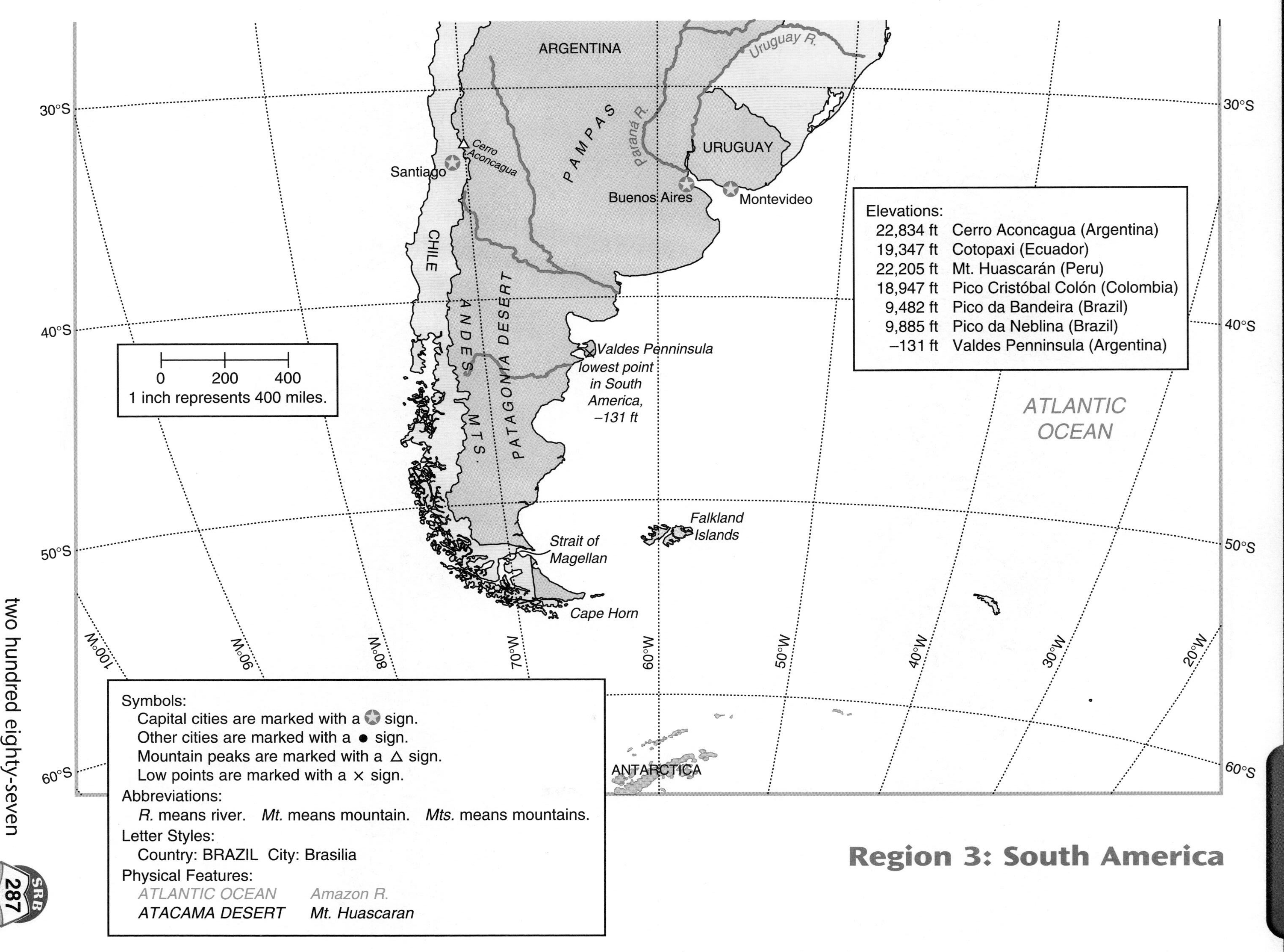

North Pole 90°N
ARCTIC OCEAN
EUROPE
Prime Meridian 0°
Arctic Circle
Baltic Sea
RUSSIA
Moscow
URAL MOUNTAINS
Ob R.
Yenisey R.
CENTRAL SIBERIAN PLATEAU
Lena R.
Angarra R.
Volga R.
Black Sea
Mediterranean Sea
AFRICA
Tropic of Cancer
Ankara
TURKEY
CYPRUS
LEBANON
ISRAEL
SYRIA
JORDAN
GEORGIA
ARMENIA
AZERBAIJAN
CAUCASUS MTS.
Mt. El'brus
Caspian Sea
Aral Sea
KAZAKSTAN
Lake Balkhash
Irtysh R.
Lake Baikal
ALTAY MTS
MONGOLIA
GOBI DESERT
UZBEKISTAN
TURKMENISTAN
KYRGYZSTAN
TAJIKISTAN
Jengish Chokusu
Euphrates R.
Tigris R.
IRAQ
Tehran
IRAN
ZAGROS MTS.
KUWAIT
Persian Gulf
QATAR
U.A.E.
Red Sea
SAUDI ARABIA
Mecca
RUB' AL-KHALI DESERT
OMAN
YEMEN
AFGANISTAN
PAKISTAN
TAKLIMAKAN DESERT
K2
KUNLUN MTS.
HIMALAYAS
PLATEAU OF TIBET
Huang (Yellow) R.
CHINA
Bhramaputra R.
Chang R.
Indus R.
New Delhi
NEPAL
Mt. Everest
BHU.
Ganges R.
Dhaka
INDIA
Calcutta
BANGLADESH
MYANMAR
Hanoi
Arabian Sea
Mumbai (Bombay)
Bay of Bengal
THAILAND
Mekong R.
LAOS
CAMBODIA
VIETNAM
Bangkok
SRI LANKA
MALDIVES
MALAYSIA
SINGAPORE
Sumatra
INDIAN OCEAN
Equator
TURKEY
CYPRUS
SYRIA
LEBANON
WEST BANK
IRAQ
ISRAEL
GAZA STRIP
Jerusalem
EGYPT
JORDAN
SAUDI ARABIA
10°E
20°E
30°E
40°N
50°N
60°N
70°N
80°N
30°N
20°N
10°N
0°
10°S
40°E
50°E
60°E
70°E
80°E
90°E
100°E

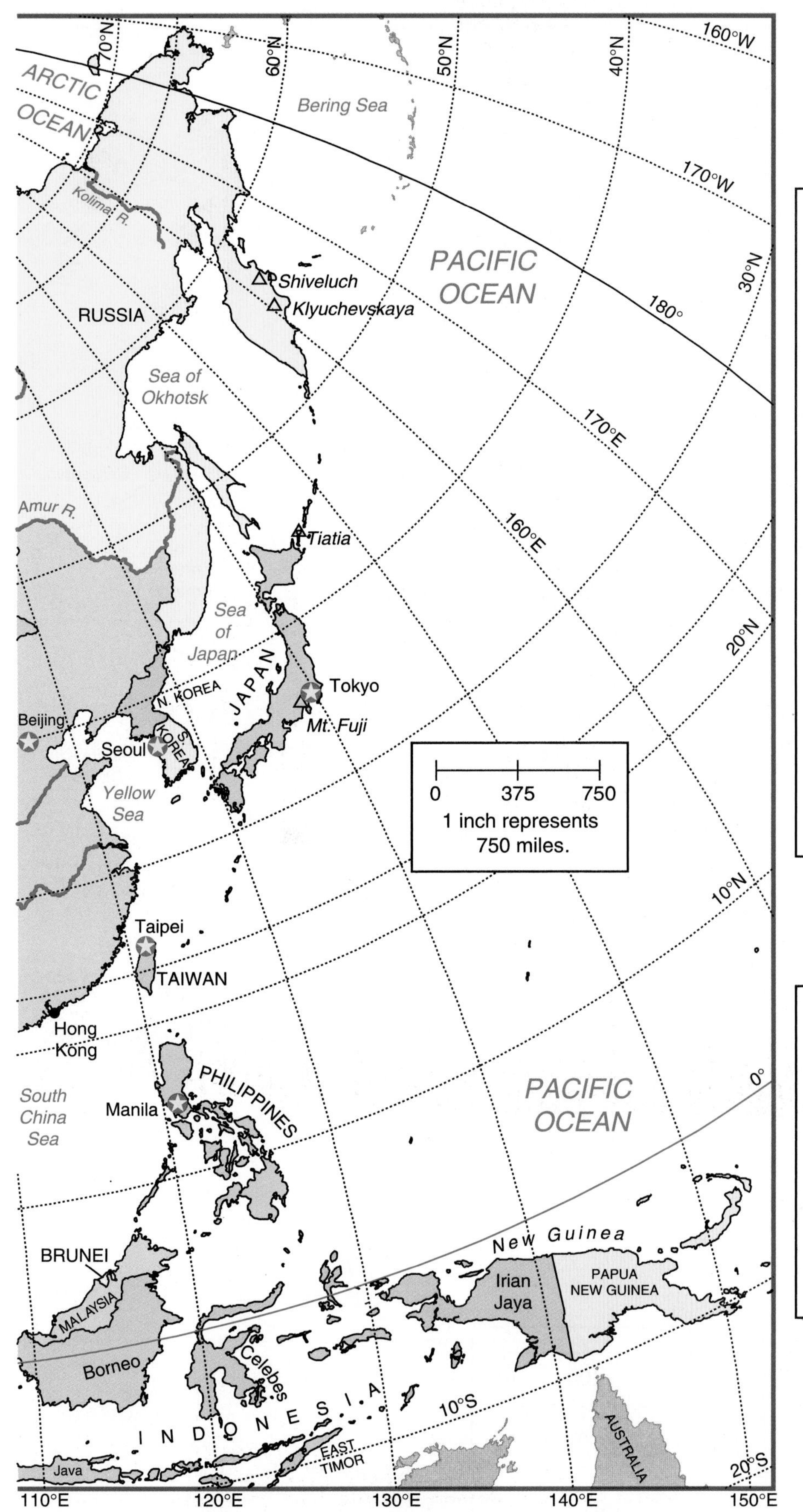

Region 4: Asia

Symbols:

Capital cities are marked with a ✪ sign.

Other cities are marked with a • sign.

Mountain peaks are marked with a △ sign.

Abbreviations:

R. means river.

Mt. means mountain.

Mts. means mountains.

U.A.E. means United Arab Emirates.

BHU. means Bhutan.

Letter Styles:

Country: CHINA

City: Beijing

Physical Features:

PACIFIC OCEAN

Ganges R.

HIMALAYAS

Mt. Everest

Elevations:

18,510 ft	Mt. El'brus (Russia)
29,028 ft	Mt. Everest (China–Nepal border)
12,388 ft	Mt. Fuji (Japan)
24,406 ft	Jengish Chokusu (Kyrgyzstan)
19,584 ft	Klyuchevskaya (Russia)
28,250 ft	K2 (China–Pakistan border)
10,771 ft	Shiveluch (Russia)
6,013 ft	Tiatia (Russia)

Region 4: Australia

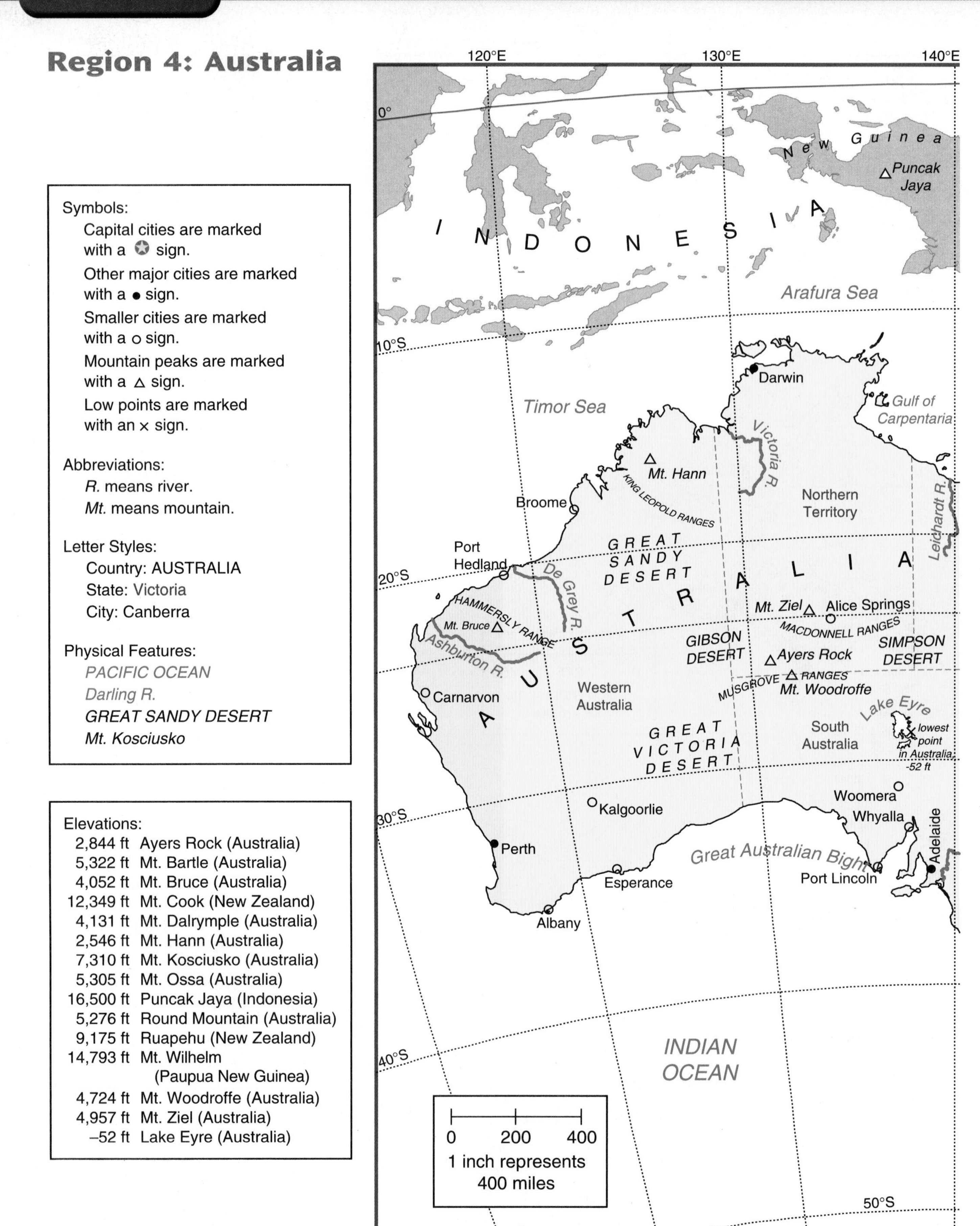

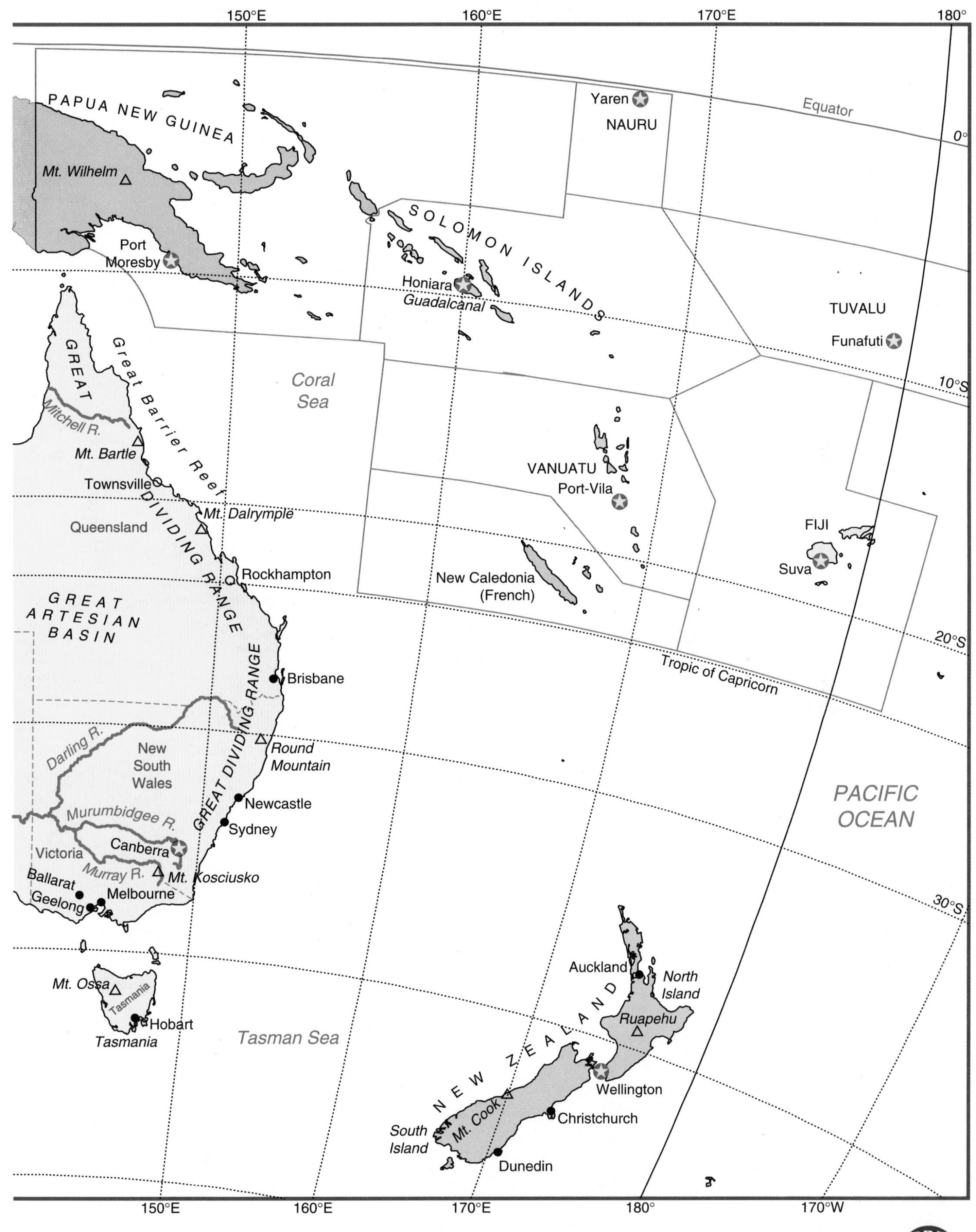
150°E
160°E
170°E
180°
Yaren
NAURU
Equator
0°
PAPUA NEW GUINEA
Mt. Wilhelm
Port Moresby
SOLOMON ISLANDS
Honiara
Guadalcanal
TUVALU
Funafuti
10°S
GREAT
Great Barrier Reef
Mitchell R.
Coral Sea
Mt. Bartle
Townsville
VANUATU
Port-Vila
Mt. Dalrymple
Queensland
DIVIDING RANGE
FIJI
Suva
Rockhampton
New Caledonia (French)
GREAT ARTESIAN BASIN
20°S
Brisbane
Tropic of Capricorn
GREAT DIVIDING RANGE
Darling R.
New South Wales
Round Mountain
PACIFIC OCEAN
Newcastle
Murumbidgee R.
Sydney
Canberra
Victoria
Murray R.
Mt. Kosciusko
Ballarat
Melbourne
Geelong
30°S
Auckland
North Island
Mt. Ossa
Tasmania
Hobart
Tasmania
Tasman Sea
Ruapehu
NEW ZEALAND
Wellington
Mt. Cook
Christchurch
South Island
Dunedin
150°E
160°E
170°E
180°
170°W

Region 5: North America

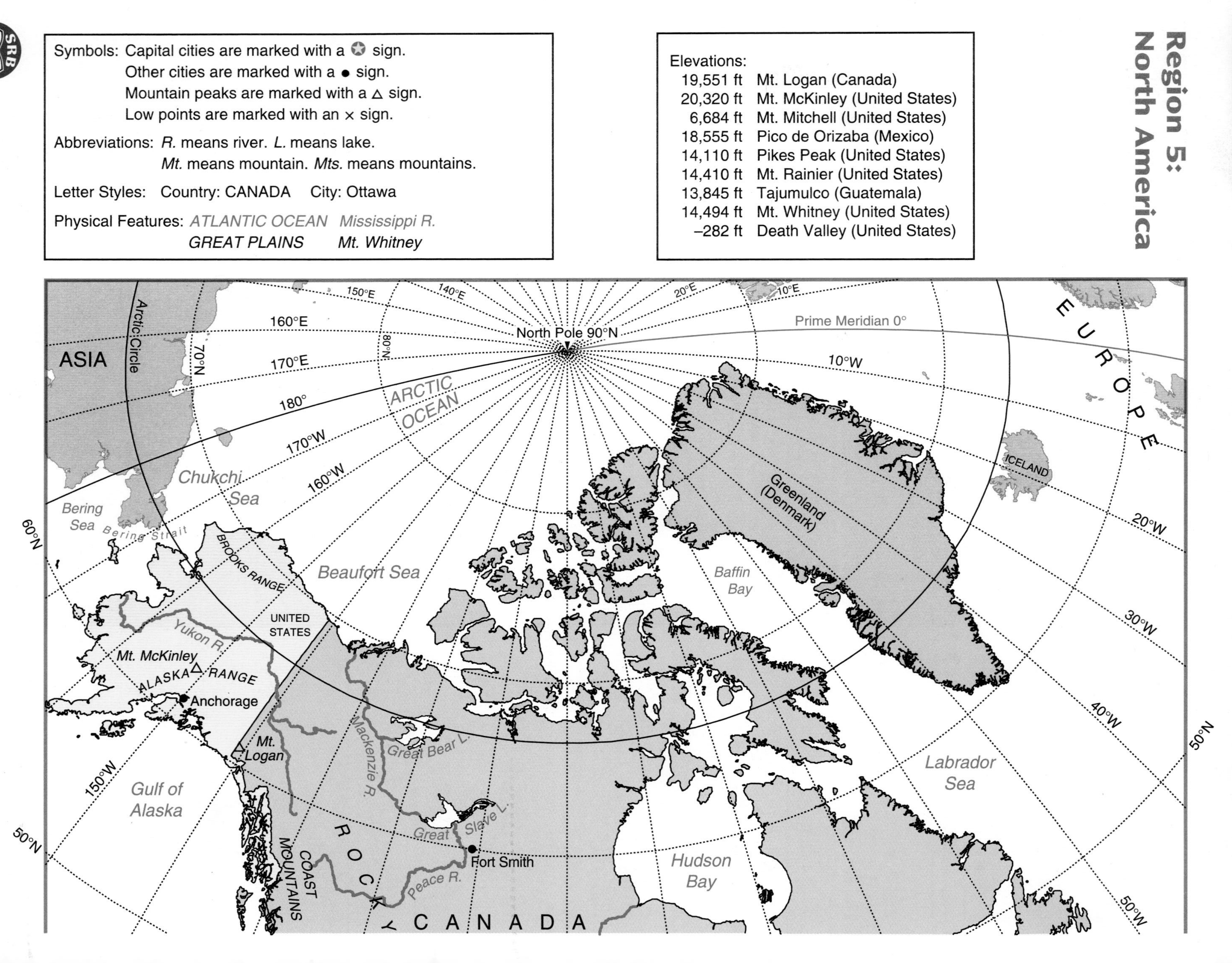

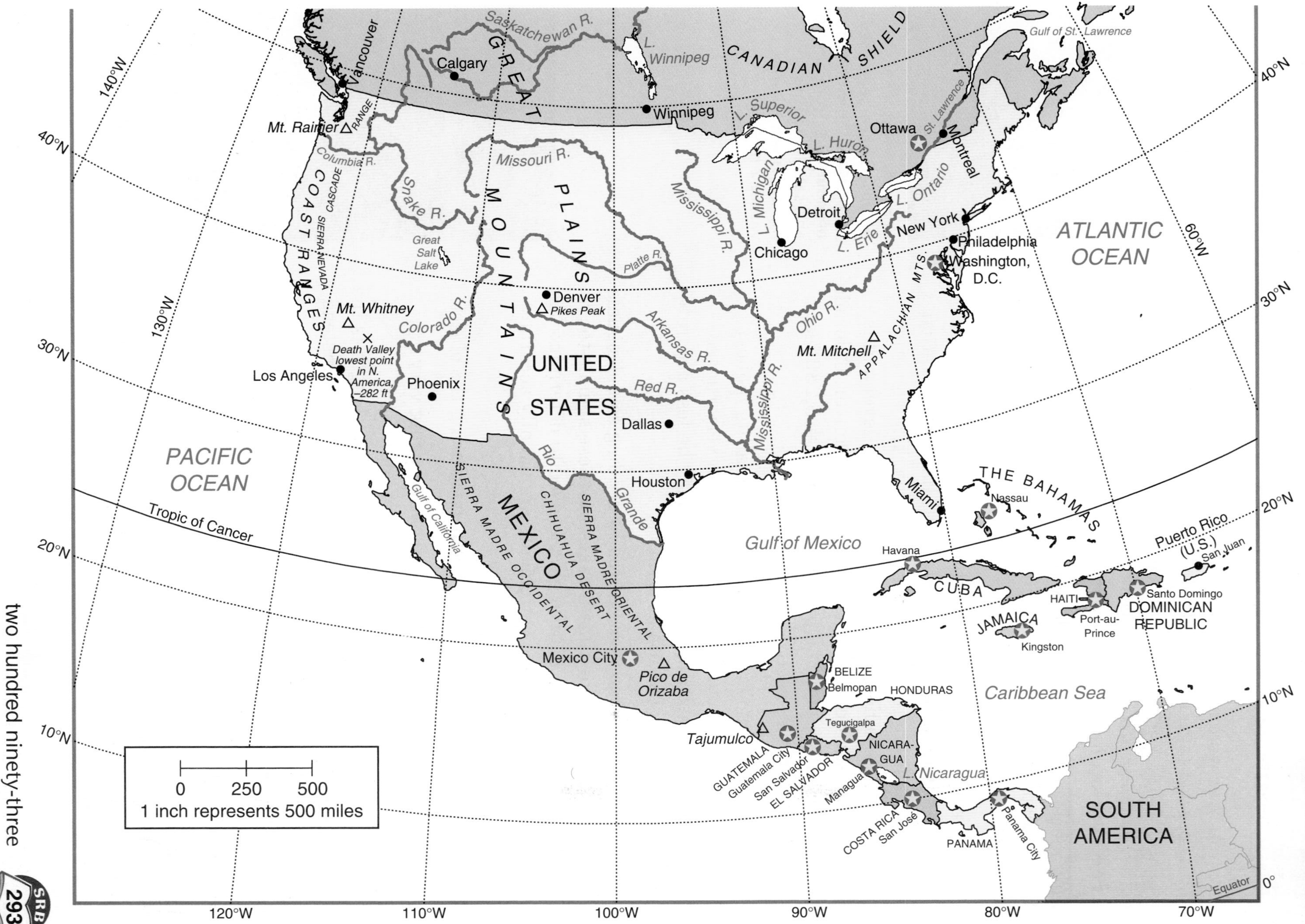

ATLANTIC OCEAN
PACIFIC OCEAN
Gulf of Mexico
Caribbean Sea
Gulf of St. Lawrence
Gulf of California
Tropic of Cancer
CANADIAN SHIELD
GREAT PLAINS
MOUNTAINS
COAST RANGES
SIERRA NEVADA
CASCADE RANGE
APPALACHIAN MTS.
SIERRA MADRE ORIENTAL
SIERRA MADRE OCCIDENTAL
CHIHUAHUA DESERT
UNITED STATES
MEXICO
SOUTH AMERICA
THE BAHAMAS
CUBA
JAMAICA
HAITI
DOMINICAN REPUBLIC
Puerto Rico (U.S.)
BELIZE
HONDURAS
NICARAGUA
GUATEMALA
EL SALVADOR
COSTA RICA
PANAMA
Vancouver
Calgary
Winnipeg
Ottawa
Montreal
Detroit
Chicago
New York
Philadelphia
Washington, D.C.
Miami
Houston
Dallas
Denver
Phoenix
Los Angeles
Mexico City
Havana
Nassau
Kingston
Port-au-Prince
Santo Domingo
San Juan
Belmopan
Guatemala City
San Salvador
Tegucigalpa
Managua
San José
Panama City
Mt. Rainier
Mt. Whitney
Death Valley lowest point in N. America, –282 ft
Pikes Peak
Mt. Mitchell
Pico de Orizaba
Tajumulco
Great Salt Lake
L. Winnipeg
L. Superior
L. Michigan
L. Huron
L. Erie
L. Ontario
L. Nicaragua
St. Lawrence
Saskatchewan R.
Missouri R.
Mississippi R.
Ohio R.
Platte R.
Arkansas R.
Red R.
Rio Grande
Colorado R.
Snake R.
Columbia R.
Equator
0°
10°N
20°N
30°N
40°N
60°W
70°W
80°W
90°W
100°W
110°W
120°W
130°W
140°W
0 250 500
1 inch represents 500 miles

Geographical Measurements

Have you ever wondered how the heights of mountains or the lengths of rivers are measured? How accurate are these measurements? Vertical measurements, such as heights and depths, are recorded as distances above or below sea level. The term **elevation** means height above sea level.

In recent years, *laser altimeters* have been used to measure the elevations of mountains from aircraft and satellites. These measures are accurate to within 2 meters.

Many measurements are made with surveying instruments. These measurements can be made more accurately by using data collected by satellites.

- The parts of a city are not all the same height above sea level. Many cities in the United States have an elevation marker somewhere in the downtown area. The elevation marker tells the height of the city above sea level and is accurate to the nearest foot. For example, Denver, Colorado, has an elevation of 5,260 feet. Since 1 mile = 5,280 feet, Denver is almost exactly 1 mile high.
- The height of a mountain is the elevation at its highest point. The land around mountains is often very rugged. So the reported heights of mountains may be less accurate than the reported elevations of cities.
- The depth of an ocean is measured by sending sound signals to the ocean floor. The time it takes for these signals to reach the bottom and bounce back is used to determine the ocean's depth. Depth measurements are usually accurate to the nearest 10 feet.

Other measurements are made without measuring the object.

- The length of a river is usually measured using very accurate maps, created with the help of satellite photographs. The instrument used to measure length is the size of a ballpoint pen but with a very small wheel instead of a ball at its tip. This instrument is moved on the map along the full length of the river. Using the map scale, the number of times the wheel rotates is converted into the actual length of the river.

River-length measurements are accurate to within $\frac{1}{5}$ of 1%, or $\frac{1}{5}$ of a mile for each 100 miles measured.

Length-of-river measurements are usually accurate to the nearest mile for each 500 miles of river. For example, the length of a 3,000-mile-long river is probably accurate to the nearest 6 miles.

Geographical Area Measurements

The heights of mountains and the depths of oceans are obtained *directly*. We find heights and depths by measuring the Earth itself.

The areas of countries and the areas of oceans are found *indirectly*. We measure very accurate maps or satellite pictures. The countries and oceans themselves are not measured.

Countries, oceans, and deserts have irregular boundaries. One way that scientists measure areas is to count grid squares. They place a transparent grid of squares on a map. Then they count the squares and parts of squares that cover the region being measured. The squares are drawn to the same scale as the map.

The shoreline of a body of water may shift greatly during different seasons of the year and over the years.

There are several reasons that it is hard to measure the following regions accurately:

Area of a country. Sometimes people disagree about the exact boundary of a country. So the area may depend on which boundary is being used.

Area of a lake, sea, or ocean. Some bodies of water have shorelines that shift greatly depending on the level of the water. So it is very hard to measure accurately the area that is covered by water.

The world's oceans are not separated from one another by shorelines. Sometimes people disagree on the boundaries between the oceans. This makes it difficult to measure the areas of oceans.

Area of a desert. Measuring desert areas is very hard. Desert boundaries may change because the climate changes. When land is cultivated, a desert boundary shifts. Also, scientists do not agree on what a desert actually is. Some define a desert as land that cannot be used for raising crops. Others define it as land that cannot be used for either crops or grazing. There are deserts that are hot and dry only part of the year. Some deserts are dry all year because it is very hot. Other deserts are dry all year because it is very cold and the water is always frozen. Very cold deserts are known as *tundras*.

Desert boundaries often change because of climate changes.

Climate and Elevation of Capital Cities

The **climate** of a city or a country refers to the average weather conditions in that place. Two kinds of weather data are shown on the opposite page: temperature and rainfall. Elevation data are also shown.

Temperature Data

Average temperatures are given in degrees Fahrenheit (°F).

- Each column lists average temperatures for a 3-month period.
- The first number is the average high temperature for that period. The second number is the average low temperature for that period.

Examples The average high and low temperatures for Cairo, Egypt, for March through May are about 81°F and 59°F. The highest temperature listed for Santiago, Chile, is 84°F. This is the average high temperature for December through February. Santiago is in the Southern Hemisphere. Countries south of the equator have summer in December, January, and February.

Santiago, Chile

Rainfall Data

Average rainfall is given in inches per month. All moisture that falls as rain or as snow is counted as rainfall. When snow falls, a sample is melted and the depth of the water is measured.

Examples The average rainfall in Monrovia, Liberia, is 30.7 inches *per month* from June through August. That's about 1 inch per day, on average. The average rainfall in Rome, Italy, is 1 inch *per month* from June through August. The total rainfall during these 3 months is about 1 in. + 1 in. + 1 in., or 3 inches.

Elevation Data

The table also lists the elevation for each capital city. A city's **elevation** is its height above sea level.

Examples The highest elevation listed is 13,166 feet for La Paz, Bolivia. The elevation listed for Amsterdam, Netherlands, is 7 feet. Amsterdam is nearly at sea level.

Amsterdam, Netherlands

Climate and Elevation of Capital Cities

Region	Capital, Country	Average Monthly Temps (°F) in Capital City (High/Low) Dec/Feb	Mar/May	Jun/Aug	Sept/Nov	Average Monthly Rainfall (in.) in Capital City Dec/Feb	Mar/May	Jun/Aug	Sept/Nov	Elevation of capital (ft)
Region 1	Algiers, Algeria	62/44	69/49	85/65	77/57	4.2	2.3	0.3	3.0	82
	Cairo, Egypt	67/50	81/59	93/71	83/64	0.2	0.1	0.0	0.0	243
	Addis Ababa, Ethiopia	69/53	72/57	66/55	68/53	0.9	3.2	8.9	2.9	7,724
	Accra, Ghana	87/78	87/79	81/75	84/76	0.8	3.7	3.5	1.8	226
	Nairobi, Kenya	77/58	77/60	72/55	76/58	2.2	4.6	0.7	2.5	5,327
	Monrovia, Liberia	87/74	88/76	81/74	84/75	2.8	10.9	30.7	23.0	59
	Rabat, Morocco	63/48	68/53	78/64	74/58	3.3	1.9	0.1	1.8	246
	Dakar, Senegal	76/66	76/67	84/76	85/76	0.1	0.0	4.7	2.3	79
	Pretoria, South Africa	81/66	75/57	67/46	78/60	4.4	2.0	0.3	2.6	4,265
	Harare, Zimbabwe	78/63	76/58	70/49	80/59	7.0	2.1	0.1	1.7	4,930
Region 2	Paris, France	44/35	57/43	73/56	59/46	2.0	1.8	2.3	2.1	315
	Athens, Greece	56/45	67/53	87/71	73/60	2.0	1.1	0.3	1.6	69
	Budapest, Hungary	38/27	60/42	78/58	58/43	1.7	2.0	2.3	2.1	607
	Reykjavik, Iceland	36/28	42/34	54/46	44/36	3.2	2.3	2.1	3.2	200
	Rome, Italy	56/40	64/48	81/65	71/55	3.3	2.4	1.0	3.9	10
	Amsterdam, Netherlands*	42/34	54/41	68/54	56/45	2.5	2.3	2.6	3.4	7
	Oslo, Norway	32/20	50/35	69/53	49/37	1.9	1.5	3.2	2.9	56
	Warsaw, Poland	34/25	54/38	72/54	53/40	1.2	1.6	2.8	1.6	351
	Madrid, Spain	52/34	65/43	87/59	69/47	1.8	1.6	0.6	1.8	1,909
	London, United Kingdom	46/35	55/39	69/52	58/43	2.7	2.2	2.1	2.9	203
Region 3	Buenos Aires, Argentina	82/63	72/53	60/42	71/51	3.8	3.5	2.2	3.3	66
	La Paz, Bolivia	56/39	57/35	55/28	58/35	4.3	1.6	0.4	1.6	13,166
	Brasilia, Brazil	81/64	81/61	80/53	83/62	10.6	5.1	0.2	5.7	3,480
	Santiago, Chile	84/53	72/45	59/38	71/45	0.1	1.0	2.7	0.6	1,555
	Bogota, Columbia	66/44	66/48	64/47	65/47	2.2	4.0	2.0	4.0	8,357
	Quito, Ecuador	66/50	66/51	67/49	67/49	4.6	5.9	1.2	4.1	9,223
	Asuncion, Paraguay	90/72	82/66	73/57	82/61	5.6	5.3	2.3	4.8	331
	Lima, Peru	78/68	76/66	67/60	69/61	0.0	0.0	0.1	0.1	43
	Montevideo, Uruguay	81/61	71/53	59/43	68/50	2.9	3.7	3.1	2.8	72
	Caracas, Venezuela	78/64	82/68	80/68	80/68	1.0	1.7	4.2	4.1	2,739
Region 4	Canberra, Australia	81/54	67/44	54/33	67/42	2.2	2.1	1.7	2.3	1,873
	Dhaka, Bangladesh	78/61	88/76	88/81	86/75	0.4	5.8	14.1	5.9	30
	Beijing, China	37/20	66/46	85/69	64/46	0.2	0.8	6.2	1.1	180
	New Dehli, India	71/50	93/71	95/81	88/68	0.7	0.5	6.2	1.9	708
	Tehran, Iran	45/33	69/53	95/75	74/58	1.5	1.1	0.1	0.5	3,906
	Tokyo, Japan	50/37	63/50	81/71	70/59	2.3	5.0	6.0	6.6	26
	Moscow, Russia	23/13	49/33	69/53	45/33	1.4	1.6	2.9	2.0	623
	Bangkok, Thailand	89/72	93/79	90/79	89/76	0.6	3.8	6.5	8.0	66
	Ankara, Turkey	38/23	59/37	80/53	64/39	1.5	1.5	0.6	0.9	3,113
	Hanoi, Vietnam	68/59	80/71	90/80	82/72	1.0	3.6	12.0	9.4	19
Region 5	Ottawa, Canada	24/9	51/34	76/58	53/38	2.6	2.7	3.2	3.0	374
	San Jose, Costa Rica	75/58	80/61	78/62	78/60	0.8	3.9	9.1	9.8	3,021
	Havana, Cuba	79/64	84/68	89/74	85/71	2.3	2.9	5.6	5.3	194
	San Salvador, El Salvador	83/66	86/69	83/69	82/68	0.2	2.3	12.2	7.6	2,037
	Guatemala City, Guatemala	73/56	78/60	74/61	73/60	0.2	2.6	8.9	5.6	4,917
	Port-au-Prince, Haiti	87/75	88/78	92/80	89/79	1.6	5.9	4.1	5.4	95
	Kingston, Jamaica	86/74	87/76	90/79	88/78	1.2	2.2	3.3	5.2	30
	Mexico City, Mexico	71/45	78/53	75/56	73/52	0.3	1.1	4.7	2.2	7,328
	Panama City, Panama	89/76	89/78	86/77	86/77	2.1	3.9	7.8	9.5	43
	Washington D.C., United States	45/30	66/46	86/68	69/51	2.8	3.4	3.8	3.1	10

*Parts of Amsterdam are as much as 13 ft below sea level.

Literacy and Standard of Living Data

The table on the opposite page lists information about TVs, radios, telephones, and cars. Each number in the table shows what you would expect to find for a group of 1,000 people.

Examples There are only 242 radios in Algeria for every 1,000 people. But there are 2,116 radios in the U.S. for every 1,000 people. That's more than 2 radios for each person in the U.S. Many people in the U.S. have more than 1 radio.

African soccer fans watching a game on TV

You can use the data table to draw graphs and compare countries.

Example This bar graph shows TV data for Region 1 (Africa). The graph shows that Algeria, Egypt, Ghana, Morocco, and South Africa all have many more TVs per 1,000 people than the other five countries have.

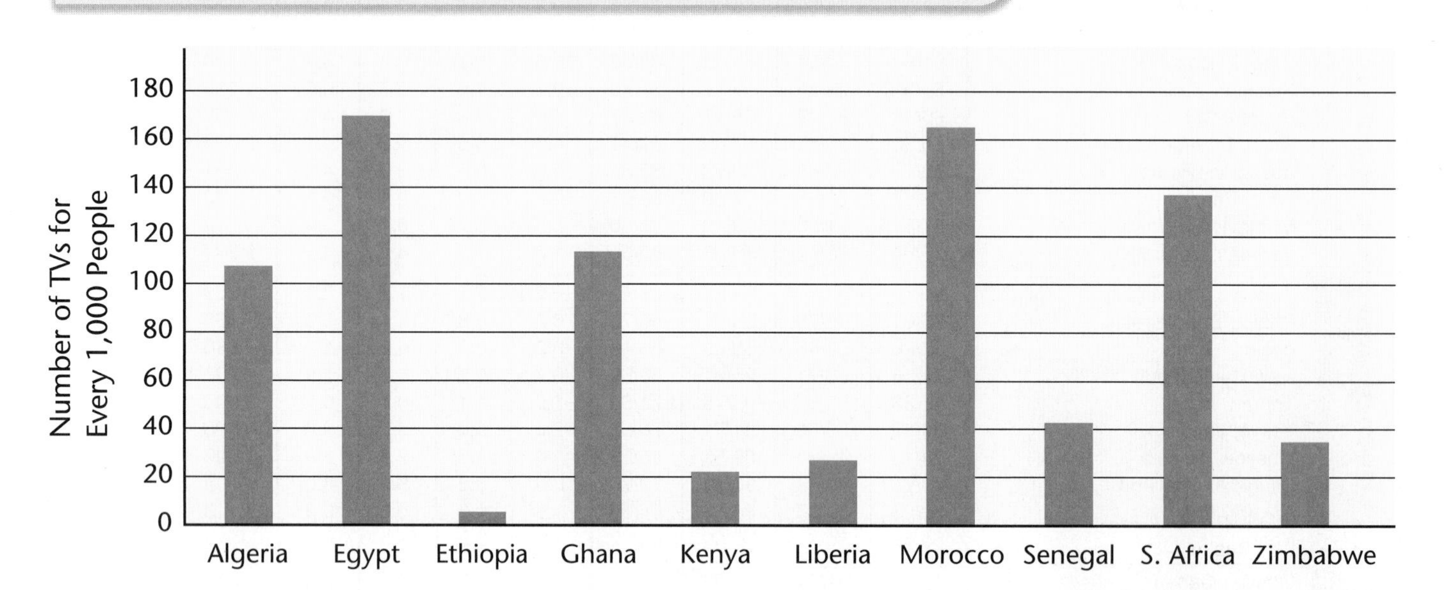

The last column in the table tells about the level of education in each country. A **literate** person is one who can read and write. The **percent of literacy** is the number of people out of 100 who are literate.

Literacy is the ability to read *and* write.

Example In Mexico, 92% of the people are literate. This means that about 92 of every 100 people can read and write.

Literacy and Standard of Living Data

Region	Country	Televisions (Per 1,000 People)	Radios (Per 1,000 People)	Telephones* (Per 1,000 People)	Cars (Per 1,000 People)	Percent Literate**
Region 1	Algeria	107	242	68	54	70
	Egypt	170	317	114	22	58
	Ethiopia	5	185	6	1	43
	Ghana	115	680	15	4	75
	Kenya	22	216	10	8	85
	Liberia	26	329	2	5	58
	Morocco	165	247	37	38	52
	Senegal	41	141	21	10	40
	South Africa	138	355	107	93	86
	Zimbabwe	35	389	24	43	91
Region 2	France	620	946	561	465	99
	Greece	480	475	488	321	98
	Hungary	447	690	369	247	99
	Iceland	505	1,075	649	541	100
	Italy	492	880	458	561	99
	Netherlands	540	980	613	375	99
	Norway	653	917	721	409	100
	Poland	387	522	318	272	100
	Spain	555	331	437	451	98
	United Kingdom	661	1,437	579	408	99
Region 3	Argentina	293	681	204	138	97
	Bolivia	118	675	69	36	87
	Brazil	333	434	211	63	86
	Chile	240	354	221	44	96
	Colombia	279	539	208	18	93
	Ecuador	213	406	114	24	93
	Paraguay	205	182	44	40	94
	Peru	147	273	65	27	91
	Uruguay	531	603	278	192	98
	Venezuela	185	296	112	57	93
Region 4	Australia	716	1,391	542	494	100
	Bangladesh	7	50	5	1	43
	China	291	342	202	5	86
	India	75	120	46	5	60
	Iran	154	265	216	14	79
	Japan	719	956	558	420	100
	Russia	421	417	247	148	100
	Thailand	274	234	102	41	96
	Turkey	328	510	274	66	87
	Vietnam	184	107	53	1	94
Region 5	Canada	709	1,038	615	519	97
	Costa Rica	229	774	253	86	96
	Cuba	248	352	51	1	97
	El Salvador	191	478	114	22	80
	Guatemala	61	79	59	45	71
	Haiti	5	53	17	12	53
	Jamaica	191	796	164	52	88
	Mexico	272	329	142	105	92
	Panama	13	299	129	74	93
	United States	844	2,116	620	757	97

*Includes telephone main lines, but not cellular telephones.

**Data are hard to measure and may vary greatly.

Population Data

The table on the opposite page lists population information for each country.

Life expectancy is the average number of years a person can expect to live. It is listed separately for males and females because women usually live longer than men.

Examples In the United States, women live an average of 80 years, and men live an average of 75 years. In Russia, women live an average of 13 years longer than men. Zimbabwe is the only country where men on average live longer than women.

Farming couple in Siberia

The **percent of people ages 0–14** is the number of people out of every 100 who are very young.

Examples In Liberia, 43% of the people are 14 or younger. That's nearly 50%, or one-half, of the people who are very young. In Italy, only 14% of the people are very young. Italy's fraction of very young people is much smaller than Liberia's fraction.

Percent urban is the number of people out of 100 who live in towns or cities. **Percent rural** is the number of people out of 100 who live in the country. These two percents add up to 100%.

Example In the United States, 80 of 100 people live in towns or cities, while 20 out of 100 people live in the country.
80% + 20% = 100%

U.S. farmer feeding cattle

The population in most countries grows larger each year. The **percent population growth** is one way to measure how fast the population is growing.

Example The population in Haiti increases by 2% each year.

For every 100 Haitians at the beginning of the year, there are 102 Haitians at the end of the year.

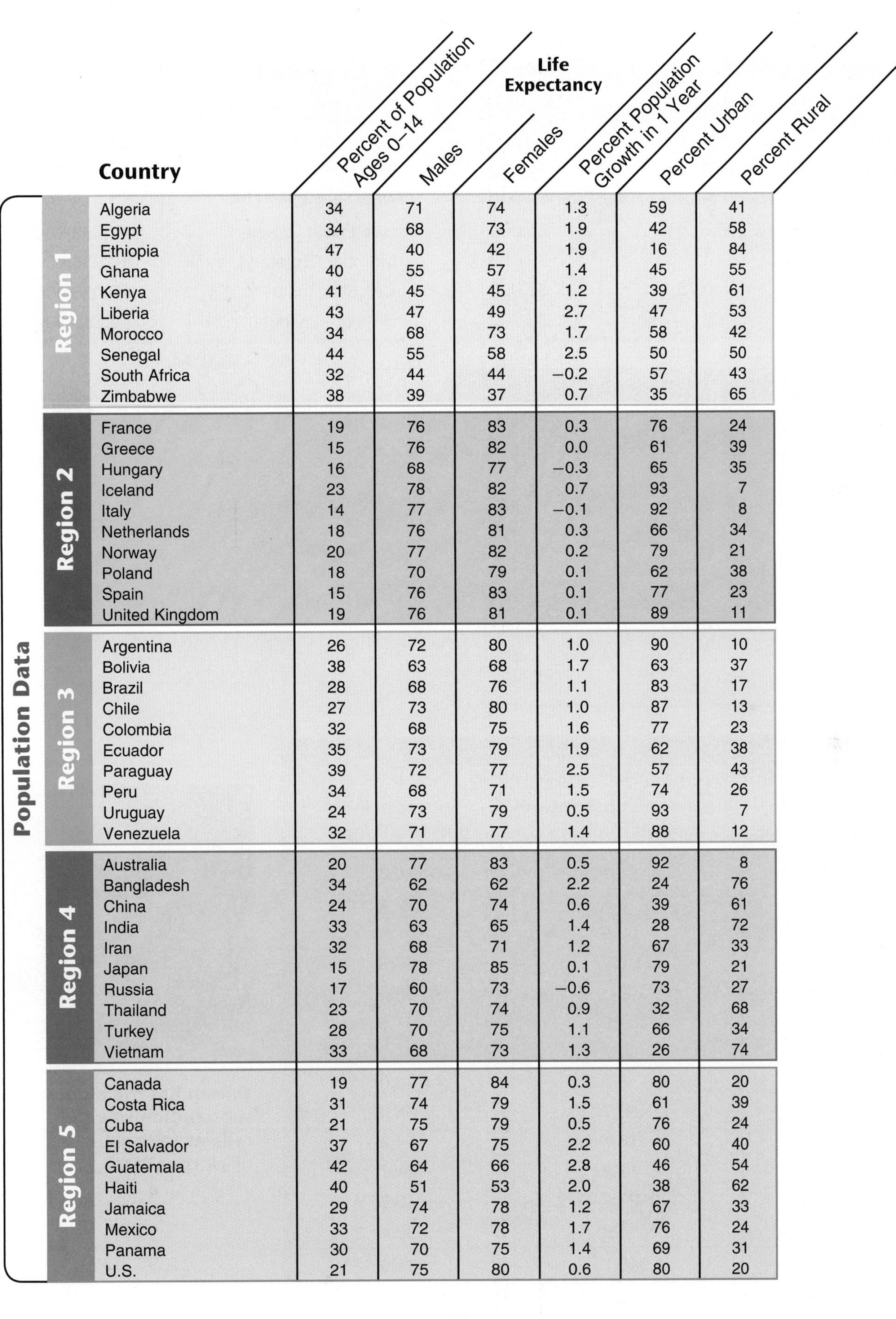

Population Data

Region	Country	Percent of Population Ages 0–14	Life Expectancy: Males	Life Expectancy: Females	Percent Population Growth in 1 Year	Percent Urban	Percent Rural
Region 1	Algeria	34	71	74	1.3	59	41
	Egypt	34	68	73	1.9	42	58
	Ethiopia	47	40	42	1.9	16	84
	Ghana	40	55	57	1.4	45	55
	Kenya	41	45	45	1.2	39	61
	Liberia	43	47	49	2.7	47	53
	Morocco	34	68	73	1.7	58	42
	Senegal	44	55	58	2.5	50	50
	South Africa	32	44	44	−0.2	57	43
	Zimbabwe	38	39	37	0.7	35	65
Region 2	France	19	76	83	0.3	76	24
	Greece	15	76	82	0.0	61	39
	Hungary	16	68	77	−0.3	65	35
	Iceland	23	78	82	0.7	93	7
	Italy	14	77	83	−0.1	92	8
	Netherlands	18	76	81	0.3	66	34
	Norway	20	77	82	0.2	79	21
	Poland	18	70	79	0.1	62	38
	Spain	15	76	83	0.1	77	23
	United Kingdom	19	76	81	0.1	89	11
Region 3	Argentina	26	72	80	1.0	90	10
	Bolivia	38	63	68	1.7	63	37
	Brazil	28	68	76	1.1	83	17
	Chile	27	73	80	1.0	87	13
	Colombia	32	68	75	1.6	77	23
	Ecuador	35	73	79	1.9	62	38
	Paraguay	39	72	77	2.5	57	43
	Peru	34	68	71	1.5	74	26
	Uruguay	24	73	79	0.5	93	7
	Venezuela	32	71	77	1.4	88	12
Region 4	Australia	20	77	83	0.5	92	8
	Bangladesh	34	62	62	2.2	24	76
	China	24	70	74	0.6	39	61
	India	33	63	65	1.4	28	72
	Iran	32	68	71	1.2	67	33
	Japan	15	78	85	0.1	79	21
	Russia	17	60	73	−0.6	73	27
	Thailand	23	70	74	0.9	32	68
	Turkey	28	70	75	1.1	66	34
	Vietnam	33	68	73	1.3	26	74
Region 5	Canada	19	77	84	0.3	80	20
	Costa Rica	31	74	79	1.5	61	39
	Cuba	21	75	79	0.5	76	24
	El Salvador	37	67	75	2.2	60	40
	Guatemala	42	64	66	2.8	46	54
	Haiti	40	51	53	2.0	38	62
	Jamaica	29	74	78	1.2	67	33
	Mexico	33	72	78	1.7	76	24
	Panama	30	70	75	1.4	69	31
	U.S.	21	75	80	0.6	80	20

Fascinating Facts

Smallest Countries by Population		
Country	Area (sq miles)	Population
Vatican City	0.17	900
Tuvalu	10	11,000
Nauru	8	13,000
Palau	177	20,000
San Marino	24	28,000
Monaco	0.75	32,000
Liechtenstein	62	33,000

Largest Cities by Population	
City, Country	Population
Tokyo, Japan	34,450,000
Mexico City, Mexico	18,066,000
New York City, U.S.	17,846,000
Sao Paulo, Brazil	17,099,000
Mumbai (Bombay), India	16,086,000
Calcutta, India	13,058,000
Shanghai, China	12,887,000

Languages with the Most Speakers		
Language	Speakers (in millions)	Countries
Chinese	873	16
Spanish	322	43
English	309	107
Hindi	180	17
Portuguese	177	33
Bengali	171	9
Russian	145	31
Japanese	122	25

Cellular Telephone Use		
Country	Number of Cellular Telephone Subscriptions (in millions)	Subscriptions per 100 People
Taiwan	25.1	110.8
Luxembourg	0.5	106.1
Italy	55.9	101.8
Iceland	0.3	96.6
Israel	6.3	95.5
Spain	37.5	91.6
United Kingdom	49.7	84.1
Greece	8.9	78.0
Japan	86.7	68.0
United States	158.7	54.3
Canada	13.2	41.7
Turkey	27.9	40.8
South Africa	16.9	36.4
World	**1,340.7**	**21.9**

Taiwan has the highest subscription rate for cellular telephone use of all countries.

World's Tallest Buildings				
Name	**Place**	**Year Built**	**Height (feet)**	**Stories**
Taipei 101	Taipei, Taiwan	2004	1,670	101
Petronas Towers, I and II	Kuala Lumpur, Malaysia	1998	1,483	88
Sears Tower	Chicago, United States	1974	1,450	110
Jin Mao Bldg.	Shanghai, China	1999	1,381	88
Two International Finance Centre	Hong Kong, China	2003	1,362	88
CITIC Plaza	Guangzhou, China	1996	1,283	80
Shun Hing Square	Shenzhen, China	1996	1,260	69
Empire State Building	New York, United States	1931	1,250	102
Central Plaza	Hong Kong, China	1992	1,227	78
Bank of China	Hong Kong, China	1989	1,209	72

For more information and ongoing updates, go to the Web site http://en.wikipedia.org/wiki/World's_tallest_structures.

Taipei 101

Sears Tower

English Channel Tunnel

World's Longest Railway Tunnels			
Tunnel	**Place**	**Year Built**	**Length (miles)**
Seikan	Japan	1988	33.5
English Channel Tunnel	UK–France	1994	31.3
Iwate-ichinohe	Japan	2002	16.0
Dai-shimizu	Japan	1982	13.8
Simplon No. I and II	Switzerland–Italy	1906, 1922	12.3
Vereina	Switzerland	1999	11.8
Kanmon	Japan	1975	11.6
Apennine	Italy	1934	11.5

Largest Oceans and Seas		
Name	Area (sq miles)	Average Depth (feet)
Pacific Ocean	64,186,300	12,925
Atlantic Ocean	33,420,000	11,730
Indian Ocean	28,350,500	12,598
Arctic Ocean	5,105,700	3,407
South China Sea	1,148,500	4,802
Caribbean Sea	971,400	8,448
Mediterranean Sea	969,100	4,926
Bering Sea	873,000	4,893
Gulf of Mexico	582,100	5,297
Okhotsk Sea	537,500	3,192

Longest Rivers		
Name	Location	Length (miles)
Nile	Africa	4,160
Amazon	S. America	4,000
Chang (Yangtze)	Asia	3,964
Huang (Yellow)	Asia	3,395
Ob-Irtysh	Asia	3,362
Congo	Africa	2,900
Lena	Asia	2,734
Niger	Africa	2,590
Parana	S. America	2,485
Mississippi	N. America	2,340

Largest Deserts		
Name	Location	Area (sq miles)
Sahara	Africa	3,500,000
Gobi	Asia	500,000
Libyan	Africa	450,000
Patagonia	S. America	300,000
Rub al Khali	Asia	250,000
Kalahari	Africa	225,000
Great Sandy	Australia	150,000
Great Victoria	Australia	150,000
Chihuahua	N. America	140,000

Largest Freshwater Lakes		
Name	Location	Area (sq miles)
Superior	N. America	31,700
Victoria	Africa	26,828
Huron	N. America	23,000
Michigan	N. America	22,300
Tanganyika	Africa	12,700
Baykal	Asia	12,162
Great Bear	N. America	12,096
Malawi (Nyasa)	Africa	11,150

Tallest Mountains		
Name	Location	Height (feet)
Everest	Nepal–Tibet	29,028
K-2 (Godwin-Austen)	Kashmir	28,250
Kanchenjunga	Nepal–India	28,208
Lhotse I (Everest)	Nepal–Tibet	27,923
Makalu I	Nepal–Tibet	27,824
Lhotse II (Everest)	Nepal–Tibet	27,560
Dhaulagiri I	Nepal	26,810
Manaslu I	Nepal	26,760
Cho Oyu	Nepal–Tibet	26,750
Nanga Parbat	Kashmir	26,660

Highest/Lowest Elevation Points		
Continent	Highest/ Lowest Point	Elevation (feet)
Africa	Mt. Kilimanjaro, Tanzania Lake Assal, Djibouti	19,340 −512
Antarctica	Vinson Massif Bentley Subglacial Trench	16,864 −8,327
Asia	Mt. Everest, Nepal–Tibet Dead Sea, Israel–Jordan	29,028 −1,312
Australia	Mt. Kosciusko, New S. Wales Lake Eyre, South Australia	7,310 −52
Europe	Mt. El'brus, Russia Caspian Sea, Russia-Azerbaijan	18,510 −92
N. America	Mt. McKinley (Denali), Alaska Death Valley, California	20,320 −282
S. America	Mt. Aconcagua, Argentina Valdés Penninsula, Argentina	22,834 −131

Temperature and Rainfall Extremes	
Hottest single days	136°F, Azizia (Alaziziyah) Libya 134°F Death Valley, California
Hottest yearly average	95°F, Dalol Depression, Ethiopia
Coldest single days	−129°F, Vostok, Antarctica −90°F, Oimekon, Russia
Coldest yearly average	−72°F, Plateau Station, Antarctica
Highest average yearly rainfall	467 in., Mawsynram, India 460 in., Mt. Waialeale, Kauai, Hawaii
Lowest average yearly rainfall	0.03 in., Arica, Chile Less than 0.1 in., Wadi Half, Sudan

Colombia has 1,700 known species of birds.

Top Countries for Mammals

Country	Continent	Number of Known Species
Mexico	N. America	491
Peru	S. America	460
Indonesia	Asia	457
Congo, Dem. Rep.	Africa	450
United States	N. America	432
Brazil	S. America	417
Cameroon	Africa	409
China	Asia	400
Colombia	S. America	359
Kenya	Africa	359

Top Countries for Birds

Country	Continent	Number of Known Species
Colombia	S. America	1,700
Peru	S. America	1,541
Indonesia	Asia	1,530
Brazil	S. America	1,500
Ecuador	S. America	1,388
Venezuela	S. America	1,340
China	Asia	1,103
Congo, Dem. Rep.	Africa	929
India	Asia	926
Argentina	S. America	897

Top Countries for Reptiles and Amphibians

Country	Continent	Number of Known Species
Colombia	S. America	1,277
Brazil	S. America	1,072
Mexico	N. America	1,014
Australia	Australia	953
Ecuador	S. America	806
Indonesia	Asia	799
Peru	S. America	736
China	Asia	630
India	Asia	599
United States	N. America	550

Top Countries for Flowering Plants

Country	Continent	Approximate Number of Known Species
Brazil	S. America	55,000
Colombia	S. America	50,000
China	Asia	30,000
Indonesia	Asia	27,500
Mexico	N. America	25,000
South Africa	Africa	23,000
Venezuela	S. America	20,000
Ecuador	S. America	18,250
Peru	S. America	17,000
Bolivia	S. America	17,000

National Flags

Region 1

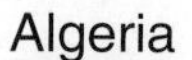
Algeria

Egypt
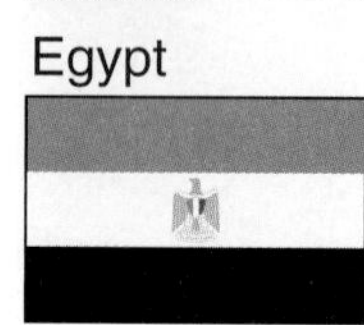
Ethiopia
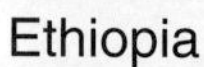

Ghana
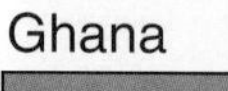

Kenya
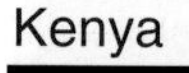
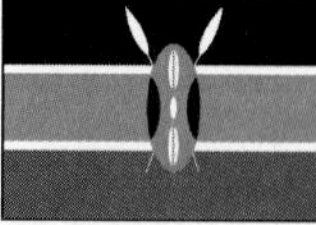
Liberia

Morocco

Senegal
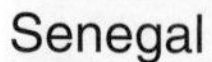

South Africa

Zimbabwe
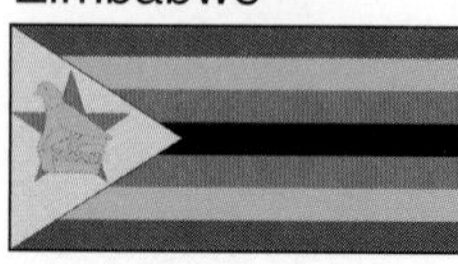

Region 2

France

Greece

Hungary
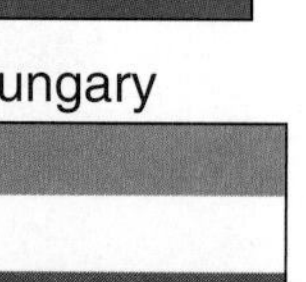
Iceland
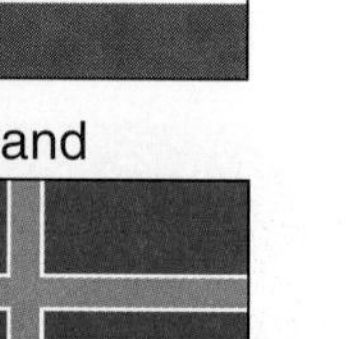
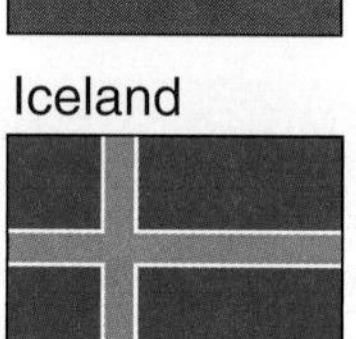
Italy
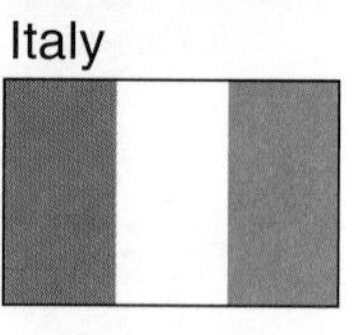
Netherlands

Norway
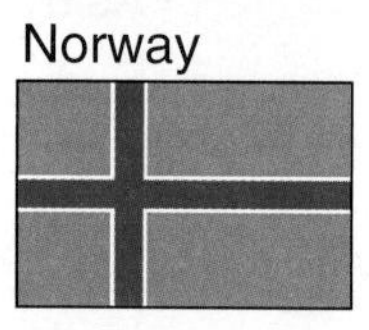
Poland
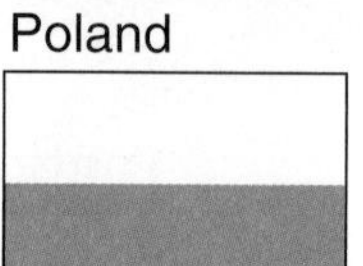
Spain

United Kingdom

Region 3

Argentina
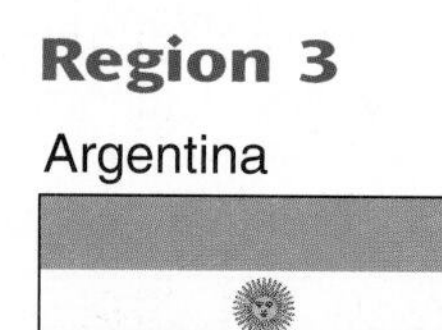
Bolivia
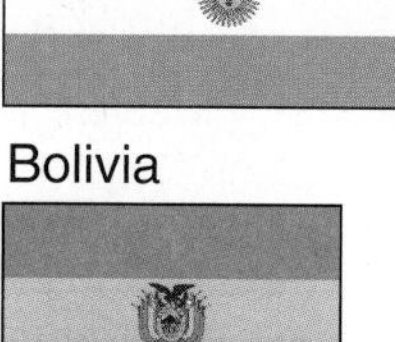
Brazil

Chile
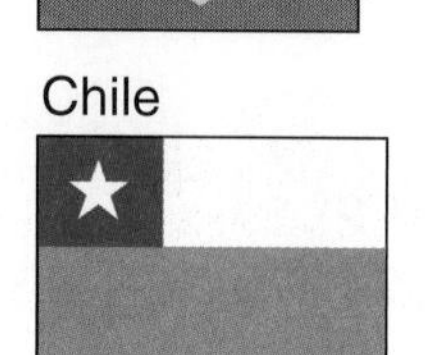
Colombia
Ecuador
Paraguay

Peru
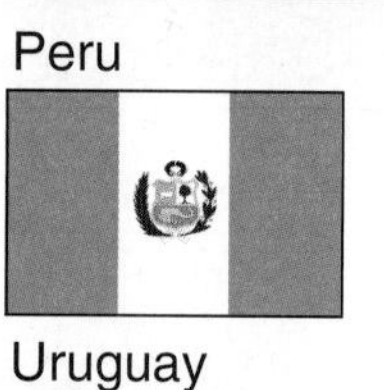
Uruguay

Venezuela
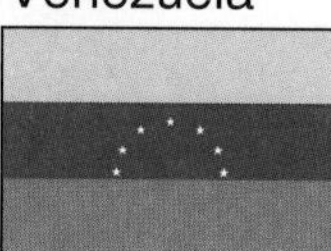

Region 4

Australia

Bangladesh

China

India
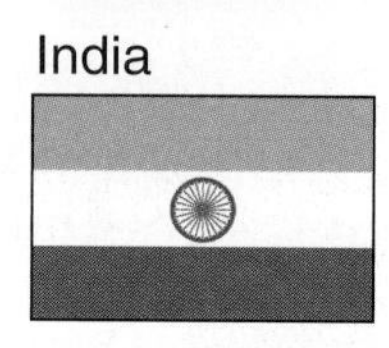
Iran
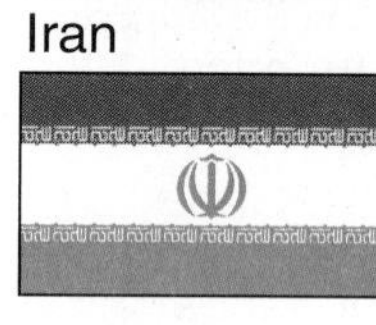
Japan

Russia

Thailand

Turkey

Vietnam
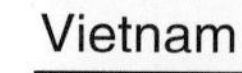

Region 5

Canada
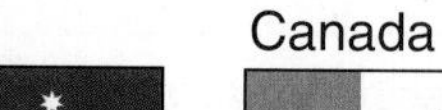
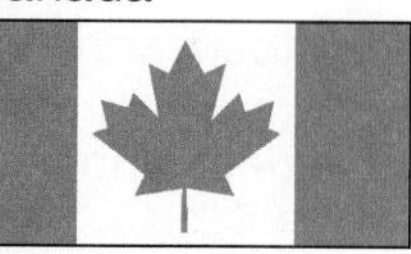
Costa Rica

Cuba

El Salvador

Guatemala

Haiti
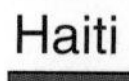

Jamaica

Mexico
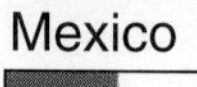

Panama

United States

Mancala

Mancala is a 3,000-year-old game of strategy. It is popular throughout Africa and Asia. The game has a variety of names, and the rules vary slightly from country to country. The game is usually played using a wooden board with 12 cups carved into it. Seeds or beans are used as counters.

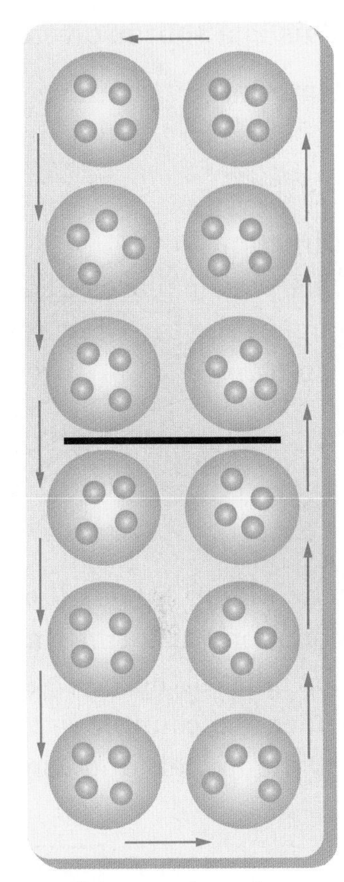

The gameboard at the start of play

Materials
- ☐ an egg carton or gameboard with 12 cups
- ☐ 48 counters, beans, or seeds

Players 2

Directions

Players sit at opposite ends of the board. Each player places four counters in each of the six cups on his or her side of the board. (See diagram.)

To begin, Player 1 picks up all of the counters from one of the six cups on his or her side. Beginning with the next cup, Player 1 drops the four counters one at a time into each consecutive cup, moving *counterclockwise* around the board.

Player 2 does the same thing with counters from one of the cups on his or her side of the board. (Players always begin by picking up counters from a cup on their side of the board).

A player *captures* counters if *both* of these conditions hold true:

1. The last counter the player drops in a cup lands on the other player's side of the board.
2. It lands in a cup with one or two counters already there.

When this happens, the player picks up all the counters in that cup. These are set aside for counting at the end of the game. Players may only capture counters from their opponent's side of the board.

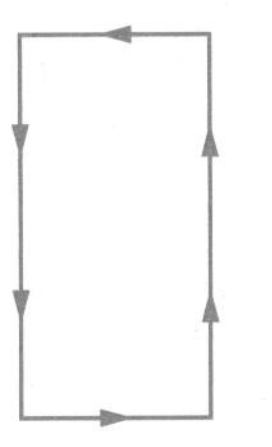

Move counterclockwise as you drop counters into the cups.

Play continues until one player has no counters left on his or her side of the board. At that time, players tally the number of counters captured, plus the number of counters remaining on their side of the board. The player with the most counters wins.

Tchuka Ruma

This is a solitaire version of *Mancala*.

Materials ☐ an egg carton or gameboard with 5 cups
☐ 8 counters

Player 1

Directions

The player places 2 counters in each of the first 4 cups. The cup on the far right remains empty. The empty cup is called the *Ruma*.

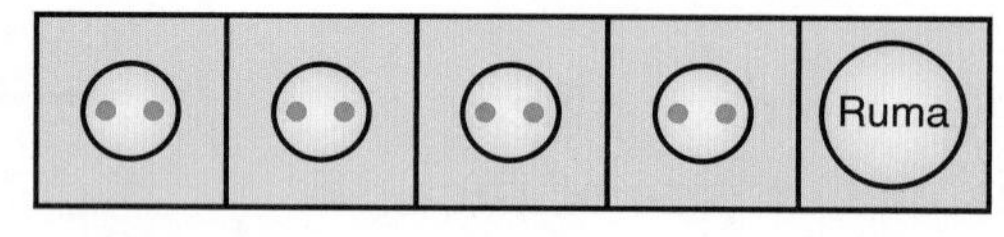

The gameboard at the start of play

The player takes all of the counters from any cup and drops them one at a time into each consecutive cup, moving from left to right. If there are still counters in the player's hand after placing a counter in the Ruma, the player goes back to the cup at the far left and continues.

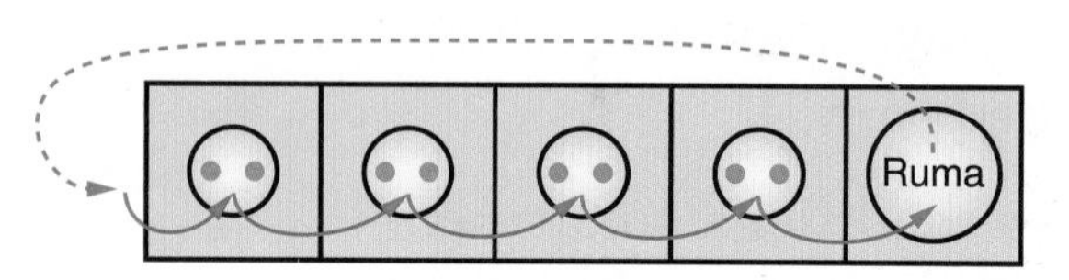

Move in the direction of the arrows as you drop counters into the cups.

If there are counters in the cup where the last counter is dropped, the player takes all of the counters (including the one dropped) and continues as before. The player drops the counters one at a time into consecutive cups, beginning with the next cup to the right. The player always moves from left to right. If there are still counters in the player's hand after placing a counter in the Ruma, the player returns to the cup at the far left and continues play.

If the last counter is dropped in the Ruma, the player can select *any* cup to begin the next move. The player takes all of the counters from this cup and drops them into other cups in the usual way.

If the last counter is dropped in a cup that does *not* have counters and is *not* the Ruma, the game ends and the player loses.

The player wins if he or she can get all of the counters into the Ruma.

Seega

This is a version of a traditional Egyptian game that is popular among young Egyptians today.

Materials ☐ *Seega* Game Mat (*Math Masters,* p. 503)
☐ 6 markers (3 each of two colors)

Players 2

Directions

Each player takes 3 markers of the same color. To begin, players place their markers on the starting lines at the ends of the game mat. (See diagram.)

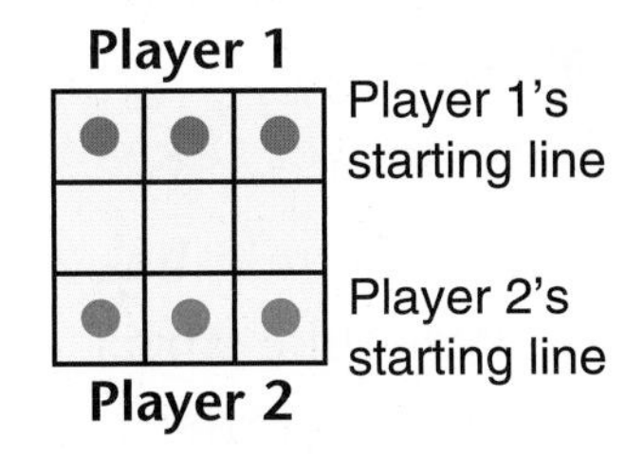

Players take turns moving one of their markers one or two squares.

- A marker can be moved to any open square that is next to it. Diagonal moves are OK.
- A marker can be moved two squares in any direction to an open square. Diagonal moves are OK, but a change in direction during the move is *not* allowed. Jumping over another marker is also *not* allowed.

Examples Moves allowed:

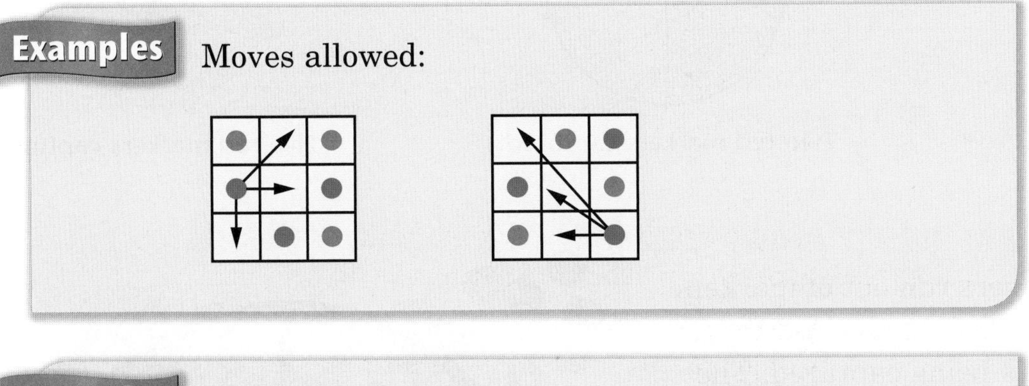

Examples Moves not allowed:

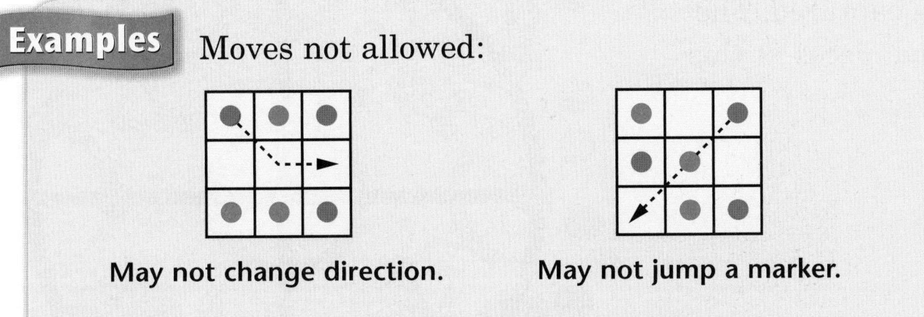

The first player to get his or her markers in a straight line is the winner. The line may be horizontal, vertical, or diagonal, but it may not be the player's starting line.

Sz'kwa

This is a Chinese children's game. Its name means "the game of four directions." In China, the game mat is often marked in the dirt or gravel, and pebbles, nuts, or shells are used as markers.

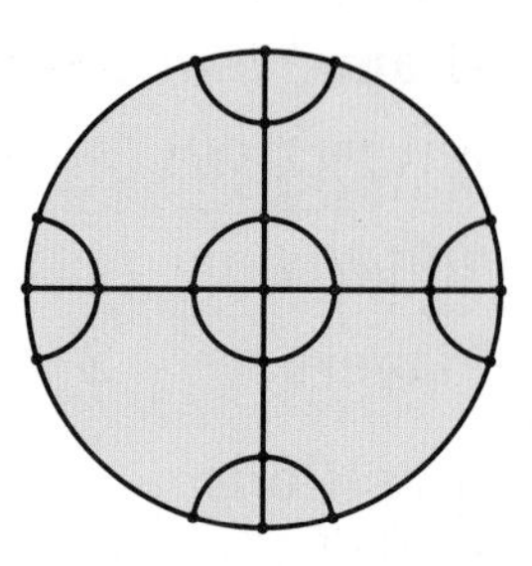

Materials ☐ *Sz'kwa* Game Mat (*Math Masters,* p. 505)
☐ 40 markers (20 each of two different colors)

Players 2

Directions

Each player takes 20 markers of the same color.

The game mat has 21 places where lines meet (called "intersections"). Players take turns. At each turn, a player places one marker on any intersection that is not already covered by a marker.

A marker is captured when it is surrounded by the opponent's markers. The captured marker is removed from the mat and kept by the opponent.

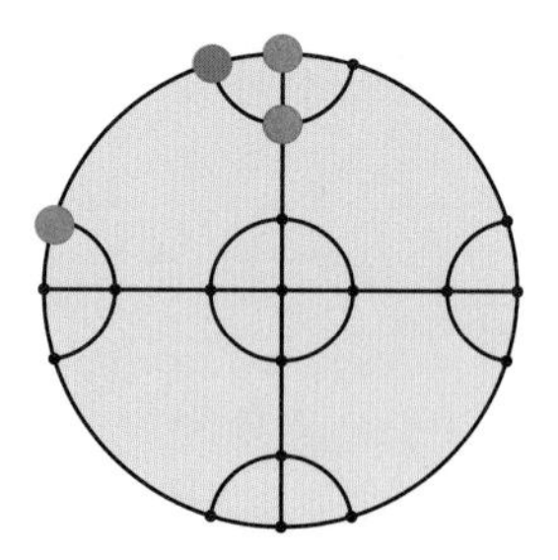

Green marker captured.

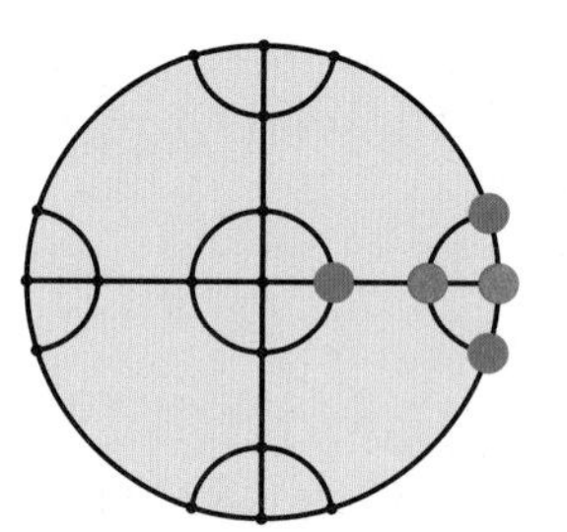

Two red markers captured.

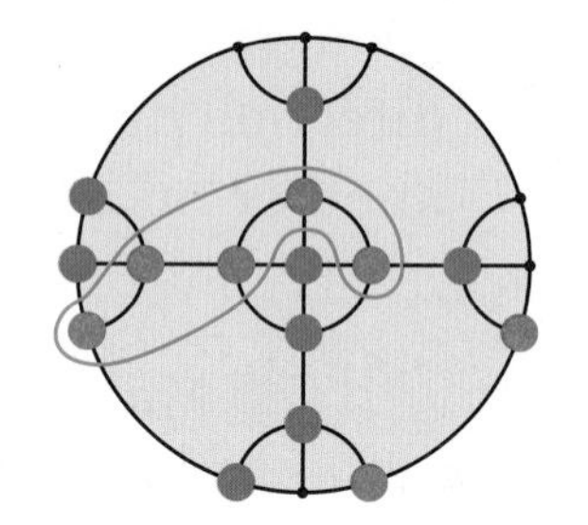

Five red markers captured.

Play continues until players run out of markers or until there is no place left on the mat to put a marker without its being captured. The player who holds more captured pieces at this time is the winner.

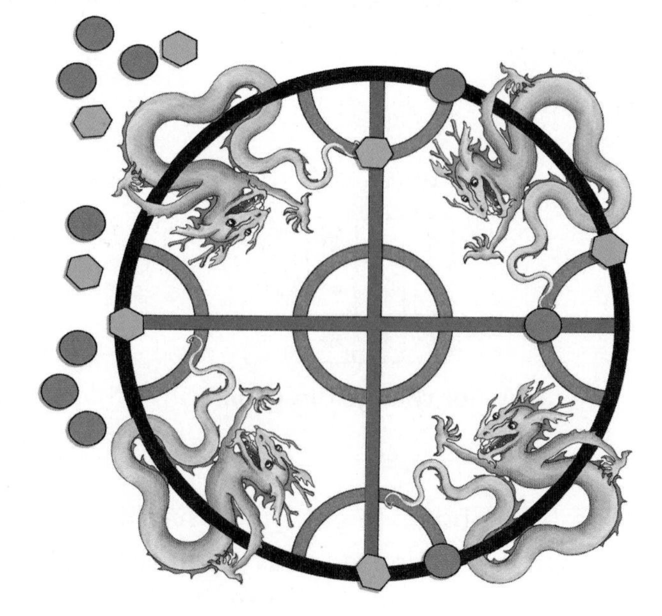

Alleyway

This game is popular in Eastern Europe. The game mat is shaped like a semicircle with 25 numbered spaces. Space 13 is left open and is called the "alleyway."

Materials ☐ *Alleyway* Game Mat (*Math Masters,* p. 456)
☐ 1 marker for each player
☐ 1 die

Players any number

Directions

Players place their markers in the Start space. They take turns rolling the die and moving their markers. A player moves his or her marker forward on the game mat by as many spaces as there are dots showing on the die.

A player's marker may land on a space already occupied by another player's marker. If that happens, the *opponent's* marker must be moved back.

- ♦ If the marker is in one of the spaces numbered 1–13, the opponent's marker must move back to the Start space.
- ♦ If the marker is in one of the spaces numbered 14–25, the opponent's marker must move back 2 spaces.

If a marker lands on another player's marker when it is moved back, then the marker it lands on must also be moved back. Use the rules given above for moving it back.

If a player's marker lands exactly on space 25, it must go back to space 14.

The winner is the first person to get *beyond* space 25.

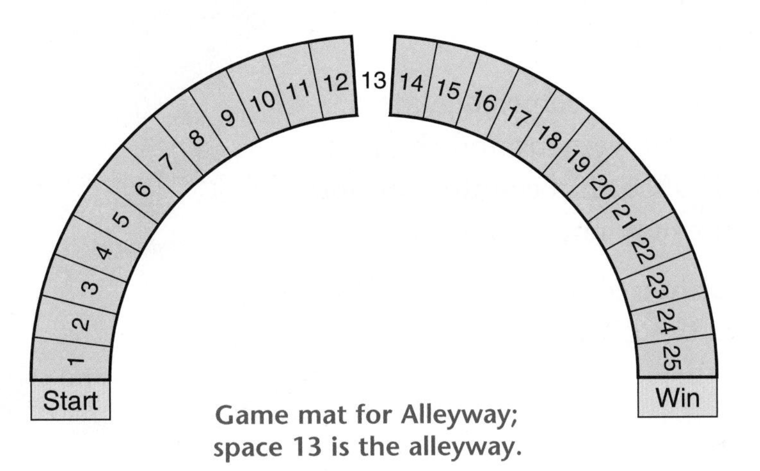

Game mat for Alleyway; space 13 is the alleyway.

Patolli

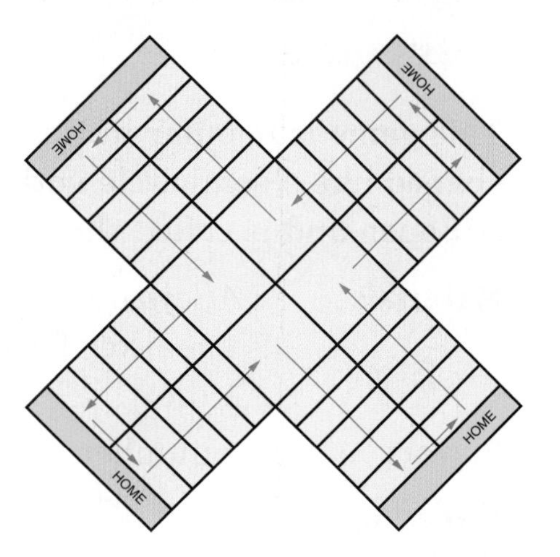

The board game *Patolli* has been played in Mexico since at least 200 B.C. The game takes its name from the Aztec word for bean—*patolli,* which means common bean or kidney bean. We do not have any record of the exact game rules. But these rules will allow you to play a game that should be similar to the ancient game.

Materials
- ☐ *Patolli* Game Mat (*Math Masters,* p. 495)
- ☐ 5 flat beans (such as limas)
  Mark one side of each bean with a dot.
- ☐ 12 counters:
  - for 2 players, 6 counters per player
  - for 3 players, 4 counters per player
  - for 4 players, 3 counters per player

  Each player should have counters of a different color from the other players.

Players 2–4

Directions

Players each place a counter on the HOME space nearest to them. Each player tosses the 5 beans. The player with the greatest number of dots showing goes first.

Players take turns tossing the beans and moving their counters.

- ♦ A counter is moved the same number of spaces as the number of dots showing on the beans. If all 5 dots are showing, the count is doubled, and the player moves the counter 10 spaces.
- ♦ A player who has more than one counter on the mat may move any one of these counters. But the player may only move *one* counter during a turn.
- ♦ If exactly one dot is showing, the player may place a new counter on his or her HOME space.

A counter is removed from the mat when it comes back to a player's HOME space after going all the way around the mat. The counter must land exactly on HOME. If a counter cannot land exactly on HOME, that counter may not be moved.

The first player to move all of his or her counters around the mat and back to HOME space wins the game.

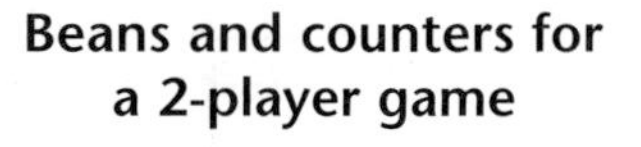

Beans and counters for a 2-player game

Sprouts

John Conway, professor of mathematics at Princeton University, invented this paper-and-pencil game in 1967.

Materials □ paper and pencil

Players 2

Directions

On a piece of paper, draw 3 dots that are widely spaced apart. You can start with more dots, but 3 dots is a good number to use when learning how to play.

Players take turns drawing a line (curved or straight) connecting any two dots, or joining a dot to itself. A player completes his or her turn by drawing another dot anywhere on the new line.

These rules must be followed when drawing the connecting lines:

- ♦ No line may cross itself.
- ♦ No line may cross any other line that has been drawn.
- ♦ No line may be drawn through a dot.
- ♦ A dot can have no more than 3 lines coming from it. A good way to keep track of this is to draw a box around any dot that has 3 lines coming from it. (See below.)

The winner is the last player who is able to draw a connecting line.

Sample Play (for an incomplete game)

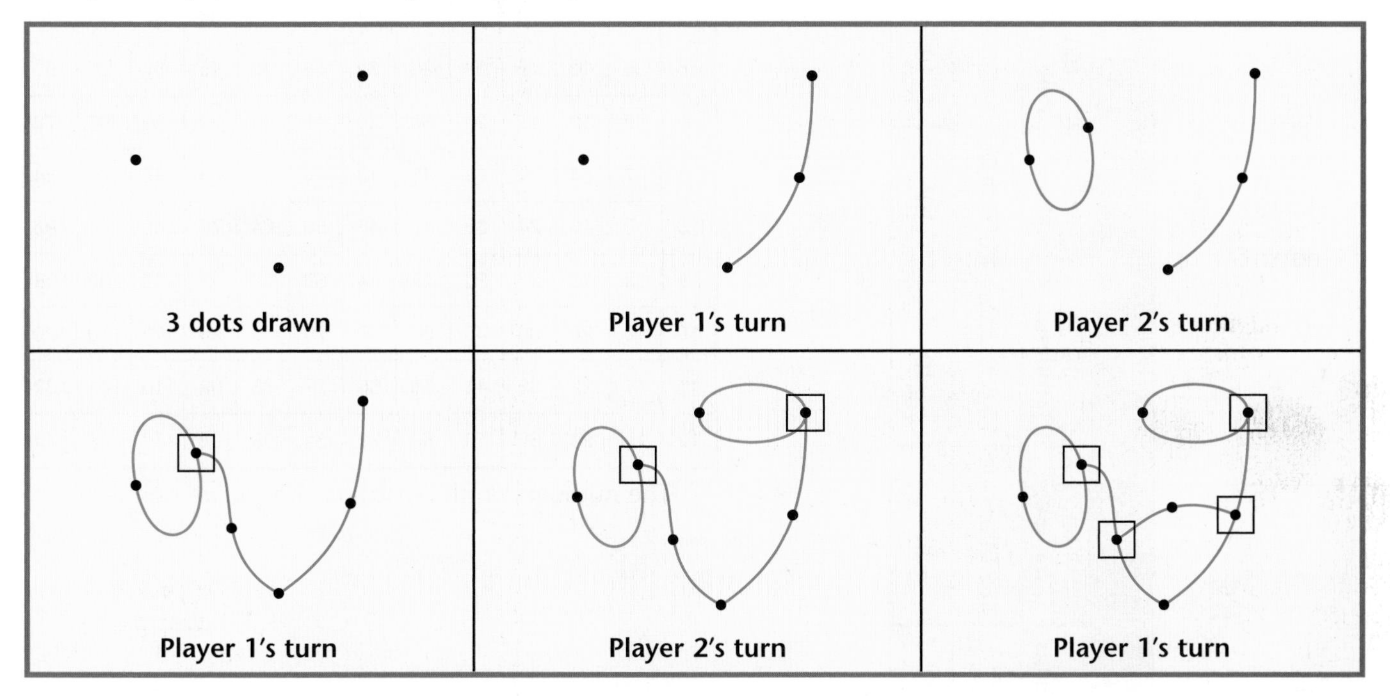

Place Value Chart

billions	100 millions	10 millions	millions	100 thousands	10 thousands	thousands	hundreds	tens	ones	.	tenths	hundredths	thousandths
1,000 millions	100,000,000s	10,000,000s	1,000,000s	100,000s	10,000s	1,000s	100s	10s	1s	.	0.1s	0.01s	0.001s
10^9	10^8	10^7	10^6	10^5	10^4	10^3	10^2	10^1	10^0	.	10^{-1}	10^{-2}	10^{-3}

Prefixes

uni-. . . . . . . .one	tera- . . . . trillion (10^{12})
bi-. . . . . .two	giga- . . . . billion (10^9)
tri- . . . . .three	mega- . . . million (10^6)
quad- . . . . . .four	kilo- . . . . . thousand (10^3)
penta- . . . . .five	hecto- . . . hundred (10^2)
hexa- . . . . . .six	deca- . . . ten (10^1)
hepta- . . . . .seven	uni- . . . . . one (10^0)
octa-. . . . . . .eight	deci- . . . . tenth (10^{-1})
nona- . . . . . .nine	centi-. . . . hundredth (10^{-2})
deca- . . . . . .ten	milli- . . . . thousandth (10^{-3})
dodeca- . . . .twelve	micro- . . . millionth (10^{-6})
icosa-. . . . . .twenty	nano- . . . billionth (10^{-9})

Multiplication and Division Table

*,/	1	2	3	4	5	6	7	8	9	10	11	12
1	1	2	3	4	5	6	7	8	9	10	11	12
2	2	4	6	8	10	12	14	16	18	20	22	24
3	3	6	9	12	15	18	21	24	27	30	33	36
4	4	8	12	16	20	24	28	32	36	40	44	48
5	5	10	15	20	25	30	35	40	45	50	55	60
6	6	12	18	24	30	36	42	48	54	60	66	72
7	7	14	21	28	35	42	49	56	63	70	77	84
8	8	16	24	32	40	48	56	64	72	80	88	96
9	9	18	27	36	45	54	63	72	81	90	99	108
10	10	20	30	40	50	60	70	80	90	100	110	120
11	11	22	33	44	55	66	77	88	99	110	121	132
12	12	24	36	48	60	72	84	96	108	120	132	144

The numbers on the diagonal are square numbers.

Metric System

Units of Length

1 kilometer (km)	=	1,000 meters (m)
1 meter	=	10 decimeters (dm)
	=	100 centimeters (cm)
	=	1,000 millimeters (mm)
1 decimeter	=	10 centimeters
1 centimeter	=	10 millimeters

Units of Area

1 square meter (m^2)	=	100 square decimeters (dm^2)
	=	10,000 square centimeters (cm^2)
1 square decimeter	=	100 square centimeters
1 square kilometer	=	1,000,000 square meters

Units of Volume

1 cubic meter (m^3)	=	1,000 cubic decimeters (dm^3)
	=	1,000,000 cubic centimeters (cm^3)
1 cubic decimeter	=	1,000 cubic centimeters

Units of Capacity

1 kiloliter (kL)	=	1,000 liters (L)
1 liter	=	1,000 milliliters (mL)
1 cubic centimeter	=	1 milliliter

Units of Weight

1 metric ton (t)	=	1,000 kilograms (kg)
1 kilogram	=	1,000 grams (g)
1 gram	=	1,000 milligrams (mg)

System Equivalents

1 inch is about 2.5 cm (2.54)
1 kilometer is about 0.6 mile (0.621)
1 mile is about 1.6 kilometers (1.609)
1 meter is about 39 inches (39.37)
1 liter is about 1.1 quarts (1.057)
1 ounce is about 28 grams (28.350)
1 kilogram is about 2.2 pounds (2.205)

U.S. Customary System

Units of Length

1 mile (mi)	=	1,760 yards (yd)
	=	5,280 feet (ft)
1 yard	=	3 feet
	=	36 inches (in.)
1 foot	=	12 inches

Units of Area

1 square yard (yd^2)	=	9 square feet (ft^2)
	=	1,296 square inches ($in.^2$)
1 square foot	=	144 square inches
1 acre	=	43,560 square feet
1 square mile (mi^2)	=	640 acres

Units of Volume

1 cubic yard (yd^3)	=	27 cubic feet (ft^3)
1 cubic foot	=	1,728 cubic inches ($in.^3$)

Units of Capacity

1 gallon (gal)	=	4 quarts (qt)
1 quart	=	2 pints (pt)
1 pint	=	2 cups (c)
1 cup	=	8 fluid ounces (fl oz)
1 fluid ounce	=	2 tablespoons (tbs)
1 tablespoon	=	3 teaspoons (tsp)

Units of Weight

1 ton (T)	=	2,000 pounds (lb)
1 pound	=	16 ounces (oz)

Units of Time

1 century	=	100 years
1 decade	=	10 years
1 year (yr)	=	12 months
	=	52 weeks (plus one or two days)
	=	365 days (366 days in a leap year)
1 month (mo)	=	28, 29, 30, or 31 days
1 week (wk)	=	7 days
1 day (d)	=	24 hours
1 hour (hr)	=	60 minutes
1 minute (min)	=	60 seconds (sec)

Decimal and Percent Equivalents for "Easy" Fractions		
"Easy" Fractions	Decimals	Percents
$\frac{1}{2}$	0.50	50%
$\frac{1}{4}$	0.25	25%
$\frac{3}{4}$	0.75	75%
$\frac{1}{5}$	0.20	20%
$\frac{2}{5}$	0.40	40%
$\frac{3}{5}$	0.60	60%
$\frac{4}{5}$	0.80	80%
$\frac{1}{10}$	0.10	10%
$\frac{3}{10}$	0.30	30%
$\frac{7}{10}$	0.70	70%
$\frac{9}{10}$	0.90	90%

The Global Grid

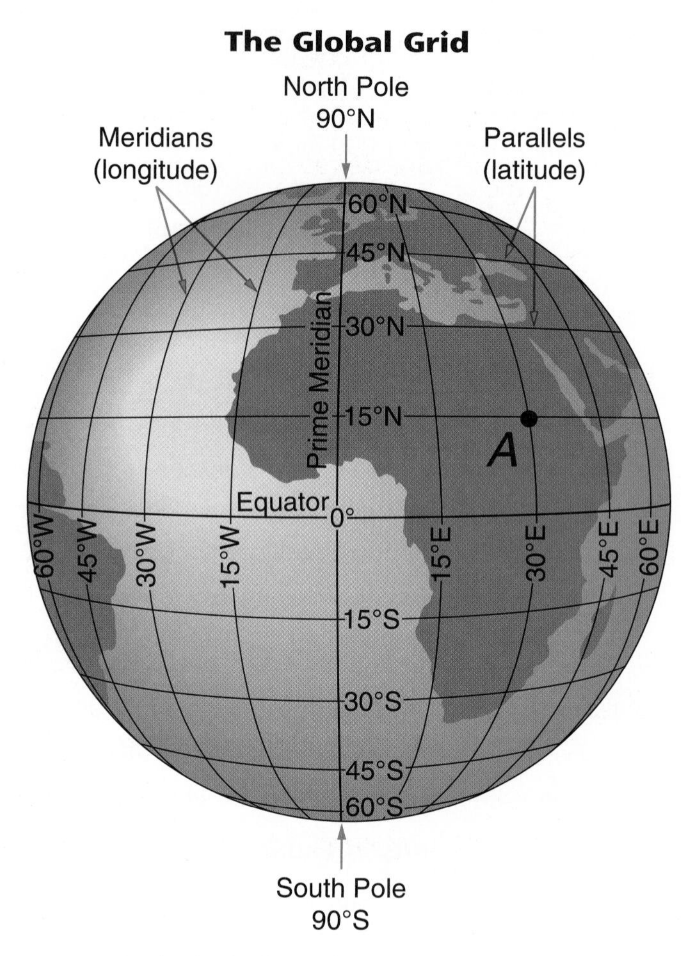

Point *A* is located at 15°N, 30°E.

Fraction-Decimal Number Line

0	0.125	0.25	0.375	0.5	0.625	0.75	0.875	1
$\frac{0}{2}$	$\frac{1}{8}$	$\frac{1}{4}$	$\frac{3}{8}$	$\frac{1}{2}$	$\frac{5}{8}$	$\frac{3}{4}$	$\frac{7}{8}$	$\frac{2}{2}$
$\frac{0}{4}$		$\frac{2}{8}$		$\frac{2}{4}$		$\frac{6}{8}$		$\frac{4}{4}$
$\frac{0}{8}$				$\frac{4}{8}$				$\frac{8}{8}$

Roman Numerals

A **numeral** is a symbol used to represent a number. **Roman numerals,** developed about 500 B.C., use letters to represent numbers.

Roman Numeral	Number
I	1
V	5
X	10
L	50
C	100
D	500
M	1,000

Seven different letters are used in Roman numerals. Each letter stands for a different number.

A string of letters means that their values should be added together. For example, CCC = 100 + 100 + 100 = 300, and CLXII = 100 + 50 + 10 + 1 + 1 = 162.

If a smaller value is placed *before* a larger value, the smaller value is subtracted instead of added. For example, IV = 5 − 1 = 4, and CDX = 500 − 100 + 10 = 410.

There are several **rules for subtracting letters.**

- ♦ The letters I (1), X (10), C (100), and M (1,000) represent powers of ten. These are the only letters that may be subtracted. For example, 95 in Roman numerals is XCV (VC for 95 is incorrect because V is not a power of ten).
- ♦ One letter may not be subtracted from a second letter if the value of the second letter is more than 10 times the value of the first. The letter I may be subtracted only from V or X. The letter X may be subtracted only from L or C. For example, 49 in Roman numerals is XLIX (IL for 49 is incorrect). And 1990 in Roman numerals is MCMXC (MXM for 1990 is incorrect).
- ♦ Only a *single* letter may be subtracted from another that follows. For example, 7 in Roman numerals is VII (IIIX for 7 is incorrect). And 300 in Roman numerals is CCC (CCD for 300 is incorrect).

The largest Roman numeral, M, stands for 1,000. One way to write large numbers is to write a string of Ms. For example, MMMM stands for 4,000. Another way to write large numbers is to write a bar above a numeral. The bar means that the numeral beneath should be multiplied by 1,000. So, $\overline{\text{IV}}$ also stands for 4,000. And $\overline{\text{M}}$ stands for 1,000 ∗ 1,000 = 1 million.

A

Addend Any one of a set of numbers that are added. For example, in $5 + 3 + 1 = 9$, the addends are 5, 3, and 1.

Algorithm A set of step-by-step instructions for doing something, such as carrying out a computation or solving a problem.

Angle A figure that is formed by two rays or two line segments with a common endpoint. The rays or segments are called the *sides* of the angle. The common endpoint is called the *vertex* of the angle. Angles are measured in *degrees* (°). An *acute angle* has a measure greater than 0° and less than 90°. An *obtuse angle* has a measure greater than 90° and less than 180°. A *reflex angle* has a measure greater than 180° and less than 360°. A *right angle* measures 90°. A *straight angle* measures 180°.

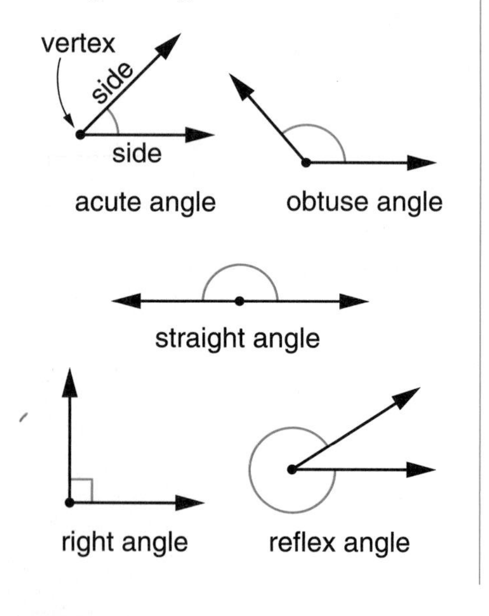

Apex In a pyramid or a cone, the vertex opposite the base. In a pyramid, all the faces except the base meet at the apex. See also *base of a pyramid or a cone.*

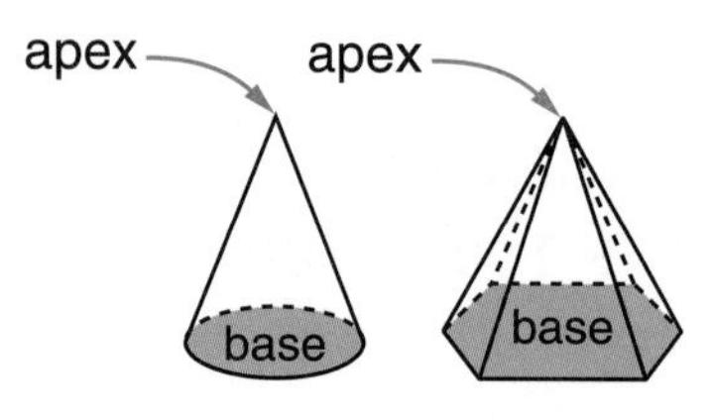

Area The amount of surface inside a closed boundary. Area is measured in square units, such as square inches or square centimeters.

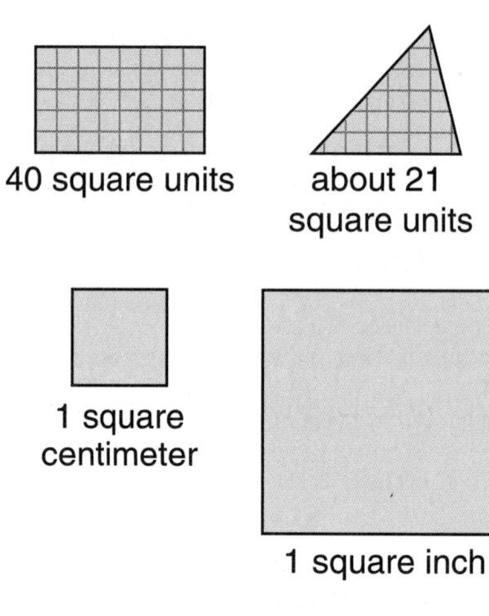

Array (1) An arrangement of objects in a regular pattern, usually in rows and columns. (2) A *rectangular array.* In *Everyday Mathematics,* an array is a rectangular array unless specified otherwise.

Associative Property A property of addition and multiplication (but not of subtraction or division) that says that when you add or multiply three numbers, it does not matter which two are added or multiplied first. For example:

$$(4 + 3) + 7 = 4 + (3 + 7)$$

and

$$(5 * 8) * 9 = 5 * (8 * 9).$$

Average A typical value for a set of numbers. The word average usually refers to the *mean* of a set of numbers.

Axis (plural: **axes**) (1) Either of the two number lines that intersect to form a *coordinate grid.*

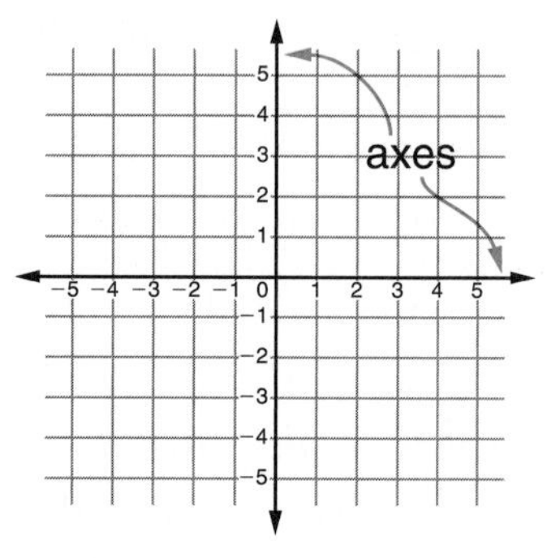

(2) A line about which a solid figure rotates.

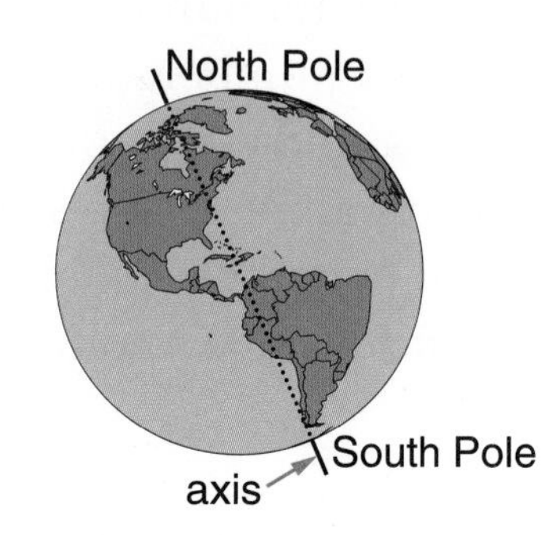

B

Bar graph A graph that uses horizontal or vertical bars to represent data.

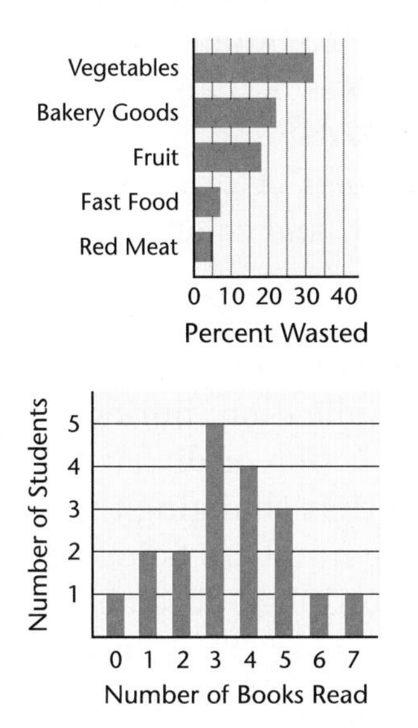

Base (in exponential notation) The number that is raised to a power. For example, in 5^3, the base is 5. See also *exponential notation* and *power of a number.*

Base of a polygon A side on which a polygon "sits." The height of a polygon may depend on which side is called the base. See also *height of a parallelogram* and *height of a triangle.*

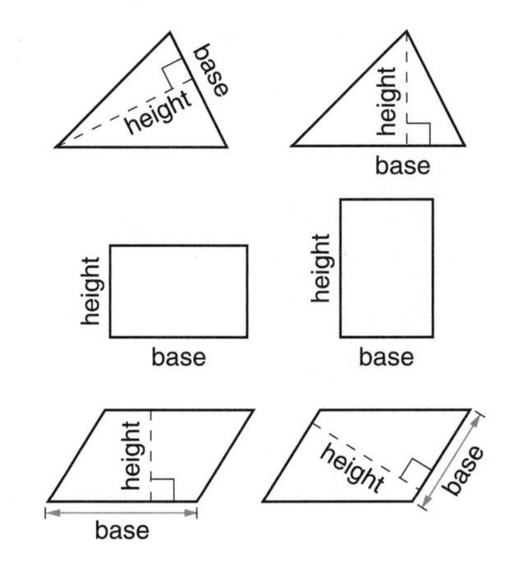

Base of a prism or a cylinder Either of the two parallel and congruent faces that define the shape of a prism or a cylinder.

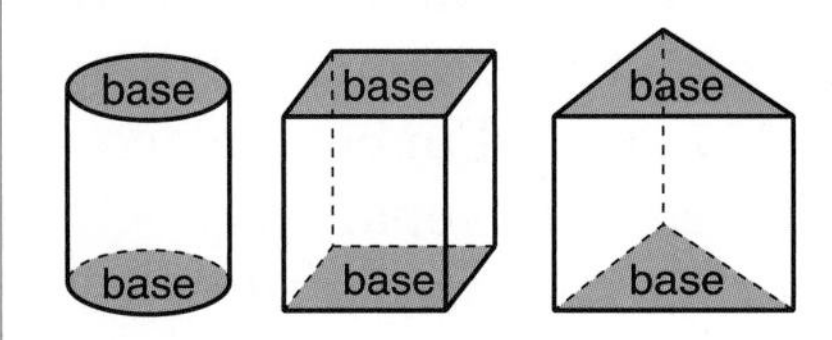

Base of a pyramid or a cone The face of a pyramid or a cone that is opposite its apex. The base of a pyramid is the only face that does not include the apex.

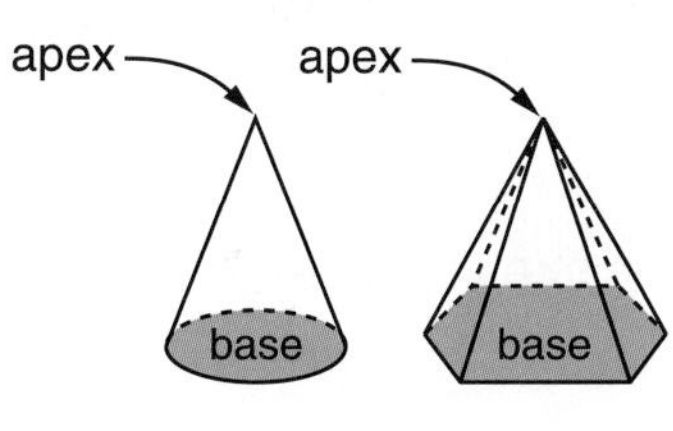

Base-ten Our system for writing numbers that uses only 10 symbols, called *digits.* The digits are 0, 1, 2, 3, 4, 5, 6, 7, 8, and 9. You can write any number using only these 10 digits. Each digit has a value that depends on its place in the number. In this system, moving a digit one place to the left makes that digit worth 10 times as much. And moving a digit one place to the right makes that digit worth one-tenth as much. See also *place value.*

Broken-line graph A graph in which data points are connected by line segments. Broken-line graphs are often used to show how something has changed over a period of time. Same as *line graph.*

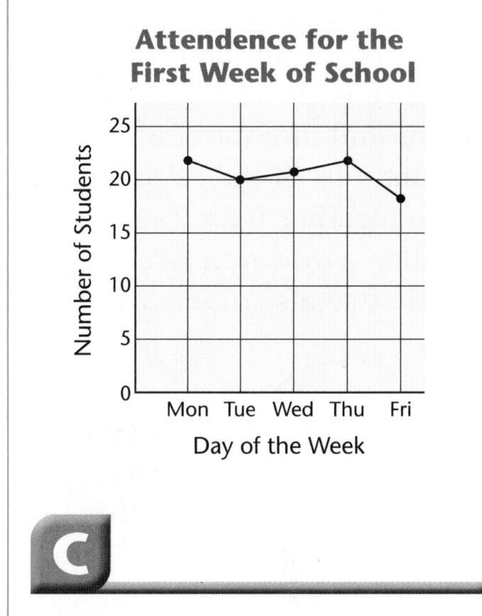

C

Capacity (1) The amount a container can hold. The *volume* of a container. Capacity is usually measured in units such as gallons, pints, cups, fluid ounces, liters, and milliliters.
(2) The heaviest weight a scale can measure.

Change diagram A diagram used in *Everyday Mathematics* to represent situations in which quantities are increased or decreased.

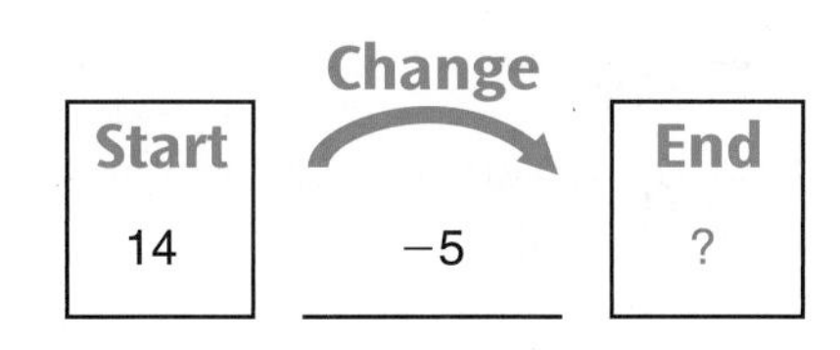

Circle The set of all points in a plane that are the same distance from a fixed point in the plane. The fixed point is the *center* of the circle, and the distance is the *radius.* The center and *interior* of a circle are not part of the circle. A circle together with its interior is called a *disk* or a *circular region.* See also *diameter.*

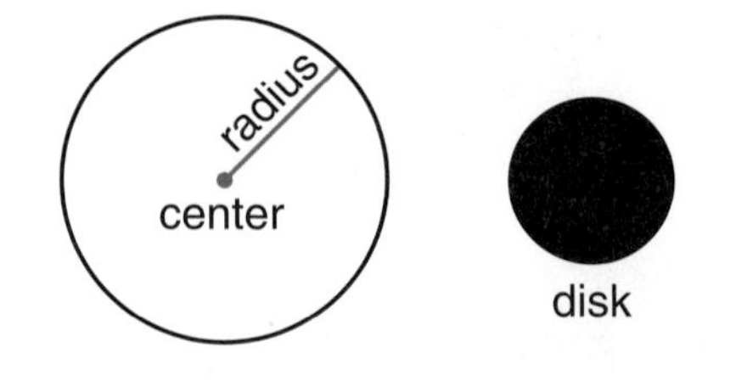

Circle graph A graph in which a circle and its interior are divided by radii into parts to show the parts of a set of data. The whole circle represents the whole set of data. Same as *pie graph.*

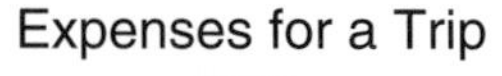

Circumference The distance around a circle; the perimeter of a circle.

Clockwise rotation A turning in the same direction as that of the hands of a clock.

Column-addition method A method for adding numbers in which the addends' digits are first added in each place-value column separately, and then 10-for-1 trades are made until each column has only one digit. Lines are drawn to separate the place-value columns.

	100s	10s	1s
	2	4	8
+	1	8	7
	3	12	15
	3	13	5
	4	3	5

$248 + 187 = 435$

Common denominator (1) If two fractions have the same denominator, that denominator is called a common denominator. (2) For two or more fractions, any number that is a *common multiple* of their denominators. For example, the fractions $\frac{1}{2}$ and $\frac{2}{3}$ have the common denominators 6, 12, 18, and so on. See also *quick common denominator.*

Common factor A counting number is a common factor of two or more counting numbers if it is a *factor* of each of those numbers. For example, 4 is a common factor of 8 and 12. See also *factor of a counting number.*

Common multiple A number is a common multiple of two or more numbers if it is a *multiple* of each of those numbers. For example, the multiples of 2 are 2, 4, 6, 8, 10, 12, and so on; the multiples of 3 are 3, 6, 9, 12, and so on; and the common multiples of 2 and 3 are 6, 12, 18, and so on.

Commutative Property A property of addition and multiplication (but not of subtraction or division) that says that changing the order of the numbers being added or multiplied does not change the answer. These properties are often called *turn-around facts* in *Everyday Mathematics.* For example: $5 + 10 = 10 + 5$ and $3 * 8 = 8 * 3$.

Comparison diagram A diagram used in *Everyday Mathematics* to represent situations in which two quantities are compared.

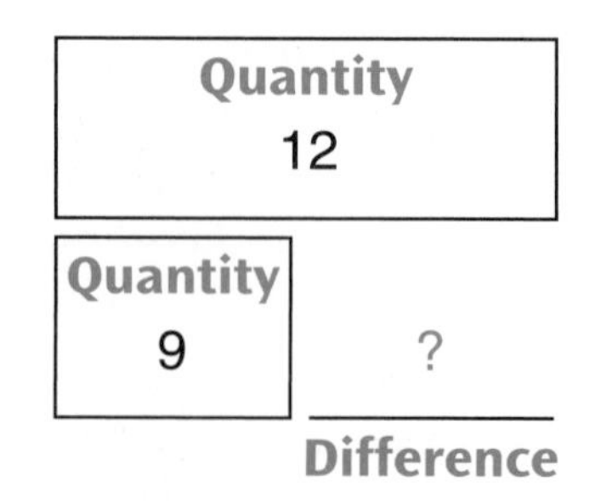

Composite number A counting number that has more than 2 different factors. For example, 4 is a composite number because it has three factors: 1, 2, and 4.

Concave polygon A polygon in which at least one vertex is "pushed in." Same as *nonconvex polygon.*

Concentric circles Circles that have the same center but radii of different lengths.

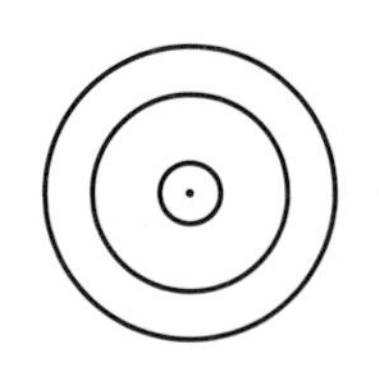

Cone A 3-dimensional shape that has a circular *base,* a curved surface, and one vertex, which is called the *apex.* The points on the curved surface of a cone are on straight lines connecting the apex and the boundary of the base.

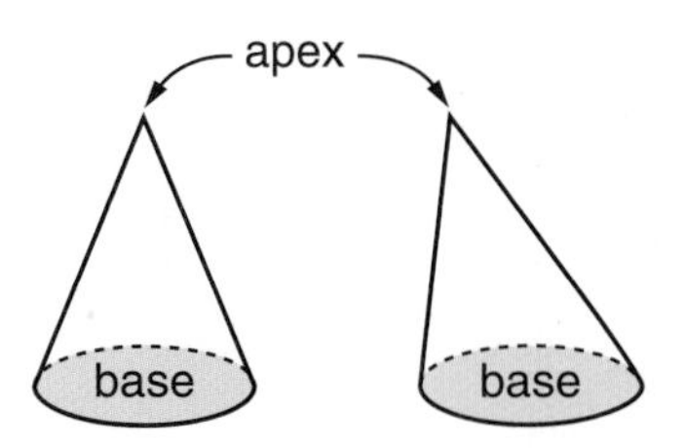

Congruent Having the same shape and size. Two 2-dimensional figures are congruent if they match exactly when one is placed on top of the other. (It may be necessary to flip one of the figures over.)

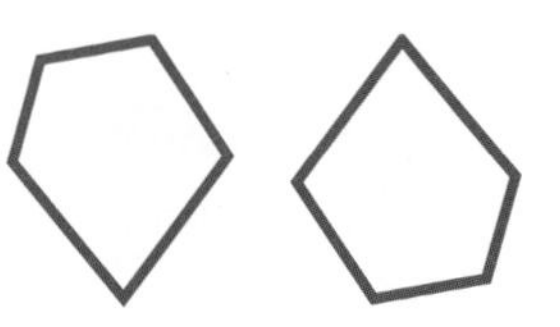

congruent pentagons

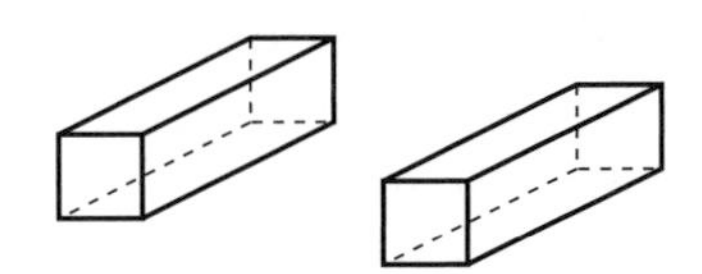

congruent prisms

Convex polygon A polygon in which all vertices are "pushed outward."

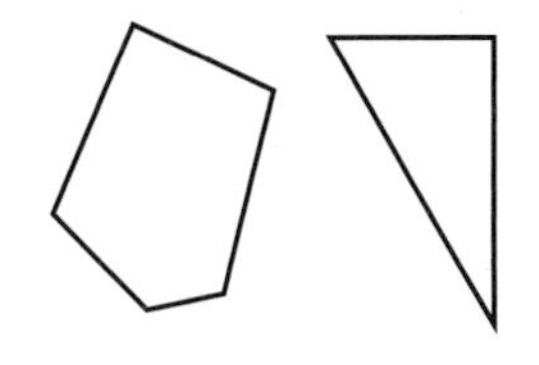

Coordinate (1) A number used to locate a point on a number line. (2) One of the two numbers in an *ordered number pair.* The number pair is used to locate a point on a *coordinate grid.*

Coordinate grid See *rectangular coordinate grid.*

Counterclockwise rotation A turning in the direction that is opposite to that of the hands of a clock.

Counting numbers The numbers used to count things. The set of counting numbers is {1, 2, 3, 4, . . .}. Compare to *whole numbers.*

Cube A polyhedron with 6 square faces. A cube has 8 vertices and 12 edges.

Cubic centimeter A metric unit of volume equal to the volume of a cube with 1 cm edges. $1 \text{ cm}^3 = 1 \text{ mL}$.

Cubic unit A unit used in measuring volume, such as a cubic centimeter or a cubic foot.

Cubit An ancient unit of length, measured from the point of the elbow to the end of the middle finger. A cubit is about 18 inches.

Curved surface A surface that is rounded rather than flat. Spheres, cylinders, and cones each have one curved surface.

Cylinder A 3-dimensional shape that has two circular bases that are parallel and congruent and are connected by a curved surface. A soup can is shaped like a cylinder.

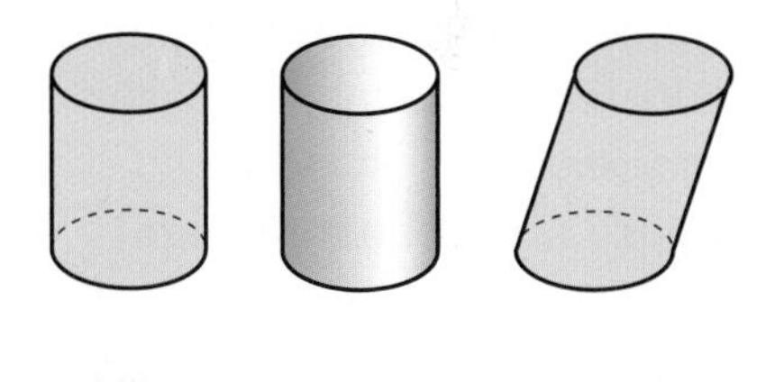

D

Data Information that is gathered by counting, measuring, questioning, or observing.

Decimal A number written in *standard, base-10 notation* that contains a decimal point, such as 2.54. A whole number is a decimal, but it is usually written without a decimal point.

Decimal point A dot used to separate the ones and tenths places in decimal numbers.

Degree (°) (1) A unit of measure for angles based on dividing a circle into 360 equal parts. Latitude and longitude are measured in degrees, and these degrees are based on angle measures. (2) A unit of measure for temperature. In all cases, a small raised circle (°) is used to show degrees.

Denominator The number below the line in a fraction. A fraction may be used to name part of a whole. If the *whole* (the *ONE,* or the *unit*) is divided into equal parts, the denominator represents the number of equal parts into which the whole is divided. In the fraction $\frac{a}{b}$, b is the denominator.

Density A *rate* that compares the *weight* of an object with its *volume.* For example, suppose a ball has a weight of 20 grams and a volume of 10 cubic centimeters. To find its density, divide its weight by its volume:

$20\text{g} / 10\text{cm}^3 = 2\text{g} / \text{cm}^3$, or 2 grams per cubic centimeter.

Diameter (1) A line segment that passes through the center of a circle or sphere and has endpoints on the circle or sphere. (2) The length of this line segment. The diameter of a circle or sphere is twice the length of its *radius.*

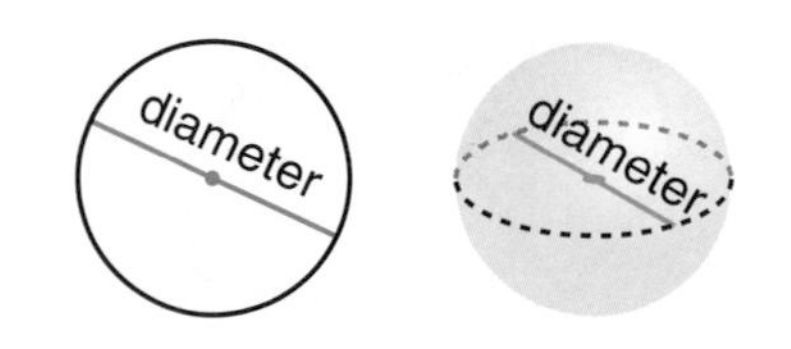

Difference The result of subtracting one number from another. See also *minuend* and *subtrahend.*

Digit One of the number symbols 0, 1, 2, 3, 4, 5, 6, 7, 8, and 9 in the standard, *base-ten* system.

Discount The amount by which the regular price of an item is reduced.

Distributive Property A property that relates multiplication and addition or subtraction. This property gets its name because it "distributes" a factor over terms inside parentheses.

Distributive property of multiplication over addition:

$a * (b + c) = (a * b) + (a * c)$, so
$2 * (5 + 3) = (2 * 5) + (2 * 3)$
$= 10 + 6 = 16.$

Distributive property of multiplication over subtraction:

$a * (b - c) = (a * b) - (a * c)$, so
$2 * (5 - 3) = (2 * 5) - (2 * 3)$
$= 10 - 6 = 4.$

Dividend The number in division that is being divided. For example, in $35 \div 5 = 7$, the dividend is 35.

Divisor In division, the number that divides another number. For example, in $35 \div 5 = 7$, the divisor is 5.

Dodecahedron A polyhedron with 12 faces.

E

Edge A line segment or curve where two surfaces meet.

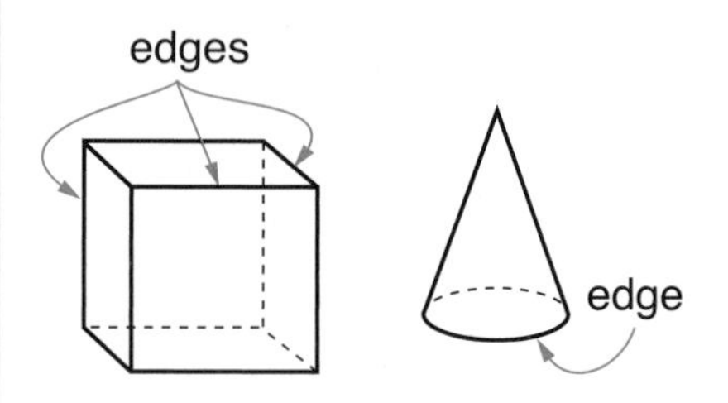

Endpoint A point at the end of a *line segment* or *ray*. A line segment is named using the letter labels of its endpoints. A ray is named using the letter labels of its endpoint and another point on the ray.

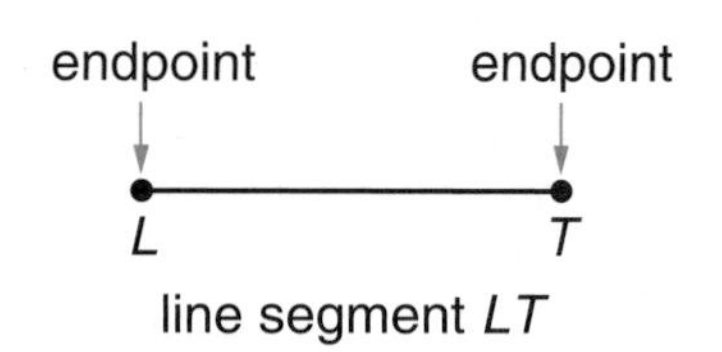

Equally likely outcomes If all of the possible *outcomes* for an experiment or situation have the same *probability,* they are called equally likely outcomes. In the case of equally likely outcomes, the probability of an *event* is equal to this fraction:

$$\frac{\text{number of favorable outcomes}}{\text{number of possible outcomes}}$$

See also *favorable outcome.*

Equation A number sentence that contains an equal sign. For example, $15 = 10 + 5$ is an equation.

Equilateral triangle A triangle with all three sides equal in length. In an equilateral triangle, all three angles have the same measure.

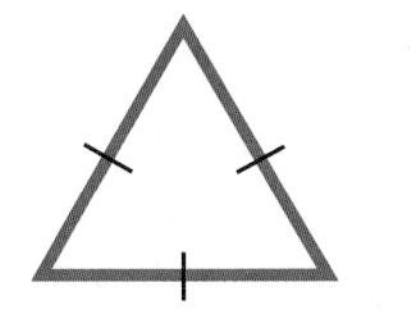

Equivalent Equal in value but possibly in a different form. For example, $\frac{1}{2}$, 0.5, and 50% are all equivalent.

Equivalent fractions Fractions with different denominators that name the same number. For example, $\frac{1}{2}$ and $\frac{4}{8}$ are equivalent fractions.

Estimate An answer that should be close to an exact answer. *To estimate* means to give an answer that should be close to an exact answer.

Even number A counting number that can be divided by 2 with no remainder. The even numbers are 2, 4, 6, 8, and so on. 0, −2, −4, −6, and so on are also usually considered even.

Event Something that happens. The *probability* of an event is the chance that the event will happen. For example, rolling a number smaller than 4 with a die is an event. The possible *outcomes* of rolling a die are 1, 2, 3, 4, 5, and 6. The event "roll a number smaller than 4" will happen if the outcome is 1 or 2 or 3. And the chance that this will happen is $\frac{3}{6}$.

If the probability of an event is 0, the event is *impossible.* If the probability is 1, the event is *certain.*

Exponent A small raised number used in *exponential notation* to tell how many times the *base* is used as a *factor.* For example, in 5^3, the base is 5, the exponent is 3, and $5^3 = 5 * 5 * 5 = 125$. See also *power of a number*.

Exponential notation A way to show repeated multiplication by the same factor. For example, 2^3 is exponential notation for $2 * 2 * 2$. The small raised 3 is the *exponent*. It tells how many times the number 2, called the *base,* is used as a factor.

Expression A group of mathematical symbols that represents a number—or can represent a number if values are assigned to any variables in the expression. An expression may include numbers, variables, operation symbols, and grouping symbols—but *not* relation symbols (=, >, <, and so on). For example: 5, $6 + 3$, $(16 \div 2) - 5$, and $3 * m + 1.5$ are expressions.

Extended multiplication fact A multiplication fact involving multiples of 10, 100, and so on. For example, $6 * 70$, $60 * 7$, and $60 * 70$ are extended multiplication facts.

F

Face A flat surface on a 3-dimensional shape.

Fact family A set of related addition and subtraction facts, or related multiplication and division facts. For example, $5 + 6 = 11$, $6 + 5 = 11$, $11 - 5 = 6$, and $11 - 6 = 5$ are a fact family. $5 * 7 = 35$, $7 * 5 = 35$, $35 \div 5 = 7$, and $35 \div 7 = 5$ are another fact family.

Factor (in a product) Whenever two or more numbers are multiplied to give a product, each of the numbers that is multiplied is called a factor. For example, in $4 * 1.5 = 6$, 6 is the product and 4 and 1.5 are called factors. See also *factor of a counting number* n.

$$4 * 1.5 = 6$$

factors product

NOTE: This definition of *factor* is much less important than the next definition.

Factor of a counting number *n* A counting number whose product with some other counting number equals *n*. For example, 2 and 3 are factors of 6 because $2 * 3 = 6$. But 4 is not a factor of 6 because $4 * 1.5 = 6$ and 1.5 is not a counting number.

$$2 * 3 = 6$$

factors product

NOTE: This definition of factor is much more important than the previous definition.

Fair Free from bias. Each side of a fair die or coin will come up about equally often.

Fair game A game in which every player has the same chance of winning.

False number sentence A number sentence that is not true. For example, $8 = 5 + 5$ is a false number sentence.

Fathom A unit used by people who work with boats and ships to measure depths underwater and lengths of cables. A fathom is now defined as 6 feet.

Favorable outcome An *outcome* that satisfies the conditions of an event of interest. For example, suppose a 6-sided die is rolled and the event of interest is rolling an even number. There are 6 possible outcomes: 1, 2, 3, 4, 5, or 6. There are 3 favorable outcomes: 2, 4, or 6. See also *equally likely outcomes.*

Formula A general rule for finding the value of something. A formula is often written using letters, called *variables,* which stand for the quantities involved. For example, the formula for the area of a rectangle may be written as $A = l * w$, where A represents the area of the rectangle, l represents its length, and w represents its width.

Fraction (primary definition) A number in the form $\frac{a}{b}$ where a and b are whole numbers and b is not 0. A fraction may be used to name part of a whole, or to compare two quantities. A fraction may also be used to represent division. For example, $\frac{2}{3}$ can be thought of as 2 divided by 3. See also *numerator* and *denominator.*

Fraction (other definitions) (1) A fraction that satisfies the definition above, but includes a unit in both the numerator and denominator. This definition of fraction includes any rate that is written as a fraction, such as $\frac{50 \text{ miles}}{1 \text{ gallon}}$ and $\frac{40 \text{ pages}}{10 \text{ minutes}}$

(2) Any number written using a fraction bar, where the fraction bar is used to indicate division. For example, $\frac{2.3}{6.5}$, $\frac{1\frac{4}{5}}{12}$, and $\frac{\frac{3}{4}}{\frac{5}{8}}$.

Frieze pattern A geometric design in a long strip in which an element is repeated over and over again. The element may be rotated, translated, and reflected.

G

Geometric solid A 3-dimensional shape, such as a prism, pyramid, cylinder, cone, or sphere. Despite its name, a geometric solid is hollow; it does not contain the points in its interior.

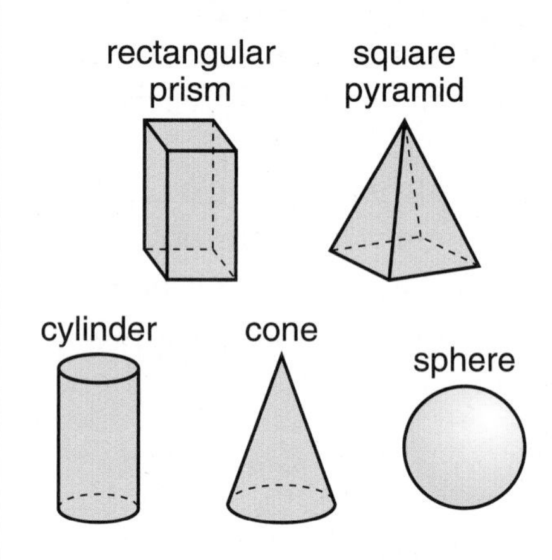

Geometry Template An *Everyday Mathematics* tool that includes a millimeter ruler, a ruler with sixteenth-inch intervals, half-circle and full-circle protractors, a percent circle, pattern-block shapes, and other geometric figures. The template can also be used as a compass.

Grouping symbols Symbols such as parentheses ( ), brackets [ ], and braces { } that tell the order in which operations in an expression are to be done. For example, in the expression (3 + 4) * 5, the operation in the parentheses should be done first. The expression then becomes 7 * 5 = 35.

H

Height of a parallelogram The length of the shortest line segment between the base of a parallelogram and the line containing the opposite side. That shortest segment is perpendicular to the base and is also called the *height.* See also *base of a polygon.*

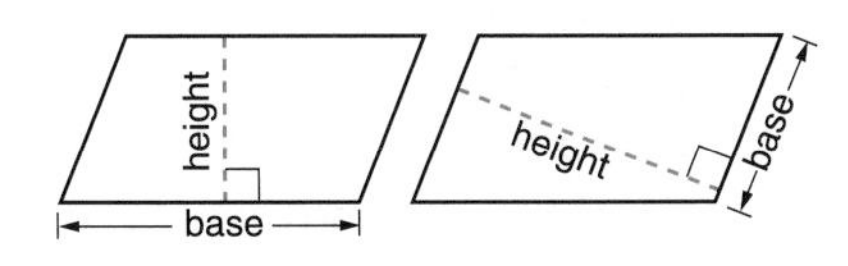

Height of a prism or a cylinder The length of the shortest line segment between the base of a prism or a cylinder and the plane containing the opposite base. That shortest segment is perpendicular to the base and is also called the *height.* See also *base of a prism or a cylinder.*

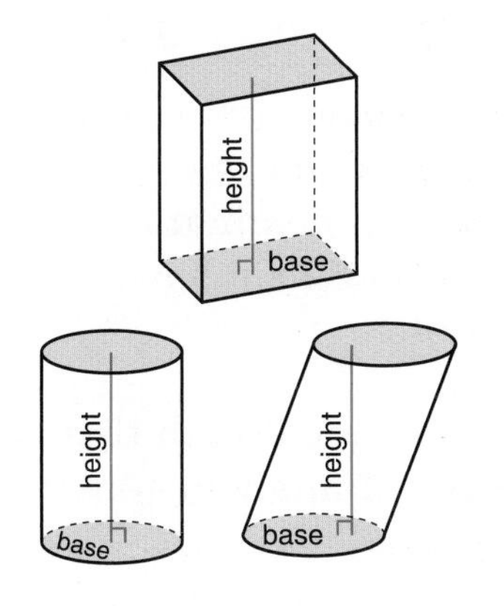

Height of a pyramid or a cone The length of the shortest line segment between the vertex of a pyramid or a cone and the plane containing its base. That shortest segment is perpendicular to the plane containing the base and is also called the *height*. See also *base of a pyramid or a cone.*

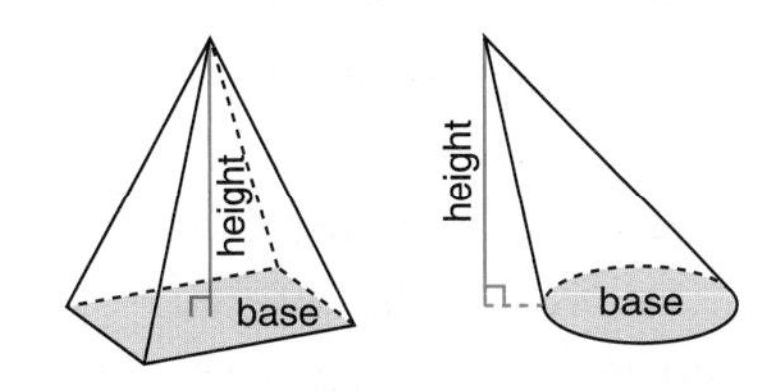

Height of a triangle The length of the shortest line segment between the line containing a base of a triangle and the vertex opposite that base. That shortest segment is perpendicular to the line containing the base and is also called the *height.* See also *base of a polygon.*

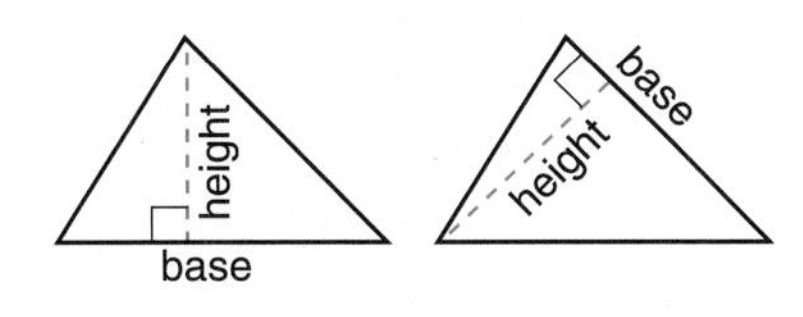

Hemisphere Half of Earth's surface. Also, half of a sphere.

Hexagon A polygon with six sides.

Hexagram A 6-pointed star formed by extending the sides of a regular hexagon.

Horizon Where the earth and sky appear to meet, if nothing is in the way. When looking out to sea, the horizon looks like a line.

Horizontal In a left-right orientation; parallel to the horizon.

Icosahedron A polyhedron with 20 faces.

Image The reflection of an object that you see when you look in a mirror. Also, a figure that is produced by a *transformation* (a *reflection, translation,* or *rotation,* for example) of another figure. See also *preimage.*

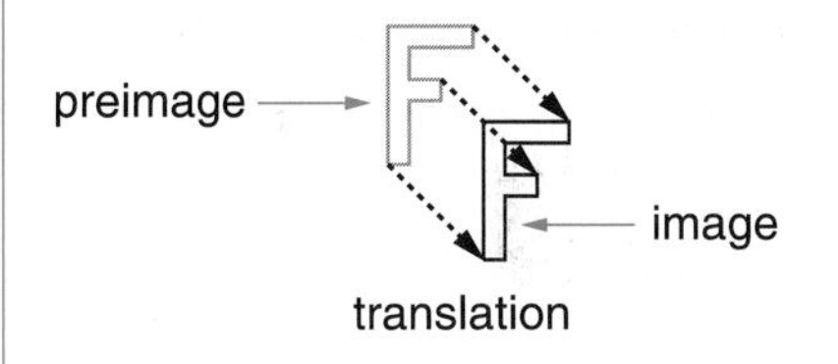

Improper fraction A fraction whose numerator is greater than or equal to its denominator. For example, $\frac{4}{3}$, $\frac{5}{2}$, $\frac{4}{4}$, and $\frac{24}{12}$ are improper fractions. In *Everyday Mathematics,* improper fractions are sometimes called "top-heavy" fractions.

Inequality A number sentence with $>$, $<$, $\geq$, $\leq$, or $\neq$. For example, the sentence $8 < 15$ is an inequality.

Inscribed polygon A polygon whose vertices are all on the same circle.

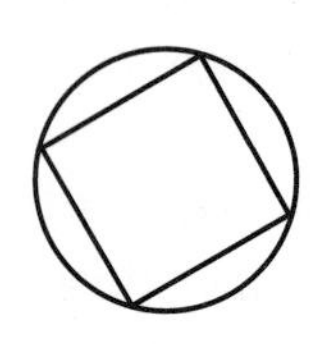

inscribed square

Integer A number in the set {..., −4, −3, −2, −1, 0, 1, 2, 3, 4, ...}; a *whole number* or the *opposite* of a whole number, where 0 is its own opposite.

Interior The inside of a closed 2-dimensional or 3-dimensional figure. The interior is usually not considered to be part of the figure.

Intersect To meet or cross.

Intersecting Meeting or crossing one another. For example, lines, segments, rays, and planes can intersect.

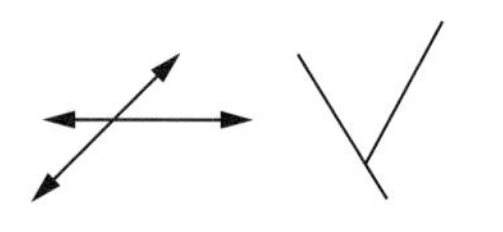

intersecting lines and segments

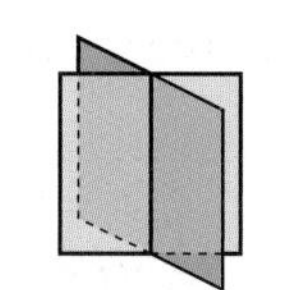

intersecting planes

Interval (1) The set of all numbers between two numbers, *a* and *b*, which may include *a* or *b* or both. (2) A part of a line, including all points between two specific points.

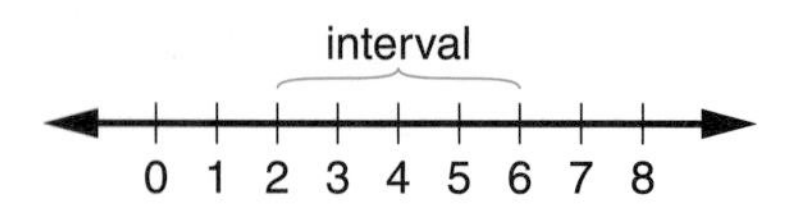

Interval estimate An estimate that places an unknown quantity in a range. For example, an interval estimate of a person's weight might be "between 100 and 110 pounds."

K

Kite A quadrilateral with two pairs of adjacent equal sides. The four sides cannot all have the same length, so a rhombus is not a kite.

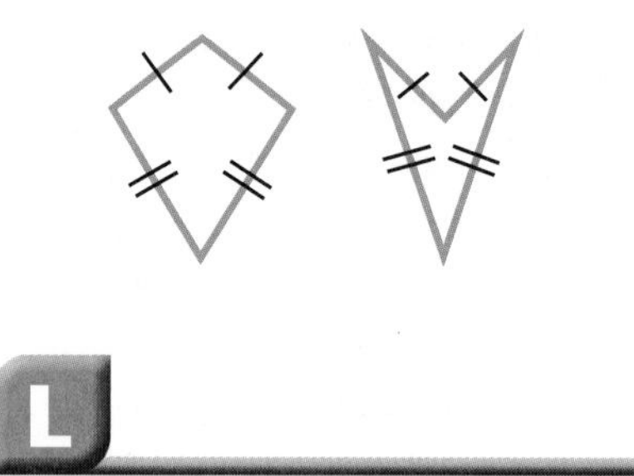

L

Landmark A notable feature of a data set. Landmarks include the *median, mode, maximum, minimum,* and *range*. The *mean* can also be thought of as a landmark.

Latitude A measure, in degrees, that tells how far north or south of the equator a place is.

Lattice method A very old way to multiply multidigit numbers.

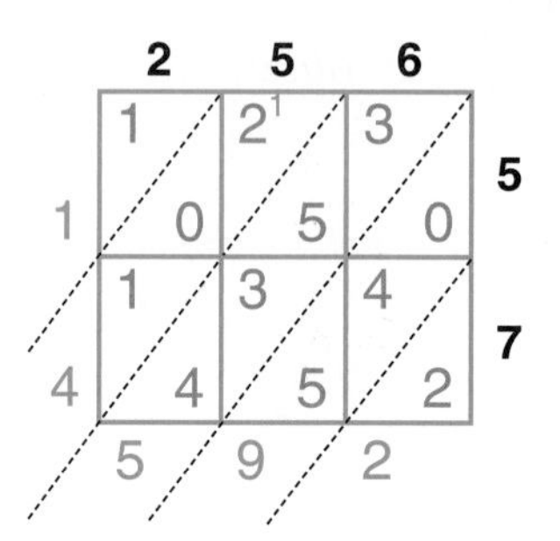

256 * 57 = 14,592

Leading-digit estimation A way to estimate in which the left-most, nonzero digit in a number is not changed, but all other digits are replaced by zeros. For example, to estimate 432 + 76, use the leading-digit estimates 400 and 70: 400 + 70 = 470.

Least common multiple (LCM) The smallest number that is a multiple of two or more numbers. For example, while some common multiples of 6 and 8 are 24, 48, and 72, the least common multiple of 6 and 8 is 24.

Left-to-right subtraction A subtraction method in which you start at the left and subtract column by column. For example, to subtract 932 − 356:

	932
Subtract the 100s.	−300
	632
Subtract the 10s	− 50
	582
Subtract the 1s.	− 6
	576

932 − 356 = 576

Like In some situations, like means *the same*. $\frac{2}{5}$ and $\frac{3}{5}$ have like denominators. 23 cm and 52 cm have like units.

Line A straight path that extends infinitely in opposite directions.

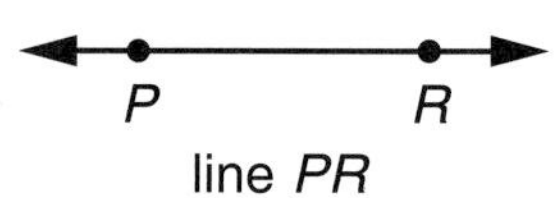

line *PR*

Line graph See *broken-line graph.*

Line of reflection (mirror line) A line halfway between a figure (preimage) and its reflected image. In a *reflection,* a figure is "flipped over" the line of reflection.

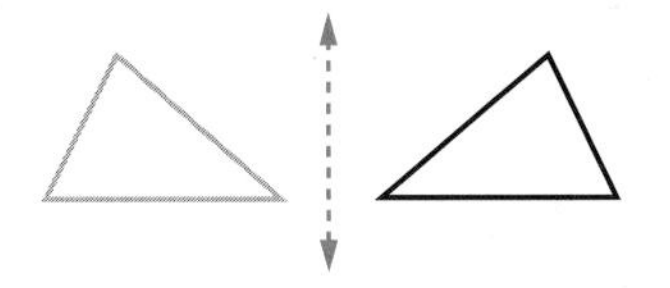

line of reflection

Line of symmetry A line drawn through a figure so that it is divided into two parts that are mirror images of each other. The two parts look alike but face in opposite directions. See also *line symmetry.*

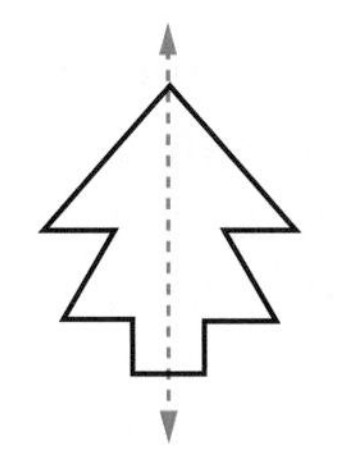

line of symmetry

Line plot A sketch of data in which check marks, Xs, or other marks above a labeled line show the frequency of each value.

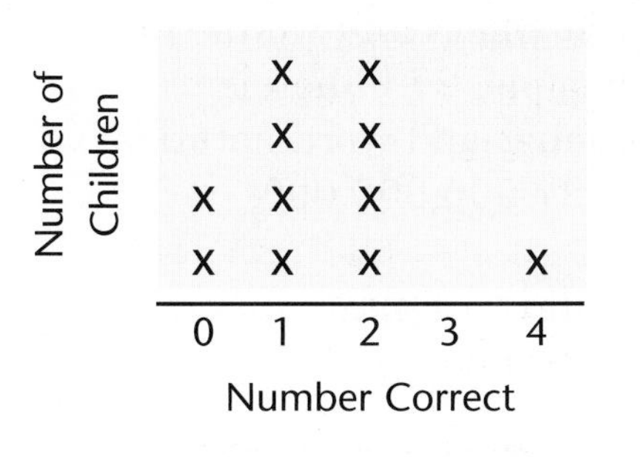

Line segment A straight path joining two points. The two points are called *endpoints* of the segment.

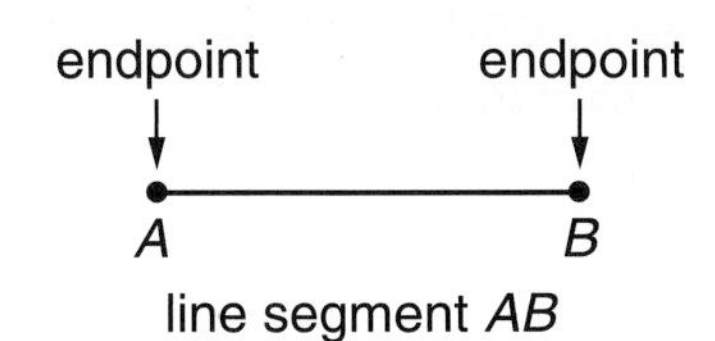

Line symmetry A figure has line symmetry if a line can be drawn through it so that it is divided into two parts that are mirror images of each other. The two parts look alike but face in opposite directions. See also *line of symmetry.*

Lines of latitude Lines that run east-west on a map or globe and locate a place with reference to the equator, which is also a line of latitude. On a globe, lines of latitude are circles, and are called *parallels* because the planes containing these circles are parallel to the plane containing the equator.

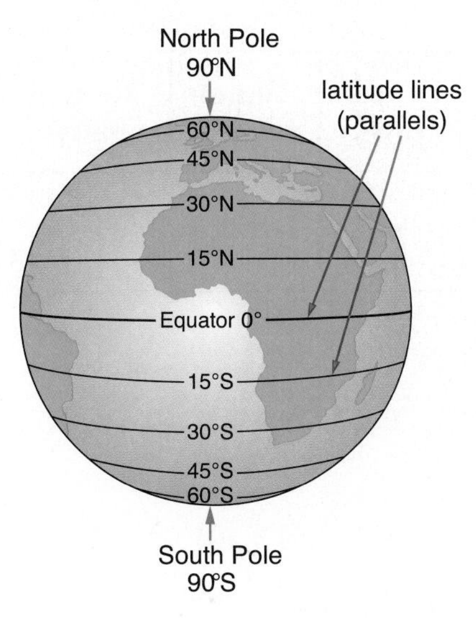

Lines of longitude Lines that run north-south on a map or globe and locate a place with reference to the *prime meridian,* which is also a line of longitude. On a globe, lines of longitude are *semicircles* that meet at the North and South Poles. They are also called *meridians.*

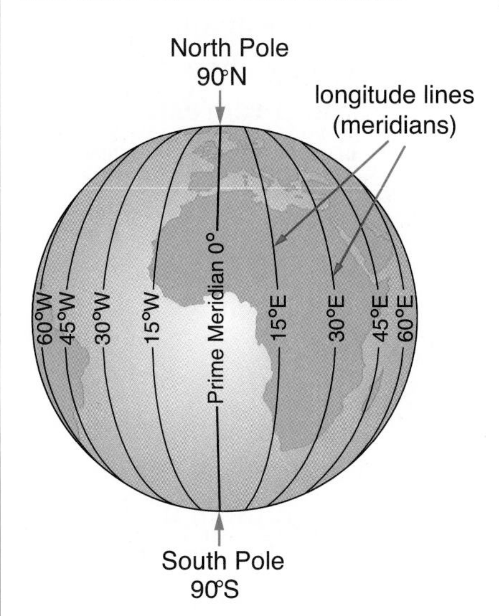

Longitude A measure, in degrees, that tells how far east or west of the *prime meridian* a place is.

Lowest terms See *simplest form.*

M

Magnitude estimate A rough *estimate.* A magnitude estimate tells whether an answer should be in the tens, hundreds, thousands, ten-thousands, and so on.

Map legend (map key) A diagram that explains the symbols, markings, and colors on a map.

Map scale A tool that helps you estimate real distances between places shown on a map. It relates distances on the map to distances in the real world. For example, a map scale may show that 1 inch on a map represents 100 miles in the real world. See also *scale.*

Maximum The largest amount; the greatest number in a set of data.

Mean The sum of a set of numbers divided by the number of numbers in the set. The mean is often referred to simply as the *average.*

Median The middle value in a set of data when the data are listed in order from smallest to largest, or from largest to smallest. If there are an even number of data points, the median is the *mean* of the two middle values.

Metric system of measurement A measurement system based on the *base-ten* numeration system. It is used in most countries around the world.

Minimum The smallest amount; the smallest number in a set of data.

Minuend In subtraction, the number from which another number is subtracted. For example, in $19 - 5 = 14$, the minuend is 19. See also *subtrahend.*

Mixed number A number that is written using both a whole number and a fraction. For example, $2\frac{1}{4}$ is a mixed number equal to $2 + \frac{1}{4}$.

Mode The value or values that occur most often in a set of data.

Multiple of a number *n* A product of *n* and a counting number. For example, the multiples of 7 are 7, 14, 21, 28, and so on.

Multiplication diagram A diagram used for problems in which there are several equal groups. The diagram has three parts: a number of groups, a number in each group, and a total number. Also called *multiplication / division diagram.*

rows	chairs per row	total chairs
15	25	?

N

Name-collection box A diagram that is used for writing equivalent names for a number.

25	
37 – 12	20 + 5
~~||||~~ ~~||||~~ ~~||||~~ ~~||||~~ ~~||||~~	5^2
twenty-five	*veinticinco*

Negative number A number that is less than zero; a number to the left of zero on a horizontal number line or below zero on a vertical number line. The symbol – may be used to write a negative number. For example, "negative 5" is usually written as –5.

***n*-gon** A polygon with *n* sides. For example, a 5-gon is a pentagon, and an 8-gon is an octagon.

Nonagon A polygon with nine sides.

Nonconvex polygon See *concave polygon.*

Number-and-word notation A way of writing a number using a combination of numbers and words. For example, 27 billion is number-and-word notation for 27,000,000,000.

Number model A *number sentence* or part of a number sentence that models or fits a number story or situation. For example, the story *Sally had $5, and then she earned $8* can be modeled as 5 + 8 = 13 (a number sentence), or as 5 + 8 (part of a number sentence).

Number sentence Similar to an English sentence, except that it uses math symbols instead of words. A number sentence *must* contain at least 2 *numbers* (or *variables* that stand for missing numbers) and 1 *relation symbol* (=, >, <, ≠). Most number sentences usually contain one or more *operation symbols* (+, −, * or ×, ÷ or /). And they often contain *grouping symbols* like parentheses. For example, 2 * (3 + 4) − 6 < 30 / 3 and 4 * *m* = 100 are number sentences.

Number story A story with a problem that can be solved using arithmetic.

Numeral A word, symbol, or figure that represents a number. For example, six, VI, and 6 are numerals that represent the same number.

Numerator The number above the line in a fraction. A fraction may be used to name part of a whole. If the *whole* (the *ONE*, or the *unit*) is divided into equal parts, the numerator represents the number of equal parts being considered. In the fraction $\frac{a}{b}$, *a* is the numerator.

O

Octagon A polygon with eight sides.

Octahedron A polyhedron with 8 faces.

Odd number A counting number that cannot be evenly divided by 2. When an odd number is divided by 2, there is a remainder of 1. The odd numbers are 1, 3, 5, and so on.

ONE See *whole* and *unit.*

Open sentence A *number sentence* which has *variables* in place of one or more missing numbers. An open sentence is usually neither true nor false. For example, 5 + *x* = 13 is an open sentence. The sentence is true if 8 is substituted for *x*. The sentence is false if 4 is substituted for *x*.

Operation symbol A symbol used to stand for a mathematical operation. Common operation symbols are +, −, ×, *, ÷, and /.

Opposite of a number A number that is the same distance from 0 on the number line as a given number, but on the opposite side of 0. For example, the opposite of +3 is −3, and the opposite of −5 is +5.

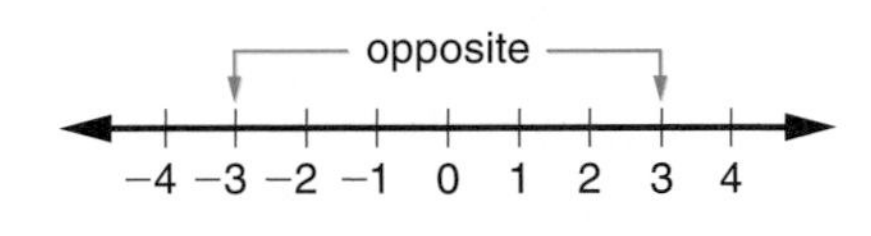

Order of operations Rules that tell in what order to perform operations in arithmetic and algebra. The order of operations is as follows:

1. Do the operations in parentheses first. (Use rules 2–4 inside the parentheses.)
2. Calculate all the expressions with exponents.
3. Multiply and divide in order from left to right.
4. Add and subtract in order from left to right.

Ordered number pair (ordered pair) Two numbers that are used to locate a point on a *rectangular coordinate grid.* The first number gives the position along the horizontal axis, and the second number gives the position along the vertical axis. The numbers in an ordered pair are called *coordinates.* Ordered pairs are usually written inside parentheses: (5,3). See *rectangular coordinate grid* for an illustration.

Origin (1) The 0 point on a number line. (2) The point (0,0) where the two axes of a coordinate grid meet.

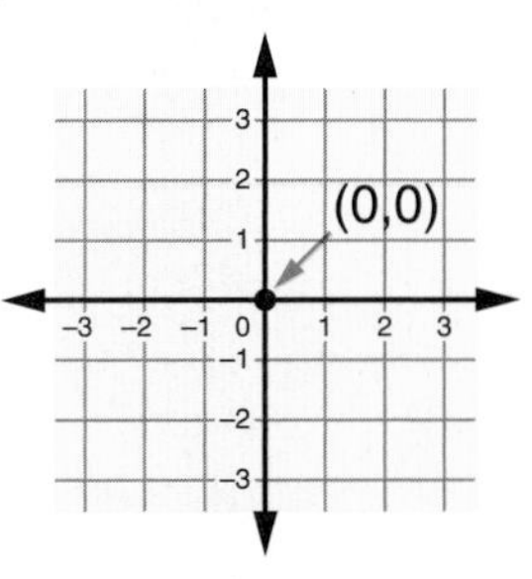

Outcome A possible result of an experiment or situation. For example, heads and tails are the two possible outcomes of tossing a coin. See also *event* and *equally likely outcomes.*

P

Pan balance A tool used to weigh objects or compare weights.

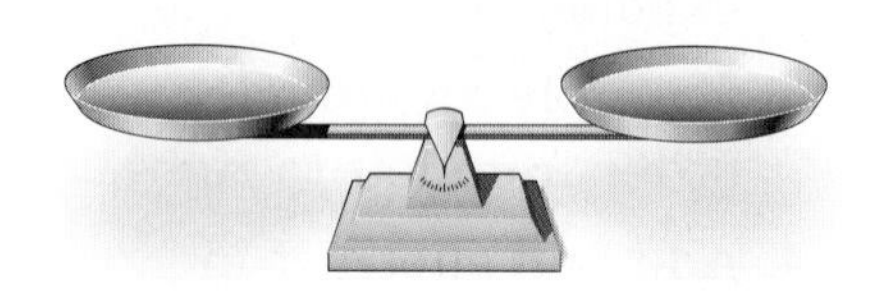

Parallel Lines, line segments, or rays in the same plane are parallel if they never cross or meet, no matter how far they are extended. Two planes are parallel if they never cross or meet. A line and a plane are parallel if they never cross or meet. The symbol || means "is parallel to."

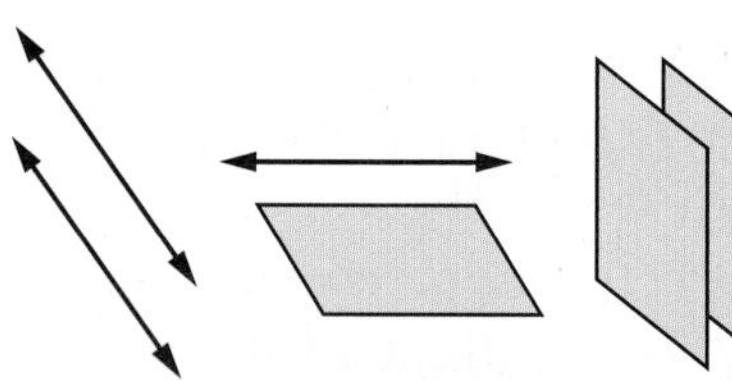

parallel lines   line parallel to a plane   parallel planes

Parallelogram A quadrilateral with two pairs of parallel sides. Opposite sides of a parallelogram are congruent. Opposite angles in a parallelogram have the same measure.

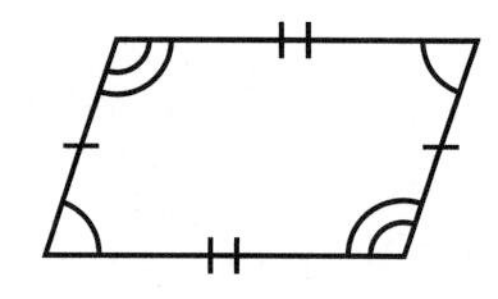

Parentheses Grouping symbols, ( ), used to tell which parts of an expression should be calculated first.

Partial-differences method A way to subtract in which differences are computed for each place (ones, tens, hundreds, and so on) separately. The partial differences are then combined to give the final answer.

		9 3 2
		−3 5 6
900 − 300	→	6 0 0
30 − 50	→	− 2 0
2 − 6	→	− 4
600 − 20 − 4	→	5 7 6

932 − 356 = 576

Partial-products method A way to multiply in which the value of each digit in one factor is multiplied by the value of each digit in the other factor. The final product is the sum of these partial products.

		6 7
		× 5 3
50 × 60	→	3 0 0 0
50 × 7	→	3 5 0
3 × 60	→	1 8 0
3 × 7	→	2 1
Add.		3 5 5 1

67 * 53 = 3,551

Partial-quotients method A way to divide in which the dividend is divided in a series of steps. The quotients for each step (called partial quotients) are added to give the final answer.

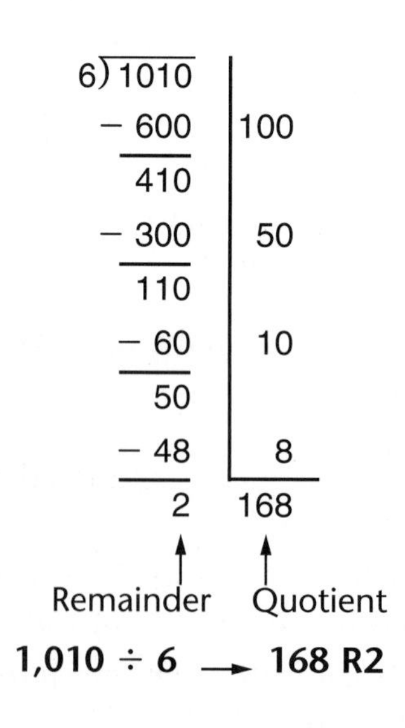

Partial-sums method A way to add in which sums are computed for each place (ones, tens, hundreds, and so on) separately. The partial-sums are then added to give the final answer.

	2	6	8
	+4	8	3
Add the 100s. →	6	0	0
Add the 10s. →	1	4	0
Add the 1s. →		1	1
Add the partial sums. →	7	5	1

268 + 483 = 751

Parts-and-total diagram A diagram used in *Everyday Mathematics* to represent situations in which two or more quantities are combined to form a total quantity.

Total 13	
Part 8	Part ?

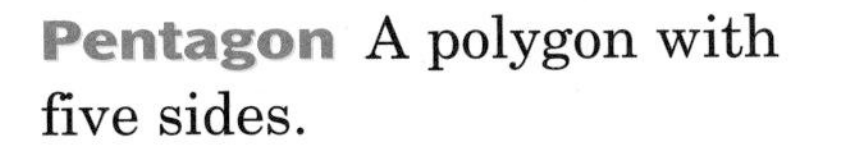

Pentagon A polygon with five sides.

Percent (%) Per hundred or out of a hundred. For example, "48% of the students in the school are boys" means that 48 out of every 100 students in the school are boys; $48\% = \frac{48}{100} = 0.48$.

Percent Circle A tool on the *Geometry Template* that is used to measure and draw figures that involve percents (such as circle graphs).

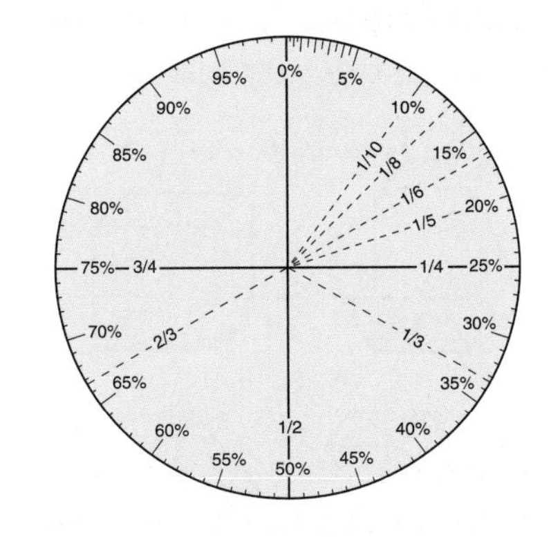

Perimeter The distance around a 2-dimensional shape, along the boundary of the shape. The perimeter of a circle is called its *circumference*.

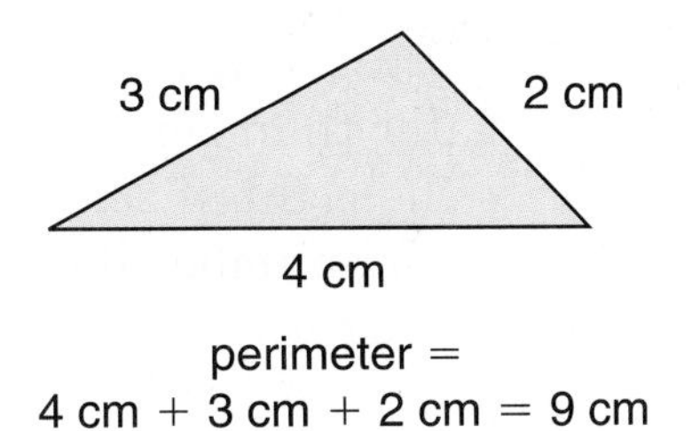

perimeter =
4 cm + 3 cm + 2 cm = 9 cm

Perpendicular Crossing or meeting at *right angles*. Lines, rays, line segments, or planes that cross or meet at right angles are perpendicular. The symbol ⊥ means "is perpendicular to."

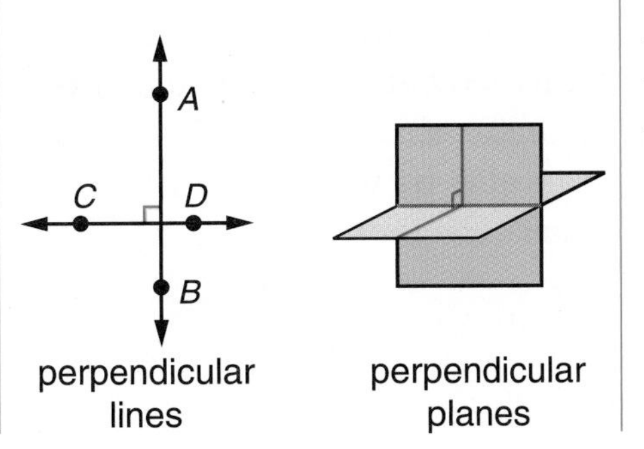

perpendicular lines

perpendicular planes

Pi (π) The ratio of the *circumference* of a circle to its *diameter*. Pi is the same for every circle and is approximately equal to 3.14. Pi is the sixteenth letter of the Greek alphabet and is written π.

Pictograph A graph constructed with pictures or symbols. The *key* for a pictograph tells what each picture or symbol is worth.

Number of Cars Washed

Friday
Saturday
Sunday
KEY: = 6 cars

Pie graph See *circle graph*.

Place value The value that is given to a digit according to its position in a number. In our *base-ten* system for writing numbers, moving a digit one place to the left makes that digit worth 10 times as much. And moving a digit one place to the right makes that digit worth one-tenth as much. For example, in the number 456, the 4 in the hundreds place is worth 400; but in the number 45.6, the 4 in the tens place is worth 40.

Plane A flat surface that extends forever.

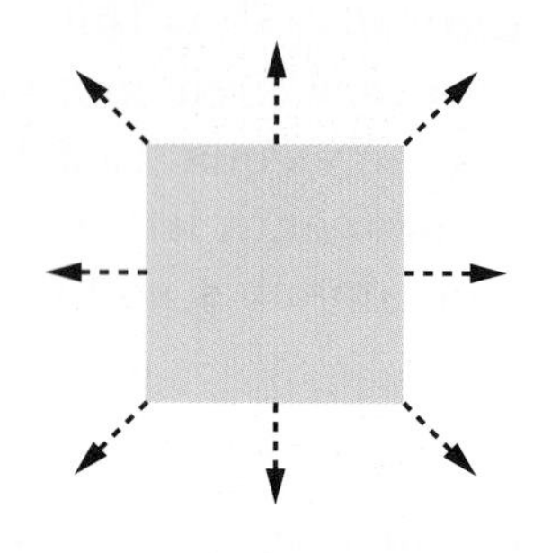

Point An exact location in space. The center of a circle is a point. Lines have infinitely many points on them.

Polygon A 2-dimensional figure that is made up of three or more line segments joined end to end to make one closed path. The line segments of a polygon may not cross.

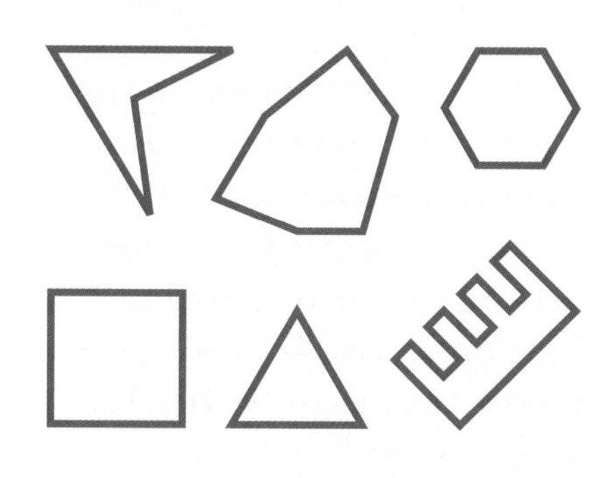

Polyhedron A geometric solid whose surfaces (*faces*) are all flat and formed by polygons. Each face consists of a polygon and the interior of that polygon. A polyhedron does not have any curved surface.

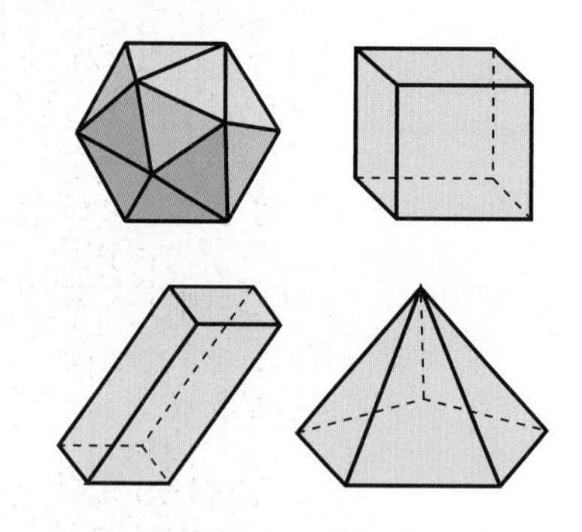

Population In data collection, the collection of people or objects that is the focus of study.

Positive number A number that is greater than zero; a number to the right of zero on a horizontal number line, or above zero on a vertical number line. A positive number may be written using the + symbol, but is usually written without it. For example, $+10 = 10$.

Power of a number The product of factors that are all the same. For example, $5 * 5 * 5$ (or 125) is called "5 to the third power" or "the third power of 5" because 5 is a factor three times. $5 * 5 * 5$ can also be written as 5^3. See also *exponent*.

Power of 10 A whole number that can be written as a *product of 10s*. For example, 100 is equal to $10 * 10$, or 10^2. 100 is called "the second power of 10" or "10 to the second power."

A number that can be written as a *product of* $\frac{1}{10}$*s* is also a power of 10. For example, $10^{-2} = \frac{1}{10^2} = \frac{1}{10 * 10} = \frac{1}{10} * \frac{1}{10}$ is a power of 10.

Precise Exact or accurate. The smaller the unit or fraction of a unit used in measuring, the more precise the measurement is. For example, a measurement to the nearest inch is more precise than a measurement to the nearest foot. A ruler with $\frac{1}{16}$-inch markings is more precise than a ruler with $\frac{1}{4}$-inch markings.

Preimage A geometric figure that is changed (by a *reflection, rotation,* or *translation,* for example) to produce another figure. See also *image.*

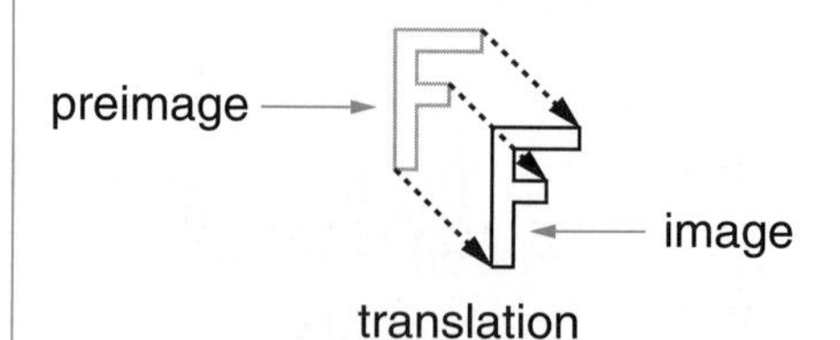

Prime meridian An imaginary semicircle on Earth that connects the North and South Poles and passes through Greenwich, England.

Prime number A counting number that has exactly two different *factors:* itself and 1. For example, 5 is a prime number because its only factors are 5 and 1. The number 1 is not a prime number because that number has only a single factor, the number 1 itself.

Prism A polyhedron with two parallel *faces,* called *bases* that are the same size and shape. All of the other faces connect the bases and are shaped like parallelograms. The *edges* that connect the bases are parallel to each other. Prisms get their names from the shape of their bases.

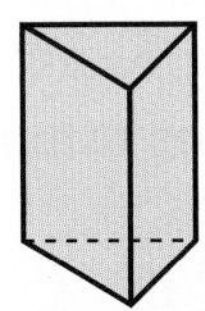
triangular prism

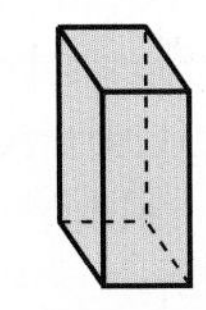
rectangular prism

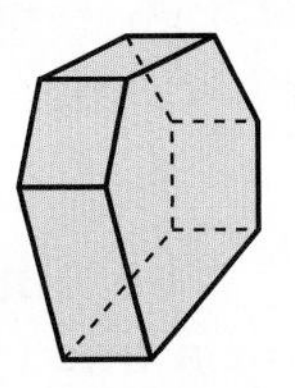
hexagonal prism

Probability A number from 0 through 1 that tells the chance that an event will happen. The closer a probability is to 1, the more likely the event is to happen. See also *equally likely outcomes.*

Product The result of multiplying two numbers, called *factors.* For example, in $4 * 3 = 12$, the product is 12.

Proper fraction A fraction in which the numerator is less than the denominator; a proper fraction names a number that is less than 1. For example, $\frac{3}{4}$, $\frac{2}{5}$, and $\frac{12}{24}$ are proper fractions.

Protractor A tool on the *Geometry Template* that is used to measure and draw angles. The half-circle protractor can be used to measure and draw angles up to 180°; the full-circle protractor, to measure angles up to 360°.

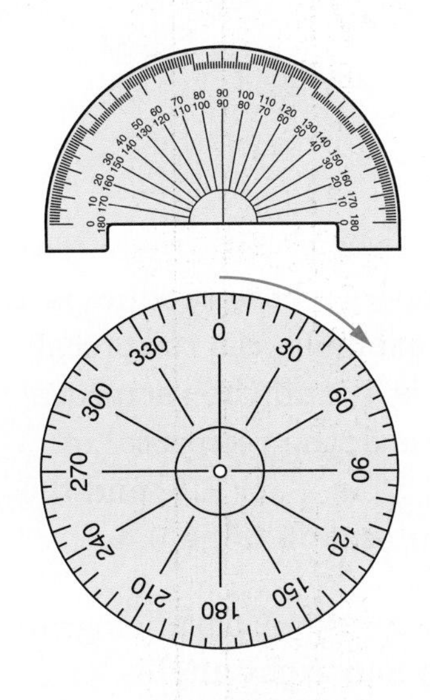

Pyramid A polyhedron in which one face, the *base,* may have any polygon shape. All of the other faces have triangle shapes and come together at a vertex called the *apex.* A pyramid takes its name from the shape of its base.

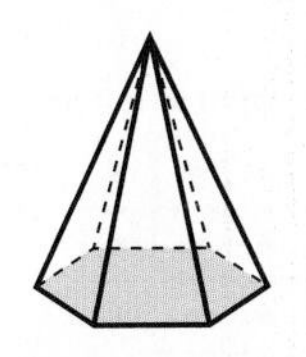
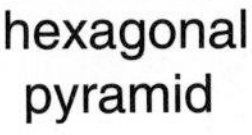
hexagonal pyramid

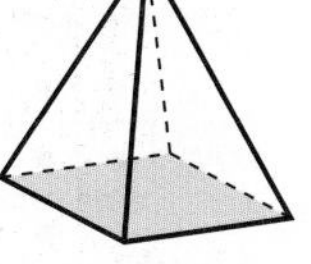
square pyramid

Quadrangle A polygon that has four angles. Same as *quadrilateral.*

Quadrilateral A polygon that has four sides. Same as *quadrangle.*

Quick common denominator The product of the denominators of two or more fractions. For example, the quick common denominator of $\frac{1}{4}$ and $\frac{3}{6}$ is $4 * 6$, or 24.

Quotient The result of dividing one number by another number. For example, in $35 \div 5 = 7$, the quotient is 7.

Radius (plural: **radii**) (1) A line segment from the center of a circle (or sphere) to any point on the circle (or sphere). (2) The length of this line segment.

Random numbers Numbers produced by an experiment, such as rolling a die or spinning a spinner, in which all *outcomes* are *equally likely.* For example, rolling a *fair* die produces random numbers because each of the six possible numbers 1, 2, 3, 4, 5, and 6 has the same chance of coming up.

Random sample A *sample* that gives all members of the *population* the same chance of being selected.

Range The difference between the *maximum* and the *minimum* in a set of data.

Rate A comparison by division of two quantities with *unlike units*. For example, a speed such as 55 miles per hour is a rate that compares distance with time. See also *ratio*.

Rate table A way of displaying rate information. In a rate table, the fractions formed by the two numbers in each column are equivalent fractions.

miles	35	70	105
gallons	1	2	3

Ratio A comparison by division of two quantities with *like units*. Ratios can be expressed with fractions, decimals, percents, or words. Sometimes they are written with a colon between the two numbers that are being compared. For example, if a team wins 3 games out of 5 games played, the ratio of wins to total games can be written as $\frac{3}{5}$, 0.6, 60%, 3 to 5, or 3:5. See also *rate*.

Rational number Any number that can be written or renamed as a *fraction* or the *opposite* of a fraction. Most of the numbers you have used are rational numbers. For example, $\frac{2}{3}$, $-\frac{2}{3}$, $60\% = \frac{60}{100}$, and $-1.25 = -\frac{5}{4}$ are all rational numbers.

Ray A straight path that starts at one point (called the *endpoint*) and continues forever in one direction.

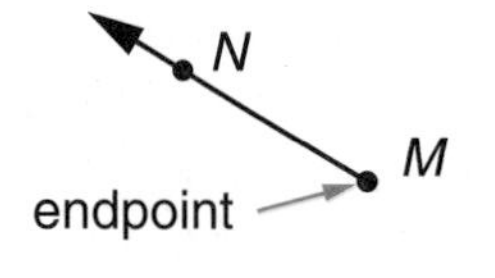

ray *MN*

Reciprocal numbers Two numbers whose product is 1. For example, the reciprocal of 5 is $\frac{1}{5}$, and the reciprocal of $\frac{1}{5}$ is 5; the reciprocal of $0.4 \left(\frac{4}{10}\right)$ is $\frac{10}{4}$, or 2.5, and the reciprocal of 2.5 is 0.4.

Rectangle A parallelogram with four right angles.

Rectangular array An arrangement of objects into rows and columns that form a rectangle. All rows and columns must be filled. Each row has the same number of objects and each column has the same number of objects.

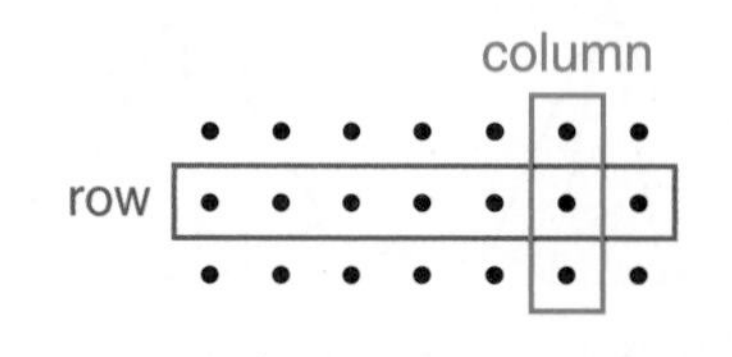

Rectangular coordinate grid A device for locating points in a plane using *ordered number pairs,* or *coordinates.* A rectangular coordinate grid is formed by two number lines that intersect at their zero points and form right angles. Also called a *coordinate grid.*

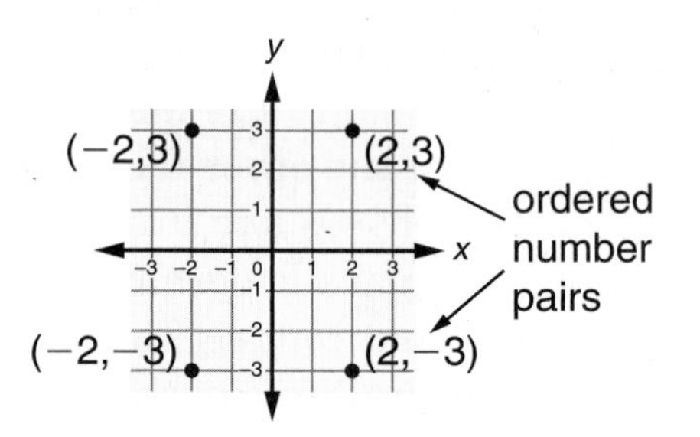

Rectangular prism A *prism* with rectangular *bases.* The four faces that are not bases are either rectangles or other parallelograms.

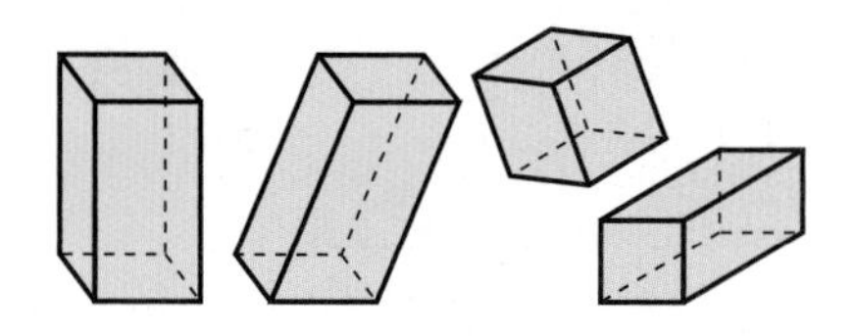

Reflection The “flipping” of a figure over a line (the *line of reflection*) so that its *image* is the mirror image of the original figure (*preimage*).

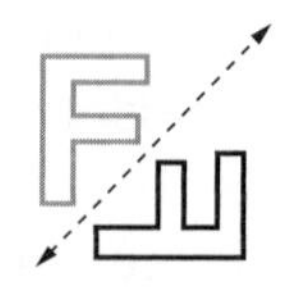

Regular polygon A polygon whose sides are all the same length and whose interior angles are all equal.

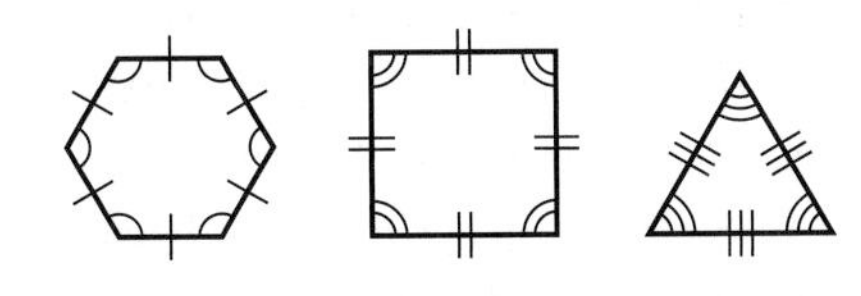

Regular polyhedron A polyhedron whose faces are congruent and formed by *regular polygons,* and whose *vertices* all look the same. There are five regular polyhedrons:

Regular Polyhedrons	
regular tetrahedron	4 faces, each formed by an equilateral triangle
cube	6 faces, each formed by a square
regular octahedron	8 faces, each formed by an equilateral triangle
regular dodecahedron	12 faces, each formed by a regular pentagon
regular icosahedron	20 faces, each formed by an equilateral triangle

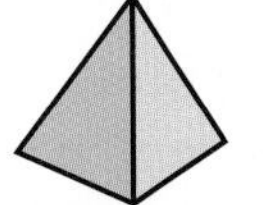

regular tetrahedron

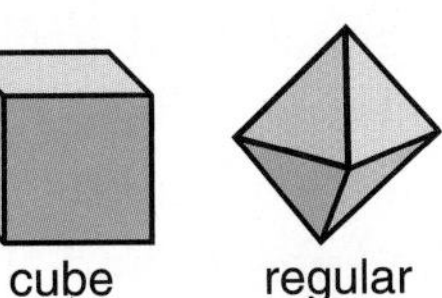

cube

regular octahedron

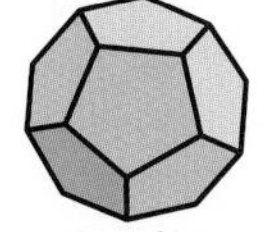

regular dodecahedron

regular icosahedron

Relation symbol A symbol used to express a relationship between two quantities.

symbol	meaning
$=$	"is equal to"
$\neq$	"is not equal to"
$>$	"is greater than"
$<$	"is less than"
$\geq$	"is greater than or equal to"
$\leq$	"is less than or equal to"

Remainder An amount left over when one number is divided by another number. For example, if 7 children share 38 cookies, each child gets 5 cookies and 3 are left over. We may write $38 \div 7 \longrightarrow 5$ R3, where R3 stands for the remainder.

Rhombus A quadrilateral whose sides are all the same length. All rhombuses are parallelograms. Every square is a rhombus, but not all rhombuses are squares.

Right angle A 90° angle.

Right triangle A triangle that has a right angle (90°).

Rotation A movement of a figure around a fixed point, or axis; a "*turn.*"

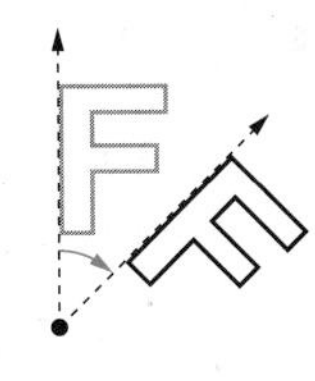

Rotation symmetry A figure has rotation symmetry if it can be rotated less than a full turn around a point or an axis so that the resulting figure (the *image*) exactly matches the original figure (the *preimage*).

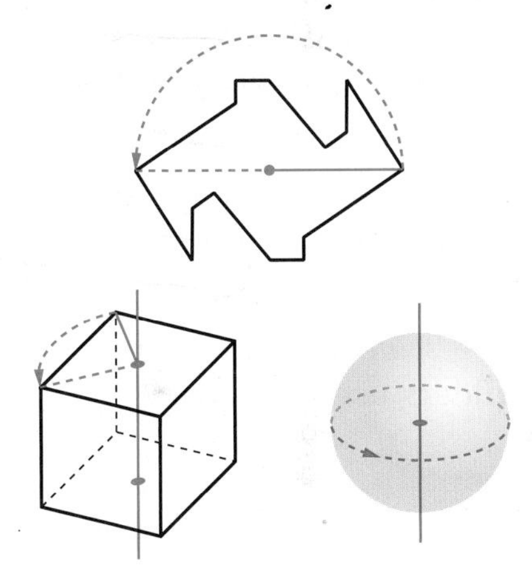

shapes with rotation symmetry

Round To adjust a number to make it easier to work with or to make it better reflect the level of precision of the data. Often numbers are rounded to the nearest multiple of 10, 100, 1,000, and so on. For example, 12,964 rounded to the nearest thousand is 13,000.

S

Sample A part of a group chosen to represent the whole group. See also *population* and *random sample.*

Scale (1) The *ratio* of a distance on a map, globe, or drawing to an actual distance. (2) A system of ordered marks at fixed intervals used in measurement; or any instrument that has such marks. For example, a ruler with scales in inches and centimeters, and a thermometer with scales in °F and °C. See also *map scale* and *scale drawing.*

Scale drawing A drawing of an object or a region in which all parts are drawn to the same *scale*. Architects and builders use scale drawings.

Scientific notation A system for writing numbers in which a number is written as the product of a *power of 10* and a number that is at least 1 and less than 10. Scientific notation allows you to write big and small numbers with only a few symbols. For example, $4 * 10^{12}$ is scientific notation for 4,000,000,000,000.

Semicircle Half of a circle. Sometimes the diameter joining the endpoints of the circle's arc is included. And sometimes the interior of this closed figure is also included.

Side (1) One of the rays or segments that form an angle. (2) One of the line segments of a polygon. (3) One of the faces of a polyhedron.

Similar Figures that have the same shape, but not necessarily the same size.

similar figures

Simpler form An equivalent fraction with a smaller numerator and smaller denominator. A fraction can be put in simpler form by dividing its numerator and denominator by a common factor greater than one. For example, dividing the numerator and denominator of $\frac{18}{24}$ by 2 gives the simpler form $\frac{9}{12}$.

Simplest form A fraction that cannot be renamed in simpler form. Also known as *lowest terms.* A *mixed number* is in simplest form if its fractional part is in simplest form.

Simplify To express a fraction in *simpler form.*

Slanted (oblique) prism or cylinder or cone A prism or cylinder in which the top base is *not* directly above the bottom base. A cone in which the apex is *not* directly above the center of the circular base.

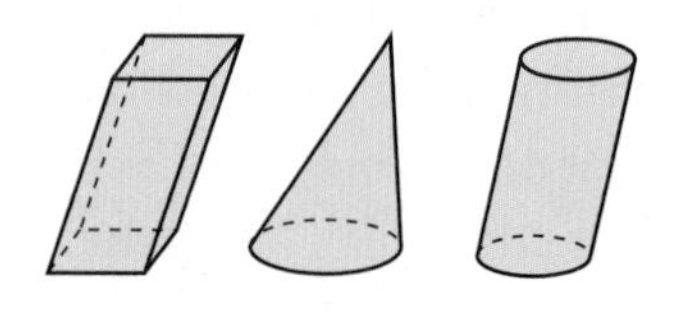

Slide See *translation.*

Solution of an open sentence A value that makes an open sentence *true* when it is substituted for the variable. For example, 7 is a solution of $5 + n = 12$.

Speed A *rate* that compares a distance traveled with the time taken to travel that distance. For example, if a car travels 100 miles in 2 hours, then its speed is $\frac{100 \text{ mi}}{2 \text{ hr}}$, or 50 miles per hour.

Sphere The set of all points in space that are the same distance from a fixed point. The fixed point is the *center* of the sphere, and the distance is the *radius*.

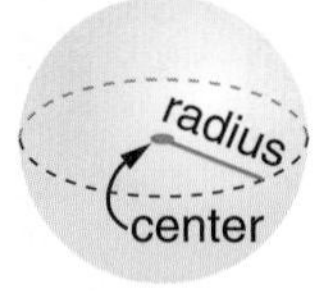

Square A rectangle whose sides are all the same length.

Square number A number that is the product of a counting number with itself. For example, 25 is a square number because 25 = 5 * 5. The square numbers are 1, 4, 9, 16, 25, and so on.

Square of a number The product of a number with itself. For example, 81 is the square of 9 because 81 = 9 * 9. And 0.64 is the square of 0.8 because 0.64 = 0.8 * 0.8.

Square root of a number The square root of a number *n* is a number that, when multiplied by itself, gives *n*. For example, 4 is the square root of 16 because 4 * 4 = 16.

Square unit A unit used in measuring area, such as a square centimeter or a square foot.

Standard notation The most familiar way of representing whole numbers, integers, and decimals. In standard notation, numbers are written using the *base-ten place-value* system. For example, standard notation for three hundred fifty-six is 356. See also *scientific notation* and *number-and-word notation.*

Straightedge A tool used to draw line segments. A straightedge does not need to have ruler marks on it; if you use a ruler as a straightedge, ignore the ruler marks.

Substitute To replace one thing with another. In a formula, to replace variables with numerical values.

Subtrahend In subtraction, the number being subtracted. For example, in 19 − 5 = 14, the subtrahend is 5. See also *minuend.*

Sum The result of adding two or more numbers. For example, in 5 + 3 = 8, the sum is 8. See also *addend.*

Surface (1) The boundary of a 3-dimensional object. The part of an object that is next to the air. Common surfaces include the top of a body of water, the outermost part of a ball, and the topmost layer of ground that covers Earth. (2) Any 2-dimensional layer, such as a *plane* or the faces of a *polyhedron.*

Survey A study that collects data.

Symmetric (1) Having two parts that are mirror images of each other. (2) Looking the same when turned by some amount less than 360°. See also *line symmetry* and *rotation symmetry.*

T

Tally chart A table that uses marks, called tallies, to show how many times each value appears in a set of data.

Number of Pull-Ups	Number of Children
0	𝍸 /
1	𝍸
2	////
3	//

Tetrahedron A polyhedron with 4 faces. A tetrahedron is a triangular pyramid.

Theorem A mathematical statement that can be proved to be true.

3-dimensional (3-D) Having length, width, and thickness. Solid objects take up volume and are 3-dimensional. A figure whose points are not all in a single plane is 3-dimensional.

Time graph A graph that is constructed from a story that takes place over time. A time graph shows what has happened during a period of time.

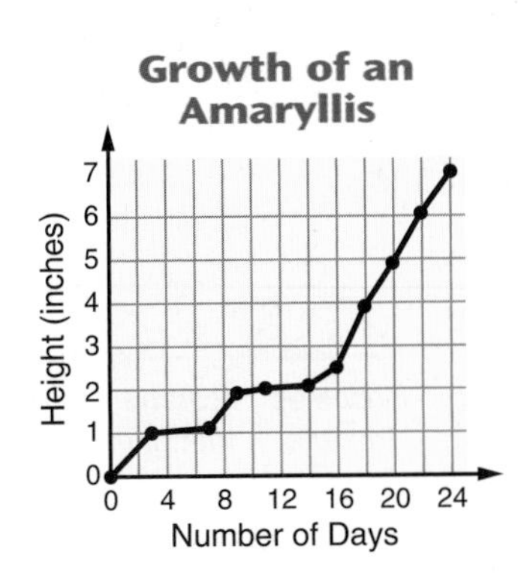

Trade-first subtraction method A subtraction method in which all trades are done before any subtractions are carried out.

Transformation Something done to a geometric figure that produces a new figure. The most common transformations are *translations* (slides), *reflections* (flips), and *rotations* (turns).

slide flip turn

Translation A movement of a figure along a straight line; a "slide." In a translation, each point of the figure slides the same distance in the same direction.

Trapezoid A quadrilateral that has exactly one pair of parallel sides.

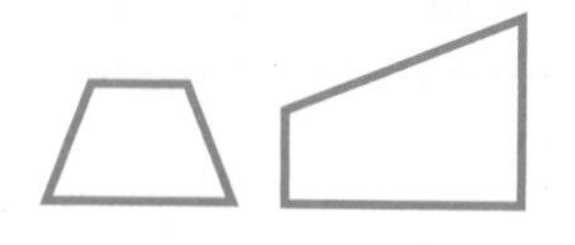

Triangle A polygon with three sides.

Triangular numbers Counting numbers that can be shown by triangular arrangements of dots. The triangular numbers are 1, 3, 6, 10, 15, 21, 28, 36, 45, and so on.

1 3 6 10

Triangular prism A prism whose bases are triangles.

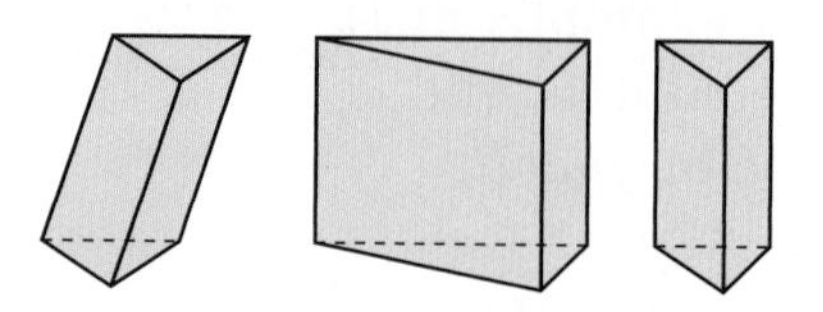

Triangular pyramid A pyramid in which all of the faces are triangles; also called a *tetrahedron.* Any one of the four faces of a triangular pyramid can be called the base. If all of the faces are equilateral triangles, the pyramid is a *regular tetrahedron.*

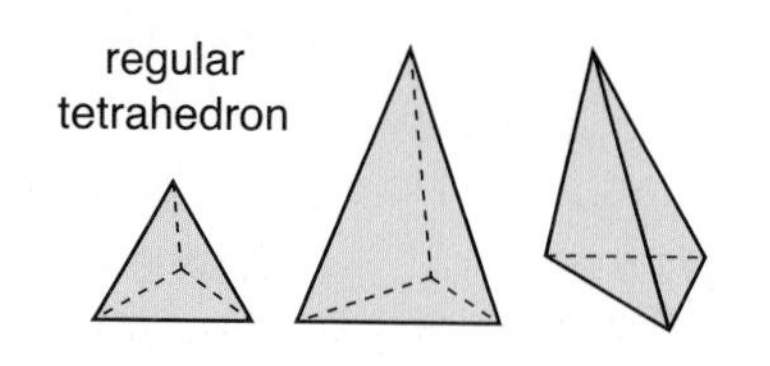

True number sentence A number sentence in which the relation symbol accurately connects the two sides. For example, $15 = 5 + 10$ and $25 > 20 + 3$ are both true number sentences.

Turn See *rotation.*

Turn-around facts A pair of multiplication or addition facts in which the order of the factors (or addends) is reversed. For example, $3 * 9 = 27$ and $9 * 3 = 27$ are turn-around multiplication facts. And $4 + 5 = 9$ and $5 + 4 = 9$ are turn-around addition facts. There are no turn-around facts for division or subtraction. See also *commutative property.*

Turn-around rule A rule for solving addition and multiplication problems based on the *commutative property.* For example, if you know that $6 * 8 = 48$, then, by the turn-around rule, you also know that $8 * 6 = 48$.

2-dimensional (2-D) Having length and width but not thickness. A figure whose points are all in one plane is 2-dimensional. Circles and polygons are 2-dimensional. 2-dimensional shapes have area but not volume.

U

Unit A label used to put a number in context. The *ONE.* In measuring length, for example, the inch and the centimeter are units. In a problem about 5 apples, *apple* is the unit. See also *whole.*

Unit fraction A fraction whose numerator is 1. For example, $\frac{1}{2}$, $\frac{1}{3}$, $\frac{1}{8}$, and $\frac{1}{20}$ are unit fractions.

Unit percent
One percent (1%).

Unlike denominators
Denominators that are different, as in $\frac{1}{2}$ and $\frac{1}{3}$.

"Unsquaring" a number
Finding the *square root* of a number.

U.S. customary system of measurement The measuring system most frequently used in the United States.

V

Variable A letter or other symbol that represents a number. In the number sentence $5 + n = 9$, any number may be substituted for *n,* but only 4 makes the sentence true. In the inequality $x + 2 < 10$, any number may be substituted for *x,* but only numbers less than 8 make the sentence true. In the equation $a + 3 = 3 + a$, any number may be substituted for *a,* and every number makes the sentence true.

Venn diagram A picture that uses circles or rings to show relationships between sets.

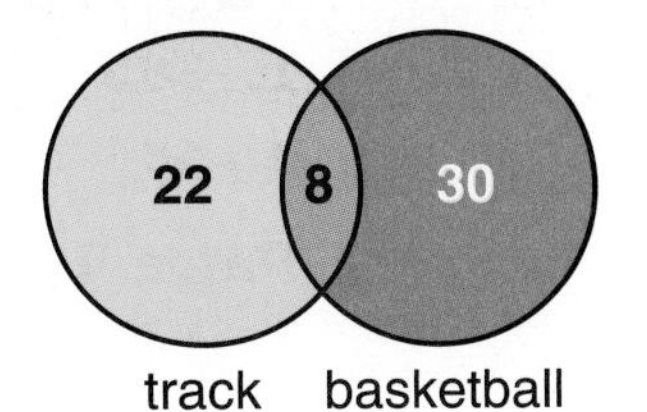

Vertex (plural: **vertices**) The point where the sides of an angle, the sides of a polygon, or the edges of a polyhedron meet.

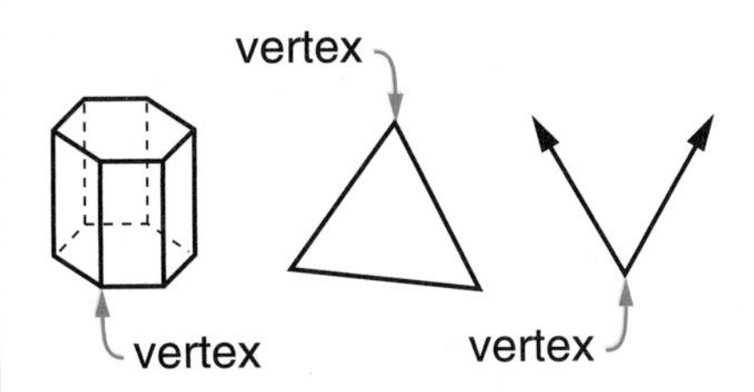

Vertical Upright; perpendicular to the horizon.

Volume A measure of how much space a solid object takes up. Volume is measured in cubic units, such as cubic centimeters or cubic inches. The volume or *capacity* of a container is a measure of how much the container will hold. Capacity is measured in units such as gallons or liters.

If a cubic centimeter were hollow, it would hold exactly 1 milliliter.
1 milliliter (mL) = 1 cm^3.

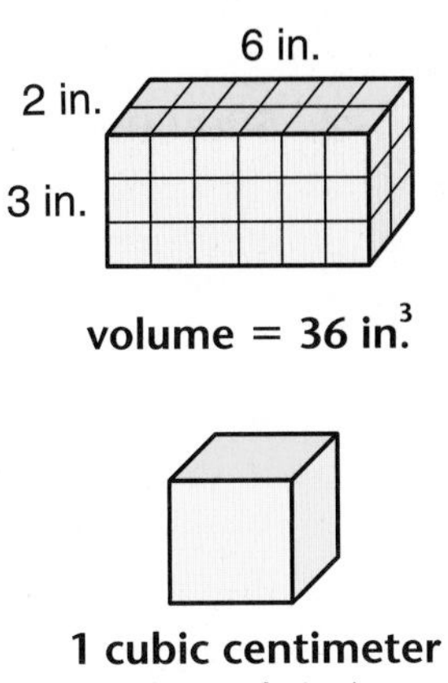

W

"What's My Rule?" problem A type of problem that asks for a rule connecting two sets of numbers. Also, a type of problem that asks for one of the sets of numbers, given a rule and the other set of numbers.

Whole (or ONE or unit) The entire object, collection of objects, or quantity being considered. The ONE, the *unit,* 100%.

Whole numbers The *counting numbers,* together with 0. The set of whole numbers is {0, 1, 2, 3, . . .}

Page 4
1. 5,000 **2.** 500,000 **3.** 50 **4.** 50,000

Page 6
1. false **2.** true **3.** true **4.** false

Page 7
1. 1, 2, 3, 4, 6, 12

2. 1, 3, 5, 15

3. 1, 2, 3, 4, 6, 8, 12, 24

4. 1, 11

Page 9
1. 12 **2.** 20 **3.** 45

Page 11
1. 579 **2.** 112 **3.** 421
4. 2,843 **5.** 135 **6.** 82
7. 99 **8.** 947 **9.** 1,572
10. 123

Page 12
1. 46 **2.** 483 **3.** 277
4. 364 **5.** 3,059 **6.** 482

Page 14
1. 98 **2.** 296 **3.** 215 **4.** 859

Page 15
1. 393 **2.** 318 **3.** 309 **4.** 737

Page 17
1. 800 **2.** 41,000 **3.** 2,100
4. 36,000 **5.** 2,400 **6.** 25,000

Page 18
1. 365 **2.** 2,709 **3.** 1,080
4. 484 **5.** 948

Page 19
1. 147 **2.** 582 **3.** 1,485
4. 2,322 **5.** 1,413

Page 21
1. 40 **2.** 47 **3.** 4,000
4. 480 **5.** 60

Page 23
1. 37 R1 **2.** 18 R2 **3.** 24 R1 **4.** 138 R2

Page 24
1. 33 R5 **2.** 76 R4 **3.** 25

Page 27
1. a. 0.7 **b.** 0.02 **c.** 0.987
d. 0.88 **e.** 0.006

2. a. $\frac{45}{100}$ **b.** $\frac{6}{10}$
c. $\frac{74}{1,000}$ **d.** $\frac{90}{100}$ or $\frac{9}{10}$
e. $\frac{9}{100}$

Page 28
1. 0.7 **2.** 0.08 **3.** 0.62
4. $\frac{154}{100}$ or $1\frac{54}{100}$ **5.** $\frac{7}{100}$ **6.** $\frac{839}{100}$ or $8\frac{39}{100}$
7. $\frac{10,362}{1,000}$ or $10\frac{362}{1,000}$

Page 31
1. a. 20 **b.** 0.20 or $\frac{2}{10}$ **c.** 0.02 or $\frac{2}{100}$
d. 0.02 or $\frac{2}{100}$ **e.** 0.002 or $\frac{2}{1,000}$
2. The value of 8 is 80.

The value of 7 is 7.

The value of 6 is 0.6 or $\frac{6}{10}$.

The value of 5 is 0.05 or $\frac{5}{100}$.

The value of 4 is 0.004 or $\frac{4}{1,000}$.

3. .248

Page 33

1. $0.68 > 0.2$ **2.** $5.39 < 5.5$

3. $\frac{1}{2} < 0.51$ **4.** $0.999 < 1.1$

Page 37

1. 5.73 **2.** 1.78 **3.** 4.02 **4.** 8.77

Page 39

1. 4 **2.** 80 **3.** 60

4. 8 **5.** 4

Page 44

1. $\frac{5}{12}$ **2.** $\frac{3}{12}$ **3.** $\frac{4}{12}$

Page 48

1. $2\frac{3}{4}$ **2.** $2\frac{1}{2}$ **3.** $\frac{7}{4}$

4. $\frac{7}{3}$ **5.** $\frac{29}{8}$

Page 50

1. a. $\frac{2}{3}$ **b.** Sample answers: $\frac{4}{6}, \frac{6}{9}, \frac{8}{12}$

2. Sample answers: $\frac{2}{4}, \frac{4}{8}, \frac{56}{112}$

3. Sample answers: $\frac{4}{24}, \frac{2}{12}, \frac{1}{6}$

Page 51

1. a. true **b.** true **c.** false **d.** true

2. a. Sample answers: $\frac{2}{10}, \frac{3}{15}, \frac{4}{20}$

b. Sample answers: $\frac{12}{60}, \frac{20}{100}, \frac{100}{500}$

Page 52

1. A: $\frac{3}{16}$ B: $\frac{7}{8}$ C: $4\frac{1}{4}$

2. a. $\frac{3}{4}$ in. **b.** $\frac{6}{8}$ in. **c.** $\frac{12}{16}$ in.

Page 54

1. $>$ **2.** $>$ **3.** $<$ **4.** $<$

Page 56

1. $\frac{2}{6}$ or $\frac{1}{3}$ **2.** $\frac{11}{12}$

3. $\frac{4}{12}$ or $\frac{1}{3}$ **4.** $\frac{7}{6}$ or $1\frac{1}{6}$

Page 57

1. $\frac{14}{20}$ or $\frac{7}{10}$ **2.** $\frac{1}{8}$

3. $\frac{4}{12}$ or $\frac{1}{3}$ **4.** $\frac{13}{12}$ or $1\frac{1}{12}$

Page 58

1. 4 **2.** $\frac{12}{5}$ or $2\frac{2}{5}$ **3.** $\frac{18}{4}$ or $4\frac{2}{4}$

4. 3 **5.** 4

Page 59

1. 7 **2.** 12

3. 10 **4.** Rita gets $8; Hunter gets $4.

Page 62

1. $\frac{1}{2}$, 0.50, 50%

2. $\frac{3}{4}$, 0.75, 75%

3. $\frac{1}{10}$, 0.10, 10%

4. $\frac{4}{5}$, 0.80, 80%

Page 71

1.

Number of Hits	Number of Players
0	////
1	///
2	////
3	//
4	/

2.

Number of Players					
	X		X		
	X	X	X		
	X	X	X	X	
	X	X	X	X	X
Number of Hits	0	1	2	3	4

Page 72

Number of Points	Number of Games
10–19	/
20–29	~~////~~
30–39	///
40–49	///

Page 73

1. 0 **2.** 4 **3.** 4

4. 2 **5.** 2

Page 74

1. min = 0; max = 4; range = 4; mode = 2, 3, and 4; median = 2.5

2. 14

Page 75

1. 154 miles

2. mean (average distance per day) = 22 miles

Page 76

1. Italy: about 42; Canada: about 26; United States: about 13
2. Italians take about 3 times as many vacation days per year.
3. Canadians take about twice as many vacation days per year.

Page 78

1. 18 children 2. 56 children

3.

Grade	Number of Children
3rd Grade	HHT HHT //
4th Grade	HHT HHT HHT ///
5th Grade	HHT HHT HHT HHT HHT /

Page 79

Cars in a Parking Lot

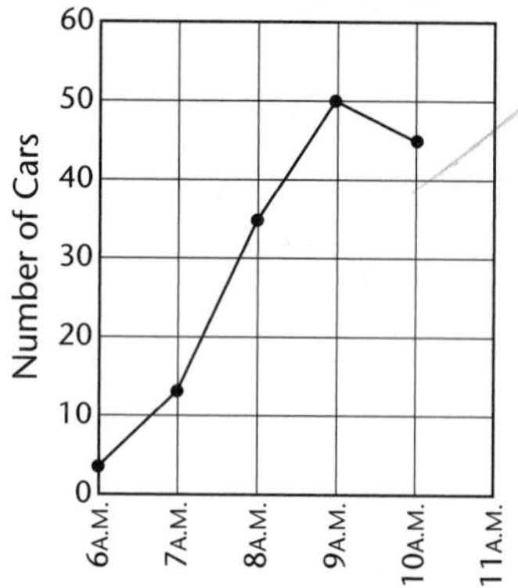

Page 80

It is more likely to snow on Monday. ($\frac{1}{6}$ is about 17%.)

Page 86

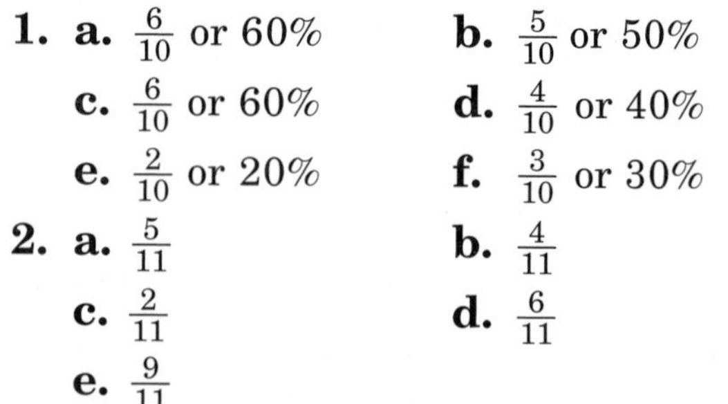

1. a. $\frac{6}{10}$ or 60% b. $\frac{5}{10}$ or 50%
   c. $\frac{6}{10}$ or 60% d. $\frac{4}{10}$ or 40%
   e. $\frac{2}{10}$ or 20% f. $\frac{3}{10}$ or 30%
2. a. $\frac{5}{11}$ b. $\frac{4}{11}$
   c. $\frac{2}{11}$ d. $\frac{6}{11}$
   e. $\frac{9}{11}$

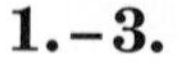

Page 91

1.–3.

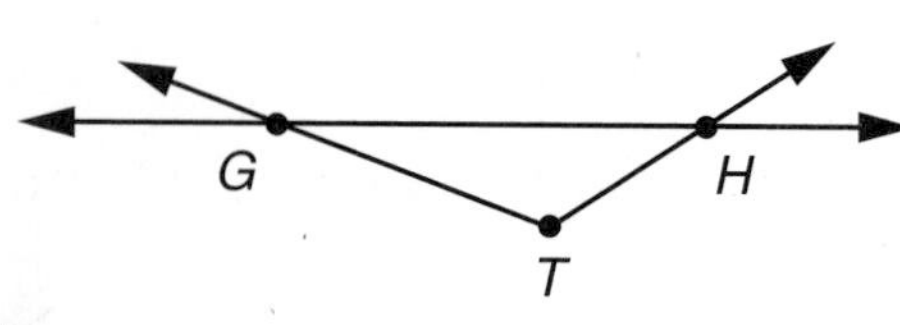

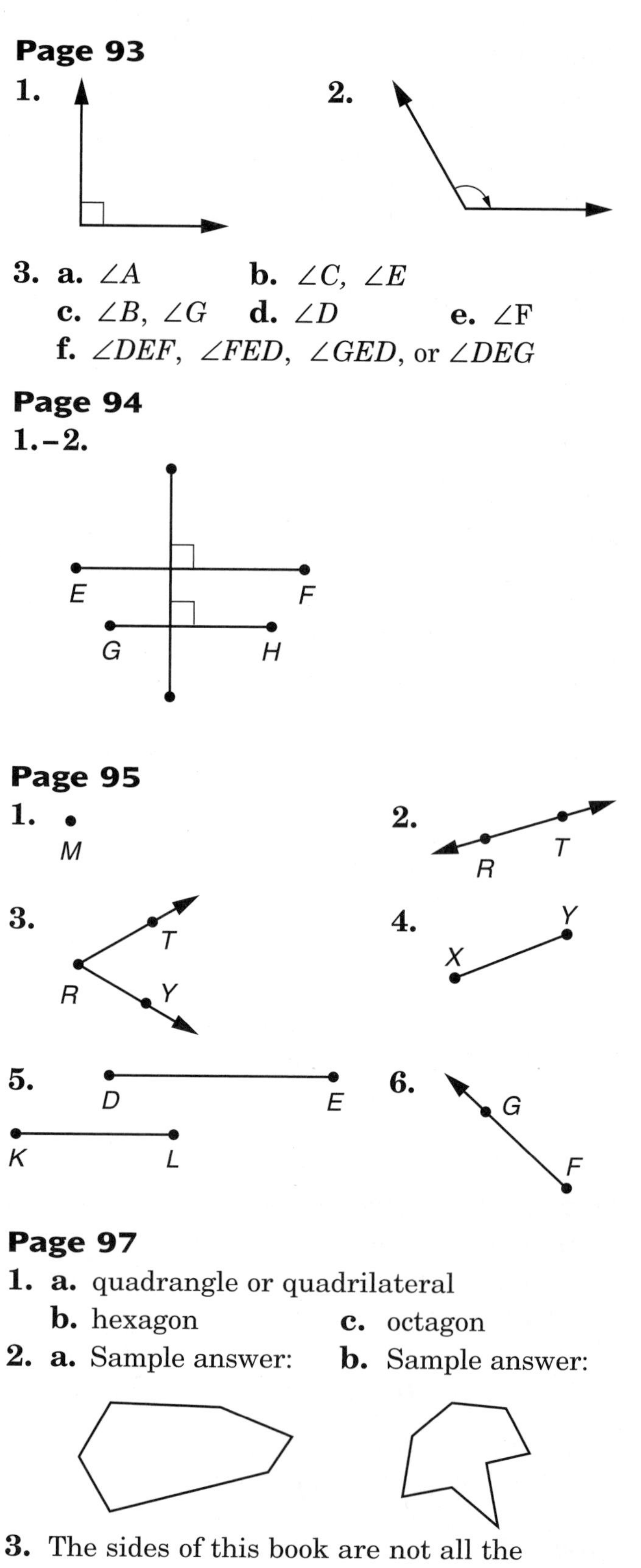

Page 93

1.

2.

3. a. $\angle A$ b. $\angle C$, $\angle E$
   c. $\angle B$, $\angle G$ d. $\angle D$ e. $\angle F$
   f. $\angle DEF$, $\angle FED$, $\angle GED$, or $\angle DEG$

Page 94

1.–2.

Page 95

1. M
2. R T
3. R T Y
4. X Y
5. D E K L
6. G F

Page 97

1. a. quadrangle or quadrilateral
   b. hexagon c. octagon
2. a. Sample answer: b. Sample answer:
3. The sides of this book are not all the same length.

Page 98

1. **a.**

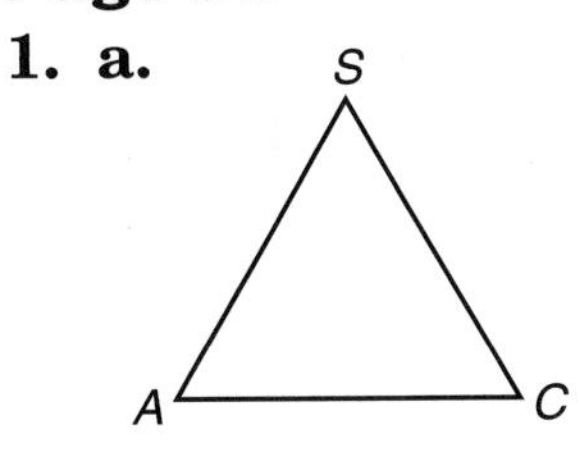

   b. *SCA*, *ACS*, *ASC*, *CSA*, and *CAS*

2.

Page 100

Sample answers:

1. All sides of a square have the same length. The sides of a rectangle may or may not all have the same length.
2. A rhombus is a parallelogram. A kite is not a parallelogram. All sides of a rhombus have the same length. The sides of a kite have two different lengths.
3. A trapezoid has exactly one pair of parallel sides. A parallelogram has two pairs of parallel sides.

Page 101

1. **a.** They each have at least 1 circular face. And each has one curved surface.
   b. A cylinder has 2 circular faces; a cone has only 1. A cylinder has 2 edges; a cone has only 1. A cone has 1 vertex; a cylinder has no vertex.
2. **a.** They each have one curved surface.
   b. A cone has a flat base, 1 vertex, and 1 edge. A sphere has no flat surface, no edge and no vertex.

Page 103

1. 5 **2.** 1 **3.** 6 **4.** 6

5. triangular prism

6. tetrahedron, octahedron, icosahedron

7. 12 **8.** 6

9. Sample answers:

   a. Their faces are equilateral triangles that all have the same size.

   b. The tetrahedrons have 4 faces, 6 edges, and 4 vertices. The octahedrons have 8 faces, 12 edges, and 6 vertices.

Page 105

You could use a, b, c, or d.

Page 107

1.

2. C

Page 109

1.

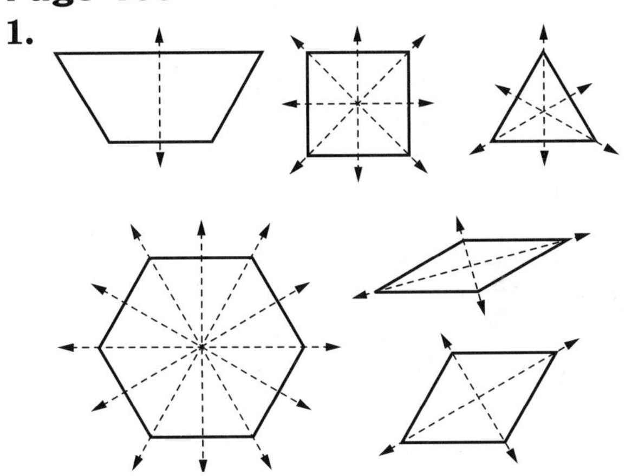

2. infinite; Any line drawn through its center is a line of symmetry.

Page 127

1. millimeter, gram, meter, centimeter

2. $\frac{1}{1,000}$ **3.** 5,000 mm

Page 128

1. 3 cm; 30 mm
2. 2.3 cm; 23 mm
3. 4.6 cm; 46 mm
4. segment 1: $1\frac{1}{4}$ in.; segment 2: 1 in.; segment 3: $1\frac{3}{4}$ in.

Page 131

1. 15 ft **2.** 64 mm **3.** 36 m **4.** 39 in.

Page 132

1. about 21 mm
2. It is slightly more than 3 * 21 mm = 63 mm
3. It is slightly more than 3 * 12 in. = 36 in.

Page 134

1. 6 units^2 **2.** 22.5 in.^2 **3.** 49 m^2

Page 135

1. 165 ft^2 **2.** 135 $in.^2$ **3.** 34.4 cm^2

Page 136

1. 6 $in.^2$ **2.** 27 cm^2 **3.** 4.8 yd^2

Page 140

1. about 180 grams; exactly 170.1 grams
2. 412 ounces

Page 142

1. 70° **2.** 270°

Page 143

1. 25° **2.** 150°
3.
4.

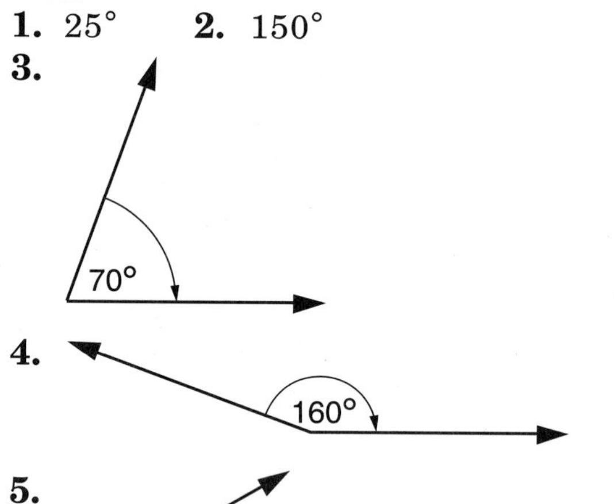

5.

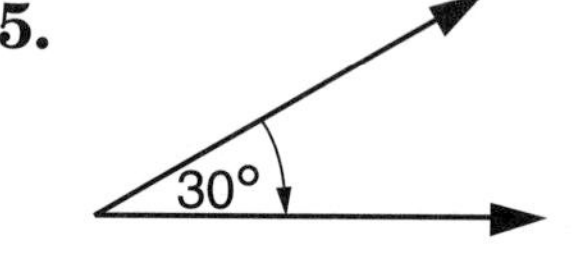

Page 144

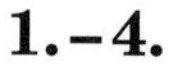

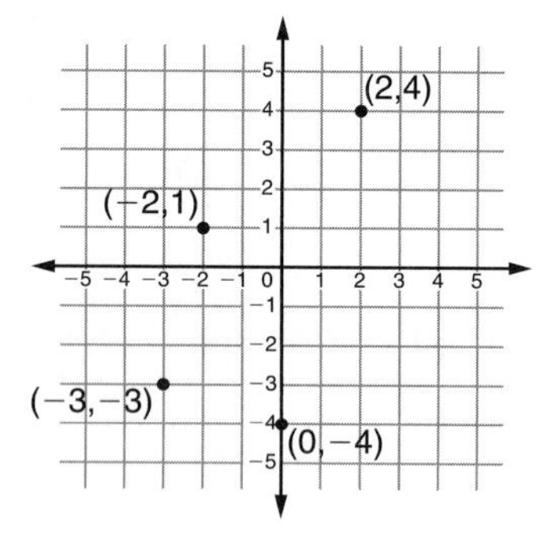

Page 148

1. true **2.** true **3.** false
4. $y = 7$ **5.** $z = 8$ **6.** $w = 6$

Page 149

1. $100 = 55 + 45$ **2.** $\frac{1}{2} = 0.5$
3. $\frac{3}{4} > \frac{1}{4}$ **4.** $3 * 50 < 200$
5. $\frac{1}{2} * 100 = 50$ **6.** $4.3 > 4.15$

Page 150

1. $x = 26$ **2.** $n = 0$
3. 1,000 **4.** $5 - (1 + 4) = 0$
5. $1 = (4 * 5) / (4 * 5)$ **6.** $10 = (3 + 2) * 2$

Page 151

1. $14 - 4 / 2 + 2 = 14$
2. $5 * 7 / (5 + 2) = 5$
3. $6 + 3 * 4 - 20 / 5 = 14$
4. $9 + 9 * 5 = 54$

Page 153

Sample answers for problems 1–3:

1.

Total			
?			
Part 4.5	Part 7	Part 5	Part 6

4.5 + 7 + 5 + 6
Ella ran 22.5 miles.

2.

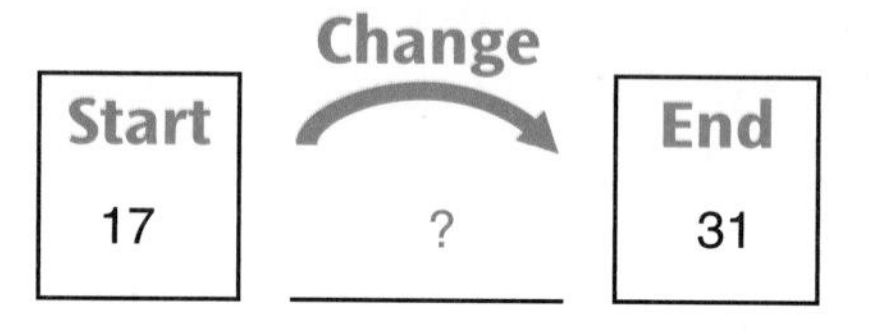

$17 + n = 31$, or $31 - 17$
The Eagles scored 14 points in the second half.

3.

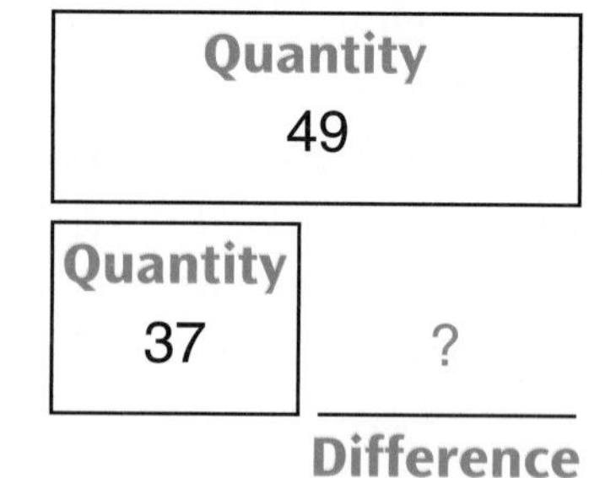

$49 - 37 = ?$, or $37 + ? = 49$
Monica had $12 more than Maurice.

Page 155

1. $x + 5 = 8$ **2.** $3 * n = 15$
3. $A = 4\ cm * 3\ cm = 12\ cm^2$
4. $A = 6\ cm * 2\ cm = 12\ cm^2$

Page 157

1. $7 * (10 + 8) = (7 * 10) + (7 * 8)$
2. $(6 * 13) + (6 * 16) = 6 * (13 + 16)$
3. $4 * (15 - 8) = (4 * 15) - (4 * 8)$
4. $(8 * 2) + (8 * 4) = 8 * (2 + 4)$

Page 159

1. 12: rectangular and even
2. 36: square triangular rectangular and even
3. 11: prime and odd
4. 21: triangular rectangular and odd
5. 16: even
6. 19: odd
7. 24: even
8. 23: odd
9. 1, 4, 9, 16, 25, 36, 49, 64, 81

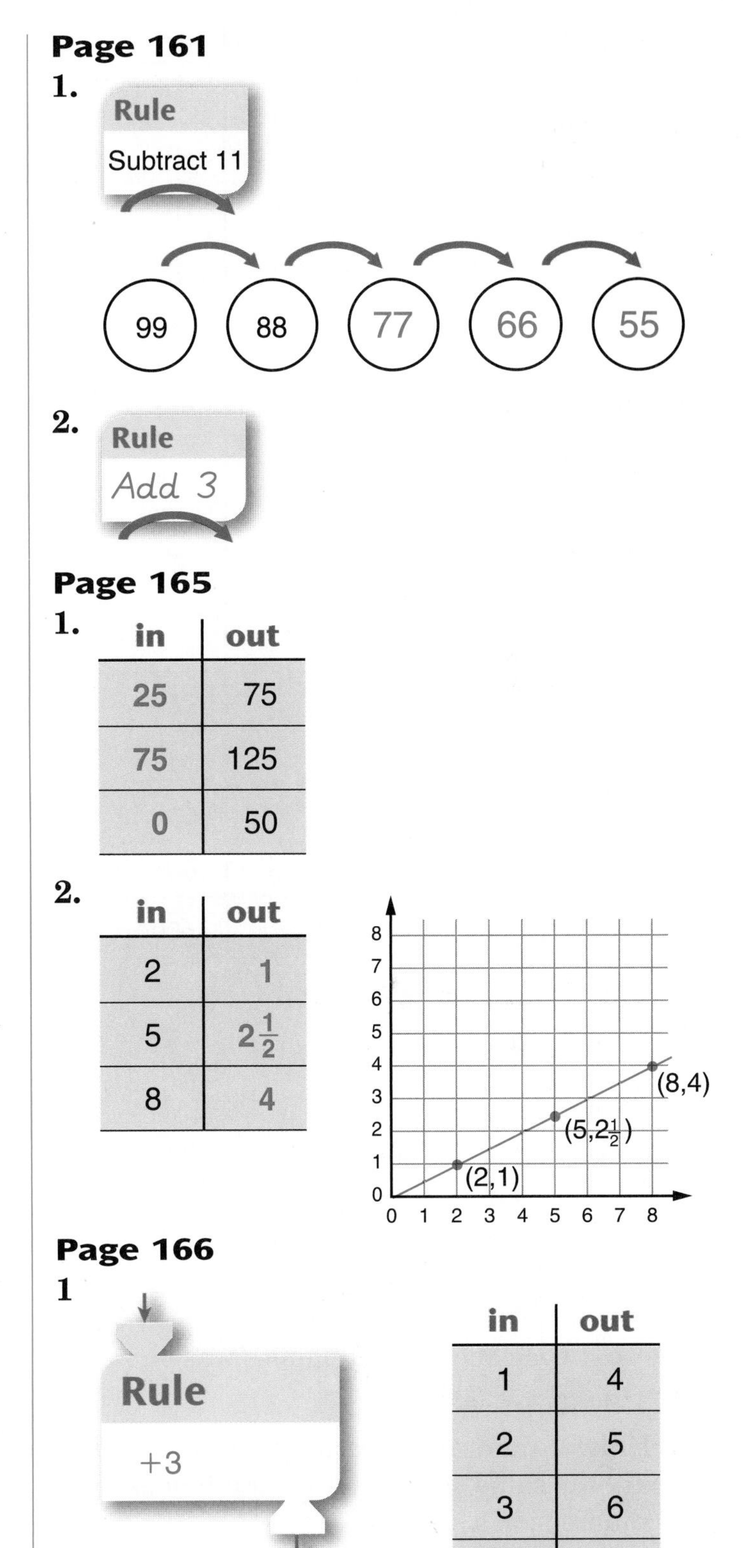

Page 161

1.

2.

Page 165

1.

in	out
25	75
75	125
0	50

2.

in	out
2	1
5	$2\frac{1}{2}$
8	4

Page 166

1

in	out
1	4
2	5
3	6
10	13
15	18

2.

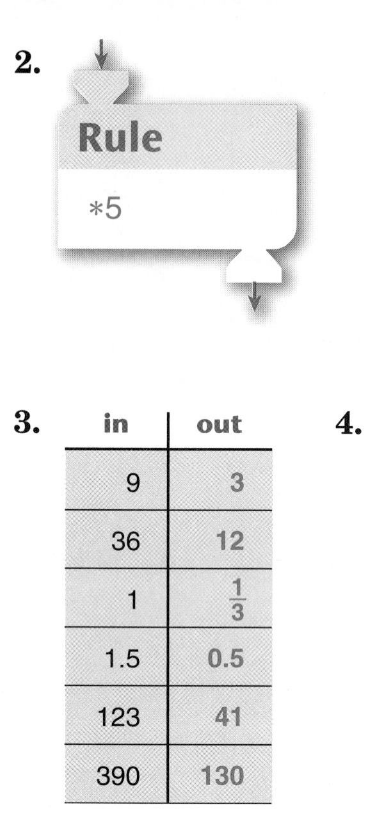

in	out
1	5
2	10
3	15
7	35
9	45

3.

in	out
9	3
36	12
1	$\frac{1}{3}$
1.5	0.5
123	41
390	130

4.

in	out
11	7
28	24
0	−4
4	0
54	50
122	118

5.

Rule
*30

in	out
20	600

Page 175

1.
2 miles	for	2 days
3 miles	for	3 days
4 miles	for	4 days
5 miles	for	5 days
	Total	14 days

The first time she walked 6 miles was on day 15.

2. Draw 2 squares that are side by side and are the same size. Call the length of each square side *s*.

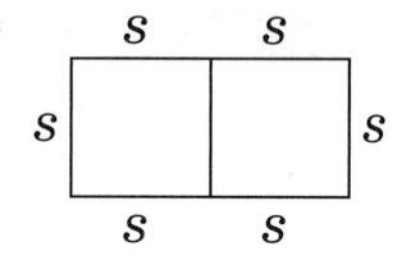

Rectangle perimeter = $6 * s = 24$ in. So s must be 4 in. The rectangle was 4 in. long and 8 in. wide.

Page 177

1. Sample answer: It took about 3 minutes to count the 100 numbers from 100,001 to 100,100. So, it would take about 30,000 minutes to say a million numbers. 30,000 minutes = 500 hr = 20.8 days or about 21 days.

2. Sample answers:

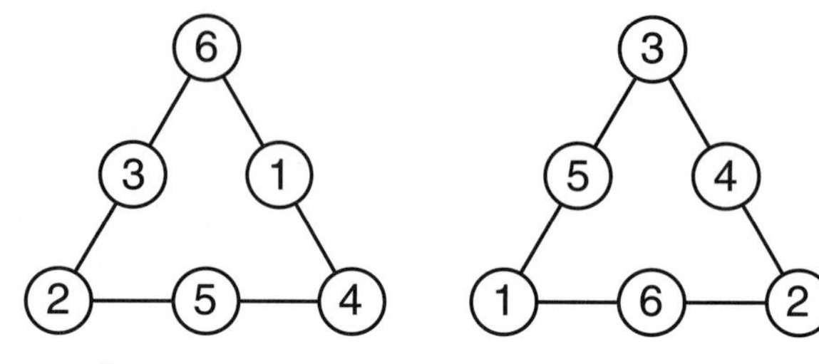

3. 7cm^2

Page 178

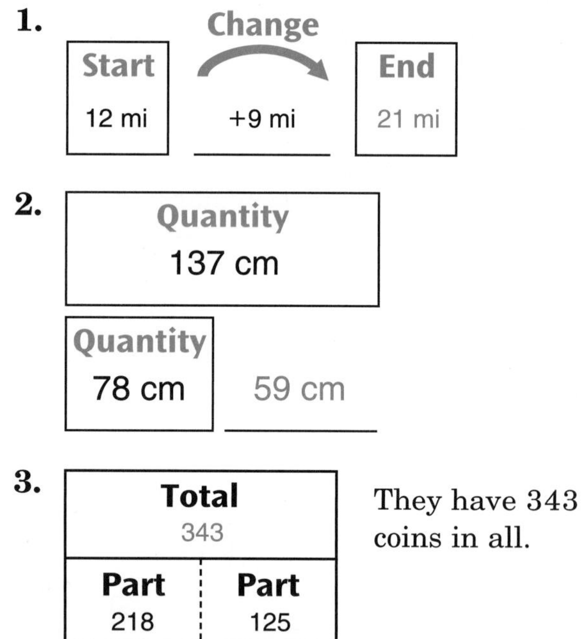

They have 343 coins in all.

Page 181

1. David is not correct. The leading-digit estimate is $600 * 30 = 18{,}000$.
2. Caroline is probably correct. There are 24 hours in a day and 7 days in a week. The leading-digit estimate is $20 * 7 = 140$ hours.

Page 183

1. 70	2. 240	3. 2,910	4. 40
5. 13,000	6. 4,000	7. 123,000	8. 1,000

Page 184

1. hundreds ($7 * 40 = 280$)
2. thousands ($8,000 / 4 = 2,000$)
3. hundreds ($4,300 - 3,700 = 600$)

Page 196

1. 35 2. 39 3. 57.6 4. 43

Page 198

1. 60 R2 2. 17 R10 3. 258 R17

Page 207

1. 0.9375 2. $\frac{113}{200}$, or $\frac{565}{1,000}$ 3. 0.7%
4. 2.91 5. 34.375% 6. $\frac{18}{25}$, or $\frac{72}{100}$

Page 209

1. 0.4 2. 522 3. 6174.614 4. 0.90

Page 214

1. 0.00049 2. 33,000,000
3. 0.000006147 or 0.0000061
4. –0.000078

Page 215

1. $6.0846 * 10^{15}$ or $6.085 * 10^{15}$
2. $5.3798 * 10^{11}$ or $5.38 * 10^{11}$
3. $1.5365 * 10^{16}$ or $1.537 * 10^{16}$
4. $2.7775 * 10^{17}$ or $2.778 * 10^{17}$
5. exactly $3.118752 * 10^{8} = 311,875,200$ hands, or about $3.1187 * 10^{8} = 311,870,000$ hands

Page 217

1. 1,018 ft^2 2. 83.9 cm

Page 220

about 450 ft^2

Page 221

$15.30; $100.30

Page 224

1. $49.75 2. $22.50

Page 226

1. 22, 30, 38, 46, 54, 62
2. 146, 130, 114, 98, 82, 66

Page 273

Los Angeles: 34° North, 118° West
Philadelphia: 40° North, 75° West
Atlanta: 34° North, 85° West

Page 24A

1. 3,591 2. 10,452 3. 139,303,013
4. 2,970,603

Page 24B

1. 55 2. 30 3. 5,986 4. 1,651

Page 24C

1. 343 2. 2,574 3. $12,656
4. 3,880

Page 24D

1. 4,199 2. 12,912 3. 96,460 4. 143,025

Page 24H

1. 62 R2 2. 392 3. 83 R7 4. 82 R6

Page 24J

1. 1,541 R3 2. 1,651 3. 699
4. 634 R5

Page 37B

1. 9.69 2. 19.572 3. 2.4644 4. 0.0063

Page 40A

1. 6.74 2. 42.21 3. 133.034
4. 29.103

Page 40B

1. 13.48 2. 0.85 3. 49.56 4. 43.95

Page 40C

1. $2,670 2. $17.88 3. $22,768
4. $450

Page 40F

1. $1.25 2. 1.35 3. 0.60 4. $9.74

Page 178B

1. Sample answers:

boxes	bars per box	total bars
7	10	*b*

There are 70 total granola bars.

2. Sample answers:

rows	band members per row	total band members
7	4	*m*

There are 28 total band members.

3. Sample answers:

hours	miles per hour	total miles
3	0.03	*t*

The snail will travel 0.09 miles. (More than the length of a football field!)

4. Sample answers:

weight at birth (lb)	scale comparison factor	weight at 1 year (lb)
6	3	*w*

Hector weighed 18 lb on his first birthday.

A

B

C

D

M

N

O

P

Q

R

S

Photo Credits

Cover (l)Tony Hamblin/Frank Lane Picture Agency/CORBIS, (r)Gregory Adams/Lonely Planet Images/Getty Images, (bkgd) John W Banagan/Iconica/Getty Images; **Back Cover Spine** Gregory Adams/Lonely Planet Images/Getty Images; **v** Burke/Triolo Productions/Getty Images; **vi vii** The McGraw-Hill Companies; **viii** Photodisc/Getty Images; **ix** Jim Cummins/Taxi/Getty Images; **1** Harry Sieplinga/HMS Images/Getty Images; **2** (t)Annette Coolidge/PhotoEdit, (c)Photodisc/Getty Images, (bl br)The McGraw-Hill Companies; **3** (t)Sharon Hoogstraten/The McGraw-Hill Companies, (b)Ryan McVay/Photodisc/Getty Images; **20** The McGraw-Hill Companies; **25** Izzy Schwartz/Photodisc/Getty Images; **28** D. Hurst/Alamy; **41** Leigh Beisch/Getty Images; **43** Photodisc/Getty Images; **57** Photodisc/Getty Images; **63** (tl)Tom Grill/CORBIS, (tr)Stephen St. John/Getty Images, (br)Photodisc/Getty Images; **64** (l)C Squared Studios/Photodisc/Getty Images, (cl)Stockbyte/Getty Images, (cr)C Squared Studios/Photodisc/Getty Images, (r)Photo Spin/Getty Images, **64-65** (tr)Stockbyte/Getty Images; **65** (tl)DAJ/Getty Images, (tr)Rubberball/Getty Images, (bl)Travel Pix/Taxi/Getty Images, (br)Neil Beckerman/Stone/Getty Images; **66** (tr)Photodisc/Getty Images, (cl) Chris Stock/Lebrecht Music & Arts/CORBIS, (cr)Lebrecht Music and Arts Photo Library/Alamy, (br)Helene Rogers/age footstock; **67** (tl)Ariel Skelley/CORBIS, (tr)worldthroughthelens/Alamy, (bl)Digital Vision/Getty Images, (br)Hans Neleman/Getty Images, **68** (tl)Manchan/Photodisc/Getty Images, (tr)Helene Rogers/Alamy, (b)C Squared Studios/Photodisc/Getty Images; **69** Edmond Van Hoorick/Getty Images; **70** Mary Kate Denny/PhotoEdit; **80** International Stock Photography/Getty Images; **81 82** The McGraw-Hill Companies; **87** Inger Hogstrom/age footstock; **88** (tr)William Leaman/Alamy, (tc)C Squared Studios/Photodisc/Getty Images, (cr)Michael Melford/Stone/Getty Images, (bl)Michael Boys/CORBIS, (bc)David Buffington/Photodisc/Getty Images, (br)Roy Morsch/CORBIS; **89** (t to b)Photolink/Photodisc/Getty Images, (2)Hemera Technologies/AbleStock.com/Getty Images, (3)Jupiterimages/Photos.com/Getty Images, (4)Bonnie Kamin/Photolibrary; **92** Martin Barraud/Stone/Getty Images; **94** Pixtal/SuperStock; **103** (t)A & L Sinibaldi/Getty Images, (c b)The McGraw-Hill Companies; **106** Glen Allison/Getty Images; **112** Steve Cole/Getty Images; **113** Vincent Besnault/Stone+/Getty Images; **119** (t)Peter Beck/CORBIS, (b)David Papazian/CORBIS; **120** (tl)Dean Conger/CORBIS, (bl)Adrian Arbib/CORBIS, (tr)Craig Aurness/CORBIS, (cr)Bryan Mullennix/Iconica/Getty Images, (b)Guenter Rossenbach/CORBIS Edge/CORBIS; **121** (tl)Peter Adams/CORBIS Edge/CORBIS, (bl)Joseph Sohm; Visions of America/CORBIS, (tr)Image Plan/CORBIS, (br) CORBIS; **122** (tl)Larry Lee Photography/CORBIS, (tr) David Muench/CORBIS, (bl)Bob Krist/CORBIS, (br)Lindsay Hebberd/CORBIS; **123** (tl tr)Roger Wood/CORBIS, (bl)James Leynse/CORBIS, (br)Bill Ross/CORBIS; **124** (tl)Louie Psihoyos/CORBIS, (tr)CORBIS, (bl)Sergio Pitamitz/CORBIS, (br)Nicholas Pitt/Digital Vision/Getty Images; **125** Brian Hagiwara Food Pix/Getty Images; **126** Siede Preis/Photodisc/Getty Images; **127** Stockbyte/Getty Images; **130** The McGraw-Hill Companies; **132** (t)Sharon Hoogstraten/The McGraw-Hill Companies, (c)C Squared Studios/Photodisc/Getty Images, (b)The McGraw-Hill Companies; **140** C Squared Studios/Photodisc/Getty Images; **147** Mitch Kezar/Stone/Getty Images; **151** Sharon Hoogstraten/The McGraw-Hill Companies; **152** Charles Orrico/SuperStock; **158** Brand X Pictures/PunchStock; **159** Mitch Tobias/The Image Bank/Getty Images; **167** (tl)Digital Stock/CORBIS, (tr)Bettmann/CORBIS, (bl)Archivo Iconografico, S.A./CORBIS, (br)Stapleton Collection/CORBIS; **168** (t)Tom Van Sant/CORBIS, (c)The Map House of London/Stockbyte/Getty Images, (b)MAPS.com/CORBIS; **169** (t)MAPS.com/CORBIS, (c)The McGraw-Hill Companies, (b)MAPS.com/CORBIS; **170** (t)Courtesy NOAA (c)Lloyd Cluff/CORBIS, (b)Joel W. Rogers/CORBIS; **171** (t)Ramin Talaie/CORBIS, (c)MAPS.com/CORBIS, (b)USGS; **172** (tr)Chad Baker/Photodisc/Getty Images, (cl)NASA/Roger Ressmeyer/CORBIS, (cr)Digital Vision/Getty Images, (br)Science and Society/SuperStock **173** Charles Smith/CORBIS; **174** The McGraw-Hill Companies; **177** Pedro Coll/age footstock; **180** The McGraw-Hill Companies; **184** agefotostock/SuperStock; **185** (t)Burke/Triolo Productions/Brand X/CORBIS, (b)Science & Society Picture Library/Contributor/SSPL/Getty Images; **186** (t)SSPL via Getty Images, (c)The Granger Collection — All rights reserved., (bl) Science & Society Picture Library/SSPL/Getty Images, (br)Sheila Terry/Photo Researchers, Inc.; **187** (t)Bettmann/CORBIS, (c) © Science Museum / Science & Society Picture Library — All rights reserved., (b)Dr. Horst Zuse,; **188** (tl)Time & Life Pictures/TIME & LIFE Images/Getty Images, (cl)Jerry Cooke/CORBIS, (cr)Bettmann/CORBIS, (br)Charles E. Rotkin/CORBIS; **189** (tl)Science & Society Picture Library/Contributor/SSPL/Getty Images, (tr)Brant Ward/San Francisco Chronicle/CORBIS, (bl) Paul Costello/Stone/Getty Images, (br)Science & Society Picture Library/SSPL/Getty Images; **190** (tl)Kim Kulish/CORBIS, (cl)Stockbyte/Getty Images, (bl)Don Farrell/Getty Images, (tr)Toshiyuki Aizawa/Reuters/CORBIS, (br)Photodisc/Getty Images; **191 192** The McGraw-Hill Companies; **227** CORBIS; **230** (t)Stockdisc/Getty Images, (b)TRBfoto/Photodisc/Getty Images; **231** Stockbyte/Getty Images; **233 240** The McGraw-Hill Companies; **262** Brand X Pictures/PunchStock/Getty Images; **265** Peter Dazeley/Stone/Getty Images; **266** (t)China Photos/The Image Bank/Getty Images, (c)Oliver Benn/Stone/Getty Images, (b)Vstock/Alamy; **267** Digital Vision/Getty Images; **268** (t to b)age fotostock/SuperStock, (2)Brand X Pictures/PunchStock, (3)Jupiterimages, (4)R. Morley/PhotoLink/Photodisc/Getty Images; **269** (t to b)flickr/Getty Images, (2)age fotostock/SuperStock, (3)PNC/Brand X Pictures/Getty Images; (4)C.I. Aguera/CORBIS; **294** (t)S. Solum/PhotoLink/Photodisc/Getty Images; (b)Jay Dickman/CORBIS; **295** (t)Adalberto Rios Szalay/Sexto Sol/Getty Images, (b)Sergei Kozak/Corbis Edge/CORBIS; **296** (t)Sexto Sol/Photodisc/Getty Images; (b)Robert Harding/Photodisc/Getty Images; **298** (t)Gideon Mendel/CORBIS; (b)Bob Daemmrich/PhotoEdit; **300** (t)Bill Bachmann/Alamy; (b)Owaki-Kulla/CORBIS; **302** RICHARD CHUNG/Reuters/CORBIS; **303** (l)Simon Kwong/Reuters/CORBIS, (c)Roger Ressmeyer/CORBIS, (r)Tim Hawkins/Eye Ubiquitous/CORBIS; **305** Carl & Ann Purcell/CORBIS; **312** The McGraw-Hill Companies.